Pulp Production and Processing

Also of Interest

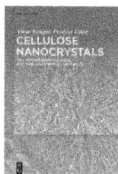

Cellulose Nanocrystals.
An Emerging Nanocellulose for Numerous Chemical Processes
Katiyar, Dhar, 2020
ISBN 978-3-11-064452-4, e-ISBN 978-3-11-064801-0

Green Pulp and Paper Industry.
Biotechnology for Ecofriendly Processing
Kumar, Pathak, Dutt (Eds.) 2020
ISBN 978-3-11-059184-2, e-ISBN 978-3-11-059241-2

Nanocellulose.
From Nature to High Performance Tailored Materials
Dufresne (Ed.), 2017
ISBN 978-3-11-047848-8, e-ISBN 978-3-11-048041-2

Downstream Processing in Biotechnology
Beschkov, Yankov (Eds.), 2020
ISBN 978-3-11-057395-4, e-ISBN 978-3-11-065480-6

Wood Research and Technology.
Cellulose – Hemicelluloses – Lignin – Wood Extractives
Editor-in-Chief: Lennart Salmén
ISSN 0018-3830, e-ISSN 1437-434X

Pulp Production and Processing

High-Tech Applications

Edited by
Valentin I. Popa

2nd Edition

DE GRUYTER

Editor
Prof. Valentin I. Popa
Technical University of Iasi "Gheorghe Asachi"
Faculty of Chemical Engineering and Environment Protection
Blvd. Mangeron 71
700050 Iasi
Romania
vipopa@tuiasi.ro

ISBN 978-3-11-065883-5
e-ISBN (PDF) 978-3-11-065884-2
e-ISBN (EPUB) 978-3-11-065921-4

Library of Congress Control Number: 2020932671

Bibliographic information published by the Deutsche Nationalbibliothek
The Deutsche Nationalbibliothek lists this publication in the Deutsche Nationalbibliografie;
detailed bibliographic data are available on the Internet at http://dnb.dnb.de.

© 2020 Walter de Gruyter GmbH, Berlin/Boston
Cover image: AVTG/iStock/Getty Images Plus
Typesetting: Integra Software Services Pvt. Ltd.
Printing and binding: CPI books GmbH, Leck

www.degruyter.com

Preface

Pulp and paper mills are ideal sites for integrated biorefinery operations for **four basic reasons. First**, they are already set up to receive and process massive amounts of delivered round wood and woods chips, and they have access to at least an equal amount of forest residuals and even a greater amount of waste paper and agriculture wastes and energy crops if needed. **Second,** these mills have basically the same existing infrastructure for warehousing and shipping out finished products around the world. **Third,** they have a well-established in-place administrative infrastructure and related human resources that can be extended to serve a biorefinery business without incurring significant new costs. **Fourth**, pulp and paper mills have operating utility support systems for processing water, electricity, steam, and waste/environmental treatment that can easily be umbrella to support biorefinery operations without major new investment. And possibly, as a strong **fifth** reason, chemical pulp mills already operate as biorefineries of sorts, producing fiber to make paper and paperboard as well as some specialized dissolving pulps to make viscose types of "bio-plastics" and rayon materials. Bioproducts can be made from pulp as cellulose derivatives and as other high-tech value-added products, as well as from spent cooking liquors including ingredients used in making coatings, adhesives, detergents, paint, varnish, ink, lubricants, waxes, polishes, gasoline additives, agricultural products, and so forth. There is also a spectrum of lignin-based products derived from refinement of black liquor. Tall oil along with turpentine could be obtained during the pulping of softwoods with kraft process. Therefore, the integrated biorefinery is a processing facility allowing extraction of polysaccharides, lignin, and other chemicals from biomass (extractives), and convert them into multiple products including materials, fuels, and high-value chemicals.

It is known that the conventional different procedures to obtain pulp and paper include mechanical and chemical treatments of fibrous raw materials. Each variant has both advantages and disadvantages being recommended in function of specific conditions. The resultant pulp yield and quality depend on the applied technology, and the by-products are used to recover chemicals and energy.

At present, deforestation is a vital problem at global level, being one of the main causes of climate change. Volume of wood used as fuel, in construction, furniture, and pulp and paper industries increases year by year. Stopping the decrease of forests surface is of imperative importance from the point of view of sustainability. Pulp and paper industry is confronted with severe lack of pulpwood, and for this reason other raw materials must be taken into consideration. This is why obtaining of chemical pulp from other raw materials is of particular importance. In these conditions some agricultural residues such as wheat straw, rapeseed stalks, and corn stoves were proposed to be used as raw materials in producing papermaking pulp. At the same time, there are many advantages of using of agricultural residues: high availability, low price, favorable chemical composition, and suitable papermaking properties of fibers.

https://doi.org/10.1515/9783110658842-202

Bleaching is the following step connected with pulp manufacturing. The process to remove the residual lignin and other oxidizable structure can be performed using different oxidants such as oxygen, chlorine dioxide, hydrogen peroxide, ozone, and peracetic acid. The use of enzymes in the bleaching process, particularly of thermostable and alkali-tolerant xylanases, is regarded as a success story of biotechnological approaches in the pulp industry. The reaction mechanism, kinetics, process parameters, and technology details of the different bleaching technology are important for optimization and to obtain good characteristics of the pulp.

On the other hand, the integration of biomass resources in the biorefining has to be associated with the new conversion environmentally friendly technologies Thus, the application of this novel bio-greening approach in the pulp and paper industry include along with fractionation of lignocellulosics; upgrading of plant extractives and resins; production of biohydrogen; and other "clean" technologies, products, and systems. In these conditions important changes in the field of research and education are expected to be produced.

At present, among the products obtained from biomass, an increased interest was observed for biochar. This product results by hydrothermal treatment of different lignocellulosic wastes existing in the pulp and paper factory (e.g., sawdust, bark, or wastes from waste paper processing) or disposable from agriculture. The interesting properties of the biochar recommend it to be applied in agriculture, bioremediation, and environmental protection.

The rapid depletion of fossil fuels has led to a huge effort to increase the use of renewable biomass. The replacement for fossil fuels comes from lignocellulosics using biorefining procedure. At the same time, some building blocks used for synthesis of biopolymers or other compounds can be obtained from a vast number of feedstocks. Thus, the most promising building blocks are hydroxymethlfurfural (HMF) and levulinic acid (LA), which could be used as intermediates for derivatives with commercial relevance for the future of biorefinery.

The knowledge and understanding of properties of cellulose depends on its basic chemistry, epimolecular, supermolecular, hypermolecular structure, interaction with water, elementary cellulose components (i.e. cellulose crystallites; wet-web strength and wet strength of materials; enthalpiometric observations of interactions in cellulosic fibrous slurries; existence of water inclusions among cellulosic chains; hydrogel structure of cellulose; H-bond ability and hydration bonding and anti-bonding concept; rheosedimentation; thermo-responsive hydrated macro-, micro-, and submicro-reticular systems of cellulose; swelling; recycling; interaction with gases and vapours).

What makes wood and nonwood fibers suitable for papermaking more than other fibers? What happens with fibers during papermaking? Is the hornification process from paper recycling a reversible one or not? Cellulosic fibers are the main constituent of all plant material, and they are a renewable source available abundantly in nature. Cellulosic fibers are made of a multilayer of a very small thread-like structure called fibrils. These fibrils can be exposed by beating/refining of fibers and provide a very large area for bonding. But the most important feature of cellulosic fibers that makes it suitable for papermaking is that fiber develops physical and chemical bonding with other fibers when it changes from wet to dry state. High tensile strength, flexibility, water-insoluble, hydrophilic character, chemical stability, and relatively colorless (white) are other particularly important features of wood and nonwood fibers, which make it suitable for papermaking. Even if papermaking is one of the most common uses of cellulosic fibers, today's applications such as paper-based microfluidic devices are a particularly attractive topic for research on high-tech applications of fibrous structures. Materials in which cellulosic fibers are used after nanofibrillation processes are also highly appreciated opening new high-tech applications for multifunctional fibrous structures.

Hydrogels represent a family of customizable three-dimensional polymeric networks with unique features, such as to absorb, swell, and release large quantities of water or biological fluids and simulate biological tissue when swollen. Polysaccharides are important candidates for the design and preparation of hydrogels, since they provide a handful of distinctive biological, physical, and chemical characteristics which favor several demanding applications, especially in the medical field. Therefore, cellulose, a key bioresource of the twenty-first century, is a versatile candidate for the construction of biomaterials with tunable network structure, receiving increasing attention in recent years due to an advantageous mixture of proper biocompatibility, hydrophilicity, and mechanical properties.

Nowadays, due to its hydrophilicity, nontoxicity and antimicrobial properties, cellulose is used in medical applications, as wound dressing, tissue engineering, controllable drug delivery system, and so on. The ability to control the material features at the nanoscale brings new and promising properties, such as high mechanical characteristics and low density, which provides the opportunity to develop new applications. The differences between nanomaterials and bulk materials are controlled by two main factors – as the increased relative surface area, so an enhanced reactivity and the quantum effects offered by the changes on mechanical, optical, thermal, and electrical properties. In addition, the combination of nanoparticles with polymers provides a route to a wide variety of old and new applications.

Water insolubility of cellulose, generally is attributed to the existence of extensive intra/intermolecular hydrogen bonds, limiting its applications. This disadvantage can be eliminated by derivatization reactions of cellulose. The introduction of functional groups (derivatization) has an important effect on macroscopic behavior; for example, solubility, stability, and viscometric/rheometric characteristics. In this respect, the synthesis of the ionic derivatives of cellulose presents an important opportunity to develop new applications.

Cellulose ester films are preferably employed in optical functional films because of its good transparency with high glass-transition temperature. Since the main chain of cellulose has low level of optical anisotropy, orientation birefringence of a stretched film can be easily controlled by chemical modification such as esterification. A small amount of crystallinity also plays an important role on the hot-stretching process. Moreover, good miscibility with various plasticizers widens the material design to control the birefringence. A novel material designs of a multi-band quarter-wave plate, one of the optical retardation films used in advanced display system, has been demonstrated using cellulose esters.

The idea to edit the second edition of this book belongs to Lena Stoll editor in the Industrial Chemistry section of De Gruyter publishing house who allow me to cooperate with Ria Senbusch, Anna Bernhard, and Dipti Dange who were involved in this project, taking the risk to bring out such a book and for ensuring the excellent quality of the publication.

We have removed few chapters from the first edition, and we have added new ones having in view new progresses in the field of cellulose as renewable and sustainable raw material.

The publishing of the book was accomplished with the contributions of renowned specialists in the pulp and paper and cellulose derivatives production from all over the world. We are grateful to these scientists for their efforts and dedication to this reference book.

The book is a very useful tool for many scientists, students, and postgraduates working in the field of pulp and paper and cellulose derivatives, aimed at opening a new era of renewable resources processed by refining. It may not only help in research and development, but may also be suitable in the line of teaching.

I hope that you as reader will enjoy the volume.

Valentin I. Popa
2020

Contents

Valentin I. Popa
Chapter 1
Biorefining and the pulp and paper industry

1.1 Introduction

The concept of biorefinery originated in late 1990s as a result of scarcity of fossil fuels and increasing trends of use of biomass as a renewable feedstock for the production of nonfood products. The term of "Green Biorefinery" was first introduced in 1997 as "Green biorefineries represent complex (to fully integrated) systems of sustainable, environmentally and resource-friendly technologies for the comprehensive (holistic) material and energetic utilization as well as exploitation of biological raw materials in form of green and residue biomass from a targeted sustainable regional land utilization."

According to US Department of Energy (DOE) "A biorefinery is an overall concept of a processing plant where biomass feedstocks are converted and extracted into a spectrum a valuable products." The American National Renewable Energy Laboratory (NREL) defined biorefinery as "A biorefinery is a facility that integrates biomass conversion process and equipment to produce fuels, power and chemicals from biomass." These definitions of biorefinery are analogous to today's integrated petroleum refinery and petrochemicals industry to produce multitude of fuels and organic chemicals from petroleum [1].

However, we think that we have a priority because we have introduced this concept in the paper [2]:

In our days, the idea that vegetable biomass represents a source of liquid fuel and of different new materials has led to the development of various research programs in this field. Our investigations in this direction are based on the following premises: (1) all kinds of vegetable biomass include almost the same components; (2) the macromolecular compounds existing in the vegetable biomass incorporate biosynthesis energy, and their conversion to useful products seems to be considered; (3) the complex and total processing technology may be modulated depending on the chemical composition of the vegetable source, as well as on the utilization of the obtained chemical compounds. The possibilities of complex processing of soft- and hardwood bark, agricultural wastes, and some energetic cultures of *Helianthus tuberosus* and *Asclepias syriaca* are exemplified.

In order to face the present state of affairs, the manifested tendency is that of adopting the existing classical technologies of carbo- and petrochemical fields in processes of converting biomass into products possessing energetic and/or chemical value. The technology of integral and complex valorization of biomass has been proposed is to be performed on several stages and modules, depending on the chemical composition of the available vegetal resources and on the corresponding field of application for the obtained products as well.

Valentin I. Popa, "Gheorghe Asachi" Technical University of Iasi, Iasi, Romania

https://doi.org/10.1515/9783110658842-001

A plant for the fractionation and refining of biomass and to use of its entire components is a "biorefinery" plant, will have to display a high level of process integration and optimization to be competitive in the near future. Forest products companies may increase revenue by producing biofuels and chemicals in addition to wood, pulp, and paper products in a so-called **integrated forest biorefinery (IFBR)**. The concept of an **IFBR** is being advanced by a number of investigators who envision converting cellulose, hemicelluloses, and lignin from woody biomass, dedicated annual crops, industrial and municipal waste in bioenergy, and basic chemicals [3–5].

A pulp mill has excellent prerequisites to be the base for a biomass-based biorefinery: large flow of raw materials (wood and annual plants), existing process equipment, and good process knowledge. The key strengths of the pulp and paper industry are the wood and biomass sourcing along with the logistic infrastructure, a sustainability existing base of integrated production, and the high efficiency and experience in combined heat and power generation. The industry has unique capabilities in handling very large volumes of biomass, and the synergies in logistics and energy integration are significant. Therefore, biorefining and bioenergy fit well into the integrated business model of forest products companies.

The current chemical pulp process use approximately 50% of the organic raw material in the production of paper pulp; the remaining 50% is combusted in the recovery boiler to produce steam. A modern energy-optimized pulp mill has a substantial excess of energy/steam. This excess can be utilized in several different ways:
– The first is to produce electrical power.
– The second is to replace the recovery boiler with a black liquor gasification unit to produce syngas.
– The third is to extract some lignin from the black liquor and sell it as a new product to be used as fuel or raw materials for biobased products.
– The fourth is to attract the other external sources of biomass or wastes and to process them with the aim of obtaining fuels and chemicals [6].

In the pulp mill of tomorrow, the hemicelluloses and extractives, dissolved in process streams, could also be extracted and used as chemical raw materials. As a consequence the chemical pulp mill could be transformed into an integrated biomass biorefinery, producing different chemicals besides traditional pulp and papers [2].

At the same time, the utilization of biomass as a renewable raw material may have the following advantages: (1) reduced dependence on imported fossil oil; (2) reduction in greenhouse gas emissions; (3) building on the existing innovation base to support new developments; (4) a bioindustry that is globally competitive; (5) the development of processes that use biotechnology to reduce energy consumption and the use of nonrenewable materials; (6) the creation of jobs and wealth; (7) the development of new, renewable materials; (8) new markets for the agriculture and forestry sectors, including access to high-value markets; (9) underpinning a sustainable rural economy and infrastructure; and (10) sustainable development along the supply chain from feed-stocks to products and their end-of-life disposal.

1.2 Possibilities of biorefining implementation into the pulp industry

Examples of fractionation technologies are several biomass pulping processes that are common practice in the pulp and paper industry (e.g., kraft pulping, sulfite pulping, soda pulping, organosolv pulping, etc.). Here, the biomass is essentially fractionated into cellulose (for paper) and black liquor, a waste stream that predominantly contains residual carbohydrates and their degradation products (e.g., from the hemicelluloses), partly degraded lignin, and inorganics from the pulping process. The main application to date of this black liquor is combustion for heat. In addition, lignin and lignin-containing residues are large side streams from the pulp and paper industry and from biorefineries that use the carbohydrate fraction of the biomass, for example, for the production of bioethanol. Globally ~ 50 Mt per year of lignin originates from the pulp and paper industry, predominantly from kraft-, soda-, and sulfite-pulping of softwood, hardwood, and agricultural residues such as straw, flax, and grasses. Only 1 Mt is used for commercial purposes including lignosulfonates from sulfite pulping and 0.1 Mt as (chemically modified) kraft lignins from kraft pulping [7].

At present, most of these sulfur-containing lignin streams are combusted for generating power and/or heat, an application with very limited added-value. These sizable amounts of lignin are in principle available for valorization into chemicals and performance products. New developments in soda-pulping technology have resulted in sulfur-free lignins from herbaceous types of biomass such as straw and grass [8, 9].

Furthermore, large amounts of (hydrolytic) lignin will be produced from future bioethanol-based biorefineries by processes such as steam explosion and organosolv pulping. The first is a thermomechanical treatment that uses sulfuric acid for the hydrolysis and steam-explosion for breaking up the fibrous biomass structure. Organosolv pulping of hardwood, grasses, and straw leads to a high-quality lignin that is essentially sulfur free. The biomass is fractionated into lignin, cellulose, and a hemicelluloses containing-side stream. Generally, the hydrolytic lignins are the main fraction in the side stream that originates from the processing of wood and agricultural residues for transportation fuels and chemical building blocks.

Taking into account the utilization of different biomass sources as raw materials in a complex integrated pulp mill producing bioproducts and biomaterials, after our opinion we have to consider the following aspects: (1) all kinds of vegetable biomass contain almost the same main compounds; (2) the macromolecular compounds existing in the biomass incorporate biosynthesis energy, and their conversion to useful products seems to be economical; and (3) the complex and total processing technology may be modulated depending on the chemical composition of the biomass sources, as well as on the utilization of the obtained chemical compounds [2].

Thus, the specific objectives of this proposal have to be the following:

1. Identification, quantification, and characterization of resources from chemical composition point of view.
2. Separation and establishment of optimal conditions for fractionation using an original scheme that allows isolating chemical compounds as a function of their structure and raw material accessible to be processed (Fig. 1.1). Conventional and nonconventional extraction procedures will be used.
3. Characterization of isolated products; comparative studies of extraction methods will be carried out; correlation of the characteristics with the possibilities of utilizing the obtained products; establishing potential applications.
4. The elaboration of sequential technological procedures to recover separated compounds with the aim of transferring them to the pilot-scale level.
5. Evaluation of the results obtained in technological transfer from the efficiency, economical, and social points of view.
6. Evaluation of the economical feasibility of applications of the proposed technologies; analysis of cost–benefit ratio.

The world distribution of phytomass evidences huge quantities still unexploited by man (about 89%), together with significant forest, agricultural, industrial, and urban wastes [4].

Wood phytomass is still being incompletely exploited (important amounts are used as fuel (about 50%) for local energetic requirements and, only to a certain extent, for chemical ones).

Agricultural phytomass is now being confronted with a new problem: "limited grounds against an ever increasing number of people," a situation that could not assure by any means new stocks (wastes excepted); consequently, there is no reason to replace valuable food products as wheat, sugar beet, and sugar cane for conversion into liquid fuels (e.g., ethanol). In such circumstances, mention should be made of the efforts to use soil inadequate for agriculture in order to energetically culture of fast-growing plants, species with high content of biological compounds or hydrocarbons. At the same time, it is known that processing of agricultural products and plants containing biological compounds results in a substantial amount of waste. This can represent a raw material allowing the separation and upgrade of different components with energy and chemical value, using technology proposed by us.

Thus from a pulp mill biorefinery, the following products can be obtained in addition to pulp and paper: phenols, adhesives, carbon fibers, activated carbon, binders, barriers, antioxidants, pharmaceuticals, nutraceuticals, cosmetics, surfactants, chelants, solvents, descaling agents, specialty polymers, and biofuels (pellets, lignin fuel, methanol, DME, ethanol, etc.). New or increased amounts of traditional products can be made from internal/or external biomass sources (Fig. 1.2).

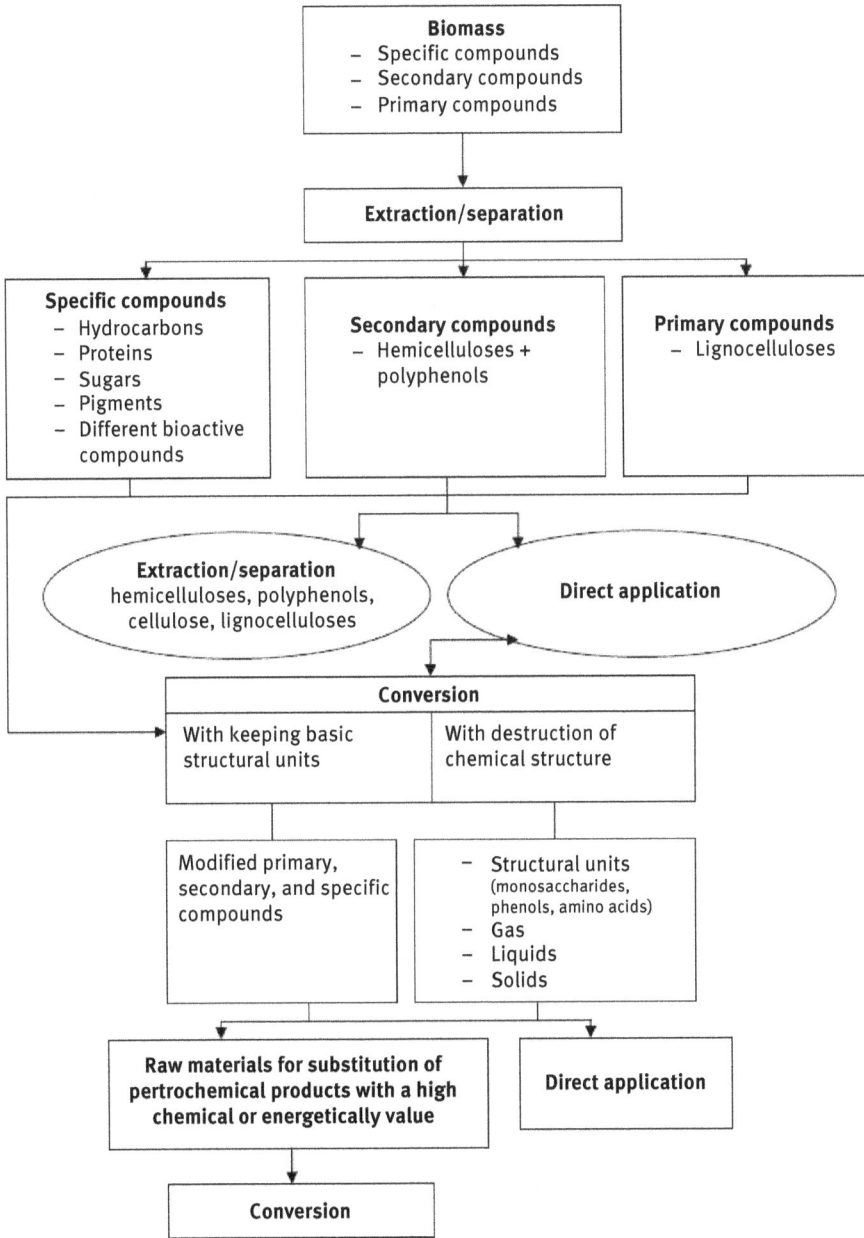

Fig. 1.1: Flow sheet of integral and complex processing of phytomass.

Three different levels can be identified: (1) a high degree of energy saving in future mills, especially chemical pulp mill, will lead to large amount of excess internal biomass, which can be transformed into the products mentioned above; (2) components

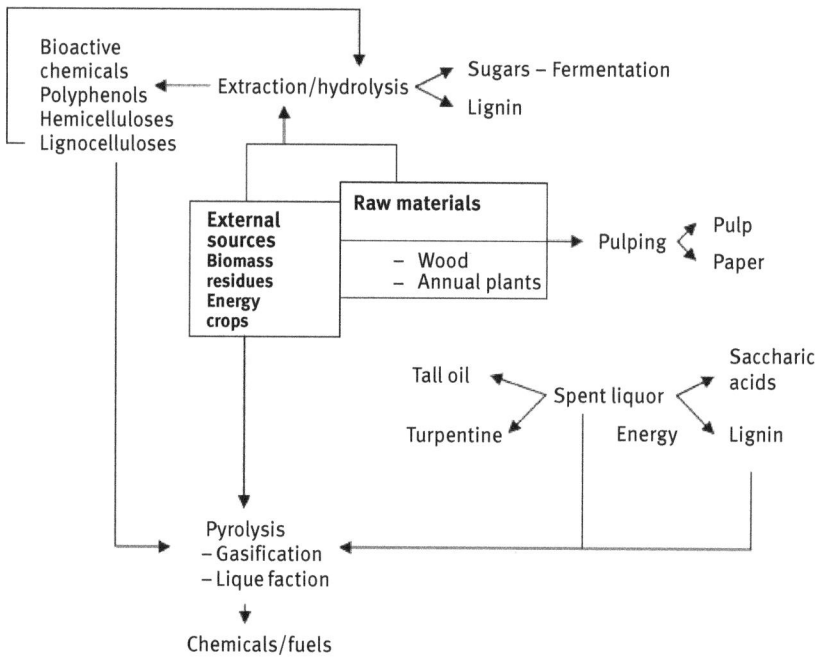

Fig. 1.2: Flow sheet for biorefining technology application using internal and external biomass sources.

in, for example, the black liquor, forest residues, and bark can be upgraded to more valuable ones and the energy balance of the mill is kept through fuel import, wholly or partly depending on the level of mill energy efficiency. The imported fuels can be biomass or other types; (3) external (imported) biomass (in some cases together with excess internal biomass) can be upgraded using synergy effects of docking this upgrading to a pulp mill.

To develop and apply new technology, we took into account the following considerations also as a result of our previous researches [2, 10, 11].

I. *All categories of phytomass contain the same compounds, arbitrarily divided into three large groups*:
    – primary compounds: cellulose and lignin;
    – secondary compounds: hemicelluloses and polyphenols; and
    – specific compounds: pigments, hydrocarbons simple sugars, alkaloids, polyphenols, other bioactive compounds, oils, proteins, and so on.

After the selective isolation of the specific and secondary compounds (performed in successive stages), the structural heterogeneity is being reduced. Thus, the residual material becomes lignocellulose (cellulose/lignin in variable ratios) characteristic of all higher plants. Consequently, *any category of available vegetal biomass may constitute a source of raw materials in its complex and integral valorification.*

II. *Compounds existing in phytomass store an important amount of energy as a result of their biosynthesis.* Thus, the biosynthesized macromolecular structures in phytomass require an amount of equivalent external energy for their cleavage into energy or chemical compounds (e.g., glucose from cellulose and phenol from lignin). That is why, depending on the available raw materials, the investigations have not been restricted exclusively to the obtaining ethanol from cellulose, via "glucose" or only to phenol separation from lignin, aiming also at the modification of the micro- and macromolecular structures existing in nature, from which valuable products can be obtained. Thus, the main objective is that all specific, secondary, and primary constituents isolated from phytomass, modified or not, should functionally substitute the classical chemical products or can represent materials with new properties.

III. *The technology of integral and complex valorization that has been proposed is to be performed on several stages and modules, depending on the chemical composition of the available resources and on the corresponding field of application for the obtained products.*

Prior to biomass harvesting, morphological elements destined for different valorifications are isolated. Then, the biomass that has been ground (containing a different content of humidity) is subjected to a stepwise processing. The technology we have thus imagined for the complex and integral processing and valorification implies two distinct stages: *extraction/separation (extraction of the specific compounds, extraction of the secondary compounds) and conversion (with or without maintaining the structural integrity of the initial compounds)*, which may be modularly applied, depending both on the species and on the chemical compounds utilized. The raw material may run through certain sequences of this flow sheet (see Figs. 1.1 and 1.2), which may be detached as single separated technology and may be applied depending on the available *amount* and *composition* of phytomass. After the separation of the specific compounds from phytomass, we can apply an **alkaline extraction**, which allows the isolation of complex consisting of polyphenols and hemicelluloses (Fig. 1.3). The idea of separating phenolic products from plants, especially from wood bark, is not new. It was proposed many years ago, yet its development has been prevented due to phenol accessibility in petrochemistry. Lately, as a consequence of the newly created economic conditions, the problem of extracting and utilizing phenolic compounds has been overcome.

The European pulp mills produce large amounts of bark as by-products, about 5.5 Mt/y. Additional amounts of bark are produced in sawmills. To date the major use of bark is a fuel, although alternative use has been sought for a long time. Thus, the bark can be used for extraction of useful chemicals and the process can be integrated in the production cycle, allowing them to be implemented industrially

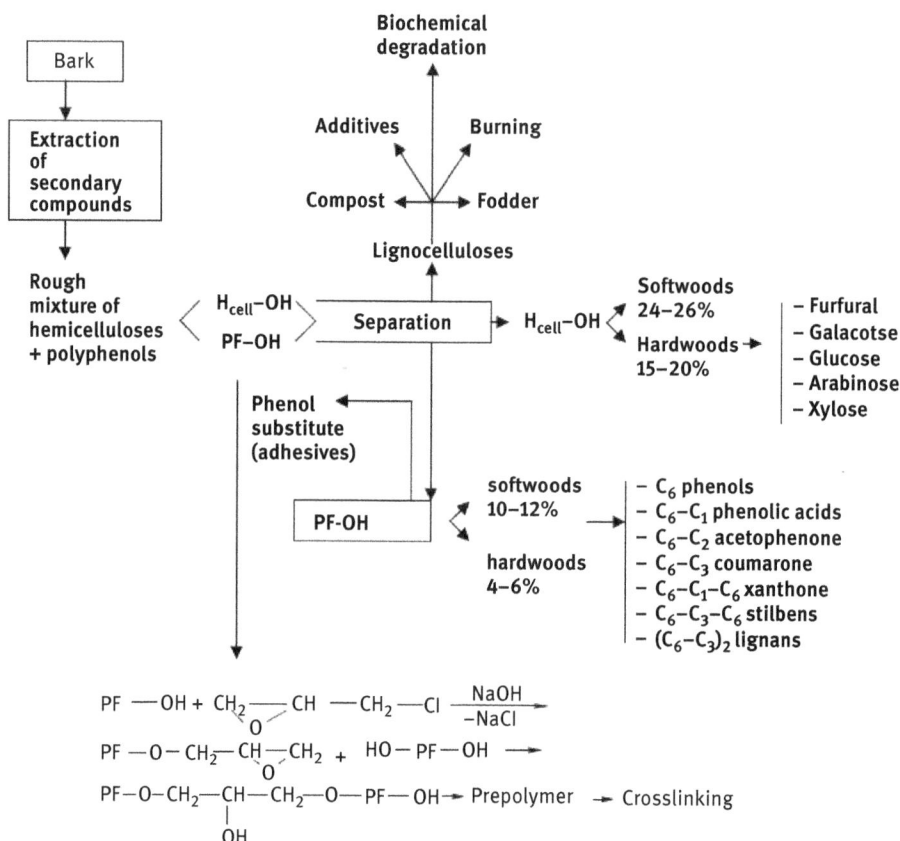

Fig. 1.3: Examples of extraction and separation of secondary compounds (isolation of hemicelluloses and polyphenols from wood bark).

(instead of being only a matter of academic interest) [5].The alkaline extracts were tested as phenol substituent in the synthesis of phenolformaldehyde resins, a bisphenol substituent in the synthesis of polyphenol epoxy resins, or as a phenolformaldehyde resin substituent [12].

Further conversion of resulted lignocelluloses could be performed by synthesis, using chemical or biochemical procedures that offer the possibilities of transforming the specific, secondary, and primary compounds while maintaining their basic structural units, or processing them by procedures presented in Fig. 1.4.

IV. *Products obtained by phytomass chemical processing, using above-mentioned methods, may structurally and/or functionally substitute certain raw materials of carbo- and petrochemical origin.*

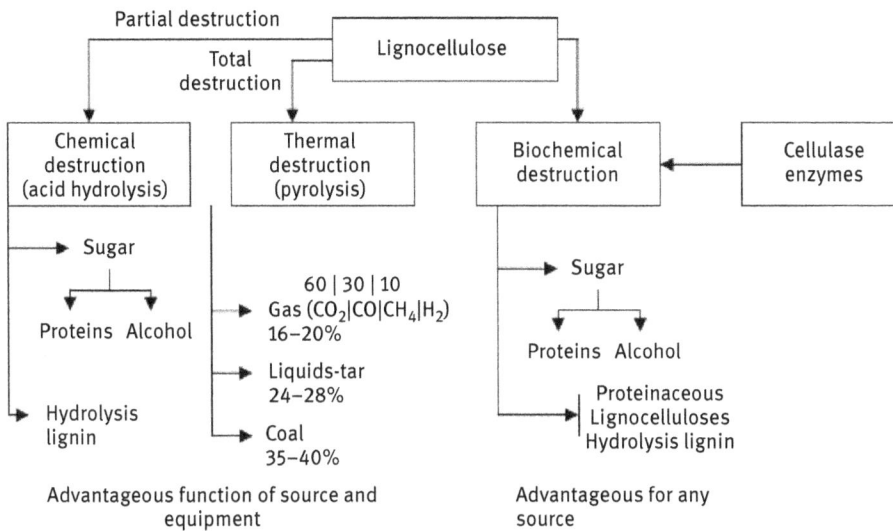

Fig. 1.4: Possibilities of lignocellulose complex valorification.

All these considerations will determine the approach to new directions of research:
- separation and direct utilization of chemical compounds isolated from biosystems;
- chemical processing of biomass and or wastes in their components by destruction, thus assuring raw materials for synthesis of polymers and chemical energy sources;
- chemical or biochemical transformation of both components and integral biomass (functionalization or functionality) for specific uses;
- elucidation of structures and functions of natural compounds in biosystems aiming at their utilization in structures with advanced properties and improving their behavior against physical, chemical and biological agents; and
- in vitro and in vivo simulation the synthesis of natural chemical compounds.

Using our proposed classification, cellulose and lignin can be considered as primary compounds in the pulp and paper industry.

1.2.1 Cellulose

The main utilization of pulp is to obtain paper and dissolving pulp (for cellulose derivatives, fibers, sponges, and films). At the same time, many chemicals can be produced from lignocelluloses, but recently there has been a new preoccupation to find special uses for cellulose. Thus, the cellulose would appear to have a great potential as nanomaterial. To fully exploit cellulose nanotechnology, research and development investment must be made in science and engineering that will determine the properties and characteristics of cellulose at the nanoscale, developing the technologies to manipulate cellulose self-assembly and multifunctionality within plants, and develop these new technologies to the point where industry can produce much more advanced and cost-competitive cellulose-based products. The properties of nanocellulose makes it an interesting material for many applications such as composites, paper and board, food, hygiene and absorbent products, emulsion and dispersion, oil recovery, medical, cosmetics and pharmaceuticals, and so on.

An expert survey with 150 respondents from 21 countries, together with other inputs, predicted biorefining to be most profitable investment over the coming 10 years. Nanotechnology will generate most attention in research, followed closely by biorefining and niche products. Nanocellulose can be produced by biorefining [13–15].

1.2.2 Lignin

Today, wood is converted into pulp, and in chemical pulp mills, steam and electricity are generated in the recovery boiler. In the pulp mill of tomorrow, lignin, hemicelluloses, and extractives dissolved in process streams will be extracted and used as chemical raw materials.

1.2.2.1 Separation of lignin

Black liquor from cooking process in kraft mills contains cooking chemicals and dissolved and degraded wood substances, with about half of the wood organic material is dissolved into the black liquor. The liquor is incinerated with recovery of inorganic cooking chemicals and production of steam. The modern chemical pulp mill produces an excess of energy in relation to its own needs, but the surplus steam often has no use and is a waste product. The export of biofuels (bark, lignin and so on) from mill would very often be more efficient than onsite incineration of all internally fuels. The production of kraft black liquor lignin amounts to about 16 Mt/y in Europe [16].

The dissolved organics consisted primarily of lignin, hemicelluloses, and degradation products of cellulose and hemicelluloses. Valuable chemical properties and fractions of highly polymeric lignin and hemicelluloses compounds are not used when the black liquor is simply burnt at the mill site for energy recovery. Many other options exist for energy production, whereas the chemical properties of the renewable lignin and hemicelluloses material are unique. The spectrum of potentially interesting products from hemicelluloses and lignin is wide, ranging from upgraded biofuels to high-value specialty chemicals. Furthermore, withdrawal of lignin and/or hemicelluloses would allow the production capacity of the pulp mill to be profitably increased if the recovery boiler is, as is often the case, the bottle-neck of the mill.

One way of utilizing the energy surplus in a modern pulp mill is to extract **lignin** from black liquor. The extraction of lignin is very flexible; lignin is energy rich (26–27 Mj/kg) and can be used as fuel to replace coal or oil in combustion and/or gasification plants; lignin may be also used as an internal fuel in the pulp mill (e.g., in the lime kiln) or sold either as fuel or raw material to be used for the production of various biobased materials/products (dispersants, various phenols, carbon fibers, etc.). Many of these components have a high market value, implying that lignin demands a high market value. It should also be kept in mind that if lignin is used as raw material, the end product will have a heat value, which, of course, should be utilized after the material has been used. Consequently, the extraction of lignin is not only a very flexible alternative, but it is also sustainable especially if the lignin is used as material before it is used as fuel. Extracting lignin is not, however, a new idea; it was proposed more than 50 years ago and a process for doing so has already been used in a few pulp mills for many years. In this process, the pH of this black liquor is lowered to about 10 and the lignin precipitates. The precipitated lignin is thereafter separated by means of filtration and a simple wash on the filter. Lignin produced using the older process was not very pure; the ash content could vary between 3 and 6%, which limited its use. Furthermore, some of the lignin was partly dissolved during the washing stage, which meant the remaining lignin was very difficult to dewater. This dissolution implies that total yield was reduced. A novel process known as the "LignoBoost" process was proposed few years ago and is a result of scientific investigation. This process has now been tested on pilot as well as demonstration scale and found to work very well: it is possible to produce lignin with a low ash content using the LignoBoost process. Although it lowers the pH of the black liquor to about 10, the separation stage is, however, quite different from the old process. The novel process starts with separation without washing. The filter cake from this stage is resuspended in low pH liquor and the lignin is filtered off and washed on a second filter. This particular procedure ensures that no noticeable amount of lignin amount will be dissolved and reprecipitated during the washing stage; as a result, the filtration resistance will be low throughout the process. The yield loss during washing will also be low. Despite the fact that there are two filters, the total filter area is normally much smaller than of the older process [17, 18].

Lignin does not possess a true repeating unit that can be selectively produced by degradation of the polymer, either chemically or enzymatically. Therefore, the future use of lignin (except as fuel) will depend on the possibilities of either its *degrading* or as *multifunctional macromolecule* [19].

Degradation of lignin can be obtained by *pyrolysis* under reductive conditions. Such methods have been tried for the production of phenols or aromatic hydrocarbons, respectively. In the latter case, substituted benzene structures can be obtained with properties suitable for *biodiesel ingredients*. Since very harsh conditions are employed in the degradation reaction, the structure of lignin seems to play a minor role, whereas the choice of reaction conditions must be done such that tar formation is minimized [20, 21].

1.2.3 Degradation products of polysaccharides

Various saccharinic acids and other monocarboxylic acids are formed from polysaccharides during alkaline pulping of wood and nonwood raw materials. Their formation depends on the cooking conditions and composition of raw materials, but typically up to 10% of the charged raw materials are converted to these acids. In addition, varying amounts of volatile acids (formic and acetic acids) and different dicarboxylic acids are formed [22].

Although dozens of compounds have been identified so far, usually a few major compounds can be recognized. Among these, the most interesting compounds for the potential applications include glycolic, lactic, 2-hyroxybutanoic, 2,5-dihydroxypentanoic (3,4-dideoxypentonic), xyloisosaccharinic, and glucoisosaccharinic acids. The isolated hydroxyl acids or acid mixtures can find applications in many different fields. The simple hydroxyl acids, glycolic, and lactic acids are currently used as industrial chemicals with several applications. The higher homolog, 2-hydroxybutanoic (which is one of the main hydroxyl acid in hardwood black liquors), could find uses in the manufacture of resins and polymers; for example, after conversion to crotonic or iso-crotonic acids. It should also be noticed that lactic and 2-hydroxybutanoic acids in the black liquors are of a racemic nature. The studies carried out have demonstrated the strong complexing capabilities of saccharinic acids, especially for many radionuclides and other materials. They have also been used for the preparation of various fine chemicals and as energy sources for aerobic and anaerobic bacteria [23, 24].

1.2.4 Black liquor as a feedstock for synthesis gas

An opportunity for biobased synthesis gas processes is integration of biofuel production with pulp and paper manufacture. For many of today's mills, this would enable the high-grade by-product energy of the biofuel plant to be fully exploited in the paper mill. The biofuel plant would also benefit from the existing infrastructure of the paper mill, in particular from that pertaining to feedstock procurement and handling.

Black liquor is a by-product from pulping, which is used today for steam and power production in the pulp mills. The black liquor is pumpable liquid, which is a significant advantage when the goal is to produce a clean syngas for use in catalytic processing. The challenge with black liquor as a fuel is that it is highly corrosive and that the cooking chemicals in the black liquor must be recovered in a form that is compatible with the pulping process. Black liquor, the spent liquor of kraft mill pulping, is a major potential biofeedstock for synthesis-gas production. However, gasifiers developed for solid biomass cannot be applied, as such, to black liquor. Significant efforts are currently being made to develop the required specialized technology. Synthesis-gas production from black liquor has the potential of being somewhat more economical than synthesis-gas production from solid biomass residues. On the other hand, the latter technology has a greater market potential, posses a smaller availability risk for a pulp mill, and is technically more certain at the present time [25].

In the gasification process, black liquor is atomized by a gas-assisted nozzle and sprayed into the reactor where it is directly gasified. Gasification of lignin offers a smart option also for other biomass sources use that can be included in pulp mill allowing as a broad range of different feedstocks to be converted into one homogeneous synthesis gas. This product can be used for electricity generation (combined power and heat gas engine, gas turbine) as well as for chemical synthesis (production of hydrogen, ammonia, methanol, methane from remethanization, hydrocarbon formation using Fischer–Tropsch reactions, etc.) by conventional technologies.

Wood biomass can be also transformed into a liquid product through pyrolysis. Other products from this process are a solid biomass residue called *char* and *noncondensable gases* mainly CO, CO_2, and light hydrocarbons. The liquid product, which is highly oxygenated, can be upgraded by removing oxygen. Application of zeolites during the upgrading removes the oxygen as water at low temperatures and as CO and CO_2 at higher temperatures, changing the yield of the produced phases. The major products in the resultant bio-oil, besides water, were 1-hydroxy-2-propanone, 2-methoxy-4-(1-propenyl)-phenol, 2-methoxy-4-methyl-phenol, and acetic acid. The phenolic products are formed in the pyrolysis of lignin, while the two other are formed of the wood carbohydrates [26, 27].

1.2.5 Biochar

Through pyrolysis of lignin along with other wastes from pulp and paper industry, carried out at temperatures that range from 250 °C to >900 °C and under limited oxygen availability, it is possible to obtain biochar. This char could be used as a solid fuel to generate process heat. It has also been demonstrated that biochar can reduce bioavailability of some heavy metals and that it has a high adsorption capacity to persistent organic pollutants and can be used for environmental protection [28, 29].

Secondary compounds: hemicelluloses and polyphenols

1.2.6 Separation of hemicelluloses

The flow of kraft black liquor carbohydrates derived by-products (including degraded sugar acids) amounts to about 12 Mt/y in Europe. In spite of the large amounts available, hemicelluloses have not yet been utilized commercially to any longer extent. The main obstacle is the difficulty of extracting them in their native form, since they are intimately associated with cellulose and lignin in the wood structure [30].

During the chemical pulping using kraft method, most of the hemicelluloses are degraded and dissolved as monomeric and oligomer sugars or sugar acids, in the black liquor. However, xylans – the main constituent of hemicelluloses in, for example, birch and eucalyptus – are preserved in a more native form to a certain extent. When isolated, the hemicelluloses have many potentially valuable properties. They could be used as paper additives, thickeners, food additives, emulsifiers, gelling agents, adhesives, and adsorbents. Some hemicelluloses have shown cholesterol-powering effects and even antitumor effects [31].

The *hemicelluloses* that normally end up in the black liquor of a hardwood kraft mill can be extracted prior to kraft pulping and used for the production of ethanol and acetic acid. The extracted liquor undergoes evaporation, hydrolysis, separation, fermentation, and distillation for the production of acetic acid and ethanol. During extraction process acetyl groups, which are side chains on the xylan hemicelluloses, are cleaved to give dissolved sodium acetate, which will be converted to acetic acid in the hydrolysis stage of the plant.

The hardwood biorefining uses green liquor in addition to AQ for wood extraction prior to modified kraft cooking to preserve both pulp yield and quality. During extraction, about 10% of the wood goes into solution. The extract mostly contains hemicelluloses derived organic compounds and has a near-neutral pH. The extracted chips require a 3% lower white liquor (as Na_2O on original oven dry wood) and a lower H factor in the subsequent kraft cook.

Galactoglucomannan (*GGM*) dominates the softwood hemicellulose. During refining of mechanical pulp, a part of the GGM is dissolved in the process water, which can be recovered from the water by ultrafiltration. The molar mass of GGM depends on the filtration methods, but is typically 20–30 kDa. GGM has been found to stabilize colloidal pitch by forming a hydrophobic layer on the surface of the pitch particles, thus reducing the risk for deposit formation during papermaking. Other possible uses of GGM can be also found, as their sorption property can be utilized together with fillers, cellulosic, and mechanical fibers. GGM can be modified by grafting it with other types of functional groups than hydroxyl groups. It is also an interesting polysaccharide as food additive, as it may be used as the so-called dietary fibers. GGM has also been studied also as film-forming constituents [32].

Specific compounds: pigments, hydrocarbons simple sugars, alkaloids, polyphenols, other bioactive compounds, oils, proteins, and so on

1.2.7 Fine and specialty chemicals

Black liquor contains compounds of potential value. Roughly half of the dissolved organic material is lignin and the rest is mainly sugar acids, other organic acids, and methanol.

Tall oil and turpentine products from kraft pulping were developed in the beginning of last century and have been produced ever since. Distillation technology has been improved during the last decades, and today, very pure rosin and fatty acids products can be manufactured. Tall oil fatty acids and tall oil rosin have found a wide variety of applications. The fragrance industry largely utilizes turpentines both sulfate turpentines and sulfite turpentines distilled from tapped oleoresin. A new use of tall oil fatty acids could be production of conjugated linoleic acids (CLAs). Nordic tall oil fatty acids are rich in linoleic acid (9,12–18:2), typically containing 40–50% linoleic acid. It is possible to make CLAs with the aid of heterogeneous catalysts and in the absence of solvents. Various antioxidant and antitumor properties have been attributed to CLAs and such preparations are marketed as dietary supplements [33].

In addition to naturally occurring constituents, chemical modification of isolated wood compounds provides new options. For example, new physiologically active compounds can be derived from wood using heterogeneous catalysis; *sitostanol* can be produced by catalytic hydrogenation from *sitosterol*, conjugated *linoleic acids* can be synthesized via isomerization of *linoleic acid*, and other lignans can be obtained from hydroxymataresinol hydrogenolysis [34].

1.2.8 Bioactive compounds

Some ideas concern new high-added value applications in the medical field, where these raw materials have not been used before, the driving force being to replace polymers/materials obtained from nonrenewable feedstocks. Xylitol and sitosterol/sitostanol are health-promoting products from wood, both of which have found their own special niche; xylitol promotes hygiene of mouth and sitosterol/sitostanol exerts a lowering effect on the cholesterol level in the human coronary flow. However, wood constituents still offer additional new opportunities for health-promoting products, as well as being a raw material of chemicals. For example, knots in spruce, that is, inner parts of the branches, have been found to contain extremely high concentration of polyphenols compared to the stem. Generally, these polyphenols are strong antioxidants. The dominating lignan, *hydroxymataresinol* (HMR), has been documented to carry favorable properties in the fight against cancer and also coronary heart diseases.

The HMR lignan is extracted from spruce knots that are separated from chips of spruce trees grown in Northern part of Finland. After purification and formulation of HMR lignan product, it has been marketed as a health-promoting food supplement since 2006.

Stilbenes are also interesting bioactive constituents in trees; these can be found in the bark of spruce, up to the level of 10%. *Piceatannol*, which is the main spruce bark stilbene, is an efficient antioxidant. The same group of antioxidants is also found in grapes and in red wine [35].

In our study lignins and polyphenols extracted from different sources of biomass using biorefining [36] have been used in model experiments to follow their actions as allelocehmicals [37].

Polyphenols extracted from spruce wood bark have been tested as plant growth regulators and the results showed that the isolated compounds exhibited similar effects to the endogenous hormones cytokinin and auxine [38]. Based on the obtained results, lignin and polyphenols were used in seeds germination, plants tissue culture, plants cultivation [38, 39], bioremediation [40] or as substrate for microorganisms development [41].

1.3 Conclusions

Plant cell wall biomass is composed of cellulose, hemicelluloses, lignin, protein, lipids, and several small molecular weight components with different ratios depending on the raw materials.

The key issue for a successful valorization of biomass to chemicals is efficient fractionation into hemicelluloses, cellulose, lignin, and the other compounds.

Thus, a plant for the fractionation and refining of biomass, and for the utilization of all its components, a "biorefinery" plant will have to display a high level of process *integration* and *optimization* to be competitive in the near future. A biorefinery is an installation that can process biomass into multiple products using an array of processing technologies. By integrating *forest biorefinery* (*FBR*) activities in an existing plant, pulp and paper mills have the opportunity to generate significant amounts of bioenergy and bioproducts and to drastically increase their revenues while continuing to produce wood, pulp, and paper products. A new generation of technologies based on thermochemical, biochemical, and chemical pathways is likely to enable the development FBR. Manufacturing new value-added by-products (e.g., fuels, bulk and specialty chemicals, pharmaceuticals, etc.) from biomass could represent for some forestry companies an unprecedented opportunity for revenue diversification. At the same time, biorefineries in the pulp and paper sector will integrate several biomass raw materials.

In addition to process technology development, product development will be essential for identifying successful new markets for biorefining products and their supply chain management. Therefore, incorporating new products, in addition to existing pulp and paper product portfolio, is a complex problem and perhaps the key to a company's successful diversification.

Biorefinery technology development will typically be implemented in retrofit and must be accompanied by careful process systems analysis in order to understand the impact on existing processes.

References

[1]   National Renewable Energy Laboratory (NREL), at http://www.nrel.gov/biomass/biorefinery.html
[2]   Simionescu CI, Rusan V, Popa VI. Options concerning phytomass valorification. Cell Chem Technol 1987, 21, 3–16.
[3]   Mao H, Gencojm JM, van Heiningen A, Pendse H. Near-neutral pre-extraction before hardwood kraft pulping; a biorefinery producing pulp, ethanol and acetic acid. Proceedings of Nordic Wood Biorefinery Conference, Stockholm, Sweden, March, 11–14, 2008, 16–29.
[4]   Popa VI, Volf I. Biomass for fuels and biomaterials. In: Biomass as renewable raw material to obtain bioproducts of high-tech value. Popa V, Volf I. eds. Elsevier, 2018, 1–37.
[5]   Volf I, Popa VI. Integrated processing of biomass resources for fine chemical obtaining: polyphenols. In: Biomass as renewable raw material to obtain bioproducts of high-tech value. Popa V., Volf I. eds. Elsevier, 2018, 113–60.
[6]   Theilander H. Withdrawing lignin from black liquor by precipitation, filtration and washing. Proceedings of Nordic Wood Biorefinery Conference, Stockholm, Sweden, March, 11–14, 2008, 36–42.
[7]   Popa VI. Biorefining and the pulp and paper industry. In: Pulp production and processing: from papermaking to high-tech products. Popa VI. ed. Smithers Rapra Technology Ltd. 2013, 1–33.
[8]   Malutan T, Nicu R, Popa VI. Contribution to the study of hydroxymethylation of alkali lignin. Bioresources, 2008, 3, 13–20.
[9]   Malutan T, Nicu R, Popa VI. Lignin modification by epoxidation. Bioresources, 2008, 3, 1371–6.
[10]   Popa VI, Volf I. Green chemistry and sustainable development. Environ Eng Manag J, 2006, 6, 545–58.
[11]   Popa VI, Volf I. Contributions to the complex processing of biomass, Proceedings of Nordic Wood Biorefinery Conference, Stockholm, Sweden, March 11–14, 2008, 229.
[12]   Simionescu CI, Bulacovschi J, PopaVI, Popa M, Nuta V, Rusan V. New possibilities of using alkaline extracts from vegetal biomass in adhesive systems for wood industry. Holzforsch Holzverw, 1988, 40,136–40.
[13]   Herrick FW, Casebier R L, Hamilton JK, Sandberg K R. Microfibrillated cellulose: morphology and accessibility. J Appl Polym Sci Polym Symp, 1983, 37, 797–813.
[14]   Wegner TH, Jones PE. Advancing cellulose-based nanotechnology. Cellulose, 2014, 13,115–8.
[15]   Popa VI. Nanotechnology and nanocellulose. Celuloză şi Hârtie. 2014, 63, 14–23.
[16]   Backlund B, Axegård P. EU-project WACHEUp, New uses of by-product streams in pulp and cork production. Proceedings of Nordic Wood Biorefinery Conference, Stockholm, Sweden, March 11-14, 2008, 50–5.
[17]   Ohman F, Wallmo H, Theiliander H. An improved method for washing lignin precipitated from kraft black liquor – The key to a new biofuel. 2007, 7, 309–15.
[18]   Tomani P, Axegård P, Berglin N, Lovell A, Nordgren D. Integration of lignin removal into kraft pulp mill and use of lignin as a biofuel. Cell Chem Technol, 2011, 45, 533–40.
[19]   Popa V.I. Lignin and sustainable development. Cell Chem Technol, 2007, 41, 591–3.

[20] Alper BS, Schuman CS. Production of chemicals from lignin.1970, Can.Pat.841708.
[21] Schabtai SJ, Zimierczak WW, Chornet E, Johnson D.Process for converting lignins into a high octane blending component, 2003, US Pat Appl. 0115792 AI.
[22] Niemelä K, Alén R. Characterization of pulping liquors. In: Analytical methods in wood chemistry, pulping, and papermaking. Sjöström S, Alén R. eds. Springer, 1999, 193–231.
[23] Tits J, Wieland E, M.H. Bradbury MH. The effect of isosaccharinic acid and gluconic acid on the retention of Eu(III), Am(III) and Th (IV) by calcite. Appl Geochem, 2005, 20, 282–96.
[24] Strand ES, Dykes J, ChiangV. Aerobic microbial degradation of glucoisosaccharinic acid. Appl Environ Microbiol, 1984, 47, 268–71.
[25] McKeough P, Kurkela E. Production and conversion of biomass-derived synthesis gas, Proceedings of Nordic Wood Biorefinery Conference, Stockholm, Sweden, March 11–14, 2008, 10–5.
[26] Aho A. Eränen K, Kumar N, Salmi T, Hupa M, Mikkola JP, Murzin DY. Catalytic upgrading vapours in a dual-fluidized bed reactor, Proceedings of Nordic Wood Biorefinery Conference, Stockholm, Sweden, March 11–14, 2008, 184.
[27] Xu C, Ferdosian F. Degradation of lignin by pyrolysis, In: Conversion of Lignin into Bio-Based Chemicals and Materials, Green Chemistry and Sustainable Technology. Xu C, Ferdosian F. authors, Springer-Verlag GmbH Germany, 2017, 13–33.
[28] Wild PJ, Huijgen WJJ, Gosselink RJA. Lignin pyrolysis for profitable lignocellulosic biorefineries. Biofuels Bioprod Bioref, 2014, 8, 645–57.
[29] Li J, Li Y, Wu Y, Zheng M. A comparison of biochars from lignin, cellulose and wood as sorbent to an aromatic pollutant. J Hazard Mater, 2014, 280, 450–7.
[30] Spiridon I, Popa VI. Hemicelluloses: structure and properties. In: Polysaccharides; structural diversity and functional versality, Second Edition, Dumitriu S, ed. Dekker/CRC Press, 2004, 475–490.
[31] Popa VI. Hemicelluloses in pharmacy and medicine. In: Polysaccharides in medicinal and pharmaceutical applications, Popa V, ed. Smithers Rapra, UK, 2011, 57–88.
[32] Mikkonen KS, Madhav P, Yadav P, Willför S, Hicks K, Tenkanen M. Films from spruce galactoglucomannan blended with poly(vinyl alcohol), corn arabinoxylan and konjac glucomannan. Bioresources, 2008, 3, 178–91.
[33] Holbom B. Bioactive extractives from wood and bark, Proceedings of Nordic Wood Biorefinery Conference, Stockholm, Sweden, March 11-14, 2008,105-110.
[34] Hupa, Auer M, and Holmbom B. Chemistry in forest biorefineries, Proceedings of Nordic Wood Biorefinery Conference, Stockholm, Sweden, March 11-14, 2008, 79–82.
[35] Holmbom B, Willför S, Hemming J, Pietarinen S, Nisula S, Eklund P, Sjöholm, R. Knots in trees a rich source of bioactive polyphenols. Materials, Chemicals and Energy from Forest Biomass. Argyropoulos DS, ed. ACS Symposium Series 954, 2007, 350–62.
[36] Bujor OC, Talmaciu AI, Volf I, Popa VI. Biorefining to recover aromatic compounds with biological properties. Tappi J, 2015, 187–93.
[37] Popa VI, Dumitru M, Volf I, Anghel N. Lignin and polyphenols as allelochemicals. Ind Crop Prod, 2008, 27, 144–9.
[38] A.Balas A, Popa VI. On characterization of some bioactive compounds extracted from *Picea abies* bark. Rom Biotechnol Lett, 2007, 12, 3209–15.
[39] Tanase C, Bujor OC, PopaVI. Phenolic natural compounds and their influence on physiological processes in plants, p.45-58, In: Polyphenols plants isolation, purification and extract preparation. Watson RR, ed. 2nd ed. Elsevier & Academic Press, 2018, 45–58.
[40] Stingu A, Volf I, Popa VI and I.Gostin I. New approach concerning the utilization of natural amendments in cadmium phytoremediation. Ind Crop Prod, 2012, 35, 53–60.
[41] Hainal AR, Capraru AM, Volf I, Popa VI. Lignin as a carbon source for the cultivation of some *Rhodotorula* species, Cell Chem Technol, 2012, 46, 87–96.

Dan Gavrilescu
Chapter 2
Pulping fundamentals and processing

2.1 Introduction to pulping

Pulping is a process by which plant material (wood, straw, grass, etc.) is reduced to a fibrous mass. Pulp refers to a suspension of cellulose fibers in water and represents the raw material for producing paper and board and obtaining cellulose derivatives. There are two major pulp grades: paper-grade pulp and dissolving pulp. The properties of paper-grade pulp refer to pulp yield, pulp brightness, and strength properties. The properties of dissolving grade pulp refer to high cellulose content and to a high reactivity toward derivatization chemicals.

According to the Confederation of European Paper Industries [1], the global production of pulp was 180.6 million tons in 2016 and the main pulp producers are North America, Europe, and Asia. The pulp and paper capacity survey (2016–2021) of the Food and Agriculture Organization of the United Nations [2] shows that in North America and Europe pulp production in the next few years will remain at relatively constant level, while pulp production in Latin America will increase due to wood abundance and low cost of labor.

A study of Haley [3] affirms that, in 2008, China overtook the United States to become the world's largest producer of paper due to the fact that this country has been rapidly expanding its paper-producing capacity. A forecast of domestic paper consumption in China shows that paper demand is projected to grow to 143 million tons in 2021 [3]. The demand for pulp in China increased at a rate of 10.3% between 2007 and 2017, and in 2017 China consumed 35% of the world produced pulp [4].

The aim of all pulping processes is to separate the fibers from the lignocellulosic materials to obtain pulp grades that must be suitable for papermaking. Cellulosic fibers can be separated from each other by mechanical, chemical, or by a combination of both treatments. According to Biermann [5], there are three main categories of pulping processes: mechanical, semichemical, and chemical.

Mechanical pulping uses mechanical energy to separate fibers from the wood matrix. There are several variants of mechanical pulping depending on the process parameters and equipment. Wood is processed either in the form of logs treated in grinders or in the form of wood chips that are converted to pulp in a refiner. Lignin or other wood components are not intentionally removed in mechanical pulping. Höglund [6] shows that in order to weaken the strength of the fiber–fiber bonds,

Dan Gavrilescu, "Gheorghe Asachi" Technical University, Iasi, Romania

https://doi.org/10.1515/9783110658842-002

steaming or/and a gentle chemical treatment of wood is performed. Classification of mechanical pulping comprises four processes:
- Groundwood pulping includes stone groundwood (SGW) and pressure ground-wood (PGW) methods. Atmospheric grinding and pressure grinding of logs using a pulp stone are the main alternatives of the groundwood pulping.
- Refiner mechanical pulping (RMP) consists of atmospheric refining of wood chips using a dedicated disc refiner.
- Thermomechanical pulping (TMP) consists of refining of steamed chips under pressure and elevated temperature using a disc refiner.
- Chemithermomechanical pulping (CTMP) involves refining of steamed and chemically pretreated chips under pressure and elevated temperature using a disc refiner.

The yield of mechanical pulp grades is high and depends on the process involved, as is presented by Sundholm [7]: groundwood pulp and RMP (96–98%); TMP (94–96%); and CTMP (90–95%). Besides the high yield, mechanical pulps exhibit a high light-scattering power, a high bulk, and a fairly high brightness. These pulps are used in producing papers with good opacity and printability at low basis weight (newsprint and magazine paper). Sundholm [7] also stated that the drawbacks of mechanical pulps refer to their high electrical energy consumption (up to 3.5 MWh/t of pulp) and to the fact that high-quality wood is required.

Semichemical pulping is performed in two steps: a mild chemical treatment of wood chips followed by a moderate mechanical refining. A partial removal of both of lignin and hemicelluloses takes place, so that the yield of pulp ranges between 60 and 80%. The chips are chemically treated at elevated temperature where about 50% of the initial lignin content of wood is removed. The softened chips are then processed in a defibrator enabling fiber separation.

According to von Koeppen [8], the most frequently employed process is neutral sulfite semichemical (NSSC) pulping, which uses cooking liquor containing Na_2SO_3 and Na_2CO_3. Sodium sulfite is the reagent for lignin sulfonation while sodium carbonate acts as a liquor pH buffer. A high cooking temperature (160–180 °C) is necessary both for both lignin sulfonation and dissolution. Hardwood is usually the best fiber source for NSSC pulp. The high lignin content makes the fibers very stiff, with the NSSC pulp being found to be the best raw material for a corrugated medium [9]. NSSC pulping generates spent liquor (usually named as red liquor) that contains dissolved lignin and hemicelluloses as well as inorganic substances. Ingruber [10] showed that the red liquor must be processed to recover the process chemicals.

Chemical pulping uses chemicals for lignin degradation and dissolving. Due to the fact that lignin is extensively removed, the fibers can be separated with little or no mechanical treatment. The process requires high temperature and pressure. Pulping chemicals are not selective regarding lignin degradation, the carbohydrates being also affected during the process. Almost half of the wood material is removed during chemical pulping and the pulp yield ranges from 45 to 55% depending on the pulping technique [11].

According to Sixta and coworkers [12] it is impossible to remove all lignin in any pulping process; hence, the cooking is intentionally stopped at a certain lignin content of the pulp. The residual lignin content represents the most important parameter that characterizes the pulp grade. The properties of chemical pulps must be compared at the same residual lignin content. In a chemical pulp mill, further delignification is achieved by bleaching of pulp.

Gullichsen [13] stated that pulping methods are classified according to the composition of cooking liquor and delignification can be performed in alkaline or acidic domain. Chemical pulping can also be performed in one or two stages.

Kraft pulping strongly dominates the world pulp production due to the following advantages:
- Kraft cooking is able to produce pulp from any kind of raw materials (wood species and nonwood plants).
- Kraft pulp exhibits very good strength properties.
- Cooking chemicals are produced via processing of the spent liquor in a recovery plant. Both steam and electrical energy are generated, so that the process is self-sufficient with respect of energy.
- The large capacity of kraft pulp mills reduces the specific consumption of materials and energy and enhances the profitability of pulp making.

Drawbacks of kraft pulping are as follows:
- Pulp yield is lower when compared with other pulping processes.
- The kraft pulp mill represents a source of polluting materials that are very different regarding their quantities and characteristics.

2.2 Mechanical pulping

During mechanical pulping, forces of various magnitude and duration are applied on wood matrix to separate the fibers. Wood is a viscoelastic material so that its behavior to mechanical forces is influenced by its moisture, temperature, load intensity, and time under load. Water plays a key role in mechanical pulping due to the fact that the softening temperature of wood main components depends on wood moisture. Salmen and coworkers [14] showed that for water-saturated lignin, the softening takes place at 80–90 °C. Mechanical energy is partially converted into heat as a result of friction between wood and pulpstone or refining plates. The fibers are removed from the wood matrix once the lignin from middle lamella has been heated to the temperature that exceeds its glass transition point.

Mechanical pulp is manufactured by mechanical defibration using two procedures: grinding of wood logs using a pulpstone (grinding process) and refining of
wood chips in a dedicated disc refiner (refining process). Figure 2.1 shows the mechanical pulping techniques.

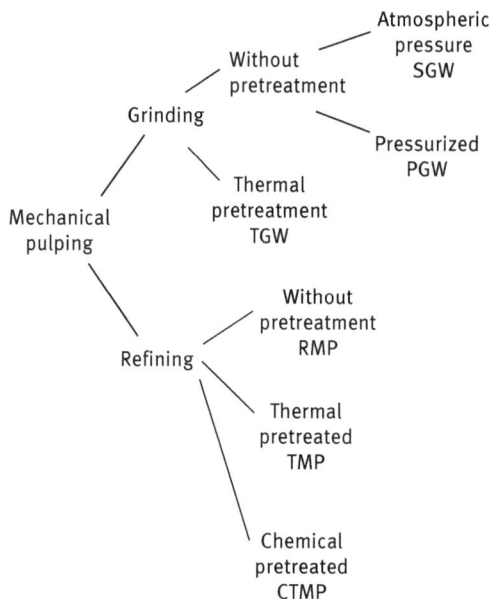

Fig. 2.1: Mechanical pulping processes.

As is shown in Fig. 2.1, the commercial grinding processes are SGW, PGW, and ther-
mogroundwood (TGW). The refiner processes are represented by RMP, TMP, and
CTMP.

A mechanical pulp mill includes the following stages: logs debarking and chipping, grinding of logs or refining of wood chips, and the screening and bleaching of
mechanical pulp.

The grinding process consists of pressing the wood logs against a rotating pulpstone in the presence of water. According to Karnis [15], the process is divided into
two consecutive stages: (i) softening of the wood and breakdown of the wood matrix and (ii) peeling and successive refining of the fibers. The pulpstone contains
grits that pass over the wood surface at very high frequencies, yielding a compres-
sion/decompression process. The grinding principle is shown in Fig. 2.2.

The wood structure is loosened due to the fatigue work performed by the grits.
At the same time, temperature increases rapidly in the grinding zone, causing the
lignin to soften. The presence of water reduces the softening temperature of lignin.

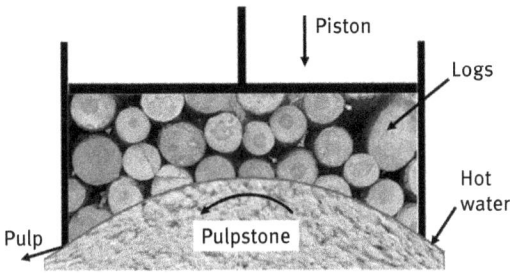

Fig. 2.2: Grinding process.

The high-frequency pressure pulsations determine deformation of the fibers and breakdown of bonds between the fibers. The fibers are peeled from the wood surface and then successively refined in the fibrous layer generated between the wood surface and pulpstone.

There are many parameters influencing the defibration of wood by grinding. Liimatainen and coworkers [16] showed that the most important are wood moisture, logs feeding speed (grinding pressure), pulpstone peripheral speed, water temperature in the grinder pit, surface shape of the pulpstone, and specific energy consumption.

The refining process uses wood chips that are fed in a dedicated disc refiner, where these are defibrated by passing through the refining zone. McDonald and coworkers [17] showed that in order to soften the wood matrix before refining, the chips are steamed and/or they undergo a mild chemical treatment. A mechanical, thermal, or chemical softening of lignin takes place.

The common refining processes are TMP and CTMP. In a TMP process, wood chips are steamed and fed into the refiner, which is pressurized, and the steam pressure corresponds to the temperature of the saturated steam. Berg [18] showed that the friction process between the fibers, and between the fibers and bar edges, causes fibers shortening and fibrillation to take place. The refining process consumes considerable energy that is partially converted into heat that increases the temperature in the refining zone between the plates, as presented by Muhic and coworkers [19]. The additional heating positively influences the chip defibration and fiber fibrillation and has an important influence on the final pulp quality. According to Miles and Karnis [20], the electrical energy demanded using refining to produce TMP pulp for magazine paper grades is 1,800 to 3,000 kWh/t, newsprint grades demand 1,800 to 2,200 kWh/t, and paperboard grades 1,000 to 1,400 kWh/t.

TMP is currently the most important mechanical pulping process and displaced older grinding processes, because similar paper properties could be achieved using a lesser amount of expensive chemical pulp, as shown by Illikainen [21] and Fernando and coworkers [22]. TMP is mostly used for producing printing papers such as newsprints, and uncoated and coated magazine papers.

Figure 2.3 presents the flowsheet of a TMP fiberline with two-stage chips refining. The chips are washed to remove sand and are steamed. The steamed chips are fed into the first refining stage where the wood material is broken into small fragments and shives. After the first refining stage, the semirefined chips feed in the second refiner for fibers separation. After refining, the pulp is screened and it enters the bleaching stage. Rejected material is processed in a two-stage refining line and after screening is combined with screened pulp. Bleached TMP pulp is sent to the paper machine.

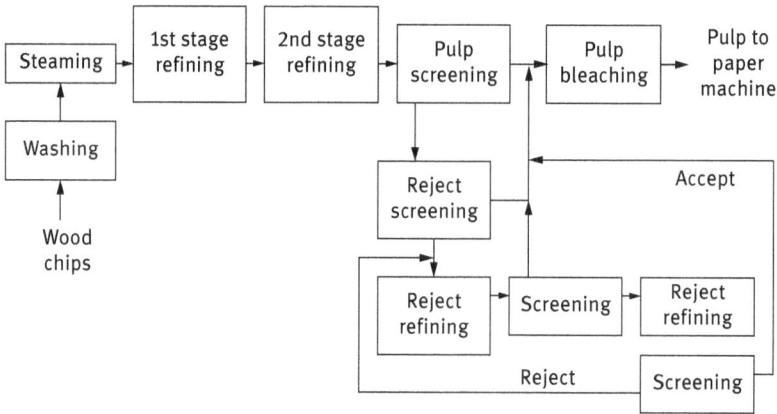

Fig. 2.3: Flowsheet of a TMP fiberline.

CTMP uses both steaming and chemical treatment for lignin softening. Chemical softening exhibits a major influence on the properties of fibers liberated from the wood matrix during the refining of chips. When wood chips are impregnated with chemicals, the softening due to swelling of the reactive middle lamellae and the primary wall regions is predominant, and this is where the defibration fracture zones are located, as was stated by Gorski [23].

During CTMP operations, the chemical treatment is carried out by the impregnation of wood chips. Sodium sulfite is the dominating reagent both in hardwood and softwood pulping. Johansson and coworkers [24] showed that the softening temperature of lignin depends on the sulfonate content of the wood, which is influenced by the sulfite dose and the temperature and duration of chemical treatment of the chips.

According to Lindholm and Kurdin [25], during the production of softwood pulp using CTMP, the parameters of the chemical treatment of the chips are sodium sulfite charge 2–4% on wood; temperature 120–135 °C; and retention time 2–15 min. The sulfonate content of the lignin ranges between 0.25 and 0.75% on pulp.

Figure 2.4 presents a typical flowsheet of a CTMP plant. The major difference between TMP and CTMP processes is the chemical impregnation stage; in this stage, chips are damped and compressed and then are expanded in a sodium sulfite solution. The expansion and condensation effects cause the absorption of the liquid phase into the wood structure.

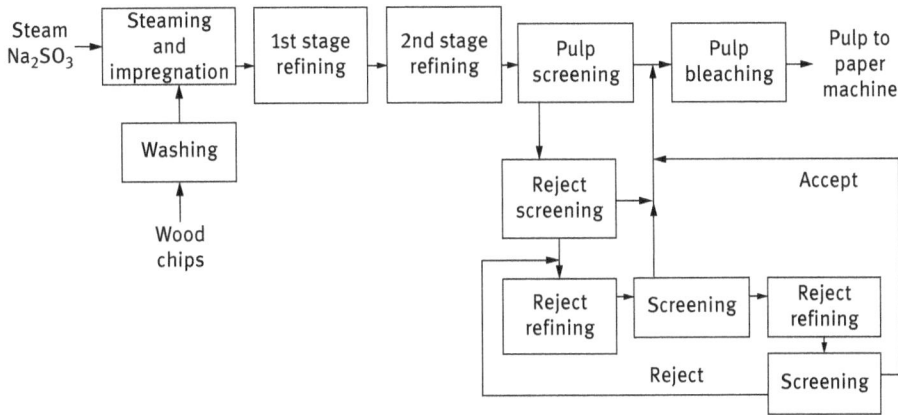

Fig. 2.4: Flowsheet of a CTMP fiberline.

The chips are refined in two stages and the obtained pulp is screened and bleached. The benefits of CTMP refer to the reduction of specific energy consumption and to an increase level of long fibers in the pulp.

The properties of the mechanical pulp depend on the pulping technique. Blechschmidt and coworkers [26] showed that the strength properties increase in the range SGW, RMP, TMP, and CTMP. Regarding the light-scattering coefficient, SGW shows the highest values, followed by RPM, TMP, and CTMP. These pulps are the raw material for the production of some of the most significant paper grades. Both softwood and hardwood species are used in mechanical pulping, and Tab. 2.1 lists the main end products.

Tab. 2.1: Paper grades achieved from the mechanical pulp of softwoods and hardwoods.

Wood species	Mechanical process	Paper grade
Softwoods	SGW, TMP CTMP	Newsprint, supercalendered, low-weight-coated (LWC) Tissue Liquid packaging boards
Hardwoods	PGW, CTMP CTMP	Tissues, printing Tissue, fluting

2.3 Semichemical pulping

The term semichemical pulping suggests that pulp is produced by a chemical treatment of chips, followed by a mechanical treatment, and both treatments exhibit a fairly equal contribution to the separation of fibers from wood matrix. During semichemical pulping, the partial remove of lignin and hemicelluloses takes place, so that the yield of pulp is 60–80%.

NSSC is the most important semichemical pulping process. The wood chips undergo partial chemical pulping using sodium sulfite liquor, followed by a treatment in a disc refiner to achieve the fibers separation. The sulfonation of the middle lamella lignin causes its partial dissolution so that the fiber–fiber bonds are weakened before the subsequent mechanical treatment. NSSC cooking liquor contains both Na_2SO_3 and Na_2CO_3, as presented by Obrocea and Gavrilescu [27]. Sodium sulfite represents the lignin sulfonation agent, while sodium carbonate acts as a buffer of liquor pH. The Na_2SO_3 consumption ranges between 6 and 9% on wood and the ratio of Na_2SO_3/Na_2CO_3 varies between 3:1 and 4:1. The pH of the liquor is 10–11 and cooking time is 0.5–2 h at 160–185 °C.

Hardwoods (beech, poplar, adler) are the best fiber source for NSSC pulp, as found by Nemati and coworkers [28]. Nonwood plants residues (straws especially) are also used, as studied by Ahmadi and coworkers [29] and by Leoponiemi [30]. NSSC pulp is used for the production of papers where stiffness is mostly essential, such as a corrugating medium. These papers show the best value for the flat crush of corrugated medium (Concora flat crush), as was found by Gavrilescu and Toth [31].

Zhao and Tran [32] studied the spent liquor of NSSC pulping (red liquor) and found that it contains lignosulfonates, hemicelluloses, organic acids, and ashes. The concentration of red liquor is quite low (5–8% dry matter) and for this reason its calorific value is small. The NSSC pulping plant is often integrated into a kraft pulp mill to facilitate chemical recovery by the so-called cross-recovery, where the red liquor is processed together with the kraft black liquor.

2.4 Chemical pulping

On a global scale, pulp is predominantly produced by chemical pulping processes, as shown in Fig. 2.5. Chemical pulp represents around two-thirds of total pulp production worldwide.

Figure 2.5 also shows that among chemical pulps, kraft pulp (bleached and unbleached) grades are the most produced if compared with sulfite pulp.

During the chemical pulping, reagents are used to split the macromolecule of lignin into fragments that are further dissolved in the cooking liquor. Fibers separation occurs when the lignin from the middle lamella is totally removed and, as a rule, no

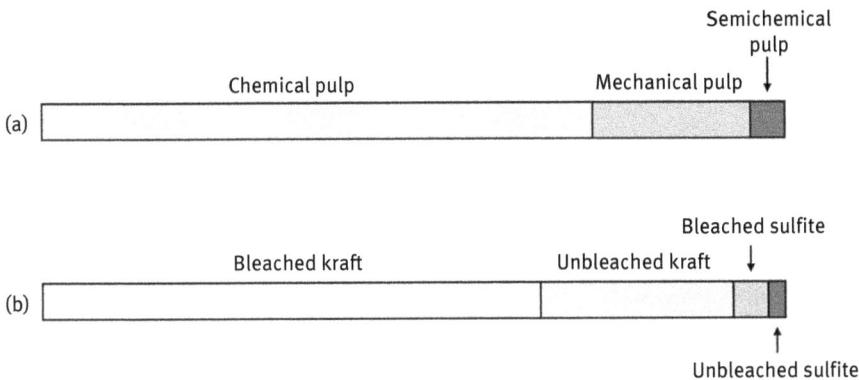

Fig. 2.5: Sharing of chemical, semichemical, and mechanical pulps (a), and the distribution of chemical pulps (b).

mechanical action is necessary. Chemical pulping requires a large quantity of reagents and a high temperature and pressure. Besides lignin, polysaccharides are also chemically modified and dissolved, so that the pulp yield ranges between 45 and 60%. An important volume of spent liquor is generated.

There are two major grades of chemical pulp: papermaking pulp and dissolving pulp. Papermaking pulp includes unbleached pulp and bleached pulp and is characterized by yield, residual lignin content, brightness, and strength properties. Dissolving pulp differs substantially from papermaking pulp and it is characterized by high alpha-cellulose content, low noncellulosic carbohydrate content, defined cellulose molar mass and molar mass distribution, low extractives and low ash content, and high reactivity toward derivatising chemicals.

The main stages of a chemical pulping process are wood preparation, cooking, pulp washing and screening, pulp bleaching, and recovery of cooking chemicals.

2.5 Kraft pulping processes

2.5.1 General description

Kraft (or sulfate) pulping represents the main cooking process, accounting for nearly 90% of the global chemical pulp production as presented by Gavrilescu and Craciun [33]. The term *kraft* refers to the fact that kraft pulp exhibits good strength properties. The cooking liquor (usually called *white liquor*) contains sodium hydroxide and sodium sulfide as the active alkali, which reacts with lignin and promotes degradation of lignin polymer to fragments that dissolve in the cooking liquor.

Figure 2.6 shows the flowsheet of kraft pulping. Wood preparation includes logs debarking, logs chipping, and chips screening. Screened chips are temporary stored in piles and then transported into the digesters house.

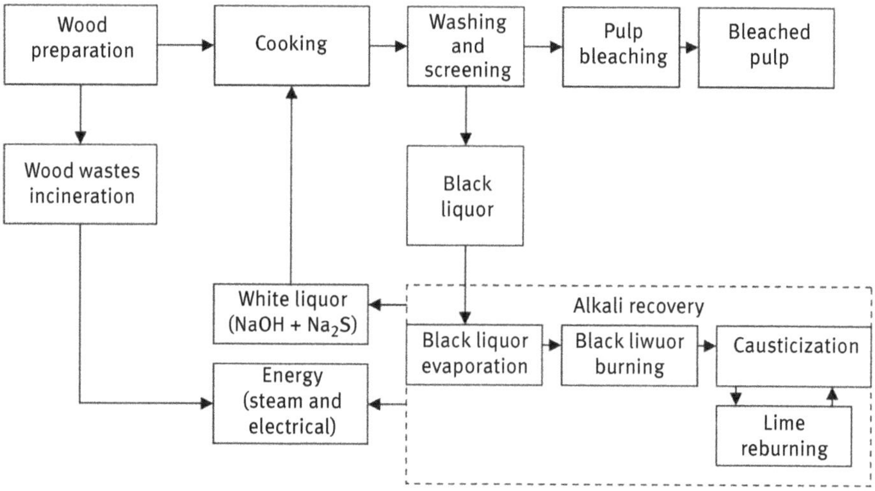

Fig. 2.6: Flowsheet of kraft pulping.

Cooking is performed in discontinuous or continuous digesters at a high temperature and pressure. After cooking, pulp is washed and screened and is finally bleached to obtain the bleached pulp, which is used for printable paper grades, while unbleached pulp is used for the production of packaging paper grades. Kraft pulping generates spent liquor (denoted as black liquor) that contains the organic compounds removed form wood and the inorganic chemicals originating from white liquor. The black liquor must be processed to recover the cooking reagents and to valorize the heat of the organic matter; the alkali recovery process includes black liquor evaporation and burning, causticizing of green liquor, and lime reburning.

White liquor is characterized by the quantitative and qualitative parameters presented in Tab. 2.2. The most important are the concentration of active alkali (AA) and sulfidity (S).

Detailed studies were made regarding the composition of kraft cooking liquor. Gierer [34] found that in the white liquor, the main ionic species are Na^+, HO^-, HS^-, and CO_3^{2-}. Other ionic species (S^{2-}, HCO_3^-, SO_4^{2-}) are of negligible importance in the electrolyte equilibria of the cooking liquor. Gierer also stated that hydroxide ions (HO^-) and hydrogen sulfide ions (HS^-) are the only species that react with lignin during kraft pulping.

The typical composition of white liquor is listed in Tab. 2.3. The major components are sodium hydroxide and sodium sulfide and the minor components are

Tab. 2.2: Parameters of white liquor.

White liquor parameter	Formula
Quantitative parameters	
Active alkali (AA)	$NaOH + Na_2S$
Effective alkali (EA)	$NaOH + 0.5Na_2S$
Total alkali (TA)	$NaOH + Na_2S + Na_2CO_3 + Na_2SO_4$
Qualitative parameters	
Sulfidity (S)	$S = Na_2S/(NaOH + Na_2S)$
Causticity (C)	$C = NaOH/(NaOH + Na_2CO_3)$
Degree of reduction of sodium sulfate (DR)	$DR = Na_2S/(Na_2S + Na_2SO_4)$

Tab. 2.3: Typical composition of white liquor (expressed as Na_2O and NaOH).

Component	Concentration (g/L)	Concentration (Na_2O; g/L)	Concentration (NaOH; g/L)
NaOH	100	100 (31/40) = 77.5	100
Na_2S	35	35 (31/39) = 27.8	35 (40/39) = 35.9
Na_2CO_3	20	20 (31/53) = 11.7	20 (40/53) = 15.1
Na_2SO_4	6	6 (31/71) = 2.6	6 (40/71) = 3.4
Active alkali concentration	–	105.3	135.9
Total alkali concentration	–	119.6	154.4

sodium carbonate and sodium sulfate. Under kraft pulping condition, sodium carbonate and sodium sulfate are not able to react with lignin.

Extensive researches have been carried out to elucidate the chemistry of kraft cooking. Potthast [35] and Sakakibara and Sano [36] showed that the reactions of lignin during kraft pulping can be divided into two main categories, degradation reactions, and condensation reactions, respectively.

Degradation reactions split the lignin macromolecule into fragments that are able to dissolve in the cooking liquor. According to Gellerstedt [37] and Gierer [38], the common degradation reactions occurring during kraft pulping include the cleavage of α-aryl ether and β-aryl ether bonds of lignin.

Condensation reactions are not desirable as they lead to the formation of alkali-stable linkages, thereby increasing the molecular size of lignin fragments, as was found by Gellerstedt and Lindfors [39]. These reactions occur if wood impregnation with white liquor is not finished and/or cooking temperature is raised too fast. According to the same authors, during a normal kraft cook, no extensive condensation involving the aromatic rings of the lignin takes place [40].

It is known that kraft spent liquor is black in color and kraft pulp is darker in comparison with pulp obtained by other pulping procedures. Robert and coworkers [41] showed that the reason for this behavior is the formation of lignin chromophores like quinonoide structures and stilbenes.

A particular reaction that occurs in kraft pulping studied by Gierer [42] involves the cleavage of methyl aryl ether bonds and generation of mercaptans represented by metylmercaptan, dimethylsulfide, and dimethyldisulfide. Mercaptans are responsible for the specific smell of the kraft pulp mill.

Sixta [43] studied the evolution of pulp yield and pulp delignification degree, and found that lignin is eliminated on a large scale during kraft pulping. Besides lignin, nonlignin components, mainly carbohydrates, are removed in some extent and, as a result, pulp yield decreases.

The reactivity of carbohydrates in kraft pulping depends on their structural features such as morphology, crystallinity, and degree of polymerization. Gentile and coworkers [44] found that hemicelluloses are degraded more than cellulose in alkaline media at high temperature.

Sjöström [45] reviewed the alkaline degradation of carbohydrates during kraft cooking and concluded that there are two basic degradation reactions: peeling and alkaline hydrolysis. In the peeling reaction, a step-by-step depolymerization occurs at the reducing end sites of the carbohydrates. The reaction generates a monosaccharide that finally transforms it into an isosaccharinic acid. During the reaction a new reducing end on the remainder carbohydrate is also formed, which can undergo further peeling reactions. The same authors also showed that the carbohydrate material lost in the peeling reaction is converted into various hydroxy acids, which consume active alkali and consequently reduce its concentration in the pulping liquor. Hemicelluloses (glucomannans and xylans) are more engaged in peeling reactions compared with cellulose. About 50–60 monomer units are peeled off from the carbohydrate macromolecule and, after that, a stopping reaction occurs.

Gustavsson and Al-Dajani [46] showed that alkaline hydrolysis of cellulose takes place at the end of the cook and determines the reduction of polymerization degree of cellulose. Due to the high temperature and strong alkaline medium, a random cleavage of cellulose macromolecule occurs. The resulting fragments may dissolve in the cooking liquor or participate in the peeling reactions. Alkaline hydrolysis of cellulose explains the severe loss of pulp yield in the final part of the kraft cook.

An important reaction of the carbohydrates in alkaline pulping is the formation of hexenuronic acids, which are formed by the alkaline degradation of 4-O-methyl-D-glucuroxylans. It was established that the presence of hexenuronic acids in the kraft pulp influences the reagents consumption during pulp bleaching and stability of pulp brightness [47].

Olm and Tistad [48] studied the kinetics of kraft pulping and showed that the dissolution of lignin and carbohydrates proceeds in three phases denoted as initial, bulk, and final delignification. Table 2.4 lists the characteristics of kraft pulping phases.

Tab. 2.4: Phases of delignification during kraft pulping.

Phases of delignification	Features
Initial	Slow delignification; lignin content is reduced by 15–25% of the initial amount, as compared to about 40% of the hemicelluloses (cooking selectivity is low)
Bulk	High delignification rate; about 70% of the lignin is dissolved as compared to 5–7% of carbohydrates (high selectivity)
Final (residual)	Slow delignification; intensive degradation of carbohydrates. Low selectivity

Initial delignification is a slow phase and only 15–25% of initial lignin is dissolved. In this stage, around 40% of hemicelluloses are dissolved so that the cooking selectivity is low. Bulk delignification is the main stage of cooking, with around 70% of the lignin being dissolved. The delignification rate is high as is the process selectivity. Final delignification begins at a delignification rate of about 90%. To avoid excessive pulp yield loss, kraft cook is often stopped before the final delignification [49].

Besides the major chemical constituents (cellulose, lignin, and hemicelluloses), wood contains a large number of low molecular organic compounds, denoted as extractives. The extractives can be separated from wood with hot water or organic solvents. Umezawa [50] found that the extractives are complex mixtures of fats, fatty acids, fatty alcohols, phenols, terpenes, steroids, resin acids, rosin, waxes, and other minor organic compounds.

During kraft pulping, extractives show a complex behavior. Some of them are volatile and accumulate in the gaseous phase of the digester; this fraction is recovered from the digester relief condensate and is called *sulfat turpentine*. Most of extractives are soluble in cooking liquor and may react with alkali. Fatty acids and resin acid esters are neutralized and recovered as tall oil soap. Lindström and coworkers [51] found that crude tall oil from pine contains 40–60% resin acids, 40–55% fatty acids, and 5–10% neutral constituents. Abietic and dehydroabietic acids cover over 60% of the resin acids, while oleic and linoleic acids predominate in the fatty acid fraction.

2.5.2 Kraft pulping parameters

Kraft pulping parameters include those factors influencing delignification rate, pulping selectivity, pulp properties, and specific consumption of raw materials and energy. MacLeod [52] showed that kraft pulping variables can be divided into key factors and minor factors. The key factors exert a significant influence on pulping results and they are active alkali charge, white liquor sulfidity, cooking temperature and cooking time, wood specie, and chip quality. The minor factors are liquor-to-wood ratio, black liquor addition, and the presence of sodium carbonate in the white liquor.

The active alkali charge represents the ratio between active alkali quantity (expressed as NaOH or Na_2O) and wood mass. Casey [53] established that the theoretical consumption of active alkali in kraft pulping is 15% NaOH on wood. In order to enhance the delignification rate, the practical active alkali charge is 50–60% higher, so that the cook is performed with an excess of active alkali. At the end of the cook, the concentration of active alkali must be 5–10 g/L NaOH to avoid the reprecipitation of dissolved lignin on fibers.

As the active alkali charge is higher, more lignin dissolves from the wood, so that this parameter determines the degree of pulp delignification. The active alkali charge ranges between 17 and 19% NaOH for obtaining hard kraft pulps and 20 and 25% NaOH for bleachable pulp grades. Softwoods require 10–12% more active alkali compared with hardwoods.

Casey [53] also showed that the active alkali concentration depends on active alkali charge and on liquor-to-wood ratio. The kraft cooking requires at least 25 g/L NaOH active alkali for fibers separation. At higher values, the delignification rate increases at the given temperature. Practically, kraft cooking requires an initial active alkali concentration of 40–60 g/L NaOH.

The sulfidity of white liquor is expressed as the percent ratio between sodium sulfide and active alkali. According to Kleppe [54], the sulfidity strongly influences the cooking rate and process selectivity. The pulping rate significantly increases if sulfidity rises up to 16%. The delignification rate continues to increase up to 35–40% sulfidity at a constant active alkali charge. As the sulfidity increases, the cooking selectivity enhances, which means that the pulp yield increases at the same pulp lignin content. A high sulfidity is advantageous for other many reasons: reduces the lime consumption for green liquor causticization and decreases fuel consumption of lime kiln, enhances the sedimentation velocity of lime particles during white liquor clarification, and raises the fluidity of the black liquor. The sulfidity ranges between 25 and 35% for cooking of hardwoods and 35 and 40% for softwoods.

Cooking temperature is the main factor influencing the delignification rate with the rate approximately doubling for an increase in cooking temperature of 8 °C. Temperature negatively influences process selectivity; in other words, at the same lignin content, increasing the temperature will decrease the pulp yield. The cooking temperature depends on the wood species, pulp grade, and cooking plant (discontinuous or continuous). Cooking of softwoods uses temperature of 170–175 °C, while cooking of hardwoods is performed at 165–170 °C.

The pulping time must be correlated with the cooking temperature in order to obtain pulp with the same properties. The *H*-factor is a kinetics model that combines cooking temperature and time into a unique parameter that measures the degree of cooking. Rantanen [55] found that the degree of delignification can be accurately estimated using H-factor if other pulping parameters such as active alkali and sulfidity remain constant. Bleachable kraft pulp grades are obtained for H-factor values in the range of 1,000–2,000.

Kraft pulp can be obtained from any kind of raw material, wood species, and nonwood plants. Wood species refer to softwoods and hardwoods, while nonwood plants are mainly represented by straw, reed, flax, hemp, kenaf, and bamboo [56].

Among wood species, softwoods are more preferred by the papermakers, as softwood pulp has the best strength properties. Hardwoods exhibit higher pulp yield, and assure good optical properties and good printability of paper. Softwoods contain more lignin, which is less reactive compared with lignin from hardwoods. For this reason, pulping of softwoods requires stronger conditions (more active alkali addition and higher cooking temperatures).

Nonwood plants able to be used in papermaking are very different regarding their morphological structure and chemical composition. According to Sanjoikari-Pakhala [57] nonwood plant fibers can be divided into four categories: grass fibers (cereal straw, reeds, bamboo, sugarcane), bast fibers (flax, hemp, kenaf, ramie), seeds fibers (cotton), and leaf fibers (sisal, acaba). Kraft pulping and other pulping methods (soda, organosolv) are used for the cooking of nonwood plants, as investigated by Leponiemi [58].

Nonwood pulp fibers are used for the production of common paper grades and specialty papers. Ashori [59] showed that bast fibers are used for papers when strength, permanence, and other special properties are needed; examples include permanent paper, fancy paper, currency, and cigarette papers.

One of the main problems in pulping of nonwood plants is their high ash concentration, especially silica. Huang and coworkers [60] established that in alkaline pulping, silica dissolves in the cooking liquor, and when the spent liquor is processed, the concentration of silica compounds increases to such an extent that it may form scale deposits.

The bark content in chips is detrimental above a certain level for many reasons: reduces the pulp yield, increases the energy and chemical consumption, and reduces the cleanliness of the pulp; hence, the bark content in chips must be restricted in accordance with the pulp grade. According to Brännvall [61], unbleached kraft pulp can tolerate certain amount of bark content, but bleachable pulp grades are more restrictive regarding chip bark content.

The chip quality refers to chip dimensions and chip dimensional uniformity both being of huge importance in all pulping processes. The chip dimensions must be as uniform as possible and the dimension limits of commercial chip are length 25–40 mm, thickness 3–8 mm, and width 15–30 mm. The most important factor among chip dimensions is chip thickness [62, 63]. Gullichsen and coworkers [64] showed that the variation of the chip thickness is the main cause of the nonuniformity of the kraft cook. A mixture of thick and thin chips leads to a pulp that contains less delignified fibers and fibers of very low lignin content.

The liquor-to-wood ratio represents the ratio between the volume of liquid phase (sum of the white liquor, black liquor, and water from wood moisture) and wood mass in the digester. During kraft cooking, it is necessary to assure the liquor phase is sufficient for complete impregnation of the chips. The significance of the liquor-to-wood ratio on the reaction kinetics of kraft pulping was studied by Gustavsson [65]. It was found that a low liquor-to-wood ratio, at a constant active alkali addition, increases the alkali concentration and raises the rate of the process. At the same time, a low liquor-to-wood ratio reduces the energy consumption for liquor heating and pumping. In the high liquor-to-wood ratio cooks, dissolved wood components could diffuse more easily into the black liquor [65]. Andersson and coworkers [49] stated that from a practical point of view, a low liquor-to-wood ratio is preferred and the values range from 2.5 to 3.5.

White liquor contains variable quantities of sodium carbonate depending on the degree of causticization. Sodium carbonate is not an effective cooking reactant, being inactive both on lignin and carbohydrates. Carbonate may have a positive effect in the kraft cook due its contribution to the alkalinity of the liquor in the final phase of cooking. However, Lundqvist and coworkers [66] found that under kraft cooking conditions, with a regular addition of white liquor, the contribution of carbonate to the alkalinity is practically zero. Sodium carbonate can precipitate calcium ions from cooking liquor yielding calcium carbonate, which forms scale deposits on heating surfaces of the pulp mill [67].

In order to assure a sufficient liquid phase for chips impregnation, a certain volume of black liquor is filled into the digester. The black liquor temperature is high and it contains the excess active alkali charged with white liquor at the beginning of the cook. The addition of the black liquor exhibits the following advantages: increases the starting temperature of the cook, reduces the white liquor charge in the digester, and increases the dry solid content of the spent liquor.

2.5.3 Kraft pulping technology

Kraft cooking is performed in discontinuous (or batch) and continuous processes. Features favoring batch kraft cooking are as follows:
- High availability – one digester down for repair does not stop pulp mill production.
- Good flexibility to switch between softwoods and hardwoods as raw materials for pulping.
- Easy to add pulp capacity of the fiberline by adding another digester.

Features favoring continuous kraft cooking, when compared with batch cooking, are as follows:
- Higher capacity fiberlines has been possible.
- Lower steam and electrical energy consumption.
- Characteristics of the pulp are more uniform.
- Higher black liquor solids and consequent lower evaporation costs in the black liquor recovery cycle.
- Lower specific investments costs.

A batch kraft cooking cycle consists in the following operations: chip filling, liquor fill, heating and cooking, and pulp discharge. Chip filling must assure an amount of wood as much as possible in order to increase the pulp yield per cook. The uniform distribution of chips across the digester is obtained by using steam as packing medium. The chips are warmed and the air is displaced from inside the chips. The cooking liquor (a mixture of white and black liquors) is charged to the bottom of the digester; once the digester was charged, the heating period starts. The cooking liquor is continuously extracted by the digester strainer and recirculated in a heater. The temperature of digester content is raised up to the cooking temperature according to cooking diagram. The digester content is kept at cooking temperature until the desired lignin content of the pulp is reached. At the end of the cook, digester content is transferred in a dedicated pulp tank.

Batch cooking uses a stationary vertical digester having a volume up to 400 m^3. A pulp mill may be equipped with a number of digesters based on the capacity of the fiberline. An indirect heating system consisting in a liquor heater and a recirculating pump is preferred. Figure 2.7 shows the conventional batch cooking system.

Batch kraft cooking is a high consumer of steam, so that effort has been made to reduce the energy consumption. Rapid displacement heating (RDH) is a modification of kraft cooking in a batch digester, which decreases the specific steam consumption. According to Sezgi [68] the purpose of the RDH method is to recover the energy of the warm black liquor and to decrease the digester heating time. The RDH process involves impregnation of steam-packed chips with black liquors of increasing temperatures. The final hot black liquor is then displaced with a mixture of hot, white and black liquors at the cooking temperature. Since the liquor is preheated, the heating time is minimal and pulp production is increased [69].

Teder [70] found that the key advantage of the RDH process is the decrease in digester steam consumption, with a reduction of 60–80% compared with conventional batch kraft pulping. The energy savings accrue from the shorter steaming time required to reach white-liquor cooking temperature.

Super batch cooking is an improvement of RDH process, the major difference being that the white liquor is added during the cooking phase. Pursiainen and coworkers [71] found that extended cooking of softwoods and hardwoods to obtain

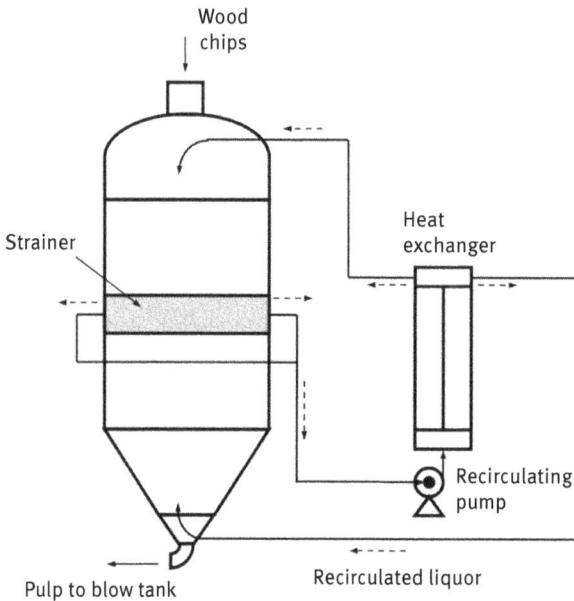

Fig. 2.7: Conventional batch cooking.

pulps with low lignin content is completely possible without negative effects on pulp quality.

A recent development of the batch process is continuous batch cooking (CBC) that combines the advantages of batch operation with the continuous preparation of cooking liquors. The impregnation liquor and the cooking liquor are prepared in the tank farm by adjusting their alkali concentration and temperature. During the cook, the liquor continuously circulates between tanks and digester. Wizani and coworkers [72] showed that the CBC advantages are due to an even alkali profile throughout the cooking stages, a reduced cooking time, and an improvement in pulp quality.

The kraft continuous pulping was developed in the 1950s and for decades became the dominant pulping process. The capacity of the pulp mills rapidly increased and now fiebrlines producing 3,000 t/day of kraft pulp using a unique digester in operation [73].

During continuous cooking chips and chemicals are continuously fed into the top of the digester and pulp is removed from the bottom. Digester is divided to zones, in which cooking phases take place. During conventional continuous cooking, the steamed chips and cooking liquor are fed by a high-pressure feeder to the top of the digester. The chips fall into the digester and the transfer circulation liquor returns to the high pressure feeder. The chips gravitationally move down the digester zones: impregnation, heating, cooking, hi-heat washing of pulp, and pulp discharge. Heat is provided by the indirect heating of cooking liquor extracted through circumferential screens [74]. The resulting black liquor is extracted through screens located about halfway up the digester. Figure 2.8 shows the principle of conventional continuous cooking.

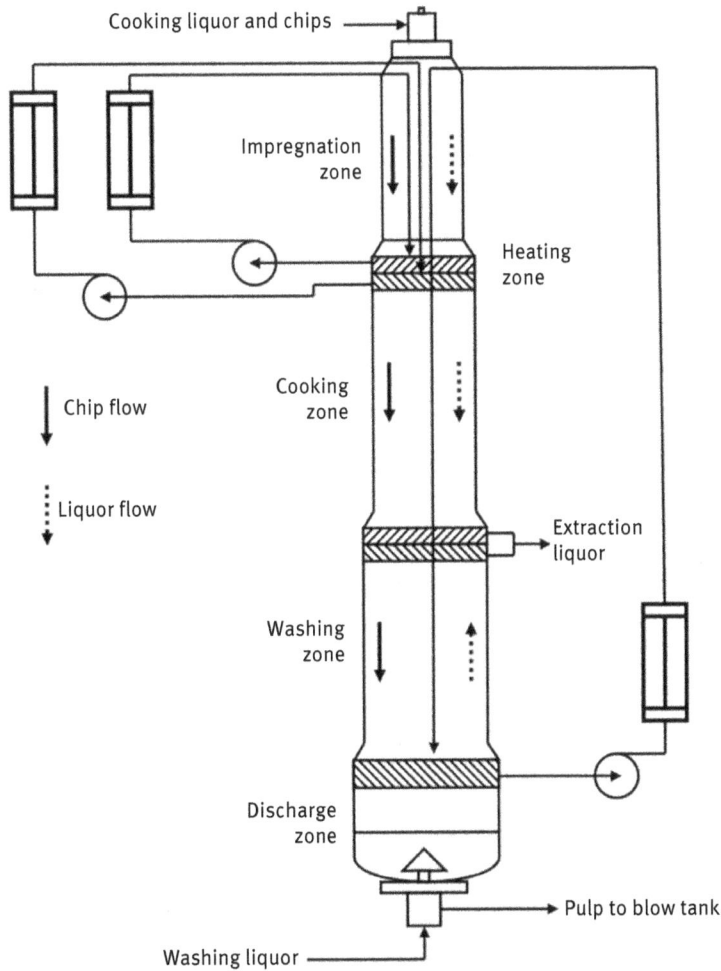

Fig. 2.8: Conventional continuous cooking.

In order to increase the capacity of pulp mills and to improve pulp properties, modifications of the conventional kraft continuous cooking have been undertaken. The basic principle is the creation of a countercurrent flow of the pulping chemicals and wood chips. Atalla and coworkers [75] showed that it is feasible to keep the concentration of active alkali lower and more uniform throughout the cooking process, and also to reduce the concentration of dissolved lignin during the cook.

During the modified continuous cooking (MCC) process, white liquor is added at several points, and the cook starts at a lower concentration of active alkali and ends at a higher concentration of active alkali, compared with conventional continuous cooking. In this way, the degradation of carbohydrates is reduced and the delignification is enhanced, as stated by Norden and Teder [76]. Extended modified continuous cooking (EMCC) uses the washing zone of the digester to further increase delignification. A volume of white liquor is added in the hi-heat washing circulation and the temperature of hi-heat washing zone is increased. Isothermal cooking is widely similar with EMCC but uses two sets of wash circulation loops with individual heaters. The delignification is extended by raising the temperature in the washing zone to levels normal for pulp cooking [77].

The latest development of the kraft process is lo-solids cooking, which aims to minimize the concentration of dissolved wood solids throughout the cooking stages, to maintain an even alkali profile and a minimal cooking temperature. Many of these objectives are similar to those of MCC and EMCC processes. According to Marcoccia and coworkers [78], Lo-Solids pulping aims at decrease the concentration of all dissolved wood solids in both the bulk and final stages of cooking, whereas earlier MCC methods were mainly designed to decrease the concentration of dissolved lignin in the final stage of cooking. During Lo-Solids pulping, multiple cooking liquor extractions are performed. White liquor and brown stock washer filtrate are added to replace the extracted liquors. An optimum profile of alkali concentration and digester temperature at every cooking stage is maintained; benefits include improved process selectivity, extended pulp delignification, and increased pulp strength, as found by Lindstrom and Lindgren [79].

2.6 Sulfite pulping

Sulfite pulping includes the following cooking variants: acid sulfite, disulfite, neutral sulfite, and alkaline sulfite. The alternatives of sulfite pulping differ by the composition and pH of cooking liquor, as listed in Tab. 2.5.

Tab. 2.5: Sulfite pulping alternatives.

Sulfite processes	pH of cooking liquor	Active species	Applications
Acid sulfite	1–2	H^+, HSO_3^-	Dissolving pulp, newsprint, tissue
Disulfite	3–5	HSO_3^-	Newsprint, tissue, printing paper
Neutral sulfite	6–9	SO_3^{2-}	Corrugated medium
Alkaline sulfite	10–13	HO^-, SO_3^{2-}	Corrugated medium, package grades

The acid sulfite pulping was the dominant cooking process until the 1940s when it was surpassed by kraft cooking. Acid sulfite pulp is brighter when compared with kraft pulp and can be used without bleaching to produce some printing paper grades. Acid sulfite pulp is weaker than kraft pulp, and not all species of wood can be cooked easily. In addition, the cooking cycle is long (up to 10–12 h) and chemical recovery of cooking chemicals, from sulfite spent liquor, is complicated or impractical. The pollution potential of acid sulfite pulping is much higher than kraft liquor because sulfite spent liquor, is reach in organic dissolved material, and due to a high sulfur dioxide loss in the air.

The cooking liquor is highly acidic and contains a mixture of sulfurous acid and sodium disulfite, the active species being HSO_3^- and H^+. During the cook, lignin macromolecule is split by hydrolysis and sulfitolysis reactions into smaller fragments that dissolve in the liquid phase. According to Sixta [80], sulfonation is the main reaction of sulfite pulping and renders the lignin sufficiently hydrophilic to dissolve in the cooking liquor. Soluble lignosulfonates are formed in the process. During acid sulfite pulping, lignin condensation reactions may be important if the disulfite ion concentration is low and the cooking liquor acidity is high. Poor chips impregnation will always increases the risk of lignin condensation that reduces the lignin reactivity and darkness of the pulp and, finally, the so-called black cook can be attained.

Gellerstedt [81] found that the main reaction of carbohydrates during acid sulfite cooking is the acid-catalyzed hydrolysis of glycosidic bonds. Hemicelluloses are degraded much more than cellulose, yielding monosaccharides that dissolve into the cooking liquor. At an elevated temperature, monosaccharides are partially dehydrated to furfural and hydroxymethylfurfural.

According to Wolfinger and Sixta [82] the rate of delignification in acid sulfite pulping is highly influenced by temperature and cooking liquor composition. Temperature elevation exerts a complex influence: it increases pH of cooking liquor, decreases sulfur dioxide solubility, and increases lignin dissolution and carbohydrates decomposition. The delignification rate increases vastly with the increasing concentration of sulfur dioxide in the cooking liquor.

Process variables in acid sulfite pulping are wood species, pH of cooking liquor and cooking base, and temperature. Acid sulfite pulping is sensitive to wood species, due its poor ability to dissolve extractives. Spruce wood is largely used while rich-extractives softwoods (pine wood) cannot be successfully cooked. Before cooking, wood must be stored for a certain period to reduce its moisture and extractives content. The most important variable of sulfite process is the pH of cooking liquor. Acid sulfite cooking is performed at pH 1–2, the pH at which all bases (calcium, sodium, magnesium, and ammonium) are fully soluble. By replacing the insoluble calcium base with sodium or ammonium base (soluble bases), a better delignification takes place and less scaling problems occur. Young [83] found that soluble bases give pulps with better yield and higher brightness. Cooking temperature determines the reaction rate and process selectivity. Acid sulfite pulping requires a temperature between 130 and 145 °C. With every increase of 10 °C cooking temperature, the delignification rate doubles, but the cook selectivity is reduced.

Disulfite pulping is carried out at a pH of 3–5 and the disulfite ion is the dominant reagent in the cooking liquor. Disulfite liquor is free from sulfurous acid (H_2SO_3). Under mild acidic conditions, the delignification rate decreases and acid hydrolysis of cellulose reduces. In order to compensate for the lower rate of lignin dissolution, disulfite pulping uses higher cooking temperatures (155–170 °C), compared with acid sulfite pulping. Cooking liquor may contain either sodium disulfite or magnesium disulfite, which are fully soluble at the pH of the liquor [84].

The major advantage over acid sulfite pulping is that the strength properties of disulfite pulp are significantly better than that of the acid sulfite pulp. The advantages over kraft pulping include a considerable higher pulp yield at the same lignin content, higher unbleached pulp brightness, and better pulp bleaching capacity.

Industrial applications include the Arbisco process that uses sodium disulfite and Magnefite that uses magnesium disulfite, as shown by Biermann [85]. Both processes are used to produce pulps for newsprint. The Magnefite process is used at pH 4.5 and 160 °C temperature with $Mg(OH)_2$ as the base. The cooking cycle is 6–8 h and pulp yield ranges 55–65% [83]. Chemical recovery is necessary due to the high cost of the base. Bailey [86] showed that Magnefite spent sulfite liquor is concentrated by evaporation and burned in a recovery boiler. Solid phase contains magnesium oxide (MgO) and the flue gases contain SO_2, which is recovered. The MgO is slurried into the water to give a slurry containing $Mg(OH)_2$, which is used to scrub SO_2 from the flue gases.

Alkaline sulfite (AS) pulping uses a cooking liquor containing a mixture of sodium sulfite and sodium hydroxide, with a pH of 10–13. This process was developed during 1980s as an alternative to kraft pulping, in order to reduce the environmental impact and to maintain pulp strength properties at the level of kraft pulp [87]. When compared with kraft pulping, AS pulping shows a very low rate of delignification and the pulp yield at the same lignin content is lower, as was found by Kettunen and coworkers [88]. In addition, the recovery of AS pulping chemicals is more difficult as compared with kraft recovery system. For these reasons, AS pulping was no applied on a commercial scale; however, interest in AS pulping increased when it was found that the addition of small quantities of anthraquinone (AQ) increased the delignification rate [89]. Moreover, if methanol is added, the delignification rate further increases and, on this basis, a new pulping method (alkaline sulfite–anthraquinone–methanol; ASAM) was developed [90]. Kordsachia and coworkers [91] found that the presence of methanol in the alkaline sulfite/AQ-system provides pulps with a highly selective delignification, yielding low lignin content with high viscosity and high strength properties.

Despite all of the above-mentioned advantages, the ASAM pulping has failed to apply at industrial scale for several reasons: the efficiency of the sodium sulfite recovery system is not economically feasible, the presence of methanol needs a dedicated recovery plant, and the concurrent kraft pulping process has been vastly improved over the last years and has diminished the advantages of the ASAM pulping [92].

2.7 Organosolv pulping

Kraft pulping is the dominant process used in producing of chemical pulp, due to its advantages related to high pulp strength properties and raw materials versatility. However, there are some limitations regarding its environmental impact. The presence of sodium sulfide in the cooking liquor creates the problem of mill odor, originating from the air emission of reduced sulfur compounds (mercaptans and hydrogen sulfide) [93, 94]. In addition, during pulp bleaching, large volumes of wastewaters which contain organic chlorine compounds are generated [95].

During 1980s and 1990s, the possibilities for substituting the kraft process were widely studied and a number of new pulping processes were developed. The new processes use organic chemicals to dissolve lignin and to separate the fibers from the wood, among their objectives being cited [96]:

- The process should be sulfur free to avoid malodorous gases generation.
- It should be capable to dissolve most of the wood lignin with very little loss of cellulose and hemicelluloses.
- It should not require higher temperatures, pressures, or pulping times than used in kraft pulping.
- It should have a simple chemicals recovery system.
- The pulp quality should be at least equal to the quality of the kraft pulp.
- The pulp should be bleached without chlorine chemicals.

The new pulping processes are grouped in the so-called organosolv pulping processes that use different types of organic compounds: alcohols (methanol, ethanol), acids (formic, acetic), phenols (phenol and cresols), esters (ethyl acetate), amines (methylamine, ethylenediamine), and ketones (methylethylketone). A classification of the organosolv pulping processes according to the basic mechanism involved in lignin dissolution is listed in Tab. 2.6.

Table 2.6 shows that several organosolv pulping methods have been developed but many of them have not progressed past the laboratory test stages. A limited number of pulping methods based on organic acids or alcohols have been tested at pilot level [107, 108]. According to Leponiemi [109], an advantage of organosolv processes is the formation of useful by-products such as furfural, lignin, and hemicelluloses.

Most problems of the organosolv processes are related to the pulp quality, which is normally lower than the quality of corresponding kraft pulps, and solvent recovery. Organic solvents are more expensive than the inorganic reactants used in kraft pulping and for this reason they have to be recovered at high recovery rates.

Only small organosolv pulp mills are in operation today, mainly producing pulp from wheat straw. A pulping process designed for the manufacture of bleached pulp, lignin and hemicelluloses from cereal straw was developed by Compagnie Industrielle de la Matière Végétale (CIMV process) [110]. According to CIMV process, wheat straw is treated at atmospheric pressure with a mixture of acetic acid/formic acid/water,

Tab. 2.6: Classification of organosolv pulping processes.

Pulping method	Main characteristics	Literature
Autohydrolysis		
Ethanol–water (no catalyst added)	42.5% ethanol, 185 °C; 55% ethanol, 195 °C	[97, 98]
ALCELL	50% ethanol, 195 °C, three displacement stages of pulping	[99, 100]
Acid-catalyzed		
Acetosolv	93% acetic acid, 0.5–3.0% hydrochloric acid, 110 °C	[101, 102]
Phenol pulping	Phenols, 100 °C	[103]
Alkaline processes		
Organocell	Methanol, anthraquinone, sodium hydroxide	[104]
Ethanol alkali	Etanol, sodium hydroxide	[105, 106]
Oxidative processes		
Milox	80% formic acid, 4% H_2O_2	[107, 108]

30/55/15 (v/v/v), for a reaction time of 3.5 h at 105 °C. In these conditions, wheat straw lignin dissolves and the hemicelluloses are hydrolyzed in oligosaccharides and monosaccharides with a high xylose content. The obtained pulp is screened and bleached. Organic acids are then recycled by concentration of the extraction liquor containing lignin and hemicelluloses. The concentrated extraction liquor is treated with water to precipitate lignin, which is recovered by high-pressure filtration.

Another organosolv process in operation that is based on full use of annually renewable fibers is Chempolis process developed at Chempolis Ltd, Oulu, Finland, by Rousu and coworkers [111]. Common nonwood materials, such as wheat and rice straw, bagasse, and different reeds and grasses are used in pulp production. Delignification is performed in a single-stage formic acid cooking, which is carried out at a temperature between 110 and 125 °C, and the cooking time can vary between 20 and 40 min; bleaching of the pulp is performed with hydrogen peroxide in several stages. The recovery of chemicals comprises the evaporation of spent liquor and distillation of the formic acid. The regenerated acid is circulated back into the cooking stage, while the dissolved solid is dried. Lignin can be further used for processing in the chemical industry or it can be burned in an ordinary power boiler. The ash from the boiler and the nutrients from pulp bleaching filtrates are returned back to the fields [112].

2.8 Conclusions and future trends

Woodpulp will be, for the next decade, the main raw material for producing paper and boards and to obtain cellulose derivatives. Among pulping processes, kraft pulping will continue to dominate global pulp production.

The large differences in wood and labor costs between world regions determine that new pulp mills will mainly be built mainly in Latin America and Asia. Modernization of existing mill will dominate in Europe and North America.

Research efforts are still necessary for a better understanding of the mechanisms involved in pulping processes in order to decrease materials and energy consumption. Reduction of the environmental impact in pulp manufacture is of huge importance for the competitiveness of pulp mills.

Global challenges, such as limited oil supply, increased concern about greenhouse gas emissions, and the decreasing competitiveness of traditional pulp and paper producers, will boost the need to convert the pulp mills into integrated forest biorefineries that will produce, besides pulp, higher value-added products such as ethanol, polymers, carbon fibers, biodiesel, and so on.

List of abbreviations

AA	Active alkali
AS	Alkaline sulfite
ASAM	Alkaline sulfite–anthraquinone–methanol
AQ	Anthraquinone
C	Causticity
CBC	Continuous batch cooking
CEPI	Confederation of European Paper Industries
CIMV	Compagnie Industrielle de la Matière Végétale
CTMP	Chemithermomechanical pulping
DR	Degree of reduction of sodium sulfate
EA	Effective alkali
EMCC	Extended modified continuous cooking
FAO	Food and Agriculture Organization of the United Nations
MCC	Modified continuous cooking
NSSC	Neutral sulfite semichemical
PGW	Pressure groundwood
RDH	Rapid displacement heating
RMP	Refiner mechanical pulping
S	Sulfidity
SGW	Stone groundwood
TA	Total alkali
TGW	Thermogroundwood
TMP	Thermomechanical pulping

References

[1]   Confederation of European Paper Industries (CEPI). Key Statistics 2017. European Pulp and Paper Industry, Brussels, Belgium, Ed. CEPI, 2017.
[2]   Food and Agriculture Organization of the United Nations (FAO). Pulp and paper capacities, survey 2016–2021. Rome, Italy, Ed. FAO, 2017.
[3]   Haley UCV. No Paper Tigers, Subsidies to China's Paper Industry From 2002–09, EPI Briefing Paper 264, Economic Policy Institute, Washington, DC, USA, 2010.
[4]   McClay B. China's Impact on Global Pulp Market, International Pulp and Recovered Paper Forum, Shanghai International Pulp Week, 14 March 2018, (Accessed June 4, 2019, at https://www.pulpmarket.ca/wp-content/uploads/CPICC-Shanghai-March-2018.pdf).
[5]   Biermann CJ. Handbook of pulping and papermaking. 2nd ed. San Diego, CA, USA, Academic Press, 1996.
[6]   Höglund H. Mechanical Pulping. In: Ek M, Gellerstedt G, Henriksson G., ed. Pulp and paper chemistry and technology, 2, Pulping Chemistry and Technology, Berlin, Germany, Walter de Gruyter, 2009, 57–89.
[7]   Sundholm J. What is Mechanical Pulping? In: Gullichsen J, Paulapuro H., ed. Papermaking science and technology, 5, Mechanical Pulping, Helsinki, Finland, Fapet Oy, 1999, 17–22.
[8]   von Koeppen A. Chemimechanical Pulps from Hardwood Using the NSSC [Neutral Sulfite Semichemical] Process. Pap. Trade J. 1986, 170, 49–51.
[9]   Worster HE. Semichemical Pulping for Corrugating Grades. In: Pulp and paper manufacture. 3rd ed. 4. Sulfite Science and Technology TAPPI-CPPA, Montreal, Canada, 1985, 130–158.
[10]   Ingruber OV. Recovery of Chemicals and Heat. In: Pulp and paper manufacture, 3rd ed. 4. Sulfite Science and Technology TAPPI-CPPA, Montreal, Canada, 1985, 244–300.
[11]   Blechschmidt J, Heinemann S. Fibrous Materials for Paper and Board Manufacture. In: Holik H., ed. Handbook of paper and board, Second, Revised and Enlarged Edition, 1. Wiley-VCH, Weinheim, Germany, 2013, 33–104.
[12]   Sixta H, Potthast A, Krotschek AW. Chemical Pulping Processes. In: Sixta H., ed. Handbook of pulp. 1. Wiley-VCH, Weinheim, Germany, 2006, 109–509.
[13]   Gullichsen J. Fiber Line Operations. In: Gullichsen J, Paulapuro H., ed. Papermaking science and technology, 6A, Chemical Pulping, Helsinki, Finland, Fapet Oy, 1999, A19–A243.
[14]   Salmen L, Lucander M, Harkonen E, Sundholm J. Fundamentals of Mechanical Pulping. In: Gullichsen J, Paulapuro H., ed. Papermaking science and technology, 5, Mechanical Pulping, Helsinki, Finland, Fapet Oy, 1999, 35–65.
[15]   Karnis A. The mechanism of fibre development in mechanical pulping. J. Pulp Pap. Sci. 1994, 20, J280–J288.
[16]   Liimatainen H, Haikkala P, Lucander M, Karojarvi R, Tuovinen O. Grinding and Pressure Grinding. In: Gullichsen J, Paulapuro H., ed. Papermaking science and technology, 5, Mechanical Pulping, Helsinki, Finland, Fapet Oy, 1999, 107–156.
[17]   McDonald D, Miles K, Amiri R. The nature of mechanical pulping process, Pulp Paper Can. 2004, 105, 27–32.
[18]   Berg J-E. Wood and fibre mechanics related to the thermomechanical pulping process. Doctoral Thesis. Mid Sweden University, Sundsvall, Sweden, 2008.
[19]   Muhić D, Sundström L, Sandberg C, Ullmar M, Engstrand P. Influence of temperature on energy efficiency in double disc chip refining. Nord. Pulp Pap. Res. J 2010, 25, 420–427.
[20]   Miles KB, Karnis A. Characteristics and energy consumption in refiner pulps. J. Pulp Pap. Sci. 1995, 21, j383–j399.
[21]   Illikainen M. Mechanisms of thermo-mechanical pulp refining. Doctoral Thesis. University of Oulu, Oulu, Finland, 2008.

[22] Fernando D, Muhić D, Engstrand P, Daniel G. Fundamental understanding of pulp property development under different thermomechanical pulp refining conditions as observed by a new Simons' staining method and SEM observation of the ultrastructure of fibre surfaces. Holzforschung. 2011, 65, 777–786.

[23] Gorski D. A TMP process: Improved energy efficiency in TMP refining utilizing selective wood disintegration and targeted application of chemicals. Doctoral Thesis. Mid Sweden University, Sundsvall, Sweden, 2011.

[24] Johansson L, Hill J, Gorski D, Axelsson P. Improvement of energy efficiency in TMP refining by selective wood disintegration and targeted application of chemicals. Nord. Pulp Pap. Res. J 2011, 26, 31–46.

[25] Lindholm C-A, Kurdin JA. Chemimechanical Pulping. In: Gullichsen J, Paulapuro H., ed. Papermaking science and technology, 5, Mechanical Pulping, Helsinki, Finland, Fapet Oy, 1999, 223–249.

[26] Blechschmidt J, Heinemann S, Suss H-U Mechanical Pulping. In: Sista H., ed. Handbook of Pulp, Wiley-VCH, Weinheim, Germany, 2006, 1069–1111.

[27] Obrocea P, Gavrilescu D. Pulping fundamentals, Rotaprint Publishers, Iasi, Romania, 1992, [in Romanian].

[28] Dillen JR, Dillén S, Hamza MF. Pulp and Paper: Wood Sources. In: Hashmi S., ed. Reference module in materials science and materials engineering. Elsevier, Oxford, UK, 2016, 1–6.

[29] Ahmadi M, Latibari AH, Faezipour M, Hedjazi S. Neutral sulfite semi-chemical pulping of rapeseed residues. Turk. J. Agric. For. 2010, 34, 11–16.

[30] Leoponiemi A. Fibres and energy from wheat straw by simple practice. doctoral dissertation. VTT Publications, Espoo, Finland, 2011.

[31] Gavrilescu D, Toth S. Corrugated Board., Sf. Gheorghe, Romania, T3 Publishers 2007, [in Romanian].

[32] Zhao L, Tran H Combustion behaviors of spent sulfite liqours. J. Sci. Technol. For. Prod. Processes 2016, 4, 50–57.

[33] Gavrilescu D, Craciun G. Kraft pulping processes. Texte Publishing House. Dej, Romania, 2012, [in Romanian].

[34] Gierer J. Chemical aspects of kraft pulping. Wood Sci. Technol., 1980, 14, 241–266.

[35] Potthast A. Chemical Pulping Processes. In: Sixta H., ed. Handbook of pulp. 1. Wiley-VCH, Weinheim, Germany, 2006, 164–185.

[36] Sakakibara A, Sano Y. Chemistry of Lignin. In: Wood and cellulosic chemistry. ed. David N-S, Hon DN-S. N-S, Shiraishi N. Marcel Dekker, New York, USA, 2001, 109–174.

[37] Gellerstedt G. Structural changes in lignin during kraft cooking. Part 2. Characterization by acidolysis. Svensk Papperstidning, 1984, 91, R61–R67.

[38] Gierer J. Chemistry of delignification. Part 1: General concept and reactions during pulping. Wood Sci. Technol., 1985, 19, 289–312.

[39] Gellerstedt G, Lindfors EL. Structural changes in lignin during kraft pulping. Holzforschung. 1984, 38, 151–158.

[40] Gellerstedt G, Lindfors E-L. On the structure and reactivity of residual lignin in kraft pulp fibres. In: Proceedings of the International Pulp Bleaching Conference. 1. SPCI Publications, Stockholm, Sweden, 1991, 73.

[41] Robert D, Bardet M, Gellerstedt G, Lindfors E-L. J. Structural changes in lignin during kraft cooking, Part 3. On the structure of dissolved lignins. Wood Chem. Technol. 1984, 4, 239–263.

[42] Gierer J. The reactions of lignin during pulping – A description and comparison of conventional pulping processes. Svensk Papperstidning. 1970, 73, 571–596.

[43] Sixta H. Kraft Pulping Kinetics. In: Sixta H., ed. Handbook of pulp. 1. Wiley-VCH, Weinheim, Germany, 2006, 185–228.

[44] Gentile VM, Schroeder LR, Atalla RH. The Structure of Cellulose: Characteristics of the Solid State. In: Atalla RH., Ed. American chemical society. ACS Symposium Series, Washington, DC, USA, 1987, 272.

[45] Sjöström E. The behaviour of wood polysaccharides during alkaline pulping processes. Tappi. 1977, 60, 151–154.

[46] Gustavsson CAS, Al-Dajani WW. The influence of cooking conditions on the degradation of hexenuronic acid, xylan, glucomannan and cellulose during kraft pulping of softwood. Nordic Pulp Pap. Res. J. 2000, 15, 160–167.

[47] Buchert J, Bergnor E, Lindblad G, Viikari L, Ek M. Significance of xylan and glucomannan in the brightness reversion of kraft pulps. Tappi J. 1997, 80, 165–171.

[48] Olm L, Tistad G. Kinetics of the initial phase of kraft pulping. Svensk Papperstidning. 1979, 87, 458–464.

[49] Andersson N, Wilson DI, Germgård U. An improved kinetic model structure for softwood kraft cooking. Nord. Pulp Pap. Res. J. 2003, 18, 200–209.

[50] Umezawa T Chemistry of Extractives. In: Wood and cellulosic chemistry. ed. David, N-S, Hon DN-S, Shiraishi N., Marcel Dekker, New York, USA, 2001, 213–242.

[51] Lindström M, Öberg P, Stenius LP. Resin and fatty acids in kraft pulp washing. Physical state, colloid stability and washability. Nord. Pulp Pap. Res. J. 1988, 3, 100–106.

[52] MacLeod M. Top ten factors in kraft pulp yield. Paperi ja Puu. 2007, 89,1–7.

[53] Casey JP Pulp and paper chemistry and chemical technology, 1, John Wiley and Sons, New York, USA, 1980.

[54] Kleppe PJ. Kraft pulping. Tappi J. 1970, 53, 35–47.

[55] Rantanen R. Modelling and control of cooking degree in conventional and modified continuous pulping processes. Academic Dissertation. University of Oulu, Oulu, Finland, 2006.

[56] Puitel AC, Marin N, Petrea P, Gavrilescu D. Lignocellulosic agricultural residues – a virgin fiber supply solution for paper-based packaging. Cellulose Chemi. Technol., 2015, 49, 633–639.

[57] Sanjoikari-Pakhala K. Non-wood plants as raw material for pulp and paper. Academic Dissertation, University of Helsinki, Helsinki, Finland, 2001.

[58] Leponiemi AA. Non-wood pulping possibilities – A challenge for the chemical pulping industry. Appita J. 2008, 61, 234–243.

[59] Ashori A. Pulp and paper from kenaf bast fibers. Fibers Polym. 2006, 7, 26–29.

[60] Huang G, Shi J X, Langrish TAG. A new pulping process for wheat straw to reduce problems with the discharge of black liquor. Bioresour. Technol. 2007, 98, 2829–2835.

[61] Brännvall E. Wood Handling In: Ek M, Gellerstedt G, Henriksson G., ed. Pulp and paper chemistry and technology, 2, Pulping Chemistry and Technology, Berlin, Germany, Walter de Gruyter, 2009, 13–34.

[62] Akhtaruzzaman AFM, Virkola NE. Influence of chip dimensions in kraft pulping, part i: mechanism of movement of chemicals into chips. Paperi ja Puu. 1979, 61, 578–580.

[63] Svedman M, Tikka P, Luhtanen M. Effects of softwood morphology and chip thickness on pulping with a displacement kraft batch process. Tappi J., 1998, 81, 157–168.

[64] Gullichsen J, Kolehmainen H, Sundqvist H. On the non-uniformity of the kraft cook. Paperi ja Puu. 1992, 74, 486–490.

[65] Gustavsson M. The significance of liquor-to-wood ratio on the reaction kinetics of spruce sulphate pulping. Master Thesis. Karlstads Universitet, Karlstad, Sweden, 2007.

[66] Lundqvist F, Olm L, Tormund D. Effects of carbonate on delignification of beech wood in kraft cooking, Nord. Pulp Pap. Res. J. 2006, 21, 290–296.

[67] Saltberg A, Brelid H, Lundqvist F. The effect of calcium on kraft delignification study of aspen, birch and eucalyptus. Nord. Pulp Pap. Res. J. 2009, 24, 440–447.

[68] Sezgi US. A combined discrete-continuous simuyltan model of an RDH tank farm. Tappi J. 1994, 77, 213–220

[69] Scheldorf JJ, Edwards LL. Challenges in modeling the RDH process: a discontinuous dynamic system. Tappi J. 1993, 76, 97–104.

[70] Teder A. Kinetics of Chemical Pulping and Adaptartion to Modified processes. In: Ek M, Gellerstedt G, Henriksson G., ed. Pulp and paper chemistry and technology. 2, Pulping Chemistry and Technology, Walter de Gruyter, Berlin, Germany, 2009, 149–164.

[71] Pursiainen S, Hiljanen S, Uusitalo P, Kovasin K, Saukkonen M. Mill-scale experiences of extended delignification with Super Batch cooking method. Tappi J. 1990, 73, 115–122.

[72] Wizani W, Krotscheck WA, Juljanski W, Bito R. CBC – Continuous batch cooking the revolution in kraft cooking. Proceedings of the Pulping Process and Product Quality Conference, Boston, MA, USA, 2000.

[73] Jansson J, Linberg T, Dahlquist E. Process Optimization and Model Based Control in Pulp and Paper Industry. (Accessed June 4, 2019, at https://pdfs.semanticscholar.org/0511/b0b4b1d776fd77f1ea0c83dcdd25aa883c7a.pdf).

[74] Laakso S. Modeling of chip bed packing in a continuous kraft cooking digester. Doctoral Dissertation, Helsinki University of Technology, Espoo, Finland, 2008, 19.

[75] Atalla RH, Reiner R, Houtman CJ, Springer EL. New Technology in Pulping and Bleaching, In: Burley J, Evans J, Younquist JA., ed. Encyclopedia of forest sciences. 2. Elsevier, Madison, WI, USA, 2004, 918–924.

[76] Nordén S, Teder A. Modified kraft processes for softwood bleached-grade pulp. Tappi J. 1979, 62, 49–51.

[77] Bachlung A, Svanberg J. inventors. Kvaerner Pulping AB assignee. US 5766413, 1998.

[78] Marcoccia B, Laakso R, McClain G. Lo-SolidsTM pulping: Principles and applications. Tappi J.. 1996, 79, 179–188.

[79] Lindstrom M, Lindgren C. inventors. Kvaerner Pulping AB assignee. US 5885414, 1999.

[80] Sixta H. Sulfite Chemical Pulping. In: Sixta H., ed. Handbook of pulp. 1. Wiley-VCH, Weinheim, Germany, 2006, 392–406.

[81] Gellerstedt G. Pulping Chemistry. In: Wood and cellulosic chemistry. ed. David N-S, Hon DN-S, Shiraishi N., Marcel Dekker, New York, USA, 2001, 859–906.

[82] Wolfinger MG, Sixta H. Modeling of the acid sulfite pulping process – Problem definition and theoretical approach for a solution with the main focus on the recovery of cooking chemicals. Lenzinger Berichte. 2004, 83, 35–45.

[83] Young RA. Wood and Wood products. In: Kent JA., ed. Handbook of industrial chemistry and biotechnology. Springer, New York, USA, 2007, 1234–1293.

[84] Lewis KN. Summer sickness of magnesium bisulfite pulp at fraser paper's edmunston mill. Master Thesis. University of New Brunswick, New Brunswick, Canada, 1994.

[85] Biermann CJ. In: Handbook of pulping and papermaking, Academic Press, San Diego, CA, USA, 1996.

[86] Bailey EL. Recent technical advances in magnesia-base cooking. Tappi. 1962, 45, 689–691.

[87] Ingruber OV, Allard CA. Alkaline sulfite pulping for force strength. Pulp and Paper Magazine of Canada, 1973, 74, T354–T369.

[88] Kettunen J, Laine JE, Yrjälä I, Virkola NE. Aspects of strength development in fibres produced by different pulping methods. Paperi ja Puu. 1982, 64, 205–211.

[89]   Kettunen J, Virkola NE, Yrjala I. The effect of anthraquinone on neutral sulphite and alkaline sulphite cooking of pine. Paperi ja Puu. 1979, 61, 685–700.

[90]   Fuchs K, Huber A, Schubert H-L. inventors, Impco-Voest-Alpine Pulping Technologies GmbH assignee. EP 0777780, 1998.

[91]   Kordsachia O, Wandinger B, Patt R. Some investigations on ASAM pulping and chlorine free bleaching of Eucalyptus from Spain. Holz als Roh- und Werkstoff. 1992, 50, 85–91.

[92]   Borgards A, Patt R, Kordsachia O, Odematt J, Hunter WD. A comparison of ASAM and kraft pulping and ECF TCF bleaching of southern pine. In: 1993 Pulping Conference, TAPPI Proceedings, Books 1–3, TAPPI Press, Atlanta, GA, USA, 1993, 629–636.

[93]   Zsu JY, Chai XS, Pan XJ, Luo Q, Li J. Quantification and reduction of organic sulfur compound formation in a commercial wood pulping process. Environ. Sci. Technol. 2002, 36, 2269–2272.

[94]   Bordado CJM, Gomes JFP. Atmospheric emissions of kraft pulp mills. J. Chem. Eng. Process. 2002, 41, 667–671.

[95]   Sillanpaa M. Studies on washing in kraft pulp bleaching academic dissertation, University of Oulu, Oulu, Finland, 2005.

[96]   Muurinen E. Organosolv pulping – A review and distillation study related to peroxyacid pulping, academic dissertation, University of Oulu, Oulu, Finland, 2000.

[97]   Kleinert TN. Organosolv pulping with aqueous alcohol. Tappi J. 1974, 57, 99–102.

[98]   Sixta H, Borgards A. New technology for the production of high-purity dissolving pulps. Das Papier. 1999, 53, 220–234.

[99]   Pye E, Lora JH. The ALCELL process – A proven alternative to kraft pulping. Tappi J. 1991, 74, 113–118.

[100] Lora JH, Aziz S. Organosolv pulping: a versatile approach to wood refining. Tappi J. 1985, 68, 94–97.

[101] Nimz H, Berg HA, Granzow C, Casten R, Muladi S. Pulping and bleaching by the acetosolv process. Das Papier. 1989, 43, V102–V108.

[102] Obrocea P, Gavrilescu D. Pulping and bleaching by the acetosolv process. Celuloza si hartie. 1988, 37, 157–159, [in Romanian].

[103] Vega A, Bao M. Fractionation of lignocellulosic materials with phenol and dilute HCl. Wood Sci. Technol. 1991, 25, 459–466.

[104] Lonnberg B, Laxen T, Sjoholm R. Chemical pulping ofsoftwood chips by alcohols. Paperi ja Puu. 1987, 69, 757–762.

[105] Aziz S, Sarkanen K. Organosolv pulping – a review. Tappi J. 1989, 72, 169–175.

[106] Gilarrahz MA, Oliet M, Rodriguez F, Tijero J. Ethanol-water pulping: Cooking variables optimization. Can. J. Chem. Eng. 1998, 76, 253–260.

[107] Sundquist J, Poppius-Levlin K. Milox Pulping and Bleaching. In: Young RA, Akhtar M., ed. Environmentally friendly technologies for the pulp and paper industry. John Willey and Sons, New York, USA, 1998, 157–190.

[108] Obrocea P, Cimpoesu G. Contribution to sprucewood delignification with peroxyformic acid. I. The effect of pulping temperature and time. Cellul. Chem. Technol. 1998, 32, 517–525.

[109] Leponiemi A. Fibres and energy from wheat straw by simple practice. Doctoral Dissertation. Aalto University, Espoo, Finland, 2011.

[110] Delmas G-H, Benjelloun-Mlayah B, Le Bigot Y, Delmas M. Functionality of wheat straw lignin extracted in organic acid media. J. Appl. Polym. Sci. 2011, 121, 491–501.

[111] Pasi R, Paivi R, Rousu E. inventors; Chempolis Oy, assignee; US 6156156, 2000.

[112] Pasi R, Rousu Paivi AJ. Sustainable pulp production from agricultural waste. Resour. Conserv. Recycl. 2002, 35, 85–103.

Adrian Cătălin Puițel, Bogdan Marian Tofănică, and
Dan Alexandru Gavrilescu

Chapter 3
Fibrous raw materials from agricultural residues

3.1 Introduction

Nonwood fibers have an extended history as raw materials used in the production of paper products, long before the wood began being utilized in papermaking [1]. The Chinese eunuch, inventor, and politician of the Han dynasty, Cai Lun is credited as the "parent of the paper," being recognized as the inventor of paper and the papermaking process in its modern version, around 105 CE, when he created for the emperor the first sheet of paper, the basic materials being the bark of trees, hemp wastes, rags of cloths, and old fishing nets [2]. Since then, even if equipment and technology of papermaking in modern times are more complex, they still make use of the same ancient technique that involves a suspension of fibers in water, draining of the water, and then drying into a thin matted sheet.

Regarding the raw materials, annual plants have been used exclusively for almost 2,000 years in the manufacture of paper products. The wood began to replace hemp, ramie, cotton and cotton cloths, old rags, and ropes only about 200 years ago, due to the increased demand for paper products, which led to a massive increase in the need for new sources of raw materials [3]. Thus, processes for pulping wood fibers made modern large-scale production possible. Of the wood species, the most used softwoods are spruce, pine, and fir, and among the hardwoods are poplar, beech, and birch.

After wood, nonwood raw materials are the second major source of raw materials for the manufacture of pulp, cellulose, and other fibrous products. The annual plants used processed in the pulp and paper industry can be grouped into four main classes [4]:
- plants specially grown for cellulosic fibers: kenaf, flax, jute, cotton, hemp;
- agricultural residues from food production: cereal straws, bagasse;
- annual technical plants, wastes from textile fibers, wastes from oil crops: cotton, hemp, flax; and
- naturally occurring uncultivated crops: different herbs, bamboo, reed, papyrus.

Adrian Cătălin Puițel, Bogdan Marian Tofănică, Dan Alexandru Gavrilescu, "Gheorghe Asachi"
Technical University of Iasi, Iasi, Romania

https://doi.org/10.1515/9783110658842-003

The primary reason that determines the use of annual plants and agricultural wastes in the pulp industry is the reduction of wood consumption for a rational and economical use in other industrial sectors [5]. Currently, wood resources are no longer meeting the consumption demand by the development of the pulp industry in most regions of the world due to the too long cycle of renewal of plant material and to deforestation [6]. On the other hand, the current technologies of the paper industry accept increased amounts of short fiber, thus being possible to expand the raw material base by capitalizing on the annual plants [7].

At present, the production of cellulosic fibers from annual plants is mainly based on the by-products from food crops: the plant material resulting from the harvesting of graminaceae (cereal straws – wheat, rye, barley, rice), the one from the processing of sugar cane (bagasse), and to a lesser extent on the crops of plants for industrialization (cotton, hemp, flax, reed) [4].

Wood accounts for approximately 47% in the production of papermaking fibers worldwide, recovered paper by 45%, and annual plants by 8%, as shown in Fig. 3.1 [8].

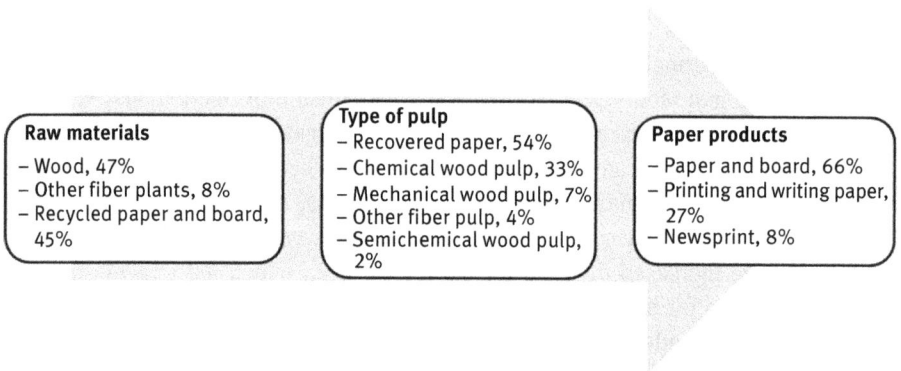

Raw materials
– Wood, 47%
– Other fiber plants, 8%
– Recycled paper and board, 45%

Type of pulp
– Recovered paper, 54%
– Chemical wood pulp, 33%
– Mechanical wood pulp, 7%
– Other fiber pulp, 4%
– Semichemical wood pulp, 2%

Paper products
– Paper and board, 66%
– Printing and writing paper, 27%
– Newsprint, 8%

Fig. 3.1: Share of raw materials for pulp and paper manufacturing worldwide in 2012.

The dynamics of the world production of fibrous materials for the manufacture of paper in the last years (Tab. 3.1) indicates the constant maintenance of the production of wood pulp, the increase of the contribution of secondary fibers, and the decrease of the production of cellulose from nonwoody plants. The orientation toward the use of the recovered paper is obvious and represents the solution for satisfying the need for increasing fibrous material in the following period. The reduction of pulp production from annual plants during the last 10–20 years is also explained by the increase of the consumption of recovered paper in countries such as China, where many small units of pulp producing plants have been closed or shifted their fiber supply for the paper and board industries [9].

Fig. 3.2 shows the evolution of pulp production from annual plants in terms of total quantities, as well as its share in world pulp production. The annual production

Tab. 3.1: World production of fibrous raw materials, million tons [9].

Fibrous raw material	2014	2015	2016	2017	2018
Mechanical wood pulp	25.2	25.3	25.7	25.2	25.2
Semichemical wood pulp	9.0	8.7	8.7	8.4	8.5
Chemical wood pulp	135.5	136.6	139.7	143.7	146.5
Pulp from fibers other than wood	14.0	13.3	12.0	11.9	11.9
Recovered fiber pulp	95.6	98.1	95.1	95.1	95.4

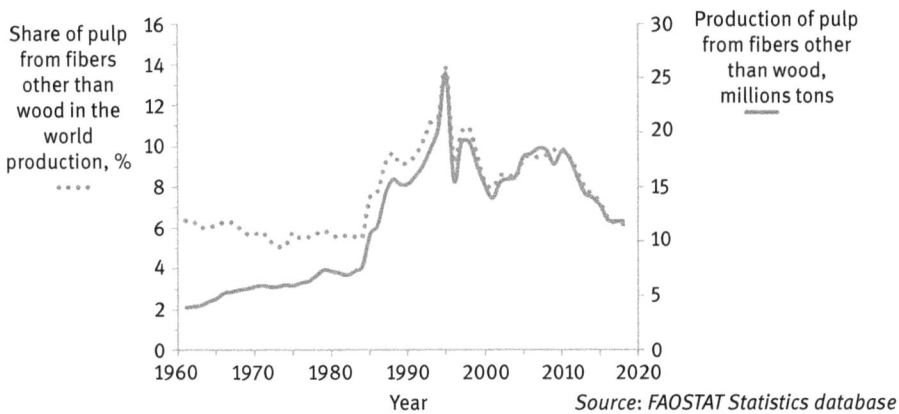

Source: FAOSTAT Statistics database

Fig. 3.2: Dynamics of world fiber production from annual plants between 1961 and 2018 [9].

of other fiber pulp also defined as pulp from fibers other than wood by Food and Agriculture Organization of the United Nations (FAO) increased continuously, reaching the maximum in 1995, with a total of approximately 25 million tons, representing 14% of the world production. Subsequently, fiber production from nonwood plants declined sharply, reaching today almost 12 million tons, respectively 6.2% of the world production.

The Asian continent, the absolute leader, accounted for a large but declining share of the global nonwood fiber pulp, producing in 2012 about 86% of the total quantity of pulps from nonwoody plants, as can be seen from Tab. 3.2 and Fig. 3.3. The explanation is related to the existence of large raw material resources such as cereal straws and other nonwoody plants, in countries such as China, India, Pakistan, Vietnam, and Thailand. Production is carried out in many small-capacity factories, equipped with simple installations. In most cases, the residual solutions are not processed in the factory, being discharged in different emissaries, under the conditions of the permissive environmental protection legislation.

Tab. 3.2: Annual production of cellulose fibers from annual plants, thousands of tons [9].

Regions		2010	2011	2012	2013	2014	2015	2016	2017	2018
Africa		229	229	218	211	211	211	211	211	181
Americas	North America	285	285	285	285	359	359	359	360	360
	Central America	38	37	10	24	192	188	188	188	188
	South America	486	382	408	401	423	405	390	391	404
Asia		15,964	15,416	13,727	11,551	11,464	10,830	9,897	9,965	10,095
Europe		1,281	1,358	1,376	1,987	1,393	1,349	927	707	646

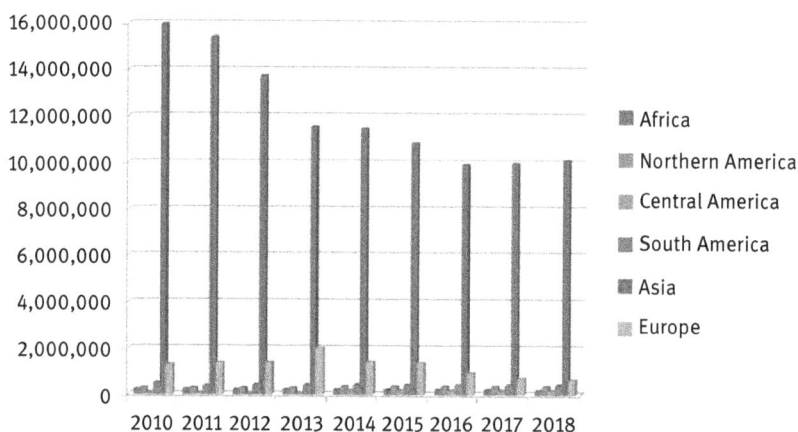

Fig. 3.3: Production of cellulose fibers from annual plants in tons, 2010–2018 [9].

China and India are the largest producers of cellulose fibers from annual plants, their production accounting over 85% of the world total (Tab. 3.3). Among the other producing countries, only Pakistan, United States of America, Italy, Colombia, Costa Rica, Argentina, Spain, Egypt, and Thailand exceed 1% of the global pulp production from annual plants in 2018 [9].

For the year 2018, in China pulp from annual plants accounts for about 35% of total cellulosic fiber production, while in India it was almost 50%. Spain, the leading European producer of cellulose fibers from annual plants, with 122,000 tons of non-wood raw materials accounts for 7% of domestic production, as shown in Tab. 3.4 [9].

Tab. 3.3: Annual production of pulp from fibers other than wood – producing countries, thousands tons [9].

Country	2010	2011	2012	2013	2014	2015	2016	2017	2018
China	12,970	12,400	10,738	8,285	7,549	6,799	5,910	5,975	6,104
India	3,070	3,020	1,995	1,995	1,995	2,270	2,920	30,200	3,020
Spain	1,113	900	900	900	828	810	414	196	122
Pakistan	370	370	370	370	370	370	370	370	370
USA	245	245	245	245	320	3320	320	320	320
Argentina	236	133	135	133	147	136	119	124	129
Italy	176	203	203	203	203	203	203	203	203
Colombia	160	160	183	178	188	197	192	188	196
Costa Rica	7,000	7,000	7,000	7,000	171	171	171	171	171
Thailand	126	131	133	133	133	116	121	117	118

Tab. 3.4: Annual production of pulp in selected countries, thousands tons [9].

Country	Fibrous raw material	2014	2015	2016	2017	2018
China	Mechanical wood pulp	865	865	865	865	865
	Semichemical wood pulp	1,710	1,710	1,710	1,710	1,710
	Chemical wood pulp	7,049	7,089	7,479	7,824	8,897
	Pulp from fibers other than wood	7,549	6,799	5,910	5,975	6,104
India	Mechanical wood pulp	504	504	504	504	504
	Semichemical wood pulp	168	168	168	168	168
	Chemical wood pulp	2,394	2,534	2,435	2,435	2,435
	Pulp from fibers other than wood	2,920	3,070	3,020	3,020	3,020
Spain	Mechanical wood pulp	90	90	356	439	439
	Semichemical wood pulp	0	0	0	0	0
	Chemical wood pulp	1,748	1,526	1,321	1,261	1,253
	Pulp from fibers other than wood	810	828	414	196	122

In recent years, the annual plant fiber production of the Asian continent has been reduced, due to decreases in China, where paper producers began to replace straw pulp with recovered paper. Today, the production of nonwood in worldwide pulp production is around 12 million tons (Fig. 3.2), but analyzing forecasts and trends in recent years, it is expected to grow again in future. Analyzing the dynamics of non-wood in world pulp production (Fig. 3.2) it can be stated that in the next years the utilization will increase continuously, at least for some years. Overall, the today ratio of 6% is similar to the same historical ratio in the 1980s before the golden decade of 1986–1995 [9].

On the short run, this rise is defined by the existing and planned expansions of the pulping capacities, while in the long run is contingent upon the regional fibrous raw material resources and the extent to which these may be economically tapped for pulp production.

Since decades, annual plants have been used as raw material in producing cellulosic fibers for papermaking. Annual plants contain the same main chemical components as wood: cellulose, hemicelluloses, and lignin. Annual plants have the same cellulose content and less lignin when compared with wood [10]. Thus, chemical pulp can be obtained from nonwoods by a mild pulping process, which consumes less energy and chemicals in a shorter cooking time. Annual plants may be cultivated and harvested every year. These special characters were the dominant direct importance for their development [11].

The annual plants have some specific problems as raw materials of pulp manufacture because harvesting is limited to a few weeks of a year and require long-term storage [12]. In addition, annual plants are planted and scattered on many small fields, an aspect that causes difficulty in their collection and transportation. A sufficient storage capacity is needed to ensure an all-year supply of pulp mill and most annual plants are attacked easily by microorganisms [13]. Transportation of wood is less expensive and less difficult than that of annual plants. Despite these drawbacks, nonwoods are still substituting wood species as an alternative resource for pulp production in the future.

3.2 Pulping processes

3.2.1 Alkaline pulping

Fibrous cellulosic materials are obtained through separation of cellulosic fibers from lignocellulosic materials using chemical or mechanical processes. Chemical processes use reagents and heat that modify and dissolve lignin into liquid phase. Mechanical processes use mechanical energy for fibers separation from lignocellulosic materials, which previously have been steamed for lignin softening. Sixta [14] shows that chemical processes allow to obtain fibrous materials (chemical pulps) with 40–52% yield and the mechanical processes lead to obtaining fibrous materials (mechanical pulps) with 85–95% yield.

Chemical processes use reagents that degrade lignin macromolecules and introduce hydrophilic groups into the resulting fragments which, thus become soluble in the liquid phase. Brannvall [15] considers that the chemical processes are not selective, besides lignin part of the polysaccharides complex passes into solution, so that about half of the lignocellulosic material results in fibers, and the other half is dissolved. Chemical processes differ according to the nature of the reagents and the conditions in which they are carried out. The agricultural wastes are processed according to the same methods used for delignification of wood: alkaline processes, acidic processes, organosolv processes, and other methods (steam explosion (SE), biopulping) [16].

Leponiemi et al. [17] consider that among chemical processes, alkaline processes are most used for the delignification of lignocellulosic materials from agriculture. The sulfate (kraft) process, which uses sodium hydroxide and sodium sulfide, is widely used to obtain chemical pulps from wood, but it is less used to obtain pulps from lignocellulosic materials from agriculture. Wheat straw kraft pulps exhibit good strength characteristics, but the presence of sodium sulfide leads to the formation of odorous compounds [18].

Soda pulping uses sodium hydroxide and can be used in the delignification of any type of lignocellulosic material. Wheat straw can be delignified using additions of 15–20% NaOH, cooking time 60–100 min at temperatures of 160–170 °C, when pulps having 40–45% yield result. Wheat straw pulp is bleached and used to obtain printing-writing papers [19].

The Soda process can be improved by adding anthraquinone, oxygen, or both. Anthraquinone addition accelerates the process and protects the carbohydrates, so that wheat straw pulps with higher yield and lower lignin content are obtained comparing with the natron pulps cooked under similar conditions [20, 21]. Tutus and Eroglu [22, 23] showed that the use of oxygen in natron-anthraquinone pulping increases the rate of the process and causes the precipitation of dissolved silicates in the spent liquor on the pulp fibers. In recent years, interest in using anthraquinone as a pulping additive has decreased as this compound is listed as a potential carcinogen [24].

The NACO process is a variant of the natron-oxygen process for obtaining pulp from straw and consists from three steps: pretreatment of straw, straw delignification, and straw pulp bleaching. In the first stage the straws are treated with 1–2% NaOH solution at 50 °C to reduce the content of silica and extractable substances. The delignification takes place in a dedicated reactor (turbopulper), with oxygen at the pressure of 6–7 bar and the 8% consistency of the pulp. The cooking temperature is 130–145 °C and the reagent is Na_2CO_3 (20–25% on o.d. straw) or NaOH (5–10%). The spent solution is processed for alkali recovery [25, 26].

SAICA process produces semichemical pulp from straw with about 70% yield from which the fluting paper for corrugated cardboard is obtained. The chipped straws are impregnated with the spent liquor followed by delignification in a continuous digester with 6% NaOH at 95–100 °C for 2.5 hours. The straw pulp is washed, refined, and screened before being sent to the paper machine. [27, 28].

Replacing NaOH with KOH as a delignifying reagent of straw results in increased process rate, reduced consumption of alkali, and lowered boiling temperature. The residual solution (black liquor) can be used to amend acidic soils. The main disadvantage is the high price of KOH and the difficulties of supplying this reagent in large quantities. Using the $KOH–K_2SO_3$–anthraquinone reagent system, straw pulp with low lignin content for pulp bleached grade is obtained [29, 30].

The neutral sodium sulfite cooking (NSSC) is part of the category of semichemical processes for obtaining fibrous materials, which combine chemical and mechanical treatments of straw. The cooking solution contains sodium sulfite and sodium carbonate, the sulfite producing lignin sulfonation, and the carbonate acting as the pH buffer [31]. Wheat and rice straws are the most widely used for the production of semichemical pulp by the NSSC, but the process can also be applied to the processing other lignocellulosic materials [32, 33]. The straw is partially cooked for weakening the linkages between fibers after which the fibers are separated by mechanical treatment of the straw in a refiner. The yield of straw semichemical pulp is 65–70% and it is used in the manufacture of papers with high rigidity as fluting paper for corrugated cardboard [34].

Semichemical pulp from agricultural wastes can also be obtained by cooking with ammonium disulfite [35]. This reagent has the advantage that the spent solution has a carbon/nitrogen ratio, which allows it to be used as an additive for humus-deficient soils [36].

The alkaline sulfite process uses a mixture of sodium hydroxide and sodium sulfite to obtain fibrous materials from wood and agricultural waste. This process is an alternative to the natron-anthraquinone pulping for obtaining bleached pulp grades from straw. The properties of straw pulp depend on the addition of reagents, the ratio of sodium hydroxide to sodium sulfite, temperature, and cooking time [37, 38].

3.2.2 Organosolv pulping

Organosolv pulping is based on the property of alcohols, organic acids, and other organic compounds to dissolve lignin from lignocellulosic materials and includes a wide range of methods [39]:
- methods using organic solvents in alkaline environment;
- methods using organic solvents in acidic environment;
- methods based on the thermal autohydrolysis of lignin by means of organic acids (acetic and formic) split from the vegetal material during the process;

- methods using delignification with sulfites and organic solvents; and
- methods based on the oxidation of lignin in an organic solvent.

Organosolv pulping is a more selective process when compared with classical pulping processes, allowing to obtain pulps having higher degree of delignification and higher yield than ordinary pulps. Of the organic solvents that can be used, the technological and economic criteria have selected two classes: low molecular weight alcohols (methanol and ethanol) and low molecular weight organic acids (formic and acetic). Technologies for the delignification of agricultural waste with organic solvents have been adopted after those used for wood delignification [40, 41].

Alcohol pulping processes use methanol or ethanol, with or without an acid catalyst. The ALCELL process uses a solution of ethanol in water in equal parts, without the presence of a catalyst. The lignin is solubilized and extracted at high temperature (195 °C), the process being favored by the acid pH (pH 4–5) that appears due to the cleavage of the acetyl groups existing in the vegetal material. The ALCELL process has an efficient solvent and lignin recovery system from the residual solution [42, 43].

The IDE (impregnation–depolymerization–extraction) process uses an ethanol solution to which small amounts of anthraquinone are added and contains three steps: impregnation of straws with a sodium carbonate solution, solubilization of lignin in the presence of water–ethanol–anthraquinone system, and extraction of lignin with fresh ethanol–water solution. Straw IDE pulp is noted for its low lignin content and high yield [44, 45].

The use of alcohols in cooking of straws has the advantage that alcohols can be easily recovered by distilling the residual solution, but the disadvantage of the high pressures that develop in the digester during the cooking. Methanol is highly toxic and all alcohols are flammable. These facts have competed to not using of alcohol based organosolv pulping at industrial scale.

The ASAM (alkaline sulfite–anthraquinone–methanol) process combines the action of sodium hydroxide, sodium sulfite, anthraquinone, and methanol for chemical modification and lignin dissolution. The charge of alkali is 15–25% and of anthraquinone is 0.05–0.1% on o.d. straw. The methanol ratio in the cooking solution is 15–30% by volumes. The cooking is carried out at 170–180 °C for 60–150 min. ASAM straw pulp has strength characteristics close to straw kraft pulp and higher yield [46–48]. The disadvantages of the ASAM refer to the toxicity of the methanol, the high pressure developing in the digester during cooking, and the need for a complex installation for recovery of reagents and methanol.

The ASAE (alkaline sulfite–anthraquinone–ethanol) process is the variant of the ASAM process that replaces methanol with ethanol for delignification of wood and agricultural waste. ASAE straw pulps exhibit strength characteristics close to straw kraft pulps and higher yields [49, 50].

Cooking of straw with organic solvents in acid medium uses formic and acetic acids that dissolve lignin at high temperature. These compounds have lower vapor pressure than methanol and ethanol, which causes much lower pressure during cooking. Another advantage of using formic and acetic acids is that the inevitable loss of reagents during cooking is partially offset by deacetylation and deformylation of hemicelluloses from lignocellulosic materials, causing the occurrence of acetic and formic acids during cooking [51].

The Acetocell process uses acetic acid with high concentration during cooking at 110 °C and atmospheric pressure to obtain pulps from agricultural lignocellulosics. In order to increase the delignification rate, hydrochloric acid is used as a catalyst in the proportion of 0.1–0.2% on o.d. raw material [52, 53]. The CIMV (Compagnie Industrielle de la Matière Végétale) process represents a development of the Acetocell process and uses a mixture of acetic and formic acids (60% and 30%, respectively), and the rest being water. The cooking of straw is carried out at 100 °C and atmospheric pressure and finally the straw pulp is bleached with peroxide [54].

The Chempolis process uses formic acid or a mixture of formic and acetic acids for delignification of wheat straw and other agricultural waste. The cooking temperature is 110–125 °C and the duration is 20–40 min. The obtained pulp is then bleached with peroxide in alkaline medium. The cooking reagents are recovered by distillation after which the residual solution is burned to produce the steam needed for the factory [55–56]. A recent development of the Chempolis process is the Formicofib technology, whereby pulp is obtained by delignification of straw with formic acid solution, followed by bleaching of a totally chlorine free process (TCF bleaching). The residual solution is evaporated and burned for energy and as by-products acetic acid and furfural are obtained [58].

The MILOX process (nomenclature derived from "MILieu pure OXidative pulping") was developed to produce bleached pulp from wood and nonwoody plants through their cooking without sulfur and pulp bleaching without chlorine. MILOX uses performic or peracetic acids obtained in situ from the respective formic or acetic acids and peroxide. Three successive extractions of the lignocellulosic material are used, the pulp having low lignin content and high brightness. MILOX pulp is comparable to kraft pulp in yield, but has lower strength characteristics [59–61]. The MILOX process has advantages related to low cooking temperature and pressure and high pulp bleaching ability by TCF bleaching. The process tolerates high ash raw materials such as nonwood plants. The most important disadvantage is the high corrosive action of formic acid at high temperature, which requires the zirconium plating of the digester.

Organosolv processes were initially used as environmental-friendly alternatives to the classic pulping processes, but subsequently the interest for them increased as procedures for the fractionation of plant biomass [62–64]. Fractionation of lignocellulosic materials using organic solvents results in three components: the fibrous fraction (pulp), the solid fraction (precipitating lignin), and the liquid fraction containing dissolved hemicelluloses. Pulp is used in the manufacture of paper or as a raw material for ethanol production, while lignin and hemicelluloses are processed to obtain chemical intermediates [65–69].

3.2.3 Other pulping processes

Steam explosion (SE) consists of treating lignocellulosic materials with high-pressure steam followed by a rapid decompression, after which the vegetal material "explodes" resulting in a mixture of separate fibers and fiber bundles [70]. SE has been proposed as an alternative to traditional thermomechanical (TMP) and chemithermomechanical (CTMP) processes for obtaining high-yield wood pulp, but it has not been extended on an industrial scale because the strength characteristics of SE pulps are inferior to TMP or CTMP wood pulps [71]. During SE hemicelluloses are depolymerized and solubilized in the liquid phase as pentoses and hexoses from which ethanol and other compounds can be obtained [72]. The solid phase (exploited chips) can be easily delignified due to its open structure to the action of reagents, which is why the SE was considered as a precooking stage [73]. SE is recommended for processing of nonwood plants, especially wheat and rice straws, resulting in pulps with strength characteristics similar to or better than TMP or CTMP pulps [74]. SE can be integrated in the biorefining process for the full valorization of lignocellulosic materials. It determines the hydrolysis of hemicelluloses and lignin and increases the accessibility of cellulose in the process of hydrolysis to glucose. Wheat straw processing by SE has been found to improve the yield of enzymatic hydrolysis of glucose during bioethanol manufacture [75, 76].

Extrusion pulping is a process similar to CTMP in which the fibers are mechanically released from the lignocellulosic materials by means of compression and shear forces. The BiVis process uses a corotating intermeshing twin-screw extruder for processing wheat and rice straws, which are initially impregnated with 0. 4–1.2% NaOH, and then processed in the extruder at 60–80 °C. The resulting fibrous material (70–80% yield) is washed and refined, being used in the manufacture of fluting paper for corrugated cardboard [77, 78]. The BiVis process has been adapted for the processing of any type of lignocellulosic material (wood and nonwood plants) in order to obtain high-yield pulp grades [79].

Biopulping uses biological methods for lignin degradation and elimination from lignocellulosic materials and represents an alternative to chemical delignification processes [80]. Biomechanical pulping uses the treatment of wood chips with fungi (*Phanerochates chrysosporium*) to reduce energy consumption at chips refining by up to 30% [81]. Treating the wheat straw [82] and rice straw [83] with white rot fungal cultures offers benefits for natron and kraft pulping by reducing the consumption of alkali or reducing the cooking time. For a fungus to be able to work commercially in biopulping and biobleaching of any vegetable material, it should display: fast growth rate and ability to grow on any kind of lignocellulosic material; preferred activity against hemicelluloses and lignin associated with low activity on cellulose; ability to degrade extractives: aggressive competition against other microorganisms that could damage cellulose; low pigmentation that may reduce pulp brightness; and good ability to sporulate in order to facilitate the inoculation of the raw material [84].

3.3 Morphological and papermaking properties of nonwood plant fibers

From anatomical point of view, annual plant fibers are constituted from narrow and elongated sclerenchyma cells. The mature fibers have the cellular walls well developed and lignified. Their main function is of plant support and protection. There are some differences between the mono- and dicotyledonous species [85]
– Monocotyledons (cereals, sugar cane corn stems) are similar to hardwoods and contain a lot of thin wall cell, parenchyma cells, vessels, and epidermal cell in a long range of dimensions.
– Dicotyledonous (kenaf, hemp, rapeseed) contain two distinct types of fibers: those located inside the stems are mainly short fibers and generally surrounded by phloem fibers that have higher lignin content and are more difficult to cook.

The nonwood plant pulps are characterized by a high diversity of fibrous elements. As a result of pulping, the obtained pulps contain high quantities of unwanted morphological elements such as parenchyma epidermis cell rich in hemicelluloses (especially pentosans) that increase the refining performances. However, the refining degree of straw pulps is up to 30 SR, while reed pulps usually have refining degree of about 25 SR [86].

The most important morphological properties for the establishment of the papermaking potential of nonwoods are as follows: fiber dimensions and their dimensional distributions, coarseness of fibers, the compaction capacity of the fiber network, the intrinsic strength, and the binding ability of fibers [87].

Fiber dimensions such as length and diameter are important parameters when compared different types from the papermaking quality perspectives. These parameters particularly influence both the fiber processing operations (beating) and also the properties of finished products (paper). The lumen diameter and the width of cell wall are also important due to their influence on fiber rigidity and strength properties of paper [88].

Fiber length is one of the most important papermaking fibers' properties. A longer fiber will be able to establish a higher number of linkages with other fibers and will further generate a stronger fibrous network compared with the situation generated in case of short fibers. The ratio between the fiber length and its diameter is considered of great importance. The L/D ratio differs as a function of fiber origin and may offer important information on the strength performances of further obtained paper. The L/D ratio of fibers suitable for obtaining good papermaking performance is 100:1. For nonwoods plant species, the medium fiber length is situated in the range 1 to 18 mm, with lower values for cereal straw and reed and with higher values in case of cotton and other textile plants.

The L/D ratio ranges from 50:1 to 1,500:1 with lower values in case of reed and higher values in case of cotton species. A comparative presentation of fiber dimensions and L/D values is listed in Tab. 3.5.

Tab. 3.5: Dimensional values for different origin papermaking fibers.

Species	Common name	Fiber length (mm)		Fiber diameter (µm)		L/D ratio	References
		Medium values	Limits	Medium values	Limits		
Gossypium sp.	Cotton	18	10–40	20	12–38	900	[85]
Triticum sativum	Wheat	1.4	0.4–3.2	15	8–34	93	
Oryza sativa	Rice	1.4	0.4–3.2	8	4–16	175	
Stipa tenacissima	Esparto grass	1.2	0.2–3.3	13	6–22	92	
Phragmites communis	Reed	1.15	0.3–2.3	19.5	12–39	60	[89]
Eucalyptus globus	Blue Gum	1.1	0.3–1.5	20	10–28	55	[85]
Fagus sylvatica	Beech	1.2	0.5–1.7	21	14–30	57	
Populus tremula	Poplar	0.9	0.2–1.6	19	13–30	50	
Quercus robur	Oak	1.1	0.5–1.6	23	14–30	50	

According to fiber length, the nonwoods virgin fiber sources may be divided into three categories:

a) long fiber length type ($L > 4$ mm) that mostly includes cotton and linen;
b) medium fiber length type ($1.5 < L < 4$ mm) including bamboo, bagasse, and kenaf; and
c) short fiber length type ($L < 1.5$ mm): straws, reed, and so on.

Coarseness of fibers or weight per unit length is an important characteristic for determining the quality of cellulosic fibers because it practically influences the behavior of fiber suspension during papermaking process: water removal on wire section, wet web and dry strength, and finally optical properties of the obtained finished papers. The effects are explained taking into account the fact that high coarseness value fibers have a low fiber wall width and have a higher specific area [90].

The standard Tappi Test Method T 234 cm-02 *Coarseness of Pulp Fibers* describes a method by which the coarseness (weight per unit length) of the fibers in a pulp may be ascertained by counting the crossings of a known weight of fibers per unit area on a prepared slide over lines of known length. The crossings are counted with a microscope and mechanical stage, by traversing the slide over a series of 5 cm distances. The coarseness of fibers is expressed in decigrex units, 1 dg = 1 mg/100 m (Tappi Test Method T 234 cm-02 *Coarseness of Pulp Fibers*). The values of fiber source depend on the raw material and pulping method. A comparative view of the values of nonwoods and wood fiber coarseness value is listed in Tab. 3.6. The variations between the two types of virgin fiber supplies sources are relatively low.

Tab. 3.6: Values for coarseness of fibers from different sources.

Species	Pulping method	Coarseness (mg/100 m)	References
Triticum sativum (wheat straw)	Kraft pulping	9.6–11.1	[91]
Phalaris arundinacea (reed)	Kraft pulping	5.6–10.2	[92]
Hibiscus cannabinus (kenaf)	Kraft pulping	15.9–17.1	[93]
Zea mays (corn stems)	Alkaline extraction	11.0	[94]
Eucalyptus grandis (eucalipt)	Kraft pulping	10.7	[95]
Populus tremuloides (poplar)	Kraft pulping	14.1–18.2	[96]
Betula pendula (birch)	Organosolv pulping	8.3–15.0	[97]
Pinus taeda (pine)	Kraft pulping	22.6–31.8	[98]
Picea abies (spruce)	Kraft pulping	21.3–21.5	[99]
Pseudotsuga menziesii (Douglas fir)	Kraft pulping	26–31	[87]

An almost linear dependence exists between the fiber length values and coarseness values, an aspect shown in Fig. 3.4.Therefore, longer fibers will be appreciated as having a higher coarseness than the short fibers. Through refining, the fibers are broken and cut; therefore the value of coarseness is reduced. A contradictory effect is that through refining most of strength properties of obtained papers increase comparative to long fiber obtained papers due to the fact that coarseness does also depend on other factor such as L/D ratio, lumen diameter, and density of cellulosic material of fibers [86].

Coarseness of fibers highly depends on the difference between the area of cross section and lumen area and implicitly on the fiber medium diameter, perimeter, width, and density of cell wall.

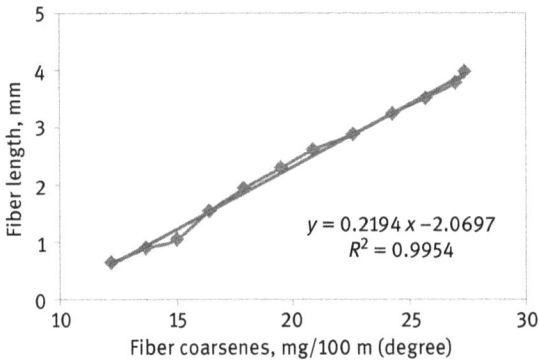

Fig. 3.4: Dependence between the length values and coarseness of fibers [86].

In order to get a better characterization of the influence of fiber dimensional elements on papermaking properties and strength values other indices have been proposed [100]:
- Felting index defined as the ratio between the length and fiber diameter;
- Rünkel index defined as the ratio between the cell wall width and lumen diameter;
- Mühlsteph index – the percentage ration between cell wall surface and fiber surface;
- fiber flexibility index a ratio between lumen diameter and fiber diameter;
- rigidity coefficient as ratio between cell wall width and fiber diameter; and
- the F factor – the ratio between fiber length and cell wall width.

The wet state fiber compaction capacity reflects the arrangement of the fibers in paper structure determining the bonding area of the fibers and the drainage performances as well as resulted paper strength properties. The most important element affecting this parameter is the cell wall width [86].

The fiber bonding capacity and the resulted relative bonding area (RBA) depend on the values of fiber bonding forces and also on their effective contact area and are directly linked with the wet state degree of compaction [86]. The RBA is the fraction of the total available fiber surface that is bonded. It is a quantity that is applied in theories of paper mechanical properties. The definition equation states that RBA = $(A_T - A)A_T$ where A_T is the total area available for bonding and A is the unbounded area. The fibers in paper bond to each other by six different mechanisms: interdiffusion, mechanical interlocking, capillary forces, Coulomb forces, hydrogen bonding, and Van der Waals forces [101]

Intrinsic strength of fibers is determined as the breaking length of paper at zero distance between the jaws – Tappi Test Method T 231 cm-07 *Zero-Span Breaking Strength of Pulp (Dry Zero-Span Tensile)*. It is a parameter considered as being of secondary importance in issues related to paper tensile stress, due to the fact that breaking of paper occurs in weaker points of its structure that are a result of fiber interbonding. Therefore in the case of unbeaten fiber obtained paper samples testing, only one-third of the existing fibers break. In case of beaten fibers obtained papers a three-forth portion of total number of fiber were broken in the tested section. These aspects represent a confirmation of the fact that intrinsic strength of fibers increases with the increase of the value of interbonding forces [86]. The most common ways of expressing the strength properties are breaking length (expressed in m or km), tensile strength (kN/m), and tensile strength index (Nm/g), determined according to standard methods at different values of basis weight (i.e., Tappi T 220 sp-06 – *Physical Testing of Pulp Handsheets, T 494 om-01 – Tensile properties of paper and paperboard using constant rate of elongation apparatus, ISO 1924-2:2008 Paper and board – Determination of tensile properties – Part 2: Constant rate of elongation method (20 mm/min)).* Table 3.7 exemplifies literature data on tensile properties of unbeaten non-wood fibers

Tab. 3.7: Tensile properties of nonwood unbeaten fiber paper sheet samples.

Source of fiber	Pulping method	Breaking length (m)	Tensile strength index (Nm/g)	References
Common reed	Kraft (17°SR)	1,500	–	[89, 102]
Wheat straw	Kraft (26°SR)	2,600	–	
Rice straw	Kraft (18°SR)	2,400	–	[103]
	Soda (17°SR)	2,500	–	
	Soda-AQ (23°SR)	3,500	–	
Vitis vinifera trimings	Soda	660	–	[104]
	Kraft	1,300	–	
Palm tree leaves	Soda-AQ	–	20.3	[105]
Palm tree fruits	Natron-AQ	–	20.8–24.0	[106]
Leucaena leucocephala	Organosolv with ethylene-glycol	3,100	–	[107]
Chamaecytisus proliferus		4,600	–	
Eucalyptus globulus	Kraft	–	22.0	[108]
Cynara cardunculus L.		–	33.2	
Arundo donax (mediteranean reed)	Kraft	–	25.2	[107]

The physical properties of the fibers are of particular importance in choosing them as raw material for paper and specialty papers or cellulose fiber-based composites. Table 3.8 displays the values of physical properties of fibers from different raw materials [4, 110]. It may be observed that the values vary in large limits according to fiber type and origin. Some properties of nonwood fibers may be comparable to rayon fibers.

Tab. 3.8: Properties of fibers from different raw materials.

Fiber source	Fiber density (g/cm³)	Elongation (%)	Tensile strength (MPa)	Young's module (GPa)
Cotton	1.55	7.0–8.0	300–600	5.5–12.6
Jute	1.3	1.5–1.8	400–800	26.5
Linen	1.5	2.7–3.2	345–1,035	69.3
Hemp	1.48	1.6	690	70
Ramie	1.5	3.6–3.8	100–938	61.4–128
Viscose	1.52	11.4	593	11
Glass fiber	2.5	2.5	2,000–3,500	70
Aramid	1.4	3.3–3.7	3,000–3,150	63–67
Carbon fibers	1.4	1.4–1.8	4,000	230–240

3.4 Conclusions

Nonwoods are the second major source of raw materials for the manufacture of pulp. The main reason that determines the use of these materials in the pulp industry is the reduction of wood consumption for a rational and economical use in other industrial sectors.

Nonwoods have the advantages of lower lignin content when compared with wood and thus the possibility of pulping in milder conditions and with lower energy consumption and chemicals. The difficulties in collection and transportation are drawbacks that limit their use on large scale.

Pulping of nonwoods may be performed by processes such as soda, kraft, sulfite, and organosolv with some modifications over the conventional wood pulping variants. The modifications, specifically for pulping of low lignin content raw materials, aim not only at increasing pulp yield and quality but also toward a facile recovery of pulping chemicals from spent liquors. Biopulping of nonwoods is at high interest in the context of reducing chemicals and energy consumption.

The pulps obtained from nonwoods are characterized by a higher diversity of fibrous elements comparing with wood pulps. The properties of the papers made from nonwood fibers vary as a function of species of raw material and on the pulping method.

References

[1]   Sixta H. Introduction, in Handbook of Pulp. editor Sixta H., Wiley-VCH Verlag, Weinheim, Germania, 2006.
[2]   Dîmboiu A. From Stone to Paper (in Romanian); Scientific Publishing House, Bucharest, 1964.
[3]   Gavrilescu D, Tofănică BM, Puiţel AC, Petrea P. Sustainable use of vegetal fibers in composite materials. Sources of vegetal fibers. Environ. Eng. Manag. J. 2009, 8, 429–438.
[4]   Gavrilescu D, Tofănică BM, Puiţel AC, Petrea PV. Vegetal fibers in composite materials – advantages and limitations, bulletin of the polytechnic institute of Iasi. Sec. Chem. Chem. Eng. 2009, 55, 85–104.
[5]   Puitel AC, Tofanica BM, Gavrilescu D, Petrea PV. Environmentally sound vegetal fiber-polymer matrix composites. Cell. Chem. Technol. 2011, 45, 265–274.
[6]   Leponiemi A. Non-wood pulping possibilities – a challenge for the chemical pulping industry. Appita. J. 2008, 61, 235–243.
[7]   Puiţel AC, Gavrilescu D, Tofănică BM. Kraft pulping – environmental impact and possibilities of reduction (in Romanian) Iaşi. Romania, Performantica Publishing House, 2010.
[8]   Tofanica BM, Gavrilescu D. Nonwood and pulp production (in Romanian). Romanian Pulp. Paper J. 2015, 64, 3–10.
[9]   FAOSTAT statistics database – FAOSTAT Statistics Division of the Food and Agriculture Organization of the United Nations, 2019, (Accessed November 1.11.2019 at http://faostat. fao.org).
[10]   Tofanica BM, Cappelletto E, Gavrilescu D, Mueller K. Properties of rapeseed (*Brassica napus*) stalks fibers. J. Nat. Fibers. 2011, 8, 241–262.
[11]   Tofanica BM, Puitel AC, Gavrilescu D. Environmental friendly pulping and bleaching of rapeseed stalk fibers. Environ. Eng. Manag. J. 2012, 11, 681–686.
[12]   Puitel AC, Moisei N, Tofanica BM, Gavrilescu D. Turning wheat straw in a sustainable raw material for paper industry. Environ. Eng. Manag. J. 2017, 16, 1027–1032.
[13]   Chesca AM, Nicu R, Tofanica BM, Puitel AC, Vlase R, Gavrilescu D. Pulping of corn stalks – assessment in bio-based packaging materials. Cell. Chem. Technol. 2018, 52, 7–8, 645–653.
[14]   Sixta H. Introduction. In: Sixta H., ed. Handbook of pulp. 2006, 1, Weinheim, Germany, Wiley-VCH, 3–19.
[15]   Brännvall E. Overview of Pulp and Paper Processes. In: Ek M, Gellerstedt G, Henriksson G., ed. Pulp and paper chemistry and technology. 2, Pulping Chemistry and Technology Berlin, Germany, Walter de Gruyter, 2009, 1–13.
[16]   Gullichsen J. Fiber Line Operations. In: Gullichsen J, Paulapuro H., ed. Papermaking science and technology. 6A, Chemical Pulping, Helsinki, Finland, Fapet Oy, 1999, A19–A243.
[17]   Leponiemi A, Johansson A, Edelmann K, Sipilä K. Producing pulp and energy from wheat straw. Appita. J. 2010, 63, 65–73.
[18]   Deniz I, Ates S. Optimisation of wheat straw Triticum drum kraft pulping. Ind. Crop. Prod. 2004, 19, 237–243.

[19]   Mollabashi G, Saraeian AR, Relasati H. The effect of surfactants application on soda pulping of wheat straw. BioResources. 2011, 6, 2711–2718.
[20]   Akgül M, Tozluoglu A. A comparison of soda and soda-AQ pulps from cotton stalks. Afr. J. Biotechnol. 2009, 22, 6127–6133.
[21]. Tofanica BM, Gavrilescu D. Alkaline pulping of rapeseed (*Brassica napus*) stalks in sulfate and soda-AQ processes. Bulletin of the Polytechnic Institute of Iasi, Sec. Chem. Chem. Eng. 2011, 57, 51–58.
[22]   Tutus A, Eroglu H. A practical solution to the silica problem in straw pulping. Appita. J. 2003, 56, 111–115.
[23]   Tutus A, Eroglu H. An alternative solution to the silica problem in wheat straw pulping. Appita. J. 2004, 57, 214–217.
[24]   Hart PW, Rudie AW. Anthraquinone – A review of the rise and fall of a pulping catalyst. Tappi J. 2014, 13, 23–31.
[25]   Fiala W, Nardi F. The naco process. Paper Technol. Ind. 1985, 26, 75–79.
[26]   Leponiemi A. Non-wood pulping possibilities – a challenge for the chemical pulping industry. Appita. J. 2008, 61, 235–243.
[27]   Delgado M., Escudero E. The "SAICA" straw pulping process. In: Becker J., ed. Small pulp and paper mills in developing countries. KomTech Gmbh, Frankfurt, 1991, 214–222.
[28]   Marín F, Sánchez JL, Arauzo J, Fuertes R, Gonzalo A. Semichemical pulping of Miscanthus giganteus. Effect of pulping conditions on some pulp and paper properties. Bioresour. Technol. 2009, 100, 3933–3940.
[29]   Wong A, Derdall G. A novel sulphite pulping and chemical recovery system for small-and medium-scale pulp mills. Pulp. Paper Canada. 1991, 92, 36–41.
[30]   Qi-pei J, Xiao-yong Z, Hai-tao M, Zuo-hu L. Cleaner production of wheat straw pulp with potash. Chem. Biochem. Eng. Q. 2006, 20, 107–110.
[31]   Gavrilescu D. Pulping Fundamentals and Processing. In: Popa VI., ed. Pulp production and processing: from papermaking to high-tech products. Smithers Rapra, Shawbury, UK, 2013, 35–69.
[32]   Rodriguez A, Moral A, Serrano L, Labidi J, Jimenez L. Rice straw pulp obtained by using various methods. Bioresour. Technol. 2008, 99, 2881–2886.
[33]   Samariha A, Khakifirooz A. Application of NSSC Pulping to Sugarcane Baggase BioResources 2011, 6, 3313–3323.
[34]   Gavrilescu D, Toth S. Corrugated Board, Sfantu Gheorghe, Romania, Ed. T3, 2007 (in Romanian).
[35]   Leminen A, Johansson A, Lindholm J, Gullichsen J, Yilmaz Y. Non-wood fibres in papermaking. Literature review. VTT Research Notes 1779. Espoo, Finland, 1996, (accessed June, 2019 at https://cris.vtt.fi/en/publications/non-wood-fibres-in-papermaking-literature-review,).
[36]   Fischer K, Schiene R. Nitrogenous Fertilizers from Lignin – A Review. In: Hu TQ., ed. Chemical modification, properties, and usage of lignin. Kluwer Academic, New York, USA, 2002.
[37]   Hedjazi S, Kordsachia O, Patt RJ, Latibari A Tschirner U. Alkaline sulfite anthraquinone (AS/AQ) pulping of wheat straw and totally chlorine free (TCF) bleaching of pulps. Ind. Crop. Prod. 2009, 29, 27–36.
[38]   Hedjazi S, Kordsachia O, Patt RJ, Latibari A, Tschirner U. Alkaline sulfite/anthraquinone (AS/AQ) pulping of rice straw and TCF bleaching of pulps. Appita. J. 2009, 62, 137–145.
[39]   Sundquist J. Organosolv Pulping. In: Gullichsen J, Fogelholm CJ., ed. Chemical Pulping 6, B of papermakingscience and technology. Helsinki, Finland, Fapet Oy, 1999, 411–427.
[40]   Muurinen E. Organosolv pulping, a review and distillation study related to peroxyacid pulping. Academic Dissertation, University of Oulu, Oulu, Finland, 2000.

[41] Boman R, Jansson C, Lindstrom LA, Lundahl Y. Non-wood pulping technology: Present status and future. IPPTA Jl. 2009, 21, 115–120.

[42] Katzen R, Frederickson R, Brush BF. The alcohol pulping & recovery process. Chem. Eng. Prog. Chem. Eng. Prog. 1980a, 2, 62–67.

[43] Katzen R, Frederickson RE, Brush BF. Alcohol pulping appears feasible for small incremental capacity. Pulp. Paper 1980b, 8, 144–149.

[44] Hultom T, Lonnberg B, Laxen T, El-Sakhawy M. Alkaline pulping of cereal straw. Cell. Chem. Technol. 1995, 31, 65–75.

[45] Ciovica S, Lonnberg B, Lonnquist K. Dissolving pulp by the IDE pulping concept. Cell. Chem. Technol. 1998, 32, 279–290.

[46] Stockburger P. An overview of near-commercial and commercial solvent-based pulping processes. Tappi J. 1993, 76, 71–74.

[47] Black P. ASAM alkaline sulfite pulping process shows potential for large-scale application. Tappi J. 1991, 74, 87–93.

[48] Obrocea P, Gavrilescu D, Stancana R. Organosolv pulping of wood alkaline-media. 1. ASAM pulping of beech wood at low-temperature. Cell. Chem. Technol. 1994, 28, 339–349.

[49] Kirci H, Bostanci S, Yalinkilic MK. A new modified pulping process alternative to sulfate method alkaline-sulfite-anthraquinone-ethanol (ASAE). Wood Sci. Technol. 1994, 28, 89–99.

[50] Kirci H, Eroglu H. Alkali sulfite anthraquinone ethanol (ASAE) pulping of cotton stalks (Gossypium hisutum L.). Turk. J. Agric. For. 1997, 21, 573–577.

[51] Sarwar JM, Nayeem RJ, Mostafizur RM. Formic acid/acetic acid/water pulping of agricultural wastes. Cell. Chem. Technol. 2014, 48, 111–118.

[52] Pan X, Sano Y. Fractionation of wheat straw by atmospheric acetic acid process. Bioresour. Technol. 2005, 96, 1256–1263.

[53] Chevanan N. Bulk density and compaction behavior of knife mill chopped switchgrass, wheat straw, and corn stover. Bioresour. Technol. 2010, 101, 207–214.

[54] Benjelloum B. The CIMV Organosolv process 2019(accessed June, 2019 at http://www.bio core-europe.org/file/1_4%20BIOCORE%20B%20Benjelloun%20CIMV%20Organasolv.pdf)

[55] Rousu PP, Rousu P, Anttila J. Sustainable pulp production from agricultural waste. Resour. Conserv. Recy. 2002, 35, 85–103.

[56] Dapía S, Santos V, Parajó JC. Study of formic acid as an agent for biomass fractionation. Biomass Bioenergy. 2002, 22, 213–221.

[57] Rousu PP, Rousu P, Rousu, E. Process for producing pulp with a mixture of formic acid and acetic acid as cooking chemical. Pat.US 6562191, 2003.

[58] Rousu P. The future is biorefining. Biorefining Concepts for Non-wood Raw Materials 2019, (accessed June, 2019 at http://www.fibrafp7.net/Portals/0/2_Paivi_Rousu.pdf).

[59] Poppius K, Laamanen L, Sundquist J, Wartiovaara I, Kauliomaki S. Bleached pulp by peroxyacid-alkaline peroxide delignification. Papery ja Puu. 1986, 68, 87–92.

[60] Sundquist J. Summary of milox research. Papery ja Puu. 1996, 78, 92–95.

[61] Seisto A, Poppius LK. Peroxyformic acid pulping of nonwood plants by the MILOX method. Part I: Pulping and bleaching. Tappi J. 1997, 80, 215–221.

[62] Nitsos C, Rova U, Christakopoulos P. Organosolv fractionation of softwood biomass for biofuel and biorefinery applications. Energies. 2018, 11, 1–23.

[63] Ramos LP. The chemistry involved in the steam treatment of lignocellulosic materials. Quimica Nova. 2003, 26, 863–871.

[64] Bozell JJ. An evolution from pretreatment to fractionation will enable successful development of the integrated biorefinery. BioResources. 2010, 5, 1326–1327.

[65] Cherubini F. The biorefinery concept: Using biomass instead of oil for producing energy and chemicals. Energy Convers. Manag. 2010, 51, 1412–1421.

[66]   Bozell JJ, Petersen GR. Technology development for the production of biobased products from biorefinery carbohydrates. Green Chem. 2010, 12, 539–554.

[67]   Boeriu CG, Fitigau FI, Gosselink RJA, Frissen AE, Stoutjesdijk J, Peter F. Fractionation of five technical lignins by selective extraction in green solvents and characterisation of isolated fractions. Ind. Crop. Prod. 2014, 62, 481–490.

[68]   Isikgor FH, Becer CR. Lignocellulosic biomass: A sustainable platform for the production of bio-based chemicals and polymers. Polym. Chem. 2015, 6, 4497–4559.

[69]   Guragai YN, Bastola KP, Madl RL, Vadlani PV. Novel biomass pretreatment using alkaline organic solvents: A green approach for biomass fractionation and 2, 3-butanediol production. Bioenergy Res. 2016, 9, 643–655.

[70]   Ahvazi B, Radiotis T, Bouchard J, Goel K. Chemical pulping of steam-exploded mixed hardwood chips. J. Wood. Chem. Technol. 2007, 27, 49–63.

[71]   Heitner C, Argyropoulos DS, Miles KD, Karnis A, Kerr RD. Alkaline sulphite ultra-yield pulping of aspen chips – a comparison of steam-explosion and conventional chemimechanical pulping. J. Pulp. Pap. Sci. 1993, 19, 58–70.

[72]   Li H, Saeed A, Jahan MS, Ni Y, Van Heiningen A. Hemicellulose removal from hardwood chips in the pre-hydrolysis step of the kraft-based dissolving pulp production process. J. Wood. Chem. Technol. 2010, 30, 48–60.

[73]   Martin-Sampedro R, Eugenio ME, Revilla E, Martín JA, Villar JC. Integration of kraft pulping on a forest biorefinery by the addition of a steam explosion pretreatment. BioResources. 2011, 6, 513–528.

[74]   Ruzinsky F, Kokta BV. High-yield pulping of switchgrass using the sodium sulphite-sodium bicarbonate system. Cell. Chem. Technol. 2000, 34, 299–315.

[75]   Zimbardi F, Viggiano D, Nanna F, Demicheme DC, Cardinale G. Steam explosion of straw in batch and continuous systems. Appl. Biochem. Biotech. 1999, 77–79, 117–125.

[76]   Fang H, James DJ, Zhang X. Continuous steam explosion of wheat straw by high pressure mechanical refining system to produce sugars for bioconversion. BioResources. 2011, 6, 4468–4480.

[77]   Cristensen H. Added Value through Fractionation of Straw. In: Andreasen L., ed. Plant based specialty products and biopolymers. Copenhagen, Denmark, Tema Nord, 1997, 102–107.

[78]   Talebizadeh A, Charani PR. Evaluation of pulp and paper making characteristics of rice stem by twin-screw extruder pulping. BioResources. 2010, 5, 1745–1761.

[79]   Clextral Group. Bivis Pulping Processes and Technology 2019 (accessed July, 2019 at https://www.striko.de/files/pdf/weitere_produkte_und_vertretungen/Papier-Extruder_Pulp__Paper_01.pdf).

[80]   Singh P, Sulaiman O, Hashim R, Rupani PF, Peng LC. Biopulping of lignocellulosic material using different fungal species: A review. Rev. Environ. Sci. Bio. 2010, 9, 141–151.

[81]   Singh D, Chen S. The white-rot fungus *Phanerochaete chrysosporium* conditions for the production of lignin degrading enzymes. Appl. Microbiol. Biot. 2008, 81, 399–417.

[82]   Bajpai P, Mishra SP, Mishra OP, Kumar S, Bajpai PK, Singh A. Biochemical pulping of wheat straw. Tappi J. 2004, 3, 3–6.

[83]   Badr El-Din SM, Kheiralla ZH, Abdel MSM, Abdel ADH. Selection of fungal isolates for biopulping of rice straw. BioResource. 2013, 8, 4969–4980.

[84]   Lopez AMQ, Silva ALSS, Santos ECL. The fungal ability for biobleaching/biopulping/bioremediation of lignin-like compounds of agro-industrial raw material. Química Nova. 2017, 40, 916–931.

[85]   Ilvessalo-Pfäffli MS. Fiber Atlas. Identification of Papermaking Fibers, Springer- Verlag, Berlin, 1995.

[86] Obrocea P, Gavrilescu D, Bobu E. Pulp and paper technology. Politechnic Institute Printing House, Iasi, Romania, 1987, 91.

[87] Biermann CJ. Handbook of pulping and papermaking. Academic Press, San Diego, 1996.

[88] Seth RS. Fiber quality factors in papermaking I. The importance of fiber length and, strength, in Material Interactions Relevant to Pulp, Paper, and Wood Industries. ed. Caulfield DF, Passaretti JD, Sobczynski SF., Material Research Society, Pittsburgh, 1990.

[89] Simionescu CrI, Rozmarin Gh. Reed chemistry (in Romanian). Technical Printing House, Bucharest, Romania, 1966

[90] Pulkkinen I, Ala-Kaila K, Aittamaa J. Characterization of wood fibers using fiber property distributions. Chem Eng Proc. 2006, 45, 546–554.

[91] Ateş S., Atik C., Ni, Y., Gümüşkaya, E. Comparison of different chemical pulps from wheat straw and bleaching with xylanase pre-treated ECF method. Turk. J. Agric. For. 2008, 32, 561–570.

[92] Finell M, Nilsson, C. Kraft and soda-AQ pulping of dry fractionated reed canary grass. Ind. Crop. Prod. 2004, 19, 155–165.

[93] Zhou C, Ohtani Y, Sameshima K, Zhen M. Selection of plant population of kenaf (Hibiscus cannabinus L.) as a papermaking raw material on arid hillside land in China. J. Wood Sci. 1998, 44, 296–302.

[94] Kadam KL, Chin CY. Brown LW flexible biorefinery for producing fermentation sugars, lignin and pulp from corn stover. J. Ind. Microbiol. Biotechnol. 2008, 35, 331–341.

[95] Savastano JH, Warden, PG, Coutts, RSP. Evaluation of pulps from natural, fibrous material for use as reinforcement in cement product. Mater. Manuf. Proc. 2004, 19, 963–978.

[96] Mansfield SD, Weineisen, H. Wood fiber quality and kraft pulping efficiencies of trembling aspen (Populus tremuloides Michx) clones. J. Wood Chem. Technol. 2007, 27, 135–151.

[97] Stener LG, Hedenberg O. Genetic parameters of wood, fibre, stem quality and growth traits in a clone test with betula pendula. Scand. J. Forest. Res. 2003, 18, 103–110.

[98] White DE, Courchene C, McDonough T, Schimleck L, Peter G, Rakestraw J, Goyal G. Effects of wood properties on loblolly pine pulp fiber and sheet characteristics. Tappi J. 2011, 10, 36–42.

[99] Sixta H. Pulp properties and applications, in handbook of pulp. ed. Sixta H., Wiley-VCH Verlag, Weinheim, Germany, 2006.

[100] Karlsson H. Some aspects on strength properties in paper composed of different pulps Licentiate thesis – Karlstad University, 2007 (accessed November 2019) at https://www.diva-portal.org/smash/get/diva2:5004/FULLTEXT01.pdf

[101] Hirn U, Schennach R. Comprehensive analysis of individual pulp fiber bonds quantifies the mechanisms of fiber bonding in paper. Sci Rep. 2015, 5, 10503.

[102] Talis F, Popescu G. Aspects regarding influence of wheat straw quality on yield and characteristics of obtained kraft pulp (in Romanian). Romanian Pulp. Paper J. 1970, 19, 305–315.

[103] Rodríguez A, Moral A, Serrano L, Labidi J, Jiménez L. Rice straw pulp obtained by using various methods. Bioresour. Technol. 2008, 99, 2881–2886.

[104] Jiménez L, Angulo V, Ramos E, De la Torre MJ, Ferrer JL. Comparison of various pulping processes for producing pulp from vine shoots. Ind. Crop. Prod. 2006, 23, 122–130.

[105] Wanrosli WD, Zainuddin Z, Law KN, Asro R. Pulp from oil palm fronds by chemical processes. Ind. Crop. Prod. 2007, 25, 89–94.

[106] Jiménez L, Serrano L, Rodriguez A, Sanchez R. Soda-anthraquinone pulping of palm oil empty fruit bunches and beating of the resulting pulp. Bioresour. Technol. 2009, 100, 1262–1267.

[107] Jiménez L, Pérez A, de la Torre MJ, Moral A, Serrano L. Characterization of vine shoots, cotton stalks, Leucaena leucocephala and Chamaecytisus proliferus and of their ethyleneglycol pulps. Bioresour. Technol. 2007, 98, 3487–3490.

[108] Abrantes S, Amaral ME, Costa AP, Duarte AP Cynara cardunculus L. alkaline pulps: Alternatives fibres for paper and paperboard production. Bioresour. Technol. 2007, 98, 2873–2878.
[109] Shatalov AA, Pereira H. Influence of stem morphology on pulp and paper properties of Arundo donax L. reed. Ind. Crop. Prod. 2002, 15, 77–83.
[110] Célino A, Freour S, Jacquemin F, Casari P. The hygroscopic behavior of plant fibers: A review. Front Chem. 2014, 1, 43.

Ivo Valchev
Chapter 4
Chemical pulp bleaching

4.1 Introduction

The main objective of bleaching is to remove encrusted substances to obtain a pure white product; therefore, the manufacturing process requires further delignification and bleaching of the fibers, as residual lignin is a major contributing factor to color. Unbleached chemical pulps still contain lignin in an amount of 2–4.5% on oven dry (o.d.) pulp depending on the wood species and process details. The bleaching process can best be regarded as a continued pulping in which a series of alternating oxidation and extraction treatments ultimately lead to an almost lignin-free fiber. During bleaching, the cleanliness of the pulp improves when the fibers of the fiber bundles, or shives, are released as the last of the residual lignin is removed from the pulp and any bark debris dissolves. The chemicals used in bleaching also effectively dissolve extractives contained in the pulp. There is a fundamental difference between bleaching of chemical pulps and mechanical pulps. In the bleaching of chemical pulps lignin is oxidized, decomposed, and finally eliminated from the pulp fibers. This results in less chromophores in the pulp. The bleaching process must be carried out as gently as possible, so that the carbohydrate is not attacked. The target brightness cannot be achieved in only one bleaching step without sacrificing pulp strength. Therefore, pulp is bleached in several steps, and the pulp is washed between them. Multistage bleaching gives the best results regarding both quality and economy, and there are alkaline and acidic bleaching stages. With only alkaline or acidic stages, the target brightness would not be attained, and so both are always used in bleaching. Using old technology, bleaching was performed with chlorine-containing chemicals: with elemental chlorine (C), hypochlorite (H), or with chlorine dioxide (D). Between stages, the dissolved lignin was extracted with alkali (E). Typical traditional bleaching sequences were CEHDED and CEDED.

During the 1980s and 1990s, environmental concern in the pulp and paper industry increased considerably, and pressure on the industry to become more environmental friendly came, to an increasing degree, from the final customers. As a result of the concern regarding chlorinated organic compounds formed during chlorine bleaching, conventional bleaching concepts were rapidly replaced by the so-called elemental chlorine-free (ECF) bleaching process, and this became the dominant bleaching technology. The complete substitution of chlorine by oxygen and chlorine

Ivo Valchev, Department of Pulp, Paper and Printing Arts, University of Chemical Technology and Metallurgy, Sofia, Bulgaria

https://doi.org/10.1515/9783110658842-004

dioxide is the key step in reducing the levels of organochlorines, measured as adsorbable organic halogens (AOX), in pulp mill effluents. The modern ECF bleaching sequences of today became established with oxygen (O), chlorine dioxide (D), and hydrogen peroxide (P) as the predominant oxidation agents. Some mills have also installed ozone (Z), peracetic acid (T), and xylanase (X) stages. Typically, a bleaching sequences using less than 4 kg chlorine dioxide/o.d. t pulp is referred to as *ECF light* and that notation is very popular. Pulp can also be totally bleached without chlorine chemicals. This kind of oxygen chemical bleaching is usually known by the abbreviation TCF – totally chlorine free. Bleaching chemicals used during TCF bleaching are oxygen-containing chemicals such as oxygen, peroxides, and ozone. Like ECF, TCF never refers to any specific bleaching sequence, but includes Q(PO), (TZQ)(PO), and A(OP)Q(PO), where Q stands for treatment of the pulp with a chelating agent and A with acid. However, very few mills are today producing only TCF-bleached kraft pulp. A bleaching stage can be defined as any treatment during the course of the bleaching, which takes place between two subsequent washings. (OO)(DQ)(PO) is thus a three-stage sequence, where (OO) is the notation of an oxygen delignification process taking place in two subsequent reactors without intermediate washing.

When the bleaching is performed mainly with chlorine-containing chemicals, the active chlorine concept is used to quantify the oxidizing power of the different substances. This means that all charges are recalculated as if only chlorine is used (Tab. 4.1). In ECF bleaching sequences, this concept still prevails.

Tab. 4.1: Active chlorine content and oxidizing equivalent (OXE) of bleaching chemicals.

Chemical (formula)	Transferred electrons (e^-/mole)	Equivalent weight (g/mole.e^-)	Active chlorine (kg/kg)	Oxidizing equivalent (OXE/kg active chlorine)
Chlorine (Cl_2)	2	35.46	1	28.20
Hypochlorite (NaClO)	2	37.22	0.95	26.86
Chlorine dioxide (ClO_2)	5	13.49	2.63	74.12
Oxygen (O_2)	4	8.00	–	125
Ozone (O_3)	6	8.00	–	125
Hydrogen peroxide (H_2O_2)	2	17.01	–	58.29
Peracetic acid (CH_3COOOH)	2	38.00	–	26.32

The oxidizing equivalent (OXE) concept suggested by Grundelius [1] is a valuable tool in comparing the efficiencies of all bleaching chemicals, where 1 OXE is equal to that quantity of substance that receives 1 mole of electrons during bleaching (Tab. 4.1). However, in reality this theoretical number should be used with caution as the same bleaching effect cannot always be expected for different chemicals even if the OXE should be the same. The treatment of a pulp with different bleaching chemicals having the same amount of OXE can lead to different brightness values [2].

4.2 Optical properties of pulp

The best known and mostly used theory for describing the optical properties of pulp and paper is the Kubelka–Munk theory. This theory was originally developed for paint films and it is the most commonly used and most widely accepted theory for estimating optical constants in pulp and paper [3–7]. The theory is based on the interrelationship between the light-scattering coefficient s, light absorption coefficient k, and reflectance factor R_∞ – brightness (eq. (4.1)):

$$R_\infty = 1 + \frac{k}{s} - \sqrt{\frac{k^2}{s^2} + 2 \cdot \frac{k}{s}} \tag{4.1}$$

where $100 * R_\infty$ is the brightness, in percent, measured at 457 nm of an optically thick or opaque sample.

The scattering coefficient s depends on the fiber area and the degree of bonding between fibers, and the indices of refraction of the fibers. A sheet of stiff, tubelike fibers scatters light more than slender, collapsed fibers. Both beating the pulp and pressing a paper sheet result in a higher degree of bonding and a denser sheet with less light-scattering interfaces. The ability of a sheet to scatter light is not or slightly affected by the degree of cooking and bleaching of the pulp. The absorption coefficient k is influenced, to some extent, by the degree of bonding between the fibers and the indices of refraction, but to a much greater extent by changes in the cooking process, the degree of bleaching, and brightness reversion. The light absorption coefficient is mainly dependent on the chromophoric groups in the pulp. During papermaking operations, it is primarily influenced only by the addition of dyes, clay, dirt, and soluble impurities such as iron compounds [8]. It is not or slightly affected by beating or pressing. In the Kubelka–Munk theory, the additivity principle is used to calculate s and k for mixtures (eqs. (4.2) and (4.3)):

$$s = s_1 \cdot g_1 + s_2 \cdot g_2 + s_3 \cdot g_3 + \cdots \tag{4.2}$$

$$k = k_1 \cdot g_1 + k_2 \cdot g_2 + k_3 \cdot g_3 + \cdots \tag{4.3}$$

where $g_1, g_2, g_3, \ldots$ are the proportions of components 1, 2, 3, $\ldots$

The additive properties of k and s can be used for predicting the optical properties for papers with different fibers and fillers. The change in brightness through a bleaching step is not proportional to the reduction in chromophore concentration. The Kubelka–Munk expression (eq. (4.1)) shows that the reflectance (brightness) is not a linear function of the chromophore concentration. At a high brightness level, the brightness is governed by only a small change in chromophore concentration, while at a low brightness level the same loss in brightness is connected with a significantly higher change in chromophore concentration. In accordance with the Kubelka–Munk theory, the following remission function (eq. (4.4)) can be used for characterizing a color change in a particular pulp or paper, brought about by an arbitrary treatment that results in a change in the absorption coefficient, but does not affect the scattering coefficient:

$$\frac{k}{s} = \frac{(1 - R_\infty)^2}{2 \cdot R_\infty} \tag{4.4}$$

For example, Tongren [9], Giertz [10], Paulsson and coworkers [11], and Li and Ragauskas [12] have used this kind of expression for the study of certain changes in pulps after aging, in the form of the postcolor number (Pc). The Pc number is the difference between k/s values before (a) and after (b) aging, as shown in eq. (3.5). The term postcolor number is used when the values are determined at a single wavelength (usually = 457 nm).

$$\text{Pc} = [(k/s)_a - (k/s)_b] \cdot 100 \tag{4.5}$$

The decrease in chromophore concentration through bleaching operations can be monitored by the relative change of k/s values $\alpha_{K/s}$ (eq. (4.6)):

$$\alpha_{K/s} = \frac{K_a/S - K_b/S}{K_a/S} = \frac{K_a - K_b}{K_a} \tag{4.6}$$

where k_a and k_b are values of the absorption coefficient before (a) and after (b) bleaching, and s is the scattering coefficient, the value of which is assumed to be constant during bleaching. The nondimensional quantity $\alpha_{k/s}$ from eq. (4.6) can be used as a kinetic variable for bleaching processes [13–16].

4.3 Residual lignin and other oxidizable structures

The amount of lignin, the types of oxidizable structures, and the metal content of the pulp entering the bleaching stages determine the consumption of bleaching agents. Froass and coworkers [17] showed that the residual lignin, compared with native lignin, is unreactive toward chlorine dioxide due to a low content of aryl ether linkages and prevalence of condensed type structures. Gustavsson [18] reports that the higher the content of β-aryl ether structures in the residual lignin after cooking, the easier it is to bleach in a QPQP sequence. However, it is difficult to establish any clear relationship between the chemical structures of the residual lignin and the bleachability of the pulp [18]. In unbleached kraft pulps, according to Lawoko [19], 85–90% of the residual lignin molecules are somewhere linked to carbohydrates, and mainly to glucomannan and xylan in network structures. The relatively faster delignification of the xylan-rich lignin–carbohydrates complex (LCC) could be explained by the hydrolysis of the predominant β-O-4 structures found in the native LCC fraction. Janson and Palenius [20, 21] observed that the lignin redeposited on the fibers at the end of the cook is much darker than the residual lignin left in the pulp. Transition metals can also be attached to lignin and extractives by complex formation. Dyer [22] also showed that pulp washing can have a significant impact on the metal content of the unbleached kraft pulp and that calcium is not likely to be a significant contributor to the chromophoric properties of pulp in this case, since the pulp is completely washed.

The kraft pulp contains not only residual lignin but also other oxidizable structures. The most important of these are hexenuronic acids that are attached to the xylan, but other nonspecified structures are also present in the unbleached pulp. In both softwood and hardwood, the xylan chain is substituted with 4-O-methylglucuronic acid groups located in the C-2 position. Under alkaline conditions, the methoxyl group can be eliminated as methanol resulting in the formation of a hexenuronic acid (HexA) group. Since the HexA group is rather stable in alkali, the xylan that remains in the pulp after the cook will contain an appreciable amount of such groups. These will influence the pulp properties and contribute to the bleachability of the pulp. According to Rosenau and coworkers, the chemical structures of HexA-derived chromophoric compounds, which make up 90% of the HexA-derived chromophores, are ladder-type, mixed quinoidaromatic oligomers of the bis(furano)-[1,4]benzoquinone and bis(benzofurano)-[1,4]benzoquinone type and the same chromophoric compounds are generated independent of the starting material [23, 24]. The HexA consumes permanganate in the Kappa number analysis of pulps and thereby appears as "false lignin" in the Kappa number measurement. Gellerstedt and Li [25] reported that typically 3–6 Kappa number units of an unbleached hardwood pulp and 1–3 Kappa number units of an unbleached softwood pulp are due to HexA. Vuorinen and coworkers [26], Buchert and coworkers [27], and Bergnor-Gidnert and coworkers [28] reported that HexA reacts with several bleaching chemicals, such as ozone, chlorine dioxide, or peracids and thus consumes these bleaching chemicals, whereas no significant degradation of HexA has been detected during the oxygen or peroxide stages. If it is present in a fully bleached pulp, HexA has been reported to decrease the brightness stability of the pulp [29-32]. Sevastyanova showed that HexA plays a dominant role in the brightness reversion in bleached kraft pulps [33]. HexA also has a strong affinity for transition metals according to Devenyns and coworkers and Vuorinen and coworkers [25, 34, 35]. However, according to Laine and Stenius [36], a high surface charge of the pulp, partly provided by the presence of HexA, leads to better paper strength properties. However, Gustavsson [18] could not reveal any clear correlation between the HexA content in the bleached pulps and the yellowing tendency. Moreover, in the case of the $D_{HT}(OP)D$ sequence, Gustavsson determined that a high HexA content, after cooking of the hardwood, has a positive effect in achieving a low bleaching chemical requirement. Furthermore, neither of the strategies to significantly reduce the HexA content in a kraft pulp by altering the cooking conditions is attractive for industrial implementation, since they would result in extensive losses in pulp yield, viscosity, and strength [18]. HexA are removed in an acid treatment (A) along the bleaching sequences, which mainly comprises acidic washing. Acidic pulping conditions, for example, at 130 °C for 2–3 h with 0.05 M sulfuric acid, or at 80 to 140 °C and pH 3.0–3.5 for 2–5 h, not only allow the removal of possible HexA side groups, but also result in the occurrence of furanoid primary degradation products, for example, 2-furancarboxylic acid and 5-formyl-2-furancarboxylic acid that lead to the formation of hitherto unidentified chromophoric compounds [37-39].

In bleached pulp, resonance-stabilized quinones are the main chromophores surviving bleaching, along with chlorine dioxide and hydrogen peroxide [40-42]. The concentration of chromophores is generally extremely low, mostly in the ppm range, and for fully bleached pulps even lower, but their resonance stabilization prevents destruction with alkaline peroxide.

Suess and Rosenau demonstrated [41, 43] that chromophores in/on cellulosics can generally be divided into two classes: primary and secondary chromophores. The primary chromophores are formed independent of the respective process chemicals, from monosaccharides, and are thus solely carbohydrate derived. Among the identified primary chromophores, hydroxybenzoquinone, hydroxyacetophenone, and naphthoquinone structures are dominant. A second class of chromophores, named *secondary chromophores*, is formed with the involvement of the process chemicals during bleaching. They are thus process specific, in contrast to the primary chromophores that are based only on the hemicelluloses material.

4.4 Oxygen delignification

The first commercial installation of an oxygen delignification high consistency (HC) system was started up in a South African mill in 1970 and the first medium consistency (MC) unit has been in operation since 1979. By 2010, the worldwide installed capacity increased significantly to about 300,000 air dry metric tons per day, representing more of the world's bleachable-grade pulp [44]. At first, the main driving forces for the installation of oxygen delignification in existing mills were environmental regulations and the complete abolition of chlorine during bleaching. The effluent from the oxygen stage could be recirculated as a countercurrent wash liquor, in the washing after the digester, and could be sent to the recovery boiler. The discharge of biochemical oxygen demand (BOD), chemical oxygen demand (COD), and AOX could thus be greatly reduced. The economic advantages of an oxygen delignification stage are the savings in operation costs through the use of lower amounts of other bleaching agents, as oxygen has a lower cost than all other oxidizing agents. Oxygen delignification removes about 40–65% of the lignin remaining in the pulp after cooking for softwood and about 35–55% for hardwood. In general, it is difficult to reach a Kappa number below 10 whatever oxygen delignification process is adopted, largely because the Kappa number does refer not only to lignin but also to structures like HexA, which are not reactive toward oxygen. The other advantage of oxygen delignification is that it is more selective in terms of yield and viscosity preservation than low Kappa kraft cooking [45]. However, reductions in pulp viscosity are usually higher compared with chlorine dioxide treatment [46]. Moreover, because high pressures are involved in oxygen delignification, the cost of an oxygen system can require a significant capital investment [44].

4.4.1 Chemistry of oxygen delignification

Oxygen is an unusual molecule in that its normal configuration is the triplet state. Oxygen has a strong tendency to oxidize organic substances, under alkaline conditions, simultaneously resulting in its stepwise reduction to water by one-electron transfer processes as shown in eq. (4.7). Depending upon the pH, this reaction yields different transients, including the superoxide anion radical ($O_2 \cdot^-$), which can combine with a hydrogen ion to form the hydroperoxy radicals ($HO_2 \cdot$), hydroperoxide anions (HO_2^-), hydrogen peroxide (H_2O_2), and hydroxyl radicals ($HO \cdot$) [41].

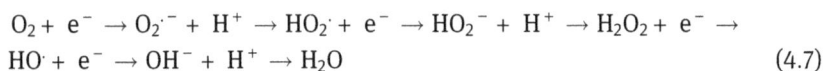

$$O_2 + e^- \rightarrow O_2 \cdot^- + H^+ \rightarrow HO_2 \cdot + e^- \rightarrow HO_2^- + H^+ \rightarrow H_2O_2 + e^- \rightarrow$$
$$HO \cdot + e^- \rightarrow OH^- + H^+ \rightarrow H_2O \qquad (4.7)$$

Free phenolic hydroxyl groups play a key role during lignin reactions [47]. When ionized by the addition of alkali, they furnish the high electron density that is needed to initiate the reaction with the relatively weakly oxidizing molecular oxygen. This, together with the weakly acidic nature of the phenolic hydroxyl groups, explains why strongly alkaline conditions are needed to achieve appreciable delignification rates. The initial step is the conversion of the ionized phenolic group to a phenoxy radical through the loss of a single electron to a suitable acceptor; this may be molecular oxygen or one of the many other radical species present in the system [48]. Once the radical is produced on the phenolic hydroxyl oxygen, resonance structures may shift the radical to the *ortho* and *para* positions relative to the carbon containing the hydroxyl group. These radicals have now become new sites available for the electrophilic attack of an oxygen molecule, which results in bridging, ring opening, and then the formation of carboxylic acids according to the scheme shown in Fig. 4.1.

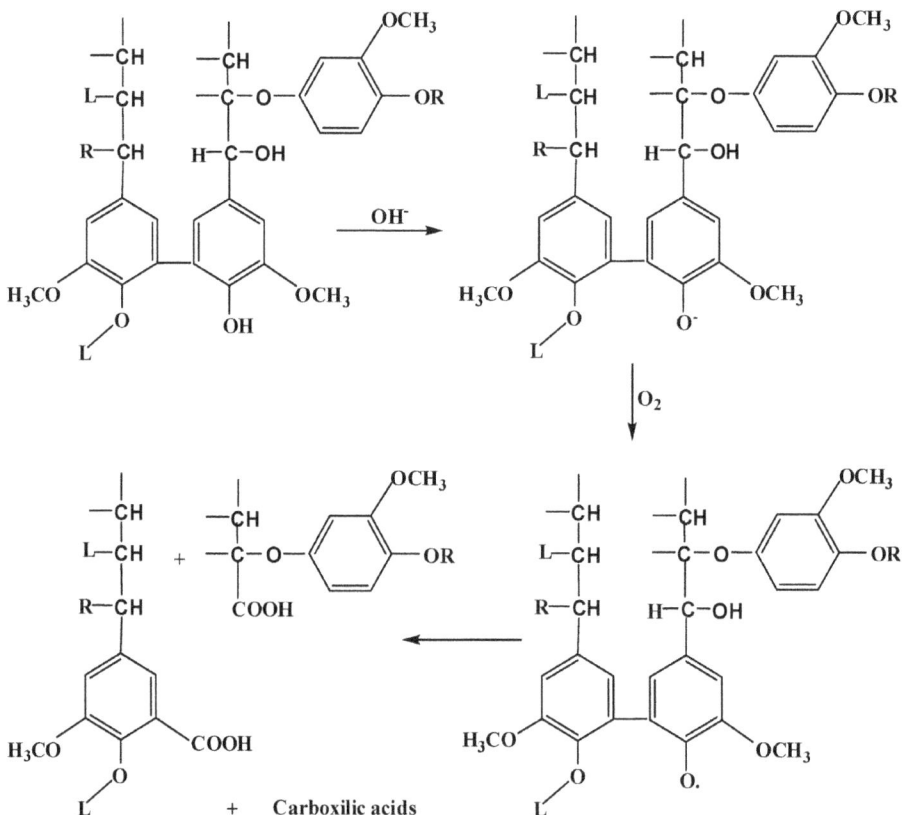

Fig. 4.1: Typical reaction of lignin with oxygen in an alkaline media.

Eventually, when the lignin undergoes a few of these steps, its aromatic structure breaks down and the number of hydrophilic oxidized groups increases; thus, the lignin fragments become water soluble and can be removed from the pulp. In addition, when the initial resonance structure moves to a carbon of an alkene side chain, oxygen will attack the radical, bridge to the adjacent carbon, and cleave the carbon–carbon linkage, leaving two carbonyl groups and eliminating part of the side chain [48]. In the oxygen delignification process, carbohydrates are attacked to a greater extent than in the chlorination and alkali extraction processes. Random chain cleavage and endwise secondary peeling reactions are two types of carbohydrate degradation reactions that occur during oxygen delignification. Random chain cleavage can occur at any point along the main chain to lower the degree of polymerization of cellulose. The viscosity loss is due to an oxidation of one or more of the hydroxyl groups located along the cellulose chain. Thereby, a carbonyl group is created and, due to the alkaline conditions, an elimination reaction occurs resulting in a cleavage of the chain into two shorter units according to the scheme (Fig. 4.2).

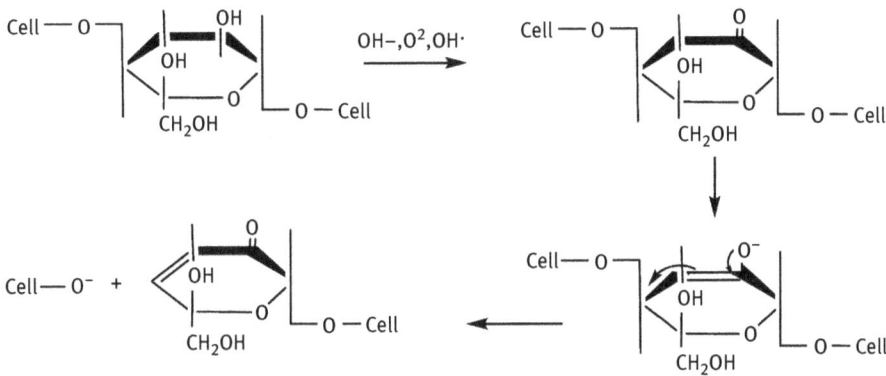

Fig. 4.2: Cellulose chain cleavage in the conditions of oxygen delignification.

The reaction that causes yield loss in an alkaline media, the peeling reaction, usually has less importance in oxygen delignification than random chain cleavage. This occurs for two reasons. First, pulps (because of their long previous exposure to strongly alkaline conditions in the digester) contain very few end units that have not been converted to the stable form by stopping reactions. The secondary reason is that oxygen itself converts reducing end groups to the stable oxidized form. However, peeling can become a problem if random chain cleavage is excessive because every chain breakage creates two new chain ends, one of which is a reducing end group. Antonsson and coworkers [49] reported that the prehistory of the pulps is a very important factor in determining the response to oxygen delignification. Sulfite pulps show the greatest yield loss during oxygen delignification, compared with those of kraft and prehydrolysis kraft pulps. It is also shown that the degree of delignification is not due to different amounts of hexenuronic acid. It is likely that lignin–carbohydrate complexes (LCC) play a very important role in limiting the reaction rate of oxygen delignification. LCC is probably native and not formed during cooking [49].

4.4.2 Process description and variables

The typical temperature in the technical oxygen delignification process is 90–100 °C, the charge of alkali is 15–25 kg/t o.d. pulp, final pH 10–11, the pressure at the reactor top is 0.4–0.6 MPa, whereas the oxygen charge is in the range of 15–25 kg/t o.d. The process is normally carried out at medium consistency in a single- or two-stage system with a retention time of 40–80 min.

Oxygen delignification is a heterogeneous process involving three phases: solid (fibers), liquid (aqueous alkali solution), and gas (oxygen). As a first step, oxygen dissolves in the aqueous phase and is then transported through the liquid to the pulp–fiber interface. The dissolved oxygen subsequently diffuses into the fiber wall and then reacts with the wood components. Agarwal and coworkers [50] determined that the intrafiber mass transfer effect does not influence the delignification rate. Hsu and Hsieh [51] also reported that the mass transfer resistance of oxygen in the gas phase is insignificant in comparison with the liquid phase resistance due to the low solubility of oxygen in the liquid phase. Therefore, to compensate for the low solubility of oxygen in water, a high oxygen pressure and fluidization of the pulp suspension are regarded as a prerequisite for oxygen delignification. The effect of pressure on the oxygen delignification of hardwood kraft pulp was studied by Nenkova and coworkers [52]. It has been established that an increase in oxygen pressure from 0.3 to 0.6 MPa leads to a Kappa number reduction of 0.5 units, a viscosity loss of about 30 dm^3/kg, and a brightness increase of nearly 2% without changing the pulp strength (Tab. 4.2). This indicates that the higher oxygen pressure increases the bleaching processes of hardwood pulp only slightly, while the results obtained from Heiningen and Ji [53], for softwood pulp, show a decrease of the Kappa number by more than 2 units without any effect on the selectivity. A higher pressure oxygen delignification would certainly be beneficial for the process, but in practice it is not economical due to the higher capital and energy cost.

The results in Tab. 4.2, obtained by varying the sodium hydroxide charge, show that a higher charge leads to more cellulose degradation (a viscosity loss of almost 100 dm^3/kg) as well as to a pulp strength loss, besides a Kappa reduction of 0.7 units [52]. Therefore, a high charge of sodium hydroxide reduces the oxygen delignification selectivity and leads to a lower pulp strength at the same degree of delignification. These results are consistent with previous reports by Yang and coworkers [54] and that obtained for softwood pulp [48]. In order to increase delignification without significantly decreasing the selectivity, Heiningen and Ji [53] suggested modification to the oxygen system design so that the alkali concentration and charge should be decoupled, similar to modern cooking systems.

It has been found that the consistency of oxygen delignification has no significant effect on the Kappa number reduction or brightness improvement (Tab. 4.2) [52]. A medium consistency oxygen system operates between 11% and 14% consistency.

Tab. 4.2: Effect of oxygen delignification conditions on the pulp properties at temperature 102 °C and reaction time 60 min.

Oxygen delignification conditions			Pulp properties				
NaOH charge (%)	Pulp consistency (%)	Oxygen pressure (MPa)	Kappa number	Viscosity (dm³/kg)	Brightness (ISO%)	Breaking length (m)	Tear index, (mN.m²/g)
2.5	10	0.3	11.0	816	47.2	8500	7.50
		0.4	10.9	809	47.6	8510	7.52
		0.5	10.7	799	48.4	8520	7.46
		0.6	10.5	787	49.5	8520	7.45
2.0	10	0.6	10.7	818	48.7	8700	7.80
2.5			10.5	787	49.5	8530	7.50
3.0			10.3	755	50.7	8400	7.70
3.5			10.0	715	52.1	8340	7.75
2.5	8.0	0.6	10.5	–	49.3	–	–
	10.0		10.5	787	49.5	8530	7.50
	12.0		10.7	–	48.7	–	–
	14.0		10.7	816	48.7	8500	7.50

At a consistency below 10%, channeling may occur in the reactor, while at about 14%, fluidization of the pulp slurry becomes difficult.

The carry-over of dissolved solids has little or no impact on the initial rapid phase of oxygen delignification; however, it will increase the overall oxygen and alkali requirements due to their preferred consumption by the dissolved organic and inorganic compounds [55]. In addition, organic unoxidized solids may negatively affect the pulp viscosity. It is theoretically possible to estimate the requirement of molecular oxygen to be close to 0.3 kg/t per oxygen delignified Kappa number unit of softwood kraft pulp, while mill measurements showed that a consumption of 1 kg O_2/t per oxygen delignified Kappa number unit can be applied, as it includes the molecular oxygen normally required for the oxidation of unoxidized carry-over with the pulp [56].

The effects of oxygen delignification temperature and reaction time on the pulp properties were studied by Nenkova and coworkers [52]. Figure 4.3 shows that a prolonged treatment and higher temperature leads to a higher lignin removal, as expected. The effect of increasing the temperature from 95 °C to 110 °C is the same as that of doubling the reaction time. When the temperature is increased from 85 °C to 110 °C,

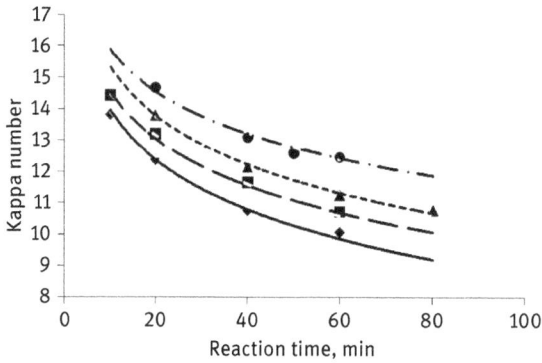

Fig. 4.3: Kinetic data of Kappa number (● 85 °C; ▲ 95 °C; ■ 102 °C; and ◆ 110 °C).

the brightness, according to the International Organization for Standardization (ISO), improves by 9% after 60 min. This is expected as a decrease in Kappa number normally leads to an increase in the brightness of the pulp.

The evolution of the pulp viscosity is shown in Fig. 4.4. It can be seen that increasing the temperature and reaction time leads to an increased viscosity loss of the bleached pulp. The behavior of the pulp viscosity in Fig. 4.4 is very similar to that of the Kappa number in Fig. 4.3. The obtained Kappa number – viscosity relationship (Fig. 4.5) shows that the selectivity (delignification – cellulose degradation) is not a function of temperature. However, the selectivity decreases upon increasing the alkali charge, as can be seen by the extra viscosity drop at the same Kappa number.

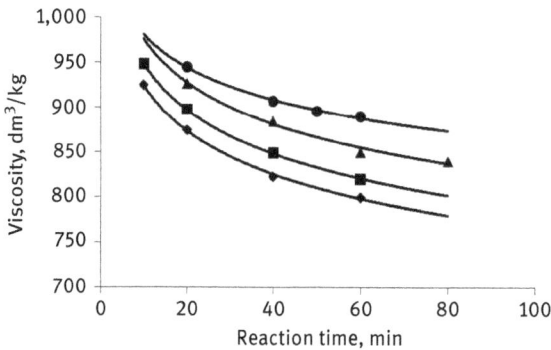

Fig. 4.4: Kinetic curves of viscosity (● 85 °C; ▲ 95 °C; ■ 102 °C; and ◆ 110 °C).

The pulp strength properties do not change considerably over the studied temperature–time interval. This result indicates that an increase in temperature can considerably accelerate the bleaching process without affecting the pulp strength properties.

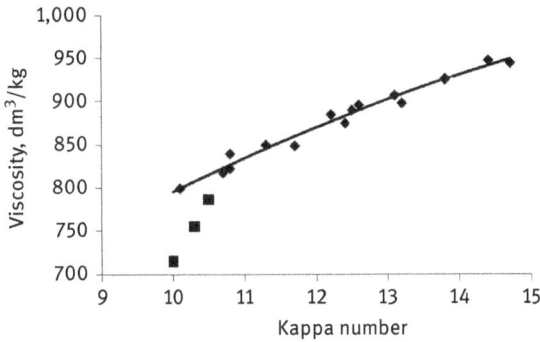

Fig. 4.5: Correlation between Kappa number and viscosity (■ 2.5–3.5% NaOH and ◆ 2% NaOH).

Therefore, it is most effective to increase the degree of oxygen delignification by optimization of the temperature–time profile [52]. Heiningen and Ji [53] reported that for softwood oxygen delignification, the selectivity does not change much from 80 °C to 100 °C, while it decreases by 30% to 40% when the temperature is raised to 110–115 °C. The lower selectivity and the higher capital cost (due to higher pressures) of the latter explain why the industry generally does not operate above 100 °C.

4.4.3 Kinetics of oxygen delignification

In an effort to optimize the efficiency and selectivity of oxygen delignification, many kinetic studies have been documented. Most kinetic models for oxygen delignification are empirical, in which the rate of delignification is considered to be proportional to the lignin content expressed as the Kappa number (K) and hydroxide ion concentration [HO$^-$], and oxygen pressure (p_{O_2}) expressed in the form of a power law (eq. (4.8)).

$$-\frac{dK}{dt} = A \cdot e^{-E/RT} \cdot [\text{OH}^-]^m \cdot p_{O_2}^n \cdot K^q \qquad (4.8)$$

The constants m, n, and q are determined empirically from experimental data, while the temperature dependence is given by the Arrhenius equation, where E is the activation energy, A is the pre-exponential factor, R is the gas constant, and T is the absolute temperature. Based on eq. (4.8), the rate of oxygen delignification has been described by using one and two region model equations by Olm and Teder [57], Hsu and Hsieh [58, 59], Iribarne and Schroeder [60], and Agarwal and coworkers [50]. The two different reaction periods, a rapid initial step followed by a long period during which the change in Kappa number is at a very slow rate, are assumed to be related to the various lignin linkages comprising its structure. The most widely accepted kinetic model of the two regions is that of Olm and Teder [57], who assumed pseudo-first-order kinetic equations in terms of the Kappa number. All those studies have an empirical character and that is why they cannot describe the complex heterogeneous oxygen delignification process. Similar empirical power law models have also been proposed for cellulose degradation. The other category of kinetic models is based on the topochemical concept that the delignification rate depends on the number of reactive sites, formed at the beginning of the process, and the growth rate of the transformed lignin from these reactive sites. Valchev and Christensen [61],and Nguyen and Liang [62] successfully verified the applicability of the topochemical equation of Avrami–Erofeev in the kinetics of oxygen delignification. The next development of the topochemical delignification model is the applicability of the modified form of the topochemical kinetic equation of Praut–Tompkins (eq. (4.9)) [52, 63]. In this kinetic model, the rate of progress of the reaction is determined by the size of the interface between the reacted and unreacted solid. According to this model, the rate of delignification v is a function of the amount of oxidized lignin that subsequently becomes soluble, α_κ, and of the amount of residual undissolved lignin at any time, $(1 - \alpha_\kappa)$:

$$v = \frac{d\alpha}{dt} = k^1 \cdot \chi \cdot \alpha_\kappa^{(\chi-1)/\chi} \cdot (1 - \alpha_\kappa)^{(\chi+1)/\chi} \tag{4.9}$$

where the degree of delignification α_κ = (Kappa$_{in}$ – Kappa)/Kappa$_{in}$ and k^1 is the rate constant. The power factors $(\chi - 1)/\chi$ and $(\chi + 1)/\chi$ determine the relative contributions by the dissolved and undissolved parts of the lignin, respectively, to the rate of delignification. Based on this topochemical mechanism, the rate of delignification depends on the size and the state of the changing reaction interface. The integrated form of eq. (4.9) used for the description of the experimental data is

$$\frac{\alpha_\kappa}{1 - \alpha_\kappa} = (k_1 \cdot t)^\chi \tag{4.10}$$

where $k_1 = k^1/\chi$ is an apparent rate constant and χ is a power factor.

After rearrangement of eq. (4.10) and with the inclusion of the Arrhenius temperature dependency of the apparent rate constant, eq. (4.11) is obtained:

$$\text{Kappa} = \text{Kappa}_{in} \left[1 - \left[1 + \left(5 \times 10^4 \cdot e^{(-5833/T)} \cdot t \right)^{-0.59} \right]^{-1} \right] \tag{4.11}$$

where t is the reaction time in min, and T is the temperature in K. The obtained ratio of the activation energy of the process to the universal gas constant, E/R, is 5,833 K. The activation energy E is 48.2 kJ/mol and depends only on the nature of the unbleached pulp. In a mill with relatively constant pulping conditions and wood supply, the activation energy is practically constant. The same is true for the coefficient χ in the Prout–Tompkins equation, which is 0.59. The pre-exponential factor, A, includes the effect of all steric factors on the process (charge of reagents, their mixing, the pressure in the system, etc.) and was found to be 5.10^4 min^{-1}. Equation (4.11) can be used for process control of the Kappa number. Inclusion of the effect of oxygen pressure and alkali in eq. (4.11) leads to eq. (4.12):

$$\text{Kappa} = \text{Kappa}_{\text{in}}\left[1 - \left[1 + \left(2.16 \times 10^4 \cdot e^{(0.175C + 0.8P - 5833/T} \cdot t\right)^{-0.59}\right]^{-1}\right] \quad (4.12)$$

where t is the time in min, T is the temperature in K, C is the alkali charge in %, and P is the oxygen pressure in MPa.

Equation (4.12) enables one to obtain the Kappa number at any moment during the oxygen delignification as a function of temperature, pressure, and alkali charge. This dependence of Kappa number may be used for simulation and control of the oxygen delignification stage.

The variation in pulp viscosity with time at different temperatures has been studied by Valchev and coworkers [63]. The dependence acquires a linear character at the 20th minute from the beginning of oxygen delignification, irrespective of temperature. This shows that the kinetics of viscosity change, which provides information on carbohydrate destruction processes, is described by a zero-order equation after the first 20 min. The zero order observed shows that the rate of destruction stays unchanged and equals the rate constant. It does not depend on the quantity of carbohydrates undergoing destruction.

4.4.4 Process technology and equipment

In a medium consistency one-stage oxygen delignification process, the pulp suspension is pumped to a reactor as presented in Fig. 4.6 [64]. Alkali is added to the pulp suspension prior to the pump. Oxygen and steam are added and thoroughly mixed into the pulp suspension in a mixer positioned immediately after the pump. In the alkaline environment, the oxygen forms a reasonably stable gas dispersion in the pulp. The mixture enters the reactor and passes upward in a plug flow while the oxygen is consumed in reactions with the lignin in the pulp. The pulp leaves the reactor and enters a blow tank (where off-gases are separated from the pulp suspension), which also serves as a standpipe for a second pump, which pumps the pulp suspension further to washing and bleaching stages.

The introduction of two-stage oxygen delignification has further increased the effectiveness of the process (Fig. 4.7) [64]. Metso introduced the two-stage OxyTrac process [65], with an oxygen pressure and temperature of 0.8–1 MPa and 80–85 °C, respectively, during the first stage of the 20–30 min retention time. During the second stage, the oxygen pressure is 0.3–0.5 MPa, the temperature is 90–100 °C, and the reaction time is 60–80 min. According to GLV [64], the optimal conditions for oxygen

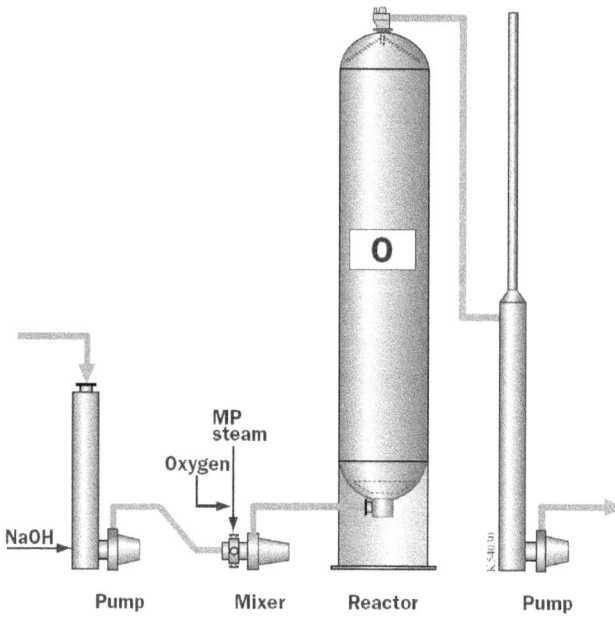

Fig. 4.6: One-stage oxygen delignification process.

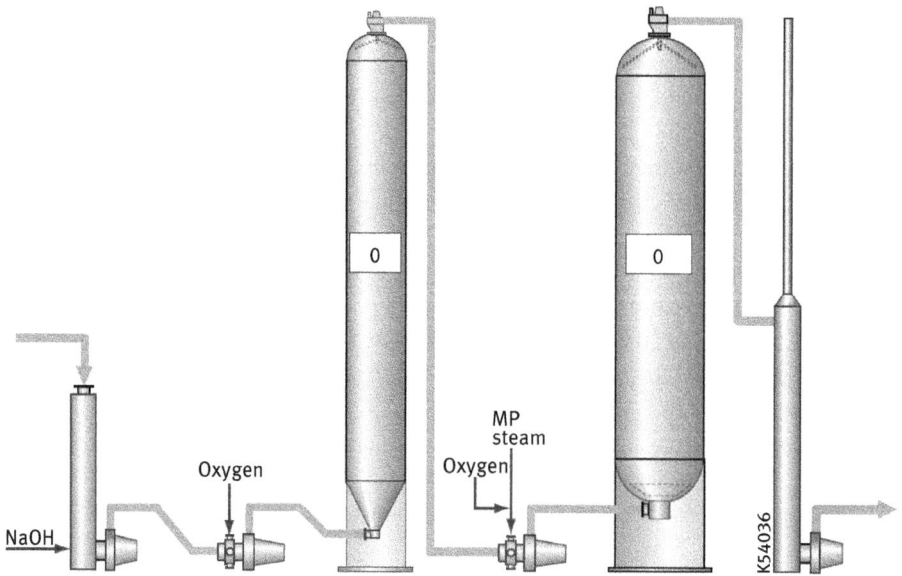

Fig. 4.7: Two-stage oxygen delignification process.

delignification during the first stage are a lower oxygen pressure, 0.3–0.35 MPa, a temperature of 80–85 °C, and a reaction time of only 5 min. The conclusion is based on the kinetics of oxygen delignification reported by Olm and Teder [57]. A high oxygen pressure of 0.6–0.8 MPa and a high temperature of 100 °C or higher during the second stage are recommended. However today, both stages are run at the highest possible pressure to boost Kappa reduction [66]. This system is applied in the industry under the name DUALOX, and its effectiveness is analogous to that of the OxyTrac process (Fig. 4.8) [64].

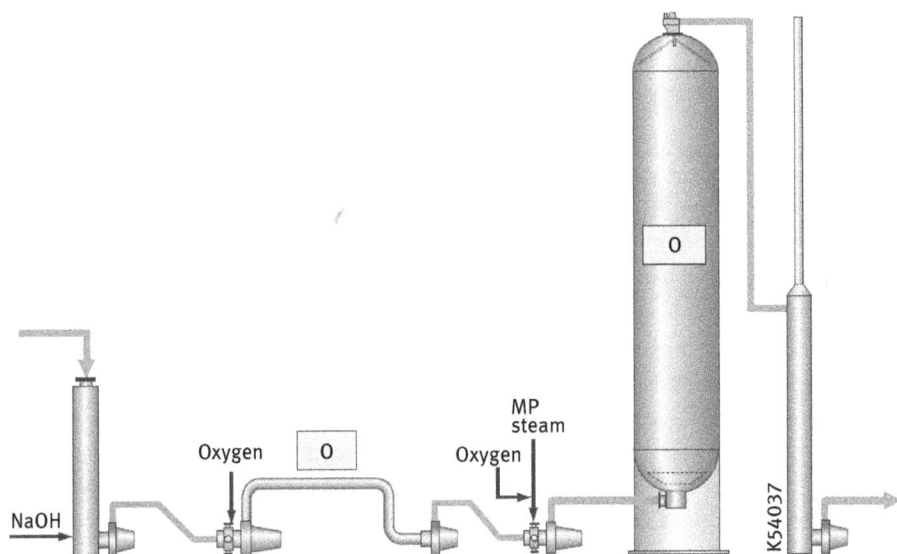

Fig. 4.8: DUALOX oxygen delignification process.

4.5 Chlorine dioxide bleaching

4.5.1 Reactions and factors in chlorine dioxide bleaching

Since turning from chlorine to ECF bleaching technologies, chlorine dioxide has become the main agent for kraft pulp bleaching. The first chlorine dioxide bleaching stage D_0 and the subsequent extraction, with the addition of oxygen and peroxide, are dominant delignification stages. Chlorine dioxide treatments, to increase the brightness, are also favorable for its stabilization and are normally final bleaching stages D_1 and D_2. Chlorine dioxide is an efficient delignification agent with high selectivity toward lignin, thus resulting in a high carbohydrate yield in the pulp [67]. Generally, chlorine dioxide reacts with phenolic lignin end groups either by oxidative ring opening, which affords muconic acid structures, or by demethylation, which affords quinone structures [68–69]. The result is a modified lignin with partly oxidized aromatic rings. Subsequently, the modified lignin is solubilized and removed in the extraction stage. The entire oxidizing power of chlorine dioxide is utilized in the acid bleach, the reduction proceeding to chlorite ions according to reaction eq. (4.13):

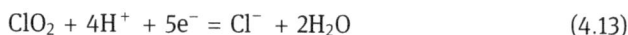

$$ClO_2 + 4H^+ + 5e^- = Cl^- + 2H_2O \qquad (4.13)$$

There are two different concepts of the performance of the postoxygen D_0 stage. The conventional operating conditions are a temperature of around 60 °C and reaction time of 30–60 min, which correspond to the complete consumption of ClO_2 [70, 71]. However, the hot chlorine dioxide (D_{HT}) concept involved in the process of bleaching hardwood kraft pulp, supported by Ragnar [70] and Lachenal and coworkers [72], is carried out at a temperature of about 90–95 °C and a prolonged reaction time despite the consumption of the ClO_2. Under these conditions, the hexenuronic acids are hydrolyzed. The reaction time of chlorine dioxide with pulp lignin is faster than with HexA. Hence, the majority of chlorine dioxide is consumed by the lignin at the beginning of the reaction, while the HexA are eliminated later through pulp acid hydrolysis at a high temperature and long exposure [73]. As a result, bleaching of the pulp is improved, the bleaching agents are saved, and at the same time the heavy metals are removed to a maximum extent without using chelating agents. Although the process may seem like a pure combination of D_0 and A, the results obtained during the D_{HT} stage show many synergies with regard to AOX discharge, chlorine dioxide consumption, and yellowing pulp properties. The final pH at the D_0 stage is very important in the bleaching processes during the next stages and for the final pulp properties. It has been proved by Dahl and coworkers [74] and Valchev and coworkers [75] that the highest intermediate brightness is achieved at pH 4, regardless of the other conditions of the D_0 stage or the type of pulp. However, the highest final brightness is obtained around or below pH 3, which can be explained by the best delignification rate, the better removal of the heavy metals from the pulp, and good strength properties of the bleached pulp [74, 75].

The effect of pH on the bleaching efficiency in the final D_1 stage, according to Valchev and coworkers [76], is shown in Fig. 4.9. The obtained results illustrate that the highest pulp brightness is achieved at a current charge of ClO_2 without preliminary acidifying, that is, final pH 6.0–7.0, but this required adjusting the final pH for individual samples. However, the higher pH values in the D_1 and D_2 stages cause a loss of pulp viscosity [74, 76]. According to Sevastyanova and coworkers [77], the hypochlorite formation seems to be responsible for the viscosity drop at a higher final pH. At the same time the HexA groups, which are usually degraded during the chlorine dioxide treatment, are present in significantly higher amounts in pulps bleached at higher pH values [77].

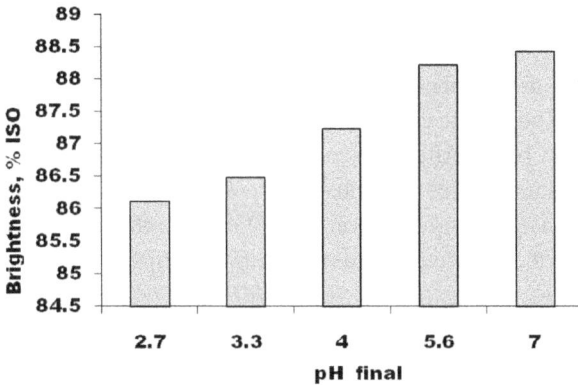

Fig. 4.9: Effect of final pH at the D_1 stage on pulp brightness.

The real effect of HexA removal can be reported after the D_1 stage, in which ClO_2 reacts very slowly with the residual lignin. The parallel reactions of ClO_2 with the products of HexA hydrolysis impair the bleaching process. Valchev and coworkers [75] presumed that the decrease in the chlorine dioxide charge and pH increase in the D_1 stage can reduce the negative effects of the HexA content in the pulp. In this way, the competitive harmful reactions are delayed and chlorine dioxide damage decreases. The lower ClO_2 charge in the D_1 stage reduces the efficiency of the D_{HT} stage to only 0.4% ISO brightness (Fig. 4.10). In the suggested optimum distribution of chlorine dioxide in the hardwood pulp bleaching sequence, the D_0 stage is favorably recommended to be carried out at a lower temperature, for a short time. According to Valchev and coworkers [75] in that case, the positive effect of the D_{HT} stage is negligible and cannot compensate for the increased power consumption and decrease in the pulp strength properties.

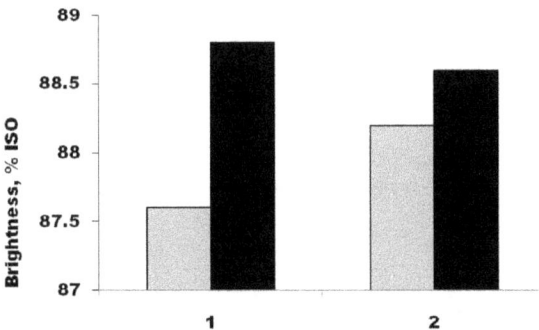

Fig. 4.10: The effect of chlorine dioxide distribution on brightness (– D_0; ■ – D_{HT}; (1) = 1.8% D_0+2% D_1 and (2) = 2.5% D_0+0.5% D_1).

4.5.2 Kinetic dependencies of chlorine dioxide bleaching

The obtained temperature–time dependencies of brightness during the D_0 stage [76] are represented in Fig. 4.11. The kinetic curves are characterized by a maximum brightness at the shorter reaction times. After maximum, the brightness decreases and is more obvious at higher temperatures. The observed reduction of brightness can be explained by structural changes and the formation of chromophoric groups in the residual lignin, as this process is better expressed at higher temperatures.

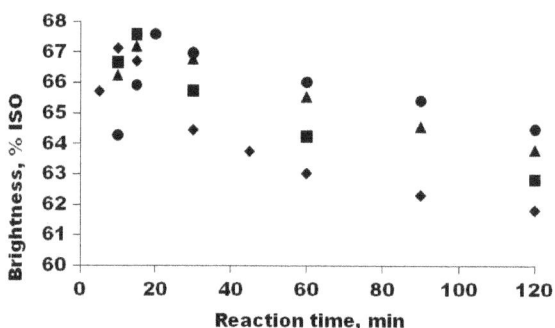

Fig. 4.11: Kinetic data of pulp brightness at the D_0 stage (♦ 90 °C; ■ 80 °C; ▲ 70 °C; and ● 60 °C).

It can be seen from Fig. 4.12 that the reaction time necessary for the total consumption of the bleaching agent is identical to the time necessary for achieving the maximal values of brightness. This means that the obtained Kappa number reduction after prolonged bleaching is probably due to the acid hydrolysis of the hexenuronic acids, but not to the lignin extraction [75, 76]. The kinetic curves of the Kappa number are represented in Fig. 4.13. It can be seen that the delignification is very fast in the first 5 min, when the Kappa number falls under 6.0; after that, the process slows

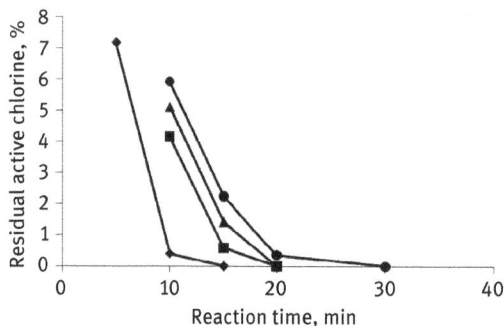

Fig. 4.12: Residual active chlorine in the D_0 stage (♦ 90 °C; ■ 80 °C; ▲ 70 °C; and ● 60 °C).

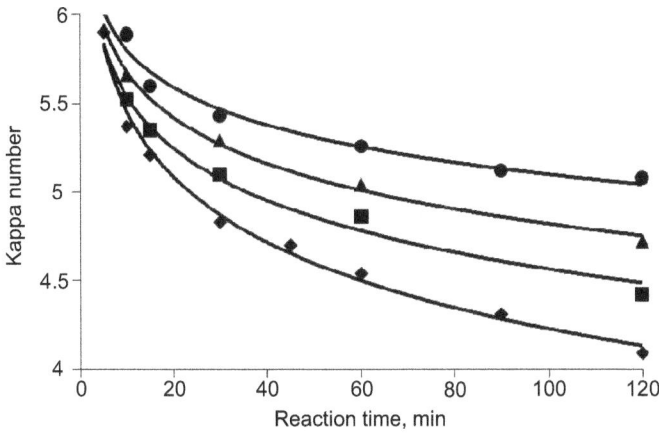

Fig. 4.13: Kinetic curves of Kappa in the D_0 stage (♦ 90 °C; ■ 80 °C; ▲ 70 °C; and ● 60 °C).

down. At a temperature of 90 °C and for a reaction time of 120 min, the Kappa number is 4. For the same reaction time at 60 °C, the Kappa number stays over 5.

The kinetics of HexA hydrolysis was studied by Valchev and Simeonova [15] in order to find the optimal conditions and mechanism of pulp bleaching, with chlorine dioxide, during the first stage – D_0. The relative change of HexA α_H is used as a kinetic variable. Various kinetic equations have been tested to describe the kinetics of the HexA hydrolysis and the best description has been achieved by the power kinetic equation, which is applied to surface chemical interactions taking place on the exponential uniform surfaces (eq. (4.14)):

$$v = \frac{d\alpha_H}{dt} = k_1 \cdot \alpha_H^{-\frac{1-\chi}{\chi}} \tag{4.14}$$

where k_1 is the apparent rate constant and χ is a coefficient of inhomogeneity.

The process of a HexA hydrolysis, run without changing the activation energy, is shown by obtained kinetic results. The process slowing down is due to decreasing of the pre-exponential factor, probably because of the exhaustion of reaction groups and difficult accessibility to them. Application of the power kinetic law, at the HexA hydrolysis in pulp bleaching with chlorine dioxide in the D_0 stage, shows that this process is a surface chemical interaction unlike the topochemical processes. One of the reasons can be a low molecular weight of the xylan basic chain and HexA distribution in the side chains. The presented kinetic model, according to Valchev and Simeonova [15], is common during the whole time interval, including the fast initial phase of the processes. It has been found by Valchev and Simeonova [15] that the Praut–Tompkins topochemical kinetic model (eq. (4.10)) successfully describes the bleaching process with respect to the light absorption coefficient during the D_1 stage. Application of a modified form of the Praut–Tompkins kinetic equation, on the basis of the kinetic variable $\alpha_{\kappa/s}$ (eq. (4.6)), can be explained by the distribution and the different accessibility of the chromophore groups during the process. A simplified equation for the determination of the running absorption coefficient has been achieved, which may be used for control of the bleaching process (eq. (4.15)):

$$\kappa = \kappa_0 \left\{ 1 - \left[1 + \left[\left(20 \times 10^6 \cdot e^{-6531/T} \right) \cdot t \right]^{-0.15} \right]^{-1} \right\}$$ (4.15)

where k_0 and k are the initial and current absorption coefficient, $\chi = 0.15$, pre-exponential factor $A = 20.10^6$ min^{-1}, and $E/R = 6,531$ K.

4.5.3 Technology of chlorine dioxide bleaching

During the D_0 stage, the pulp suspension is pumped to a reactor. Optionally, sulfuric acid is normally added to the pulp suspension prior to the pump. Chlorine dioxide is added and thoroughly mixed into the pulp suspension in a mixer positioned immediately after the pump. The mixture enters the reactor and passes upward in a plug flow, while the chlorine dioxide is consumed in reactions with the lignin in the pulp. Chlorine dioxide bleaching is sometimes carried out in upflow–downflow reactor combinations, where a smaller upflow section is used. This reactor has some operational advantages as it can absorb variations in production rates. The pulp leaves the reactor and enters a second pump, which pumps the pulp suspension further to the washing and bleaching stages. The typical retention time for the D_0 stage is 20–60 min at a temperature of 50–70 °C, while for the D_1 and D_2 stages, the retention time is 1–3 h at a temperature up to 90 °C. The normal retention time for the hot chlorine dioxide process, D_{HT}, is 2 h. The temperature is increased to about 90 °C by the addition of steam in a steam mixer, placed before the chlorine dioxide mixer. The preferred material of construction for the equipment in the chlorine dioxide stage is titanium.

4.5.4 Reactions in the extraction stages

The basic idea of the extraction stage E is to extract the oxidized lignin that has been formed in the chlorine dioxide bleaching stage, since the solubility of lignin is low under acidic conditions. The performance of the extraction stage can be somewhat boosted by the addition of a small amount of oxygen in the EO stage, a small amount of hydrogen peroxide – the EP stage or both oxygen and hydrogen peroxide in the EOP stage. The major reactions encountered in an E stage are the neutralization of carboxyl groups, thereby strongly increasing the water solubility of the oxidized lignin, and the elimination of organically bound chlorine and the formation of a chloride ion. Oxygen and H_2O_2 in the E stages will always either improve the brightness of pulp or make it possible to reach the same brightness with less chemicals. Kinetics of the extraction stages, which have been studied by Valchev and Simeonova [15] on the basis of the degree of delignification α_κ, are better described by the exponential kinetic equation (eq. (4.16)). This equation is applied to processes that take place on uniformly heterogeneous surfaces:

$$v = \frac{d\alpha_\kappa}{dt} = v_0 \cdot e^{-a \cdot \alpha_\kappa} \tag{4.16}$$

where α_κ is the degree of delignification, v_0 is an initial rate, and a is a coefficient connected with the inhomogeneity of the surface.

The obtained simplified kinetic expression (eq. (4.17)) could be used for the fast and precise delignification control:

$$\alpha_\kappa = \frac{1}{a} \cdot ln(a_\kappa \cdot A_0) + \frac{1}{a} \cdot \left(ln\, t - \frac{E_0}{RT} \right) \tag{4.17}$$

The extraction stages are either of the upflow or upflow–downflow type. Typical data for an upflow–downflow (EOP) stage are temperature 80–90 °C, retention time up to 30 min, pressure 0.3–0.5 Mpa in the top of the upflow tower, and a 60–120 min retention time at atmospheric pressure in the downflow tower; the charge of sodium hydroxide is normally 10–25 kg/ton of pulp.

4.6 Peroxide bleaching

4.6.1 Chelating stages

Due to environmental pressure, there is an increasing use of hydrogen peroxide as a total or partial substitute for chlorine-based bleaching agents with ECF and TCF sequences. However, to achieve satisfactory brightness all types of transition metal ions have to be removed from the pulp. Transition metals decompose hydrogen peroxide to oxygen and water via the intermediate formation of hydroxyl and superoxide radicals. The amount of transition metals in the pulp depends on the wood species and the soil where the wood was grown. Normally, manganese and iron ions dominate, while other metals like copper and cobalt are present only in trace quantities.

Strong acidic stages or treatment with chelants are the standard procedures for successful peroxide bleaching. During standard ECF bleaching, chelants are not required. Transition metals are sufficiently removed by the acidic conditions of the D_0 stage. The pulp cleaning is effective at a final pH <3.0 and the process improves upon an increase in temperature [75]. The highest final brightness can be achieved by the implementation of a preliminary acid treatment of the stage A pulp, with subsequent washing [75]. The strongly acidic conditions during stage A have the disadvantage of removing not only metals like manganese but also magnesium, which provides viscosity protection. Magnesium is therefore normally charged into the pulp to improve pulp quality [74]. During TCF bleaching, the removal of transition metal ions is more important because hydrogen peroxide is applied at much higher charges. The concentration of manganese in the pulp, proceeding into a pressurized PO stage, should not exceed 2 ppm according to Andtbacka [78]. During TCF bleaching, metal removal at the mill scale is typically carried out at the Q stage with chelants at a moderate pH, a temperature between 70 °C and 90 °C, and a retention time of about 1 h.

4.6.2 Reactions and factors in hydrogen peroxide bleaching

Peroxide bleaching is activated with alkali. At higher pH, the equilibrium of eq. (4.18) is pushed to right side to the formation of the reactive perhydroxyl anion:

$$H_2O_2 + OH^- \leftrightarrow OOH^- + H_2O \qquad (4.18)$$

The perhydroxyl anion formed according to eq. (4.18) is primarily responsible for the bleaching effect in pulp and leads to an increase in brightness [79]. Therefore, the bleaching process will certainly be accelerated with higher charges of alkali. On the other hand at higher alkalinity, more peroxide is consumed in side reactions. In addition, high alkali charges have an increasing effect on extraction, and lead to a higher pulp viscosity and simultaneous negative effect on the effluent load and yield. A higher temperature accelerates the bleaching process and the decomposition of hydrogen peroxide into hydroxyl radicals, which in turn leads to the generation of the superoxide anion radicals. The radical species contribute not only to the oxidation of lignin, but also to a certain extent to the oxidation of the polysaccharides and leads to a drop of the pulp viscosity. Therefore, in ECF bleaching, the mill practice is to keep the temperature level below 90 °C during the peroxide stages. The peroxide bleaching effect most likely occurs by one of two mechanisms: a lignin retaining part, where the perhydroxyl anion reacts with the chromophores resulting in colorless carboxylic acid groups, and a lignin degrading part where radical species, originating from the hydrogen peroxide, oxidize, and depolymerize the lignin molecule. Unlike oxygen, hydrogen peroxide in an alkaline medium does not attack phenolic structures. The main reaction pathway of the perhydroxyl anion is nucleophilic addition to enone and other carbonyl structures [80–82]. In this way, the chromophore groups are removed. The two radical species, hydroxyl and superoxide, increase the extent of lignin degradation. The hydroxyl radicals introduce radical sites into the substrate. These substrate radicals may be oxidized, may disproportionate, or may combine with superoxide radicals. The superoxide radicals, after coupling with substrate radicals, bring about fission of the C–C bonds and thereby open aromatic rings and cleave ring-conjugated double bonds. In addition, they are intermediates in the formation of hydrogen peroxide and hydroxyl radicals.

4.6.3 Process variables and technology of peroxide bleaching

In ECF bleaching, hydrogen peroxide is widely used in the EOP extraction stages following chlorine dioxide treatment and is one of the most effective possibilities for the increase of the final pulp brightness [81]. The influence of hydrogen peroxide charge during the EOP stage on the intermediate and final pulp brightness, according to Valchev and coworkers [75], is represented in Fig. 4.14. A significant improvement of the bleaching process selectivity in the EOP stage was obtained by adding up to 0.5% hydrogen peroxide. The higher peroxide charge leads to a relatively lower bleaching effect and can be used to reach a desired brightness, if necessary. In the suggested optimal bleaching conditions, 1 kg peroxide replaces approximately 2.5–3 kg ClO_2 [75].

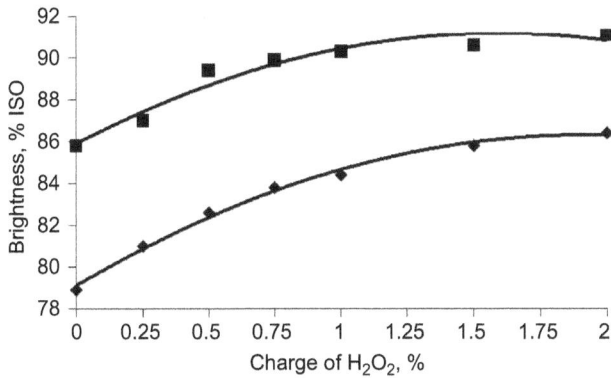

Fig. 4.14: Effect of H_2O_2 charge on the pulp brightness (♦ after EOP and ■ after D_1).

The temperature in an EOP stage is between 75 °C and 90 °C and the retention time 60–120 min, while the pH is typically about 11 at the start of treatment and about 10 at the end. A higher pH may result in alkali-induced decomposition of hydrogen peroxide. The charge of sodium hydroxide is normally 10–25 kg per ton of pulp. The equipment used for the EOP extraction stages is very similar to EO extraction equipment. Pressurized hydrogen peroxide bleaching during the PO stage is a very efficient tool for increasing brightness, although the Kappa reduction capability of hydrogen peroxide is decreased, partly because hydrogen peroxide, in contrast to chlorine dioxide and ozone, is not reactive toward HexA. An extensive use of hydrogen peroxide often leads to a somewhat increased yellowing of the fully bleached pulp, especially in the case of hardwood kraft pulp. Extensive use of hydrogen peroxide, as in a PO stage, normally requires a chelating agent treatment (Q) prior to the peroxide stage. Pressurized peroxide bleaching expands the reaction rate by the increase of the temperature, peroxide concentration, and improving peroxide diffusion [80, 81]. Using the pressure peroxide methods substantially reduces the retention time and increases the brightness ceiling compared with conventional process [81, 82].

Valchev and coworkers investigated the influence of oxygen utilization during the bleaching processes [13]. It has been established that the initial stage is very important for the bleaching process. The Kappa number of the pulp decreases very quickly in the first 10 min, while in the next stage the delignification is slower. The Kappa number value decreases upon increasing the temperature, but there is not a positive effect on Kappa number reduction at temperatures higher than 100 °C in the presence or absence of oxygen. This is probably caused by the forced decomposition of peroxide under these conditions.

Oxygen affects the delignification more significantly, especially in the initial stage of the process, but the delignification continues slowly and after full peroxide consumption. Higher brightness (1.5–2%), extended delignification, and higher peroxide consumption (5–7%) are reached using pressure peroxide bleaching compared with atmospheric bleaching. Correlations between Kappa number peroxide consumption and brightness show that higher peroxide consumption, in the presence of oxygen, corresponds to an increasing brightness and all experimental data are described with one and the same correlation. However, this increase of peroxide consumption leads to a further extended Kappa number reduction. Consequently, the effect of oxygen during peroxide bleaching is mainly expressed as an improvement of the delignification process [13]. Typical parameter values for a PO stage are normally temperature of 90–105 °C, retention time of about 2 h, pressure at top 0. 3–0.5 MPa, and a final pH of 10.0–11.0. The charge of hydrogen peroxide is usually 10–30 kg/ton and depends on the Kappa number of the incoming pulp. The charge of sodium hydroxide is directly proportional to the charge of hydrogen peroxide and is normally 20–30 kg/ton of pulp. The equipment for pressurized peroxide bleaching is very similar to that used in oxygen delignification equipment.

The pulp bleaching history has a great influence on the performance of a final hydrogen peroxide bleaching stage, regarding the bleachability and selectivity parameters, and bleached pulp properties. The final peroxide stage completes bleaching and extracts all remaining potential chromophores. This stage requires temperature and alkalinity to be efficient. The reaction with chromophores or precursors of chromophores is rather fast. A benefit of using peroxide in the final stage of a bleaching sequence is the improved brightness stability. The advantage of the application of hydrogen peroxide can be attributed to the destruction of carbonyl groups and quinoid structures remaining in the pulp after the D stage.

The identified remaining chromophores in fully bleached and aged pulp have a common structural element, hydroxy- and dihydroxyquinones. These compounds are resonance stabilized and have a peculiar reactivity because they can form a dianion under alkaline conditions. They are the main chromophores surviving the ECF bleaching [37, 83]. Traces of quinone compounds are detected in pulp bleached with a final D stage, but no such products were found in pulp bleached with a final peroxide stage [84]. The destruction of quinoid structures is one of the main reactions of alkaline hydrogen peroxide in bleaching processes. For the development of brightness, hydrogen peroxide requires alkaline bleaching conditions. The perhydroxyl anion is a strong nucleophile and cleaves side chains in the residual lignin, thus destructing chromophores. Lignin degradation takes place in addition via radicals resulting from the decomposition of hydrogen peroxide. The disadvantage of these radicals is their unselectivity, the side reactions are cellulose oxidation, and the subsequent cleavage of the polymer chains [85].

According to Valchev and coworkers [86], the comparison of the results of pulp postbleaching with xylanase, peracetic acid, and hydrogen peroxide shows that the greatest effect on pulp brightness is achieved by the peracetic acid treatment (Fig. 4.15).

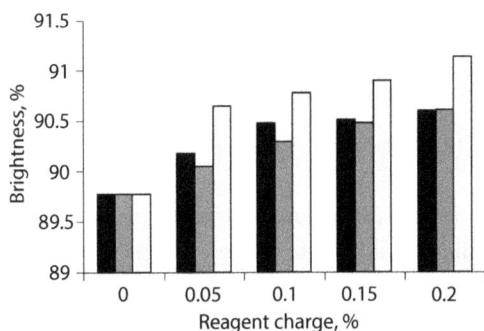

Fig. 4.15: Postbleaching effect on the pulp brightness (■ – X; ▪ –P; – Paa).

The peroxide treatment leads to the lowest Pc number, but the selectivity of the pulp bleaching with this reagent is also at its lowest (Fig. 4.16). On the other hand, the decrease in viscosity, observed during peroxide bleaching, does not affect the pulp strength. Typical data for postperoxide bleaching is temperature about 80 °C, retention time up to 120 min under atmospheric conditions, peroxide charge 1–3 kg/ton of pulp, and a final pH of 8.0–10.0. The application of pulp postbleaching does not require additional capital investment, and the effect achieved justifies the expense for bleaching reagents.

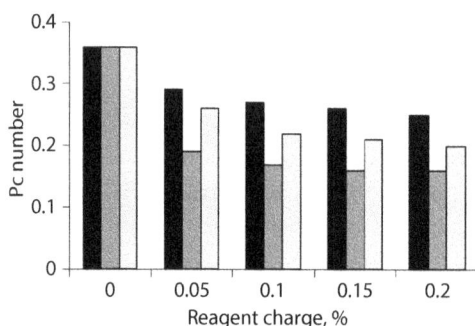

Fig. 4.16: Postbleaching effect on the Pc number (■ – X; ▩ – P;   – Paa).

4.6.4 Kinetics of peroxide bleaching

The kinetics of the EOP stage, pressure delignification OP, and atmospheric peroxide stage P were studied by Valchev and coworkers [13, 15] on the basis of the variation of the Kappa number with reaction time at different temperature values. It has
been found that the best description was achieved by the exponential kinetic equation (eq. (4.16)), which takes place on uniformly heterogeneous surfaces. The obtained values of activation energy E are independent of the degree of increasing
delignification. The pre-exponential factor A decreases in the process according to
eq. (4.19):

$$\ln A = \ln A_0 - a \cdot \alpha \tag{4.19}$$

where a is a coefficient of inhomogeneity of the surface and A_0 is an initial pre-
exponential factor.

Consequently, a higher rate of delignification during pressure peroxide bleaching,
compared with atmospheric bleaching, could be due to an increase in the number of
interactions between lignin and the reagent species. The pressure peroxide delignification is faster than peroxide bleaching at atmospheric pressure and the effect of oxygen on the reaction rate corresponds to a nearly 10 °C difference of temperature. The
kinetics of peroxide and pressure peroxide bleaching, with respect to absorption coefficients, has also been investigated by Valchev and coworkers [13, 15], and it has been
established that the modified equation of Prout–Tompkins (eq. (4.10)) gives the best
description of the process. The relative change of absorption coefficient $\alpha_{k/s}$ was calculated according to eq. (4.6). It is well known that eq. (4.10) is successfully applied to
topochemical reactions of the chain mechanism; a similar equation can be used in the
case of diffusion controlled heterogeneous processes. Investigation of the processes of
peroxide and pressure peroxide bleaching shows that the kinetics of delignification is most successfully described by an exponential kinetic equation (eq. (4.16)).
The activation energy E is independent of the presence of oxygen and does not

change during the process. This means that one and the same type of interactions take place. The kinetics of the bleaching process, with respect to the absorption coefficient, is described by the modified topochemical equation of Prout–Tompkins. The values for activation energy E and pre-exponential factor A established in this case confirm the assumption that the effect of oxygen is determined by the value of the pre-exponential factor and does not depend on the activation energy. The topochemical kinetic mechanism, which describes the change of the absorption coefficient, is determined by the course of two different processes: removal of the lignin from the pulp and transformation of the chromophore groups in the solid phase. These features of pressure peroxide bleaching are confirmed by the obtained correlations. The effect of oxygen action on peroxide bleaching is expressed with some increase in the rate of bleaching processes without changing their kinetic mechanism [13, 15].

4.6.5 Peracetic acid (Paa) in pulp bleaching

Paa is a highly selective bleaching chemical, whose use ensures good strength properties of the pulp can be maintained. Today, Paa occupies a niche in TCF bleaching and is also recommended as a final treatment step to boost the brightness of the TCF pulp [87, 88]. Its application is necessary due to the need to modify the residual lignin to allow its destruction during hydrogen peroxide bleaching and to improve the economics of TCF sequences. During pulp bleaching, Paa is consumed by lignin, hexenuronic acids, oxidation reactions, and spontaneous decomposition, catalyzed by transition metals [89, 90]. In neutral or slightly acid pH, the pulp carbohydrate degradation is mainly dependent on the concentration of peracetic acid [91]. At low pH, acid hydrolysis is probably the main reason for carbohydrate depolymerization; at neutral pH, decomposition of Paa may produce highly reactive oxygen species, such as hydroxyl radicals, formed by the homolytic cleavage of O–O, which causes the degradation. Peracetic acid is less sensitive to transition metals than hydrogen peroxide. The study of Paa, beyond its use in TCF bleaching, has shown that Paa removes HexA with an effectiveness equal to chlorine dioxide and ozone [92]. As a final stage, Paa exerts a bleaching effect, but does not necessarily improve the pulp brightness stability. Most likely, the reactions it is involved in hydroxylation, as well as oxidation, generate quinones, which cause chromophore formation [83]. Peracetic acid application in the last stage of eucalyptus pulp bleaching caused a slight decrease in pulp viscosity, Kappa number and HexA content, but had no significant effect on pulp reversion [93].

Paa can be used as an equilibrium solution with a mixture of peracetic acid, acetic acid, and hydrogen peroxide, or as pure distilled Paa. In order not to waste the hydrogen peroxide content, the equilibrium Paa treatment has to be followed by the peroxide stage without intermediate washing. Following the addition of caustic soda, the unused hydrogen peroxide content in the pulp is reacted in the subsequent P step. This results in an improved delignification and a higher brightness, which is achieved in the following alkaline peroxide stage. Following the investigations of Khristova and coworkers and Simeonova and coworkers [94, 95], the alkali charge in peracetic acid bleaching has to be selected at an initial pH above 10 in order to finish the process at pH 5.0–6.0. The final pH, around 5, being used for peracetic bleaching is also in accordance with Zawadski's investigations [96]. At that pH, the chromophore groups are easily destroyed and the bleached pulp has a higher brightness. With the increase of pH above 10.8, a sharp decrease of the pulp brightness is observed (Fig. 4.17) [86], which is due to neutralization of the Paa. On the other hand, the performance of bleaching in a neutral and low acid medium does not allow the peroxide to react, which also negatively affects the pulp brightness. The bleaching reagents are entirely consumed in the high-alkali conditions, whereas in an acid medium the peroxide content in the product barely reacts. The comparison of pulp post bleaching with xylanase, peracetic acid, and hydrogen peroxide shows that the greatest effect on pulp brightness is achieved by the Paa treatment (Fig. 4.16). The Paa does not affect cellulose under the applied conditions of the last bleaching stage [86]. An additional advantage of Paa in comparison with other bleaching reagents is its antibacterial action, as well as the reduced consumption of chemicals in the subsequent paper production as a result of the cleaning effect on the pulp fibers.

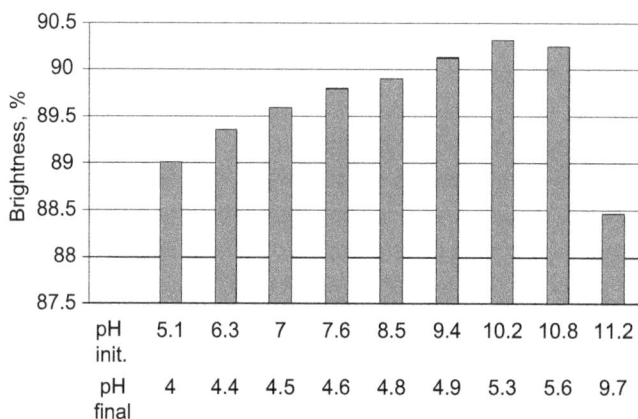

Fig. 4.17: Effect of pH on the pulp brightness after the post peracetic acid stage (initial brightness 89.2% ISO, peracetic acid charge 0.15%).

It has been found by Valchev and coworkers [97] that the modified topochemical Prout–Tompkins equation most accurately describes the kinetics of the Paa bleaching, with respect to the relative change of the light absorption coefficient. The kinetic investigation shows that the peracetic bleaching, as a final step, is an energetic uniform process. The reaction rate depends on the number of chromophore structures and the presence of steric and diffusion difficulties, which are included in the preexponential factor. For economic reasons, only distilled peracetic acid is of interest, its bleaching effect being equal or slightly better than that for the equilibrium type. Distilled Paa does not need the pH to be changed from the normally slightly acidic regime to an alkaline one, compared with bleaching using hydrogen peroxide [83]. Tetraacetylethylenediamine (TAED) has also been investigated as a bleaching activator [94]. TAED can react with peroxide over a range of pH conditions; in alkaline conditions, one mole of TAED reacts with two moles of hydrogen peroxide to form two moles of the active bleaching species of the peracetate anion.

4.7 Ozone bleaching chemistry and technology

Ozone is an extremely powerful oxidant with an oxidation potential of 2.07 V. Ozone has been investigated, on a laboratory scale, for use in the bleaching of chemical pulps for a long time, but for both economic and pulp quality reasons, the first implementation of ozone on an industrial scale was not until 1990, when the first installation of an ozone bleaching plant came on stream in Lenzing. Ozone reacts very rapidly with the pulp and the reaction takes place, to a lesser or greater extent, in the mixers. The importance of highly efficient mixing cannot therefore be overestimated. Since ozone decomposes spontaneously into oxygen at high temperatures, ozone stages are run at a low temperature. The stability of ozone in water increases with decreasing pH. This is one of the important reasons why ozone bleaching is performed under acidic conditions, preferably in the pH range of 2–3. Ozone decomposes via the catalytic action of transition metal ions, present in the pulps and bleaching liquors, giving rise to hydroxyl, hydroperoxyl, and superoxide anion radical intermediates [98, 99]. The reactions of ozone with lignin involve an initial electrophilic attack by the oxidant, followed by the loss of oxygen, which results in hydroxylation of the aromatic ring. In a subsequent step, ozone reacts with the aromatic ring, and finally, this is cleaved [100]. Carbohydrate decomposition during ozone bleaching depends on pH, temperature, and the concentration of transition metal ions [101]. The ozone stage leads to the introduction of carbonyl groups and gylcosidic bond cleavage [81]. Thus, alkaline extraction after ozone treatment decreases the molecular weight of the carbohydrates by splitting the alkali-sensitive linkages. As expected, the alkaline treatment of ozone-bleached pulps, in general, reduces fiber strength. An advantage of the ozone treatment is that the extractives are removed very efficiently. Another advantage is its ability to efficiently remove HexA in the pulp, which improves the brightness stability of the final bleached pulp and reduces the ClO_2 demand in the bleach plant. Ozone bleaching is normally only used for the bleaching of hardwood kraft pulps and for sulfite pulps. Unfortunately, ozone delignification is accompanied by a concomitant degradation of the polysaccharide fraction. In general, the tearing strength of ozone-bleached softwood kraft pulps is found to be 10–20% lower compared with conventionally bleached pulps of the same provenience [102]. It has been reported that the efficiency of a TCF bleaching sequence is significantly improved when an ozone stage is arranged between two hydrogen peroxide stages [103]. The investigations of Davies and coworkers [83] have shown that the highest brightness and extremely low reversion are obtained after the combination of a small amount of ozone with a subsequent peroxide treatment.

Typical parameter values for the medium consistency ozone Z stage are a retention time of a few seconds, pH 2–3, temperature about 50 °C, pressure 0.8 MPa, and an ozone charge of 3–5 kg/ton. In a medium consistency Z stage, the pulp suspension is pumped into two consecutive mixers. Ozone is added and thoroughly mixed into the pulp suspension in a mixer positioned immediately after the pump (Fig. 4.18) [64].

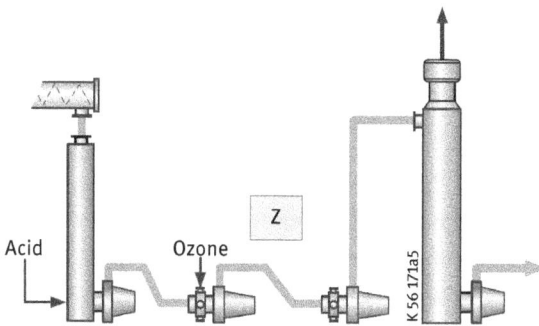

Fig. 4.18: Medium consistency ozone process.

Unlike the MC process, high consistency ozone bleaching is carried out under atmospheric conditions, where the pulp meets the gas in a screw reactor. A slightly longer retention time is utilized than in an MC system, but the time is still only about 1 min.

4.8 Xylanases and laccases in pulp bleaching

The use of xylanases for kraft pulp bleaching has been one of the greatest success stories of enzymes in the pulp and paper industry. It is known that a number of pulp mills continue to use xylanase enzymes to increase the bleaching efficiency. The xylanase treatment of pulp promotes increased efficiency of the subsequent delignification and bleaching processes. Thereby, a 25% saving of bleaching reagents has been achieved, without investment in expensive equipment. The investigations into a pulp enzyme treatment during the bleaching process began in the mid-1990s [104–106]. The effect of hemicellulases on the pulp brightness is attributed to the removal of xylan from the fiber surface, as well as to the degrading of xylan–lignin complexes in the inner layers of the pulp matrix. The elimination of hemicelluloses makes the pulp fiber structure more accessible to a subsequent bleaching with chemical reagents, such as hydrogen peroxide, chlorine dioxide, and ozone, and focuses their action to only the lignin chromophore groups [106, 107]. This improved lignin removal and the removal of the bleach-consuming HexA might explain the mechanism of xylanase prebleaching [108]. The pulp treatment by xylanase also leads to a reduction in the content of heavy metals, such as Cu, Fe, and Mn, most likely due to existing bonds between HexA and xylan [109]. According to Valchev and coworkers [14], the xylanase treatment increases the brightness of the peroxide-bleached pulp by about 4% ISO and allows approximately a 1 unit Kappa number decrease upon a 10% lower hydrogen peroxide consumption (Fig. 4.19). The overall xylanase treatment effect is two times higher than the oxygen effect on the peroxide bleaching.

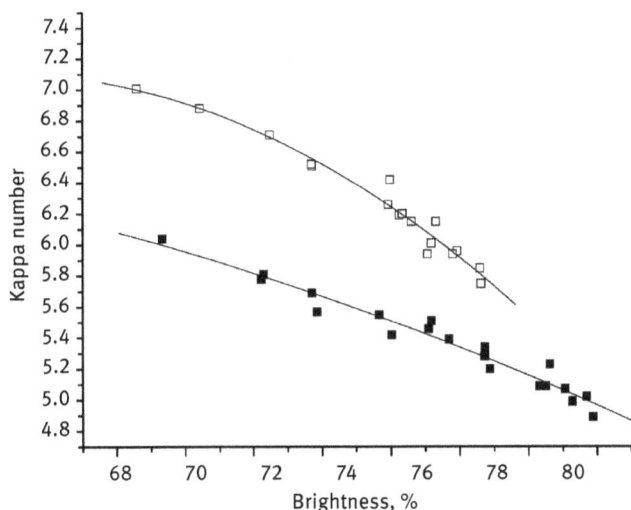

Fig. 4.19: Correlation between the Kappa number and brightness. (■ – enzyme treated pulp and □ – untreated pulp).

There is also variability in the stage of the process where the xylanase treatment is performed. It could be before or after oxygen delignification, or after the bleaching sequence. During ECF bleaching sequences, the best location for the xylanase treatment is in the storage towers of the unbleached pulp. After a continuous xylanase treatment of hardwood pulp under industrial conditions, a 21% reduction of chlorine dioxide is registered [76]. Typical data for a xylanase prebleaching treatment is temperature about 55–65 °C, retention time up to 120 min, enzyme charge 0.3–1 XU per gram of pulp, and a pH of 8.0–9.0. Because the kraft process results in pulp that is alkaline and at higher temperatures, enzymes that do not require adjustment of temperature or pH are better suited to the process [110].

It has been shown that a xylanase posttreatment of hardwood kraft pulps, during a final stage, resulted in a significantly reduced yellowing [111]. The enzyme treatment primarily resulted in a degradation of HexA in the pulp, reducing the amount of this structure by up to one-third.

Valchev and Tsekova [112] do not indicate a relation between HexA content and pulp brightness during xylanase post treatment. According to them, the effect on the bleached pulp may be explained by the hydrolysis of the xylan, located on the surface, and extraction of the stabilized quinone chromophoric structures retained by it. The obtained correlation between the pulp brightness and the enzyme action shows that an optimum bleaching effect is achieved at a low degree of xylan conversion and at a low enzyme charge, which does not significantly affect the pulp yield and the strength properties (Fig. 4.20) [86, 112].

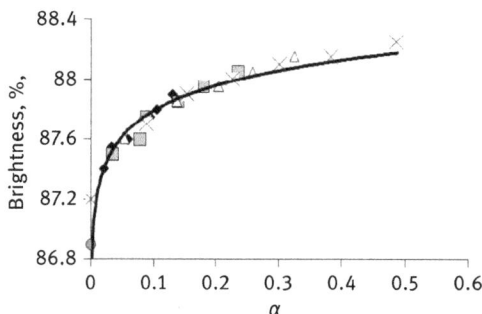

Fig. 4.20: Correlation between the brightness and degree of xylan conversion (pulpzyme HC charge 0.05%).

Kinetic studies of xylan decomposition with different xylanases show quite good results when the modified topochemical Prout–Tompkins equation(4.10) is used [113–115]. The nonhomogeneous distribution of xylan, as well as the difference in its structure throughout the pulp matrix, makes the classical Michaelis–Menten rate equation inadequate to describe the kinetics of enzyme-catalyzed decomposition. The application of a modified Prout–Tompkins equation to the action of xylanase on pulp indicates that the process is connected with a continuous change of reaction area, which is determined by the xylan chain structure and the lignin–carbohydrate bonds. The activation energy does not change during the course of the process, which is an indicator of an interaction with a specific type of xylan. The amount of remaining unreacted xylan in the pulp has a decisive effect on the rate of its removal [112, 116].

While the efforts into the use of xylanase products are directed toward their in-dustrial-scale implementation, the research is aimed at the development of new en-zyme technologies with a direct delignification effect.

Laccases (E.C.1.10.3.2.), a group of multicopper oxidases, are the most promising oxidoreductive lignin-degrading enzymes [117]. Laccases oxidize the lignin structure into radicals, which are spontaneously degraded. However, due to the steric hindran-ces, laccases alone are not very efficient for delignification. Furthermore, their oxida-tion potential provides the oxidation of lignin's phenolic end groups only. It is recognized that their substrate specificity can be extended to nonphenolic end groups, as well as by adding oxidized low molecular mass compounds (mediators) [118]. 1-Hydroxybenzotriazole (HBT) can act as a mediator. Laccase oxidizes this me-diator, forming an HBT radical during the initial stage of the process [119]. The media-tor is small enough to penetrate the lignin network and perform oxidations of the nonphenolic end groups. The application of a laccase-HBT mediator system provides an efficient, selective, and completely chlorine-free bleaching stage [119]. It has been discovered that pulp treatment with a laccase–mediator system (LMS) leads to high lignin extraction (40–60%) in the subsequent bleaching stages. In addition, the car-bohydrate hydroxyl groups, present in the pulp, remain unchanged and neither a de-crease in viscosity nor changes in the content of hexeneuronic acids are observed. Moreover, the enzyme treatment does not directly affect the degree of pulp bright-ness. Generally, oxygen is required during normal delignification processes. The small size molecules of the oxidized mediator provide a relatively uniform lignin de-polymerization [120]. The investigations of Valcheva and coworkers [121] show the highly selective delignification effect of LMS. A Kappa number decrease of 32% is pro-duced after chlorine dioxide bleaching, while it reaches only 17% after oxygen bleaching. It is evident that the enzyme effect on delignification is more pronounced in the case of chlorine dioxide bleaching [122]. It has been found that the exponential kinetic equation (eq. (4.16)), valid for processes taking place on uniformly nonho-mogeneous surfaces, provides a good interpretation of the LMS action, when using the kinetic variables of the degree of delignification α_k and the relative change in the light absorption coefficient $\alpha_{k/s}$ [121]. The combination of LMS and ClO_2 leads to the domination of the delignification reaction, while the combina-tion of LMS and O_2 leads to a decrease in the content of chromophoric groups [16]. Some problems related to enzyme stability and the high cost of LMS still exist, but the results obtained provide reasonable optimism for the future application of this type of enzyme system in the pulp bleaching process.

4.9 Conclusions

After the rapid development of pulp bleaching technology in the 1990s, there followed a period in which the individual bleaching processes were mainly optimized, without this being bound to considerable changes in the multistage bleaching sequences. No essential amendments to the environmental legislation are expected in the coming years, meaning that the main driving force in the development of pulp bleaching technologies will be an economic one.

Since the wood raw material is the main expenditure for a pulp mill, it is more logical to focus on achieving a high pulp yield. An extended and selective oxygen delignification stage would give both a high overall yield and less environmental impact due to the lower requirement for bleaching chemicals. As a result of in depth kinetic studies, the established conditions of the oxygen delignification processes should probably be changed so that in the oxygen system design, the alkali concentration and charge are decoupled, similar to modern cooking systems, to create a more uniform delignification rate throughout the reactor.

Chlorine dioxide will continue to be the main bleaching reagent for kraft pulp bleaching. Through optimization of the bleaching conditions and temperature–time dependences of the processes in the individual chlorine dioxide bleaching stages, consumption of the bleaching reagent may be considerably reduced and a high final brightness and high carbohydrate yield may be achieved in the pulp.

New efficient catalysts of peroxide bleaching, which will improve the selectivity of the bleaching process, will be sought. The development of production methods for cheap peracetic acid will considerably extend its use in the multistage bleaching sequences.

The development of more thermostable, alkali-tolerant enzyme systems, which do not give significant yield losses, might be interesting in pulp treatment in order to break up LCC.

High final brightness without additional capital investment may be achieved through pulp post bleaching. In the integrated pulp and paper mills, the place of that bleaching will probably be relocated prior to the paper machine.

List of abbreviations

A	Acid treatment
AOX	Adsorbable organic halogen
BOD	Biochemical oxygen demand
C	Chlorine
COD	Chemical oxygen demand
D	Chlorine dioxide
D_{HT}	Hot chlorine dioxide
E	Alkaline extraction

ECF   Elemental chlorine free
EO   Alkaline extraction and oxygen
EOP   Alkaline extraction, oxygen, and peroxide
EP   Alkaline extraction and peroxide
H   Hypochlorite
HBT   1-Hydroxybenzotriazole
HC   High consistency
HexA   Hexenuronic acid
LCC   Lignin–carbohydrates complex
LMS   Laccase–mediator system
MC   Medium consistency
O   Oxygen delignification
o.d.   Oven dry
OXE   Oxidizing equivalent
P   Peroxide
Paa   Peracedic acid
Pc   Postcolor number
PO   Pressurized peroxide
Q   Chelating agent
T   Peracids
TAED   Tetra acetyl ethylene diamine
TCF   Totally chlorine free
X   Xylanase treatment
Z   Ozone

References

[1]   Grundelius R. Oxidation equivalents, OXE – an alternative to active chlorine. Tappi J. 1993,
      76(1), 133–135.
[2]   Ragnar M, Leite M. Bleaching of cellulose pulp in a first chlorine dioxide bleaching step.
      Kvaerner pulping, 2004, Patent WO 2004/079087M.
[3]   Kubelka P, Munk F. Zeits. F. Tech. Physik. 1931, 12, 593–601.
[4]   Steele F. The optical characteristics of paper. II. A precision opacimeter. Paper Trade J. 1935,
      101(17), 31–35.
[5]   Steele F. The optical characteristics of paper. III. The opacifying power of fibres and fillers.
      Paper Trade J. 1935, 104(8), 157–158.
[6]   Judd D. Optical specification of light-scattering materials. Paper Trade J. 1938, 106(1), 39–46.
[7]   Van den Akker J. Scattering and absorption of light in paper and other diffusing media.
      A note on the coefficients of the Kubelka-Munk theory. Tappi. 1949, 32(11), 498–501.
[8]   Giertz H. Some optical consequences of the consolidation of paper. In Proceedings of the
      Transactions of the Symposium on Consolidation of the Paper Web. Cambridge, England,
      1966, 928–948.
[9]   Tongren J. A reflectance method for the study of discoloration of artificially aged papers.
      Paper Trade J. 1938, 107(8), 34–42.
[10] Giertz H. Yellowing of pulp. Finnish Paper Timber Journal. 1945, 27, 359–361.

[11] Paulsson M, Lucia L, Ragauskas A, Li C. Photoyellowing of untreated and acetylated chemithermomechanical pulp under argon, ambient, and oxygen temperatures. J. Wood Chem. Technol. 2001, 21(4), 343–360.

[12] Li C, Kim D, Ragauskas A. Brightness reversion of mechanical pulps. XIX. Photostabilization of mechanical pulps by UV Absorbers: Surface photochemical studies using diffuse reflectance technique. J. Wood Chem. Technol. 2004, 24(1), 39–53.

[13] Valchev I, Valcheva E, Dimitrov I. Kinetics of the pressured peroxide bleaching of hardwood pulp. Part I. Effect of oxygen on the peroxide bleaching. Cellul. Chem. Technol. 2003, 37, 131–139.

[14] Valchev I, Dimitrov I, Valcheva E, Veleva S. Kinetics of the pressured peroxide bleaching of hardwood pulp. Part II. Effect of enzyme treatment on pulp. Cellul. Chem. Technol. 2004, 38, 67–77.

[15] Valchev I, Simeonova G. Kinetic models for investigation and control of the bleaching processes. 2005. International pulp bleaching conference. Stockholm, Sweden, 14–16 June, 2005, 272–274.

[16] Radeva G, Valchev I, Valcheva E. Comparative kinetic analysis of a laccase–mediator system treatment of pulp after oxygen delignification and chlorine dioxide bleaching. Cellul. Chem. Technol. 2009, 43, 7–8. 317–323.

[17] Froass P, Ragauskas A, McDonough T, Jiang J. Relationship between residual lignin structure and pulp bleachability. Int. Pulp Bleaching Conference, Washington D.C., Proceedings, 1996, 1, 163.

[18] Gustavsson C. On the interrelation between kraft cooking conditions and pulp composition: Doctoral Thesis. Royal Institute of Technology. Department of Fibre and Polymer Technology, Stockholm, Sweden, 2006.

[19] Lawoko M. Lignin Polysaccharide Networks in Softwood and Chemical Pulps: Characterisation. Structure and Reactivity. Doctoral Thesis. Royal Institute of Technology. Department of Fibre and Polymer Technology, Sweden, 2006.

[20] Janson J, Palenius I. Aspects of the color of kraft pulp and kraft lignin. Pap. Puu. 1972, 54(6), 343–352.

[21] Janson J, Palenius I, Stenlund B, Sagfors P. Differences in color and strength of kraft pulps from batch and flow cooking. Pap. Puu. 1975, 57(5), 387–396.

[22] Dyer T. Elucidating the formation and chemistry of chromophores during kraft pulping. Thesis for the degree of doctor of philosophy. Institute of Paper Science and Technology, Atlanta, Georgia, USA, 2004.

[23] Rosenau T, Potthast A, Zwirchmayr N, Hettegger H, Plasser F. et al, Chromophores from hexeneuronic acids: identification of HexA-derived chromophores. Cellulose. 2017, 24, 3671–3687.

[24] Rosenau T, Potthast A, Zwirchmayr N, Hosoya T, Hettegger H. et al. Chromophores from hexeneuronic acids (HexA): synthesis of model compounds and primary degradation intermediates. Cellulose. 2017, 24, 3703–3723.

[25] Vuorinen T, Teleman A, Fagerström P, Buchert J, Tenkanen M. Selective hydrolysis of hexenuronic acid groups and its application in ECF and TCF bleaching of kraft pulps, International Pulp Bleaching Conference. Washington, D.C., USA, 1996, 1, 43–51.

[26] Gellerstedt G, Li J. An HPLC method for the quantitative determination of hexenuronic acid groups in chemical pulps. Carbohydr. Res. 1996, 294(1), 41–51.

[27] Buchert J, Teleman A, Harjunpää V, Viikari L, Vuorinen T. Effect of cooking and bleaching on the structure of xylan in conventional pine kraft pulp. Tappi J. 1995, 78(11), 125–130.

[28] Bergnor-Gidnert E, Tomani P, O. Dahlman. Influence on pulp quality of conditions during the removal of hexenuronic acids. Nordic Pulp and Paper Res. J. 1998, 13(4), 310–316.

[29] Buchert J, Tenkanen M, Ek M, Teleman A, Viikari L, Vuorinen T. Effects of pulping and bleaching on pulp carbohydrates and technical properties. In Proceedings of the International pulp bleaching conference. Washington, D.C., USA, 1996, 1, 39–42.

[30] Buchert J, Bergnor E, Lindblad G, Viikari L, Ek M. Significance of xylan and glucomannan in the brightness reversion of kraft pulps. Tappi J. 1997, 80(6), 165–171.

[31] Siltala M, Winberg K, Henricson K, Lonnberg B. Mill scale application for selective hydrolysis of hexenuronic acid groups in TCFz bleaching of kraft pulps. In Proceedings of the International pulp bleaching conference. Helsinki, Finland, 1998, 1, 279–286.

[32] Granström A, Eriksson T, Gellerstedt G, Rööst C, Larsson P. Variables Affecting the Thermal Yellowing of TCF-bleached Birch Kraft Pulps. Nordic Pulp and Paper Res. J. 2001, 16(1), 18–23.

[33] Sevastyanova O. On the importance of oxidizable structures in bleached kraft pulps. Doctoral Thesis. Royal Institute of Technology. Stockholm, Sweden, 2005.

[34] Devenyns J, Chauveheid E. Uronic acid and metals control. In Proceedings of the 9th International Symposium on Wood and Pulping Chemistry (ISWPC). Montreal, Que, Canada, 1997, M5: 1–4.

[35] Devenyns J, Chauveheid E, Martens H. Uronic acids and metals control. In Proceedings of the International pulp bleaching conference. Helsinki, Finland, 1998, 1, 151–157.

[36] Laine J, Stenius P. Surface properties of bleached kraft pulp fibres and their comparison with the properties of the final paper products. In Proceedings of the 8th International Symposium on Wood and Pulping Chemistry (ISWPC). Helsinki, Finland, 1995, 589–596.

[37] Johansson M, Samuelson O. Epimerization and degradation of 2-O-(4-O-methyl-a-D-glucopyranosyluronic acid)-D-xylitol in alkaline medium. Carbohydr. Res. 1977, 54, 295–299.

[38] Clavijo L, Cabrera M, Kuitunen S, Liukko S, Rauhala T, Vuorinen T. Changes in a eucalyptus kraft pulp during a mild acid treatment at high temperature. Papel. 2012, 73, 59–64.

[39] Teleman A, Hausalo T, Tenkanen M, Vuorinen T. Identification of the acidic degradation products of hexenuronic acid and characterization of hexenuronic acid-substituted xylooligosaccharides by NMR spectroscopy. Carbohydr. Res. 1996, 280, 197–208.

[40] Suess H. and C. Leporini. 2007. Best practice for highest and very stable brightness, consequences for a short sequence. Paper presented at ABTCP. Sao Paulo. Brazil. Accessed April 05, 2012. http://h2o2.evonik.com/product/h2o2/en/Pages/default.aspx

[41] Suess H, Davies D, Dietz T. Pushing the brightness ceiling of "Difficult" softwood kraft pulps. Paper presented at PAPTAC. Montreal, Canada, 2008. (Accessed April 05, 2012, at http://h2o2.evonik.com/product/h2o2/en/Pages/default.aspx)

[42] Rosenau T, Potthast A, Kosma P, Suess H, Nimmerfroh N. Chromophores in aged hardwood pulp – their structure and degradation potential. Paper presented at ISWFPC, Durban, South Africa, 2007. (Accessed April 5, 2012, at http://h2o2.evonik.com/product/h2o2/en/Pages/default.aspx)

[43] Rosenau T, Dietz T, French A, Henniges U, Potthast A. Structures of chromophores isolated from different cellulosics, molecular mechanisms of cellulose aging and yellowing and implications for bleaching. In Proceedings of the 16th International Symposium on Wood and Pulping Chemistry (ISWPC), Tianjin, China, 2011, 525–528.

[44] Genco J, Van Heiningen A, Miller W. Oxygen Delignification, The Bleaching of Pulp. In: P. Hart, A. Rudie., eds. The bleaching of pulp. TAPPI Press, Atlanta GA, USA, 9, 2012.

[45] Parsad B, Kirkman A, Jameel H, Gratzl J, Magnotta V. Mill closure with high kappa pulping and extended oxygen delignification. Tappi J. 1996, 79(9), 144–152.

[46] McDonough T. In: C.W. Dence, D.W. Reeve., eds. Pulp bleaching principles and practice, Tappi Press, Atlanta, 1996, 213.

[47] Shin S, Schroeder L, Lai Y. Understanding Factors Contributing to Low Oxygen Delignification of Hardwood Kraft Pulps. J. Wood Chem. Technol. 2006, 26(1), 5–20.

[48]   Violette S. Oxygen delignification kinetics and selectivity improvement. Thesis for the Degree of Doctor of Philosophy, The University of Maine, USA, 2003.

[49]   Antonsson S, Lindstrom M, Ragnar M. A comparative study of the impact of the cooking process on oxygen delignification. Nordic Pulp and Paper Res. J. 2003, 18(4), 388–394.

[50]   Agarwal S, Genco J, Cole B, Miller W. Kinetics of Oxygen Delignification. Journal of Pulp and Paper Science. 1999, 25, 10: 361–366.

[51]   Hsu C, Hsieh J. Effects of Mass transfer on Medium Consistency Oxygen Bleaching Kinetics. Tappi J. 1985b, 68(11), 126–130.

[52]   Nenkova S, Valchev I, Simeonova G. Possibility for increasing the effectiveness of oxygen delignification of hardwood pulp. J. Pulp Pap. Sci. 2003, 29(10), 324–327.

[53]   Van Heiningen A, Ji Y. Southern pine oxygen delignified pulps produced in a Berty throughflow reactor: How to obtain the highest degree of delignification while maintaining pulp yield and quality. Tappi J. 2012, 11(3), 9–17.

[54]   Yang R, Ragauskas A, Jameel H. Oxygen degradation and spectroscopic characterization of hardwood kraft lignin. Ind. Eng. Chem. Res. 2002, 41(24), 5941–5948.

[55]   Iijima J, Taneda H. The effect of carryover on medium-consistency oxygen delignification of hardwood kraft pulp. J. Pulp Pap. Sci. 1997, 23(12), J561–J564.

[56]   Ragnar M, Ala-Kaila K. On the demand for oxygen in oxygen delignification of chemical pulp. IPW/Papier. 2004, 8, T146 –T149.

[57]   Olm L, Teder A. The kinetics of oxygen bleaching. Tappi J. 1979, 62(12), 43–46.

[58]   Hsu C, Hsieh J. Oxygen bleaching kinetics at ultra-low consistency. Tappi J. 1987, 70(12), 107–111.

[59]   Hsu C, Hsieh J. Reaction kinetics in oxygen bleaching. AIChE J. 1988, 34(11), 116–122.

[60]   Iribarne J, Schroeder L. High-pressure oxygen delignification of kraft pulps 1. Kinetics Tappi J 1997, 80(10), 241–250.

[61]   Valchev I, Christensen P. Kinetics of Mg(OH)2-oxygen delignification of sulphite pulp – Part I. Common kinetic dependencies. In Proceedings of the 7th International Symposium on Wood and Pulping Chemistry (ISWPC). Beijing, China. 1993, 1, 197–205.

[62]   Nguyen K, Liang H. Kinetic model of oxygen delignification. Part 1– effect of process variables. Appita J. 2002, 55(2), 162–165.

[63]   Valchev I, Valcheva E, Christova E. Kinetics of oxygen delignification of hardwood kraft pulp. Cellul. Chem. Technol. 1999, 33(3–4), 303–310.

[64]   GLV Group. Pulp Technologies. (Accessed April 5, 2012, at http://www.glv.com/Pulp_Paper/ Pulp_Technologies/Chemical/BusinessCateg.aspx)

[65]   Firmennachrichten-ausland. (Schweden Sunds Defibrator – Umweltfreundliches Bleichverfahren erhält 1997 den Product Excellence Award). Wochenblatt für Papierfabrication. 1998, 126(8), 380.

[66]   Ragnar M. A Compact Way to Extend Oxygen Delignification. In Proceedings of the 7th International Conference New Available Technologies. Stockholm, Sweden, 2002, 31–35.

[67]   Barroca M, Marques P, Secom M, Castro A. Selectivity Studies of Oxygen and Chlorine Dioxide in the Pre-Delignification Stages of a Hardwood Pulp Bleaching Plant. Ind. Eng. Chem. Res. 2001, 40, 5680–5685.

[68]   Joncort M, Froment P, Lachenal D, Chirat C. Reduction of AOX formation during chlorine dioxide bleaching. Tappi J. 2000, 83(1), 144–148.

[69]   Brogdon B, Lucia L. New Insights into Lignin Modification during Chlorine Dioxide Bleaching Sequences (III): The Impact of Modifications in the (EO) versus E Stage on the D1 Stage. J. Wood Chem. Technol. 2005, 25(3), 133–147.

[70] Ragnar M. Modification of the DO-stage into D* makes 2-stage Bleach Plant for HW Kraft Pulp
      a Reality. In Proceedings of the International pulp bleaching conference. Portland, USA,
      2002, 1, 237–244.
[71] Dyer, T. and A. Ragauskas. Developments in Bleaching Technology Focus on Reducing
      Capital, Operating Costs. Pulp Pap. 2002, 3, 49–53.
[72] Lachenal D, Chirat C, Viardin M. High Temperature Chlorine Dioxide Delignification.
      A Breakthrough in ECF Bleaching of Hardwood Kraft Pulps. Tappi J. 2000, 83(8), 96.
[73] Eiras K, Colodette J. Eucalyptus Kraft Pulp Bleaching with Chlorine Dioxide at High
      Temperature. J. Pulp Pap. Sci. 2003, 29(2), 64–69.
[74] Dahl O, Ninimäki J, Tirri T, Kuopanportti H. Bleaching softwood kraft pulp: the role of chlorine
      dioxide dosage and final pH in the D stages. Pap. Puu. 1997, 79(3), 560–564.
[75] Valchev I, Simeonova G, Nenkova S, Gaidarov J. The evolution of hardwood kraft pulp
      bleaching – A comparison of options. Cellul. Chem. Technol. 2005, 39(1–2), 105–114.
[76] Valchev I, Simeonova G, Nenkova S, Valcheva E, Gaidarov J. Evolution of ECF hardwood
      bleaching for better chemical efficiency. Balkan Pulp and Paper News. 2004, 5(12), 25–28.
[77] Sevastyanova O, Forsström A, Wackerberg E, Lindström M. Bleaching of eucalyptus kraft
      pulps with chlorine dioxide: Factors affecting the efficiency of the final D stage. Tappi J. 2012,
      11(3), 43–53.
[78] Andtbacka S. The importance of washing in oxygen delignification and TCF bleaching.
      Pulp&Paper Canada. 1998, 99, 3, 57 –60.
[79] Andrews D, Singh R. The Bleaching of Pulp. ed. R.P. Singh. 3rd ed. Tappi Press, Atlanta,
      1979, 211–253.
[80] Gierer J, Imsgard F. The reactions of lignin with oxygen and hydrogen peroxide in alkaline
      media. Svensk Papperstidning. 1977, 80(16), 510–518.
[81] Gellerstedt G, Pettersson I, Sundin S. Chemical aspects of hydrogen peroxide bleaching.
      In Proceedings of the International Symposium on Wood and Pulping Chemistry. Stockholm,
      Sweden, 1981, 1, 120–124.
[82] Gierer J. Formation and involvement of superoxide (O_2/HO_2) and hydroxyl (OH) radicals
      in TCF bleaching processes: A review. Holzforschung. 1997, 51(1), 34–46.
[83] Davies D, Dietz T, Suess H. A comparison of options to improve brightness stability
      of chemical pulp. Pulp Pap. Canada. 2009, 110, 8, 25:31.
[84] Jääskeläinen A, Saariano A, Matousek P, Parker A, Towrie M, Vuorinen T. Characterization
      of residual lignin structures by UV Raman spectroscopy and the possibilities of Raman
      spectroscopy in the visible region with Kerr-Gated fluorescence rejection. In Proceedings
      of the 12th International Symposium on Wood and Pulping Chemistry (ISWPC). Madison,
      USA, 2003, 1, 139–142.
[85] Filho C, Suess H. Hydrogen peroxide in chemical pulp bleaching – an overview. Paper
      presented at Iberoamerican Congress on Pulp and Paper Research CIADICYP. Sao Paulo,
      Brazil, 2002.
[86] Tsekova P, Gaidarov J, Valchev I, Blyahovski V. Progress in post bleaching for highest
      brightness. In Proceedings of the 16th International Symposium on Wood and Pulping
      Chemistry (ISWPC). Tianjin, China, 2011, 761–765.
[87] Ruohoniemi K, Heiko J, Laakso I, Martikainen S, Väyrynen V, Jäkärä J. Experience in the use
      of peracetic acid in ECF and TCF bleaching. In Proceedings of the International pulp bleaching
      conference. Helsinki, Finland, 1998, 1, 145–150.
[88] Jakara J, Paren A, Autio P. The use of peracetic acid as a brightening agent. In Proceedings
      of the 53rd Appita Annual Conference. Rotorua, New Zealand, 1999, 2, 463–467.
[89] Yuan Z, Ni Y, Van Heiningen A. 1997. Kinetics of the peracetic acid decomposition. Can.
      J. Chem. Engg. 1997, 75(1), 37–41.

[90]   Yuan Z. Peracetic acid bleaching of softwood kraft pulp, Thesis for the Degree of Doctor of Philosophy. University of New Brunswick, Canada, 1997.

[91]   Jääskeläinen S, Poppius-Levlin K. Carbohydrates in peroxyacetic acid bleaching. In Proceedings of the International pulp bleaching conference. Helsinki. Finland. 1998, 2, 423–428.

[92]   Bergnor-Gidnert E, Tomani P, Dahlman O. Influence on pulp quality of conditions during the removal of hexenuronic acids. Nordic Pulp and Paper Res. J. 1998, 13(4), 310–316.

[93]   Barros P, Silva V, Hämäläinen H,. Colodette J. Effect of last stage bleaching with peracetic acid on brightness development and properties of eucalyptus pulp. BioResources. 2010, 5(2), 881–898.

[94]   Khristova P, Tomkinson J, Valchev I, Dimitrov I. Totally chlorine-free bleaching of flax pulp. Bioresour. Technol. 2002, 85(1), 79–85.

[95]   Simeonova G, Valchev I, Nenkova S. Hardwood kraft pulp bleaching by peracetic acid. J. of UCTM. 2003, 38(1), 151–156.

[96]   Zavadzki M. Quantified determination of quinone chromophore during ECF bleaching of kraft pulp. Thesis for the Degree of Doctor of Philosophy. Institute of Paper Science and Technology. Atlanta, Georgia, USA, 2004.

[97]   Valchev I, Simeonova G, Blyahovski V. In Proceedings of the 15th International Symposium on Wood and Pulping Chemistry (ISWPC). Oslo, Norway, 2009.

[98]   Pan G, Chen C, Chang H, Gratzl J. The effect of pH, temperature, buffer systems and heavy metalions on stability of ozone in aqueous solution. J. Wood Chem. Technol. 1984, 4(3), 367–387.

[99]   Staehelin J, Hoigné J. Decomposition of ozone in water in the presence of organic solutes acting as promoters and inhibitors of radical chain reactions. Environ. Sci. Technol. 1985, 19(12), 1206–1213.

[100] Gierer J. The chemistry of delignification. A general concept. Part II. Holzforschung. 1982, 36(1), 55–64.

[101] Ni Y, Kang G, Van Heiningen A. Are hydroxyl radicals responsible for degradation of carbohydrates during ozone bleaching of chemical pulp?. J. Pulp Pap. Sci. 1996, 22(2), J53–J57.

[102] Lindholm C. Effect of pulp consistency and pH in ozone bleaching. Part 6. Strength properties. Nordic Pulp and Paper Res. J. 1990, 5, 1, 22–27.

[103] Lierop B, Berry R, Roy B. High brightness bleaching of softwood kraft pulps with oxygen, ozone and peroxide. J. Pulp Pap. Sci. 1997, 23(9), 428–432.

[104] Viikari L, Rauna M, Kantelinen A, Linko M, Sundquist J. In Proceedings of the 3rd International Conference Biotechnology in the Pulp and Paper Industry, Stockholm, Sweden, 1986, 67–69.

[105] Viikari L, Rauna M, Kantelinen A, Linko M, Sundquist J. In Proceedings of the 4th International Symposium on Wood and Pulping Chemistry. Paris, France, 1987, 1, 151–154.

[106] Viikari L, Kantelinen A, Sundquist J, Linko M. Xylanases in bleaching: From an idea to the industry. FEMS Microbiol. Rev. 1994, 13(2–3), 335–350.

[107] Deneault C, Leduc C, Valade J. The use of xylanases in kraft pulp bleaching: a review. Tappi J. 1994, 77(6), 125–131.

[108] Davis M, Rosin B, Landucci L,. Jeffries T. Characterization of UV Absorbing Products Released from Kraft Pulps by Xylanases. In Proceedings of the Tappi Biological Science Symposium, 1997, 435–443.

[109] Buchert J, Viikari L. Significants of pulp metal profil on enzyme-aided TCF bleaching. Paperi Ja Puu – Paper and Timber. 1995, 77(9), 582–587.

[110] Qingzhi M, Wang Q, Wang C, Feng N, Zhai H. Application of pure, thermostable, alkali-tolerant xylanase in bleaching of oxygen-delignified pine kraft pulp. Tappi J. 2015, 14(11), 689–694.

[111] Simeonova G, Sjödahl R, Ragnar M, Lindström M, Henriksson G. On the effect of a xylanase post-treatment as a means of reducing the yellowing of bleached hardwood kraft pulp. Nordic Pulp and Paper Res. J. 2007, 22(2), 172–176.

[112] Valchev I, Tsekova P. Xylanase post-treatment as a progress in bleaching processes. Appita J. 2010, 63(1), 53–57.

[113] Valchev I, Yotova L, Valcheva E. Kinetics of Xylanase treatment of hardwood pulp. Bioresour. Technol. 1998, 65(1), 57–60.

[114] Valcheva E, Veleva S, Valchev I, Dimitrov I. Kinetic model of xylanase action of kraft pulp. React. Kinet. Catal. Lett. 2000, 71(2), 231–238.

[115] Valcheva E, Valchev I, Yotova L. Kinetics of enzyme action of Cartazyme NS-10 prior to bleaching of kraft pulp. Biochem. Eng. J. 2001, 7(3), 223–226.

[116] Dimitrov I, Valchev I, Valcheva E. Topochemical kinetics of xylanase action on kraft pulp, Biocatalysis and Biotransformation. Biocatal. Biotransform. 2005, 23(1), 33–36.

[117] Osiadacz J, Al-Adami A, Bajraszwska D, Fisher P, Peczynska-Croch W. On the use of Trametes versicolor laccase for the conversion of 4-methyl-3-hydroxyanthranilic acid to actinocin chromophore. J. Biotechnol. 1999, 72(1–2), 141–149.

[118] Widsten P. Kandelbauer. Laccase applications in the forest products industry: A review. Enzyme Microb. Technol. 2008, 42(4), 293–307.

[119] Crestini C, Argyropoulos D. Structural Analysis of Wheat Straw Lignin by Quantitative 31P and 2D NMR Spectroscopy. The Occurrence of Ester Bonds and α-O-4 Substructures. J. Agric. Food. Chem. 1997, 45(4), 1212–1219.

[120] Camarero S, Ibarra D, Martinez A, Romero J, Gutierrez A, Del Rio J. Paper pulp delignification using laccase and natural mediators. Enzyme Microb. Technol. 2007, 40(5), 1264–1271.

[121] Valcheva E, Veleva S, Radeva Gr, Valchev I. Enzyme action of the laccase-mediator system in the pulp delignification process. React. Kinet. Catal. Lett. 2003, 78(1), 183–191.

[122] Valchev I, Radeva G, Valcheva E. Effect of laccase-mediator system on the oxygen delignification and chlorine dioxide pulp bleaching. J. UCTM. 2005, 40(2), 107–110.

Emmanuel Koukios, Lazaros Karaoglanoglou, Dimitrios Koullas,
Nikolaos Kourakos, Anna Moutsatsou, Sofia Papadaki,
and Ioannis Panagiotopoulos

Chapter 5
Recent advances in processing of biomass feedstocks for high added value outlets through bio-greening pathways

5.1 Introduction and methodology

The bioresource technology unit (BTU) is a research and innovation group operating since the mid-1980s within the Organic Technologies Laboratory of the Chemical Engineering School at the National Technical University of Athens. The BTU work is driven by the vision of the emergence of a new generation of technologies aiming at the conversion of feedstocks of biological origin to industrial, energy, and other products and services. According to the BTU approach, science, research, process, and product development, as well as demonstration and diffusion of innovations are considered as "levers" for sociotechnical change.

The strategy of the group is characterised by the following attributes [1]:
- *greening* oriented, aiming at cleaner processes, products, and services
- *systemic*, consisting of the identification of critical issues for investigation within well-defined frames, that is, those of integrated material and energy systems
- *interdisciplinary*, including science and engineering disciplines, mainly chemcal, biological, mathematical, social, and economic
- *networked*, with networks operating at regiomal, national, European, amd global levels.

In the following paragraphs, we will summarize the scope, background, and approach of the BTU research tean activities regarding the utilization of biomass through "greener" pathways, classified in seven technological areas.

Emmanuel Koukios, Lazaros Karaoglanoglou, Dimitrios Koullas, Nikolaos Kourakos, Sofia Papadaki, Ioannis Panagiotopoulos, Bioresource Technology Unit, School of Chemical Engineering National Technical University of Athens, Greece
Anna Moutsatsou, Bioresource Technology Unit, School of Chemical Engineering National Technical University of Athens, Greece; National Gallery, Athens, Greece

https://doi.org/10.1515/9783110658842-005

5.1.1 Bio-feedstocks

All types of biomass, such as plant, animal, and microbial wastes, residues and coproducts, as well as new and old industrial and energy crops, are assessed with emphasis
on lignocellulosic biomass. The assessment criteria include biomass potential and its
availability, in both quantitative and qualitative terms, the latter covering the physicochemical and other properties of the particular biomass resources. To map and evaluate the landscape of the possible optimal use of biofeedstocks, appropriate models are
constructed linking biomass technical, logistic, economic, social, and environmental
factors [2]. The BTU experience with such bioresource models ranges from local and
regional studies to national and European ones.

5.1.2 Biorefineries

The concept of biorefining occupies a central position in the BTU work, as it concerns the development of new type of interface between bioresources and biopro-
cesses and bioproducts. So, throughout the group's history, a lot of effort has been
dedicated in order to define, analyze, promote, and assess the role and value of
building feasible and sustainable biorefineries. Key topics of these activities include
the typology of biorefineries, the strategic emphasis on refining lignocellulosic
types of biomass (e.g., straws and other woody parts of agricultural plants), and the
development of biomass fractionation processes, with the use of physicochemical
pretreatment and upgrading stages. BTU was one of the partners in putting together
the first biorefinery pilot plant on the island on Bornholm, Denmark, in the frame
of a pioneering European project [3].

5.1.3 Plant fibres

The science and technology of plant fibres are of a high significance for the biomass
field; this is due on the one hand to the high availability of lignocellulosic resources
for novel biobased applications, and on the other hand on the long and rich experience of the established fibre industry that could be used as a model for the emerging biobased ones. So far relevant BTU activities have focused on the substitution
of wood fibres in paper- and board-making by nonwood ones; the development of
"clean" (without S and Cl – with oxygen and organic solvents) pulping and bleaching processes; novel, "smart" plant fibre characterization protocols, for example,
with the use of fluorescence spectroscopy, and the degradation and conservation of
old paper [4].

5.1.4 Bioethanol

This biomolecule occupies one of the key positions in the BTU "greening" approach, as it could serve as a "bridge" between our fossil fuel – and internal combustion engines – dominated economy, and a solar – and other renewable energies – emerging one. The BTU's emphasis has been on the conversion of lignocellulosic biomass from both agro-residues and energy crops, with the following particular strategic topics of interest [5]: (a) putting together the optimal biomass pretreatments taking into account the role of cellulose crystallinity and lignin content; (b) optimizing the water-ethanol separation and other critical factors affecting the process energy balance; and (c) using a multi-criteria assessment approach combining technical, economic, energy and environmental aspects.

5.1.5 Biohydrogen

If the bioethanol molecule can be part of the "bridge" away from the fossil-fuel based economy, and a step towards a sustainable bioeconomy, it is with the biohydrogen molecule that we can have a good glimpse of such an emerging sustainable energy and materials society and economy. The BTU relevant activities include both biochemical and thermochemical conversion process pathways, as well as their combinations, for example, a thermochemical pretreatment step followed by a thermochemical one or a biochemical conversion stage opening the way to a thermochemical one. The range of potential biofeedstocks is also broad covering cellulose-, sugar- and starch rich types of biomass. The resulting product vectors include coproducts that could enhance the feasibility and sustainability of the whole system [6].

5.1.6 Thermo-refining

According to the BTU philosophy, the thermochemical refineries of biomass feedstocks will gradually replace the presently dominant petrochemical refineries of fossil carbon resources. In order to facilitate such strategic transition, the group's research has focused on some key points for the feasibility and sustainability of the emerging biobased industries [7]. They include the prediction of the performance of the biofeedstocks under various high-temperature process regimes, including possible specific problems and opportunities, the development of appropriate pretreatments, with emphasis on the role of biomass inorganics (ash), and the key role of biomass densification and production of upgraded solid biofuels as a transition fuel linking the wood with the agricultural industries.

5.1.7 Integrated biosystems

The various types of biomass utilization systems tend to interact strongly with each other and with other technological and socioeconomic systems at local, regional, European, and global levels. In order to efficiently manage the resulting complexity, the BTU research portfolio has been extended to include new topics, such as the modeling of biosystems; the prediction of the dynamics of such complex systems; the development of decision support tools and expert systems to enrich and improve the strategic management of business, research, and administration; as well as the promotion of future-oriented approaches, linked to the formulation of suitable policy environments, aiming as sustainable bioeconomy targets [8].

5.2 Dyeing of chemical pulp with plant-derived dyes

This research shows the potential of plant materials, in particular the residues and by-products of agricultural and forestry activities, to be used satisfactorily as dye raw materials for dyeing of chemical paper. An alternative use of plant dyes in addition to their well-known and widely used applications, such as dyeing textile fibres, food, beverages, and cosmetics is suggested. It is important to emphasize that the coloring of pulp and in particular chemical pulp, which is the raw material for most paper products, from plant based dyeing raw materials is a field that has not been extensively explored despite its wide application. Chemical pulp makes up about 70% of the total paper mass production and it is clearly superior to mechanical pulp because it gives paper with high mechanical strength and whiteness, while it does not tend to get yellowish hues when exposed to air and mainly to light. It is used in the production of a wide range of products such as printing paper, packaging materials, functional food packaging materials, household paper, and personal care and hygiene papers. The application of plant derived dyes to these products is of particular importance as they not only make them easily recyclable due to the reduced resistance of dyes to leaching but also enhance the final product with unique properties such as antimicrobial and antifungal activity due to the presence of a variety of natural chromophore compounds such as tannins and naphthoquinones.

In this way, the pulp produced is environmentally friendly, as the use of synthetic petrochemical dyes is avoided, which exhibit significant environmental effects not only during their production stages but also in their application and final disposal. In addition, the application of plant-based dyes to chemical pulp facilitates its subsequent recycling process, and in particular the bleaching process, where hot water flushing can replace or drastically reduce the application of strong chemicals and especially chlorides [9].

The application of plant derived dyes to chemical pulp hides has some weaknesses such as achieving repeatability of the dye effect. This is due to the peculiarity of the plant dye raw material, whose color properties depend on a number of independent factors such as its cultivation conditions, climate, manner, and period of collection. Also, the final dyeing effect is influenced by the management and processing of the plant material such as its shredding (plant matter granulation), its drying, and extraction. In addition, dyeing conditions are a critical factor in achieving the desired dyeing effect. This report focuses on the critical factors of successful dyeing associated with the selection and pretreatment of dyeing plant materials.

The selection of dyeing plant raw materials usually starts from oral testimonies and bibliographical research on dyeing plants in the European area. For the final selection of most suitable plant-based raw material important role is played by

– The abundance with which they are found in European flora.
– Their unique characteristics due to their origin, as they are plant materials, which are nonedible plant materials, residues/by-products of agricultural and forestry activities.
– The holistic utilization of plant materials in a wide range of applications beyond chemical pulp dyeing such as food technology, cosmetology, and pharmacology, since the plant parts of many dyeing plants other than chromophore compounds contain numerous bioactive antioxidants such as antioxidants.
– Also, to a large extent the dyes themselves are multiactive substances that exhibit beyond their dyeing capacity of antioxidant and anti-inflammatory activity.

According to the literature, the application of plant-based dyes, whose dyeing ability has been tested in the present study, gave satisfactory dyeing effects when tested on other substrates, such as cotton, woolen and synthetic fibers. Application of plant-based dyes whose dyeing ability has been tested and given satisfactory dyeing effects on other substrates such as cotton fibres, woollen, and synthetic fibres. The dyeing ability of most dye plant raw materials is based on the presence of water-soluble compounds such as flavonoids, more importantly quercetin and anthocyanins, as well as quinoid bromides such as fluororubin. For this reason, water is suggested as an ideal extraction medium. The choice of water as an extraction medium ensures a safe and environmentally friendly process for the production of plant dyes, drastically reducing costs and finally the ability to utilize the plant-based residues in a fairly wide range of applications. In addition, given that the paper mill manages huge quantities of water, the production of dye aqueous extracts may possibly take place at the same site where the paper dyeing process is carried out, as it allows the dyeing aqueous extract to be introduced directly into the production line of paper.

In order to optimize the dyeing effect, the pretreatment modes and conditions of the dyeing plant materials before they are brought to the extraction stage are of particular importance:

I. The study of the effects of the drying methods and conditions of the plant raw material

The types of drying proposed for the drying of plant raw materials are usually the following: natural drying (open air drying), air drying, and freeze-drying [10]. These types of drying are proposed on the basis of several criteria, most importantly ease of application, low installation costs, and operating costs, as well as the final quality of the product. Natural drying is chosen because of its ease of application and extremely low cost. Air-drying is quite efficient giving end products with satisfying organoleptic characteristics and has relatively low installation and operating costs. Finally, freeze-drying is costly but gives products with excellent properties as their structure is significantly affected [9, 10]. The choice of the optimum method varies from case to case and is a field of study for each plant material individually, as it varies with the colorants they contain, the initial moisture content, the porosity of the material, and its various physical characteristics.

II. The study of the effect of the granulometry of the extracted dry plant raw material. The granulometry of a material has a significant effect on the performance of its extract, as it determines the wettability of the material. In order to ensure high extraction efficiency, it is necessary to rapidly immerse the material to be extracted to its interior.

III. The study of the effect of the change in the ratio of the weight of the extracted amount of the dry plant raw material to the weight of the dyed dry chemical pulp (M/P ratio) on the color

Increasing the value of the M/P ratio causes an increase in the value of the color difference (DE) to some limit value beyond which the value of total DE remains constant. Therefore, a further increase in the value of the M/P ratio beyond the limit value does not lead to a further change in the color of the paper sheets produced by the dyed pulp, as the fibers appear to be no longer able to retain more dye. The saturation limit value is different for each type of dyeing material and enables the prediction of which ratio of M/P can achieve the strongest dyeing effect. This results in significant savings in resources and energy, by avoiding unnecessary consumption of extra amounts of dye, as the overall DE remains relatively unchanged beyond the saturation threshold.

5.3 Forest chemicals: innovative pine oleoresin products

Pine oleoresin is an abundant renewable resource mainly composed by diterpenic resin acids of the general formula $C_{19}H_{29}COOH$. It is obtained either by tapping living pine trees or as a by-product of pulp industry. The crude oleoresin is processed to yield turpentine, an essential oil, and nonvolatile rosin among other products. These two main primary products of pine resin are considered as very valuable raw materials and they are applied in a wide range of industrial activities. However, a sharp decline in the production of pine oleoresin by tapping living pine trees in Western Europe has been observed in recent years.

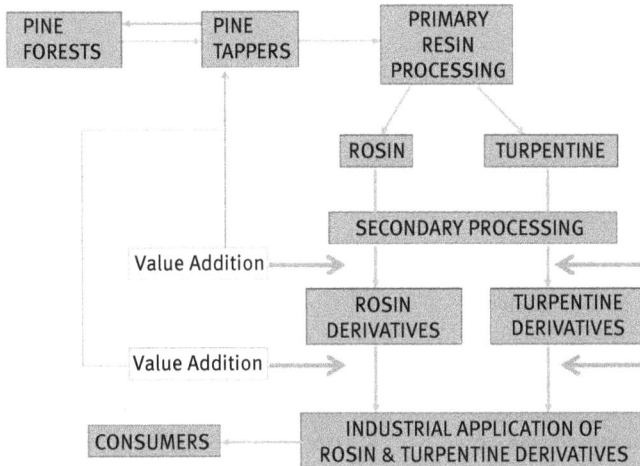

Fig. 5.1: Pine oleoresin processing flow chart and value addition opportunities [11].

Developing a new product portfolio, close to the feedstock production site, could provide the necessary incentives to the local population, encouraging their involvement in the pine tapping activity. Participatory processes, where the relevant stakeholders will attend to all the crucial decision-making steps, were followed in order to ensure that the outcomes will fulfil the necessary socioeconomic sustainability criteria [12]. Furthermore, data were collected through a market survey, to listen to the "Voice of Customer," as an initiating step for the product development process, according to Juran's Quality Handbook. This was followed by the identification of the critical product specifications and preliminary experiments carried out to explore the effect of multiple factors on these specifications.

The three major outcomes of the followed participatory process and market survey are the following: (I) Innovative rosin product development would probably benefit the existing industries but not necessarily the income of pine tappers. Therefore, it is more likely that it would increase the import of rosin. Given that none of the existing industries proceed in the further processing of turpentine, this fraction is potentially available for further exploitation. (II) Value added turpentine product development in a local unit close to the production site, quite probably will have an impact on local population's income. (III) Although in long term a wide range of product portfolio should be aimed, in short term a product of mass consumption with relatively simple production processes will facilitate the start-up steps of the project.

Following these outcomes turpentine was selected as the major raw material to be focused on. The three main application fields of turpentine which were examined in depth along with the pros and cons of each product category are presented in Tab. 5.1. Taking into consideration this table, and the fact that presently available turpentine in the target region is not enough for making further processing feasible, the wood treatment product category has been selected for the development of a new product.

Tab. 5.1: Comparative assessment of potential turpentine end products [11].

	Positive impact +	Negative impact −
Wood treatment	– Simple production process – Local production-consumption – Easy to promote products – Turpentine is the only organic solvent which is considered as environmentally friendly	– Relatively lower value addition
Insecticides	– New trends in insecticides – Utilization in local activities – Multiple utilization of the same product (flavor or insecticides)	– Time consuming biological experiments – More complex production process – Necessity of cooperation with a larger producer
Flavor and fragrances	– High value addition – Versatile product development	– More complex production process – More difficult promotion – Low possibility for local production-promotion – Necessity of cooperation with a larger producer

The protective effect of turpentine on wood is well known, however, in recent years other, economically more feasible, mainly petroleum-based agents have replaced its utilization in wood treatment. The main advantage of turpentine in comparison with these agents is its renewable and biodegradable nature. Turpentine, according to the current literature, can be utilized either in wood varnish and polishing agents or in products aiming to clear and revive already existing wood varnishes. In both cases turpentine is usually combined with other renewable raw materials like drying oils, rosin and its derivatives, vinegar, and bee wax.

Within this framework, samples with several levels of turpentine, boiled linseed oil, vinegar, rosin, and maleic-modified rosin has been prepared. Technical specifications of these samples like odor, water repellent efficiency, VOCs, drying time, phase separation, viscosity, polishing ability, easiness of use, and their specialization according to the wood species were explored using timber from different tree species like iroko, oak, and beech.

Following the preliminary tests, four combinations of raw materials were selected for further tests. This selection was based on the current literature, market information, and empirical estimation of their appearance on the wood surface and handling properties.

Tab. 5.2: Potential formulas for wood protection products [11].

	Turpentine	Linseed oil	Vinegar	Maleic-modified rosin
Sample 1	33% v/v	33% v/v	33% v/v	–
Sample 2	60% v/v	20% v/v	20% v/v	2% w/v
Sample 3	50% v/v	50% v/v	–	–
Sample 4	50% v/v	50% v/v	–	4% w/v

*All the samples presented in the Tab. 5.2 also contain about 2% v/v Ca drier, which reduce the drying time of the product.

Two of these samples (1, 2) are intended to be used on already existing varnishes, to clean and revive them, whereas the other two samples will be applied on bare timber. These samples were subjected to water repellence and VOC tests together with two commercially available solvent containing wood treatment products (Com 1 and 2) – see Figs. 5.2 and 5.3.

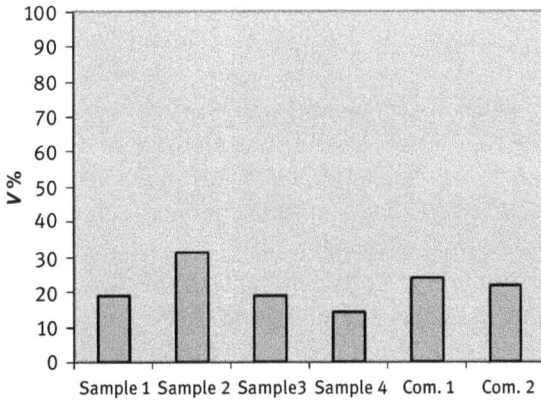

Fig. 5.2: Water repellent efficiency of sample 3 and sample 4 on beech, iroko and oak timber [11].

The tentative outcomes can be summarized into the following [11]:
- The utilization of rosin and maleic-modified rosin in the recipe had an increasing effect on the viscosity, drying time, gloss, and water repellence.
- The species of the wood samples seem to play a significant role in the efficacy of the wood treatment preparation. Therefore, products tailored according to

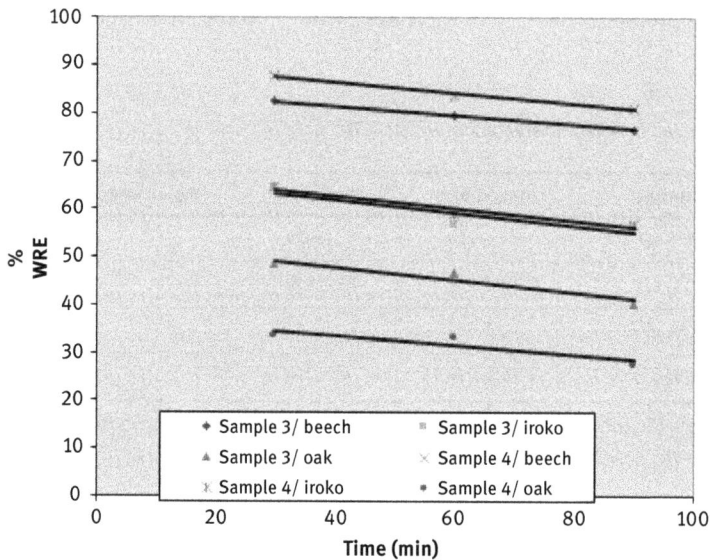

Fig. 5.3: Volatile organic compounds emission [11].

the wood type which they will be applied on, would optimize the efficacy of the products.

- When rosin is used in a varnish cleaning and polishing preparation it improves the gloss of the preparation but it also causes the darkening of the existing varnish; the utilization of maleic-modified rosin prevents this darkening.
- Using a big proportion of linseed oil causes also the same problem. However, a recipe which optimize the proportion of linseed oil, rosin, turpentine, and vinegar in a way that the final products give high polishing and cleaning results without darkening the existing coating can be prepared.
- The addition of vinegar to the recipe cause phase separation, this fact should be taken into consideration in the storage and utilization of the product; the addition of dryers improves significantly the drying time of the preparation.

5.4 Hydrothermal fractionation of lignocellulosic wastes

Lignocellulosic wastes and residues can be hydrothermally fractionated to vectors of valuable products and coproducts. For example, wheat bran contains starch and C5 polysaccharides (as part of their hemicelluloses). So, whenever wheat bran is considered as a wheat grain industry left-over and is not used, it can be recycled for further exploitation by hydrothermal treatment [4].

The process concept followed is shown in brief in Fig. 5.4 Starch is produced at a first stage, specifically before depolymerization reactions occur. After the separation of the starch fraction, the residual material is hydrothermally treated, so that hydrolysis of the (nonstarchy) C5 polysaccharides to C5 carbohydrate oligomers and monomers takes place [13]. Depending on the treatment parameters (mainly time) the production of soluble xylan oligomers or xylose monomers can be maximized, and the production of lignocellulosic (LC) solids can be affected, for example, minimized. Since autohydrolysis is targeted, the utilization of additional chemical reagents is usually avoided.

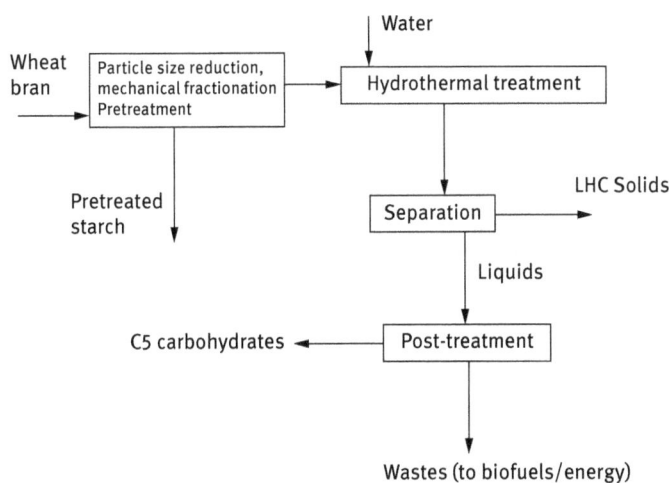

Fig. 5.4: Hydrothermal fractionation of lignocellulosic wastes.

As it can be seen in Fig. 5.5, the solid residues of such a process can be valuable feedstocks from which raw materials for several uses can be recovered. In particular, wheat bran can be a resource for various applications, such as sorbents to clean the oil spills, or to produce activated carbon, or biofuels and energy. Apart from these main uses, cellulosic fibres can be recycled from wheat bran. From the latter even raw materials for composites can be produced [14].

5.5 Optimizing the utilization cycle of leafy biomass

Besides lignocellulosic wastes, leafy biomass constitutes another type of available biomass for industrial and energy utilization. In this section we will introduce the concept of leafy biomass refining, using sugar beet leaves as the appropriate case study [14].

	Cellulose recovery, cellulosic material, paper	Composites materials	Skimmers and sorbents for spill cleanup	Energy production	Others (activated carbon, feedstock)
Eucalyptus solid residue	✓✓✓	✓✓	✓✓	✓	✓
Spent grain solid residue		✓✓	✓✓✓	✓	✓✓
Wheat bran solid residue	✓	(✓)	✓✓	✓✓✓	✓✓
Corn cobs solid residue		✓✓✓	✓✓	✓✓	✓

Fig. 5.5: Assessing the utilization of solid residues from the hydrothermal fractionation of various wastes and residues [14].

Sugar beet is an extremely important crop for sugar production from the beet roots in Europe at industrial level. However, there are other possible uses of this crop to consider. In particular, the sugar beet leaves contain protein that can be recovered, and, therefore, a global exploitation of the crop can be achieved. As a result, sugar beet leaves fractionation protocol has been investigated, a brief version of which can be seen in Fig. 5.6:

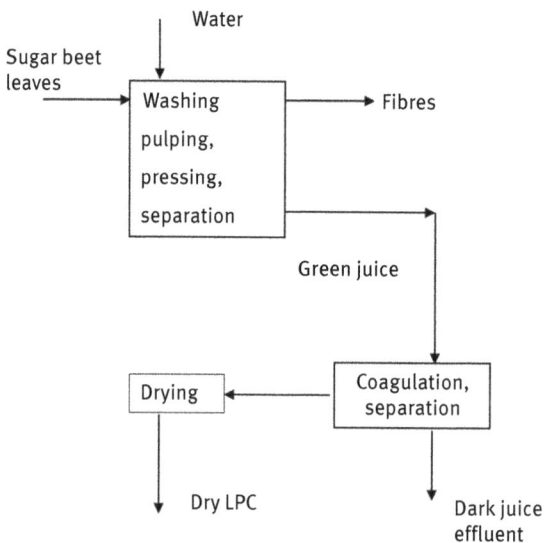

Fig. 5.6: SBL fractionation protocol (in brief).

As shown in Tab. 5.3. it was found that the later the harvesting, the higher the percentage of sugar beet leaf (SBL) dry matter (DM). In addition, there is an optimum harvesting time with regards to leaf nitrogen content (% dry matter).

Tab. 5.3: SBL characteristics.

Harvesting month	Leaf DM (%)	Leaves N in DM (%)
July	6.70	3.15
August	8.28	4.61
September	9.28	4,27
October	10.47	3.90

It seems (Tab. 5.4), that earlier (i.e., first) harvesting results in higher fibre production, while later (i.e., second) harvesting leads to greater amounts of dark juice production. No matter the harvesting time, the leaf protein concentrate being approximately the same (approximately 14–15%). Moreover, it is noticeable that the production of dark juice is minimized.

Tab. 5.4: SBL DM fractionation efficiency (%).

Harvesting month	LPC(%)	Fibres	Dark juice
July	15.2	60.80	24.00
August	14.21	39.02	46.77
September	15.08	49.57	35.36
October	14.96	46.12	38.92

An optimal nitrogen recovery in the leaf protein concentrate is achieved during the early harvesting as shown in Tab. 5.5, where SBL nitrogen fractionation efficiency (i.e., distribution of 100 g Leaf N_2 in the SBL fractions) is depicted.

5.6 Optimal feedstocks for biohydrogen production

Biorefineries are based on the integration of the biomass conversion processes to produce power, fuels, and chemicals. Hydrogen is a very interesting fuel within the biorefinery, because it has zero CO_2 emissions when burnt and it is the prime energy carrier to be used in fuel cells for electricity generation. In particular, the use

Tab. 5.5: SBL N_2 fractionation efficiency (distribution of 100 g leaf N_2 in the SBL fractions).

Harvesting month	LPC (%)	Fibre (%)	Dark juice (%)
July	33.0	39.0	28.0
August	23.7	31.3	45.0
September	25.5	41.9	32.6
October	26.8	43.2	30.0

NOTE: This is an original research, a part of which was presented in an ERASMUS Programme in Foggia, Italy, and another part supported a NTUA post graduate biomass course [14], and thus the contribution of these audiences is acknowledged.

of biomass for biohydrogen production [15] is expected to be substantially developed in the future [16–17]. Biohydrogen has key advantages compared to other hydrogen-producing technologies because the particular bioprocess is relatively efficient, it has low energy demands and it can use a variety of biomass feedstocks, including sugary, starchy, and lignocellulosic raw materials [18].

Biomass pretreatment is needed to make the available carbohydrates from sugary, starchy, and particularly lignocellulosic raw materials accessible for hydrogen fermentation. Depending on the type of biomass, pretreatment type and severity varies. For example, in case of sugary biomass, most of the sugars are readily fermentable, so the pretreatment is relatively simple. On the other hand, in case of lignocellulosic biomass which has C5 and C6 sugars, the pretreatment is more complex, usually requiring heat and chemicals. Typical lignocellulosic biomass pretreatment methods include dilute-acid pretreatment, steam pretreatment and alkaline pretreatment. Among various lignocellulosic raw materials, agricultural residues have advantageous use for biohydrogen production partly because they typically require simpler pretreatment (see Fig. 5.7) [19].

Wheat straw and barley straw are typical agricultural residues that can be used as feedstocks for biohydrogen production [20–22]. The fermentability of wheat straw is relatively low. This is partly related with the release of inhibitors which typically takes place during dilute-acid pretreatment, and with other, as-yet-unknown inhibiting compounds. Barley straw shows good fermentability. A comparison of alkaline pretreatment and dilute-acid pretreatment of barley straw has recently gained attention. Generally, the optimization of the pretreatment of barley straw as a raw material for biohydrogen production is based on various parameters, such as sugar yield, release of inhibitors, use of chemicals, possible recycling of chemicals, and investment costs. Dilute-acid pretreatment is performed with lower amount of chemicals but results in relatively lower sugar concentrations. The release of fermentation inhibitors depends

Agricultural residues

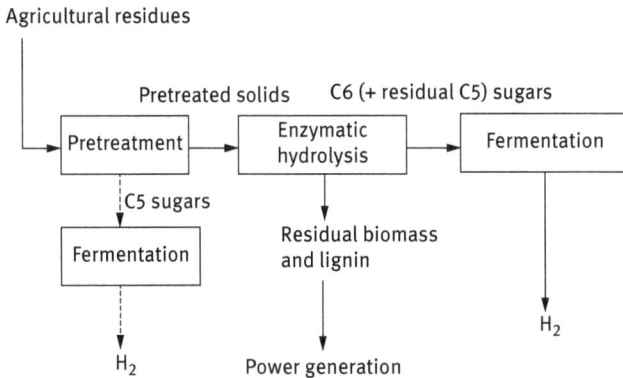

Fig. 5.7: A schematic outline of a pretreatment process for biohydrogen production from agricultural residues. The dashed arrows on C5 sugars represent the potential of C5 sugars to be used for hydrogen fermentation (reprinted from Panagiotopoulos and Koukios, 2018).

on the severity of the pretreatment. The maximum concentration of fermentable monomeric sugars is typically observed after the pretreatment at the highest pretreatment severity. However, this corresponds to a low conversion of hemicellulose of the straw and the largest formation and release of inhibitors [18]. In terms of fermentability for thermophilic hydrogen production, it seems that the application of a pretreatment temperature of up to 170 °C is acceptable, when the initial sugar concentration in the fermentation medium is 10–15 g/L. On the other hand, alkaline pretreatment, with relatively higher solid/liquid ratios, leads to higher sugar concentrations, specifically when a dedicated pretreatment equipment, such as a conical screw reactor, is used. Increasing the solid/liquid ratio leads to increased sugar concentrations but relatively low sugar yields. The fermentability of the alkaline pretreated substrates is good, particularly when the sugar concentration is lower than 25 g/L.

Commercial production of hydrogen from lignocellulosic biomass is expected to take place in 2030–2040. Among other lignocellulosic raw materials, agricultural residues, such as wheat straw and barley straw, are expected to be widely used because they are less recalcitrant and, thus, require simpler pretreatment. The selection of the optimal feedstocks for biohydrogen production is expected to be mainly based on the criteria of sugar yield, fermentability and availability.

5.7 Methods to evaluate paper-based works of art

Works of art on paper, such as watercolour paintings, are nowadays gaining popularity, commercial value, and prestige in the art community. Many museums are reorganizing and reassessing their collections in order to promote them. Nevertheless, authenticity control of paintings on paper is not covered as a separate subject by the fundamental or current relevant literature. Furthermore, in the common practice of museums this kind of studies is usually based on stylistic criteria, as well as on visual methods of examination by art historians. If physicochemical analysis is carried out, its protocol is defined mainly by investigation of pigments anachronisms, like in cases of panel and canvas paintings. At the same time, art forgers have used several methods to imitate the appearance and condition of original artistic paper supports [23].

The aim of the research that has been done by our group [24] was the development of an integrated, reliable, and easy-to-use methodology for the characterization of paper substrates in the framework of a noninvasive control of the authenticity of watercolour paintings, with application to works of Greek painters of the end of nineteenth century and the beginning of the twentieth century. Regarding the analytical protocols their selection was based on the accessibility of equipment at museum laboratories and mainly the noninvasive character of the whole process.

Thus, it was proved that noninvasive visible light microscopy and fluorescence light microscopy (including in-situ fluorescence microscopy) must constitute the basic part of the methodology formed [25]. The particular protocol for the microscopic observation of artistic paper substrates includes quantitative analysis of captured images, extraction of colour-related data from microscopic images, and finally graphical representation and statistical analysis of the results. This protocol was shown to be adequate in order to differentiate the naturally aged from the nonaged paper substrates, even if the latter were subjected to a forging technique, other than artificial ageing. In the next step, the usefulness of attenuated total reflection (ATR) spectroscopy was shown to be multiple (see Fig. 5.8)

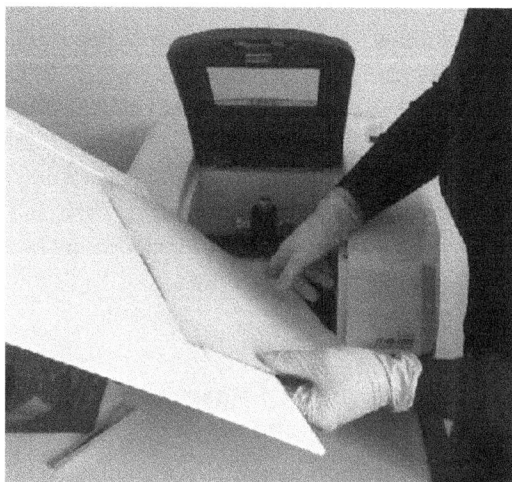

Fig. 5.8: Noninvasive analysis of original watercolour painting by means of ATR spectroscopy.

ATR is a sampling technique used in conjunction with infrared spectroscopy, which enables samples to be examined directly in the solid or liquid state without further preparation. In particular, it was found out that it is possible to differentiate naturally aged substrates from artificially aged ones subjected to dry heating at temperatures above 100 °C, while it is possible to differentiate some naturally aged papers from artificially aged ones under suitable temperature–humidity conditions.

According to our experimental results, the integrated protocol for noninvasive authenticity control of nineteenth-century paintings on paper, based on the information that the paper substrate may provide, is presented in Fig. 5.9.

The results from such investigations should be considered as indications to be further assessed in conjunction with observations deriving from the physicochemical investigation of the painting techniques and materials, as well as with the style, origin, and historical framework of the artists and their artworks (e.g., [26]).

In the framework of future development of the present methodology, the main aim of the research was the collection and processing of experimental results of, as many as possible, original watercolour paintings of the nineteenth century or more recent. Through the examination of a great amount of original watercolour paintings it would be made possible:

– to create a wide data-base for the safer classification of paintings under investigation, for example, through the more precise determination of the oxidation index threshold able to differentiate naturally aged from artificially aged artistic papers; and
– to further investigate the chronological limit of natural ageing for which such a differentiation could be possible.

5.8 Educating engineers for greening project skills

The need to transform our educational practices is an issue that employs the academic community and the broader society. The emergence of new techno-economic dimensions and new fields such as green technologies, sustainability, bioeconomy, smart cities, internet of things (IoT), artificial intelligence (AI), machine learning (ML), additive manufacturing, or as often mentioned 3D printing, intelligent robots, chatbots and digital assistants, and blockchain are the new normals in our lives. All these bring a wide-ranging change, which has also a direct impact on education issues.

5.8.1 Major trends

Watching [27] the waves of technological change according to A. Toffler and the cycles of the economy as mentioned by Kondratiev, we find that we are on the verge of

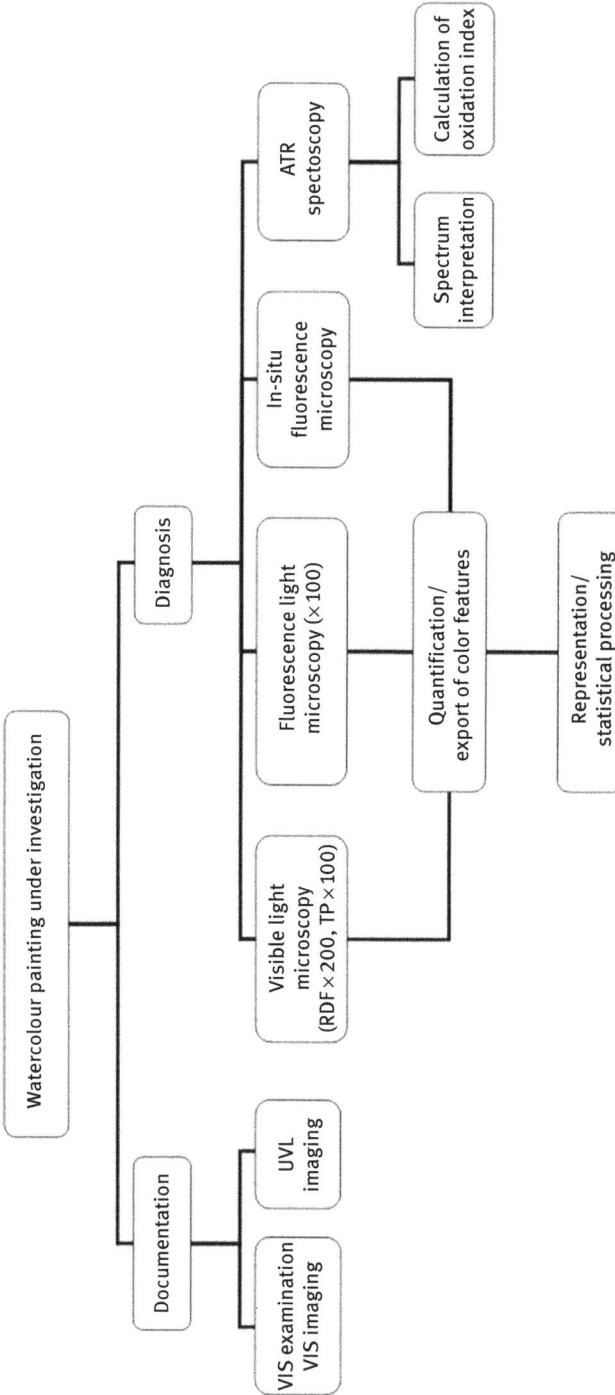

Fig. 5.9: Integrated protocol for noninvasive authenticity control of nineteenth-century paintings based on paper substrate [24].

change, at the crossroads of technology, production, economy, and society transformation. All these changes arise the need to reapproach and realign our education approach and consequently our didactic methodologies and tools. The flowing working reality, the production transformations, and the emergence of the info-nano-bio society, lead to gaps in knowledge of entire generations. Nowadays we are talking about STEAM (science, technology, engineering, art, and mathematics) and digital gap. The digital divergence of generations (as shown by the internet readiness index) is a small confirmation of the above.

Till now our education model comes from the first industrial revolution and the industrial schools. The traditional teacher centered model seems now obsolete. The depreciation rates of knowledge are exponential. The internet, the social networks, the ability to access a huge number of resources, the amount of information, the MOOCs (massive open online courses) offered by Universities, and other agencies are now part of today's student life. We can describe the new role of teacher as a coach and facilitator for the students.

There is an increased demand on student-centered teaching, collaborating methods of teaching, interactive living labs, adaptive and interactive blended learning environments. Furthermore, one of the fundamental educational requirements today is to enhance motivation and involvement and achieve an increased active student's participation in classes.

After a period of study, pilot implementations, and adaptation, our research team tried to incorporate a range of modern methodologies and tools in everyday teaching practice. This integration was gradual and targeted at both the undergraduate and postgraduate courses and laboratories.

5.8.2 Learning styles

In particular, one of the pillars of our educational approach and methodology was the adaptation of teaching materials and presentations based on the learning styles of our audience. Our goal was to adapt the educational material and presentations to the particular learning characteristics of the participated students. Customized learning, maybe, is the optimal way that suits exactly the needs of any individual student. Relevant literature indicates that there is an expanding demand for this [27].

Our research tool was the index of learning styles (ILS) developed by Felder and Solomon, one of the most prevalent and widely used models in the specific area. This tool uses four relevant axes, complementary to each other: active vs. reflective, sensitive vs. intuitive, visual vs. verbal, and sequential vs. global. In the results area, we can see from our research samples that a rather strong orientation to visual and sensitive was recording. Even more a slight turn to active learners' type is obvious. Furthermore, we make a comparison of our findings with other

European technical universities research findings such as Aalborg, in Denmark. We notice that the differences between the two research samples are very few.

After the analysis of the results we have started to modify all presentation materials, enriching our slides with diagrams, incorporating video clips, and animated introductions, in order to be more attractive, motivate the audience, and increase their course engagement.

5.8.3 Student participation – digital curation

The next step in changing our teaching practice was enhancing the active participation of our students, mainly postgraduate ones, in the digital curation trial.

Nowadays, on daily basis, we indicate the increasing tendency of saving, uploading on the cloud, and transferring educational material through any kind of digital devices. Given that, for our students the digitalized educational material is almost everywhere. Additionally, reusing any kind of digital material is common place, achievable and supported by the use of metadata. The potential of direct data and information "mining," accelerates the direct access of any student and trainee to material of digitalized format. In another point of view, the digitalized material can be actively protected, whereas, in the meantime, its discoverability is enhanced. The benefits of these two factors, are reflected on the extensive reuse of the digital material during its lifetime which exponentially increases its added-value. For many researchers this is the practical definition of content curation (CC) and diffusing information function. The question arises, following the relatively thorough approach and the clarification of the function of CC, concerns how and in which manner it can help and enhance our didactic function in the class.

In order to move our students into the center of educational function and practice, leverage their ability not only to pathetically receive lectures and material but to try and find complimentary texts and manipulate with all these contents, we use a well-known CC tool, the Scoop.it. The platform was chosen based on the large number of their members society, the ease of use, and the ability to connect with widespread social networks such as Facebook and LinkedIn.

We ask students to upload their preferred material and mainly write comments on this in order to record their criticism. Simultaneously, the platform allows members of the course, the Scoopers community, and all connected social networkers to comment, edit, and supplies with new bookmarks the original post. In this way we had a double cycle of leveraging the participation and also provide the critical thinking of the class.

The acceptance of these action was proved by the very strong percentage of participation (near 80% positively disposed). Accordingly, clear positive are also the feedback which was recorded by the number of visits by third observers [27].

5.8.4 Enhancing analytical capacity – mind maps

Another issue was the strengthening of the analytical capacity, skills, and development of students critical thinking. In this dimension cognitive maps were chosen to be used as a tool. The presentation of complex concepts and their analysis in the breakdown as well as the synthesis of a complex of different and diverse parts was our aim. Thus, our students are invited to analyze a complex conceptual schema in the field of knowledge modules as well as to represent a complex concept by using cognitive maps.

In order to document the student's behavioral intention to use/acceptance of mind maps as a supportive learning medium and tool, we use the technology acceptance model (TAM) by Davis et al. [28]. The overall usefulness received from our questionnaires using the Linkert scale has been mentioned in Fig. 5.10.

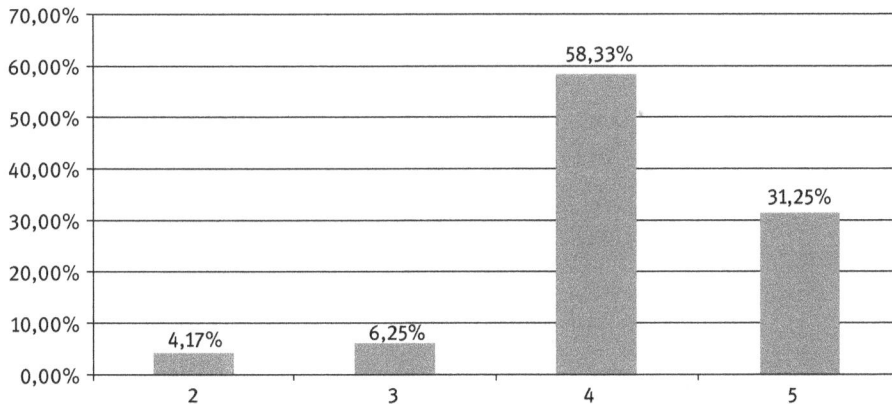

Fig. 5.10: Overall, "I think that the use of mind maps is useful in my studies."

This positive feedback from our students led us to incorporate this approach into our lessons in the future.

5.8.5 Promoting team work – wikis

Enhancing collaborative learning and promoting the ability of our students to work as a team was our fourth goal. Here the use of wikis was considered both appropriate and mandatory.

It is worth noting that, in addition to the many positive points of the use of wikis, the international bibliography records a sufficient number of cases of nonuse for a wide range of reasons. Our team has chosen to create a general-purpose wiki for the course with the ability for each participating student to create their own page and post their content there. Of course, there was always the possibility of commenting and filling in for the approval of another page's material. The effort was aimed at promoting cooperation and noble competition among students.

Our findings on participation cannot be described as encouraging. In particular, we observed a 40% low rate of participation among our students [27]. Additionally, the rates for the frequency of the wiki visits and the number of interventions/enrolments on the part of students were lower. The workload and lack of time were, in the vast majority, the "arguments" of students for the lack of petitions and observations as well as for the overall use of the educational tool.

5.9 Concluding remarks

In their so far productive life – of approximately one generation – the BTU activities have resulted in a number of specific outcomes and outputs, including 20 completed doctoral theses; more than 200 completed diploma theses (equivalent to a master's degree); education of more than 1,000 university students with courses at their senior level of engineering degree studies on biomass subjects; more than 100 research and technology projects with external financial support, more than 50% of which European and international; these have led to ca. 150 peer-reviewed publications and more than 300 conference and other papers that have received so far ca. 3,000 citations, that is, an H-factor of ca. 30 for the group leader; all these activities were possible by extensive and systematic networking with more than 100 research and innovation expert groups in 30 countries across four continents.

A particular aspect of the BTU activities that is linked with the group's philosophy and mandate is its active participation in large projects of strategic interest and potential significant effects for the future of bioeconomy at various levels of action; some examples follow:

- national networks: RENES (renewable energies); SELL (soft energies at the local level) conferences; FORESIGHT (forward looking inputs to policy making); HEBIS (Hellenic biotechnology society); ELETAS (technology evaluation and assessment society);
- european initiatives: SAST – MONITOR (technology assessment), EUREC (renewable energy centres), BIORAF (biorefinery); European Foresight; RISOE (forward looking inputs to policy making; KBBE – bioeconomy advisory group; STAR (regional bioeconomy initiative);
- global coordination: EXTERNE (external costs of renewables); European – Eastern Asia initiative; ICTPI (international conferences on technology policy and innovation).

References

[1]   Koukios E. Knowledge-based greening as a new bioeconomy strategy for development: Agroecological utopia or revolution? In: Monteduro M, Buongiorno P, Di Benedetto S, Isoni A., eds. Law and agroecology: A transdisciplinary dialogue. Springer-Verlag, Berlin Heidelberg, 2015, 439–450.

[2]   Koukios E, Karaoglanoglou L. Assessment of biomass potential for sustainable energy and fuels production in European regions. In: Barz M, Dinkler K., eds. Technical and economic aspects of renewable energies – know-how transfer as development opportunity for Southern Europe. Berlin, MBV, July 2016.

[3]   Papatheofanous M, Billa E, Koullas D, Monties B, Koukios E. Optimizing multisteps mechanical-chemical fractionation of wheat straw components. Ind. Crops. Prod. (Elsevier) 1998, 7, 249–256.

[4]   Avgerinos E, Billa E, Papatheofanous M, Koullas D, Koukios E. Developing molecular strategies for the delignification and characterisation of annual plant fibres. C.R. Biologies (Elsevier). 2004, 327(9–10), 927–933.

[5]   Tsoutsos T, Doudakmani O, Tournaki S, Koukios E. Study of the alcohol-producing sector – The bioethanol system. Greek Review of Agrarian Studies, 1992, 3(2), 100–114 (in Greek).

[6]   Koukios E, Daouti-Koukios I, Koullas D, Avgerinos E. Critical parameters for optimal biomass refineries: The case of biohydrogen. Clean Technol. Environ. Policy J. 2010, 12(2), 147–151.

[7]   Arvelakis S, Koukios E. Critical factors for high temperature processing of biomass from agriculture and energy crops to biofuels and bioenergy. WIREs Energy Environ. (Wiley Interdisciplinary Reviews). 2012, 2(4), 441–455 – https://doi.org/10.1002/wene.28.

[8]   Koukios E. Technology management for biobased economy: Mapping, dynamics, policies – The case of Greece In: Trzmielak D, Gibson D., eds. International cases on innovation, knowledge and technology transfer, University Press, Lodz, 2014, 129–145.

[9]   Papadaki S, Krokida M, Economides D, Koukios E Dyeing of chemical pulp with natural dyes. Cellul. Chem. Technol. 2014, 48(3–4), 385–393.

[10]   Papadaki S, Krokida M, Economides D, Vlyssides A, Koukios E. 2013 Effect of drying methods on dyeing capacity of various plants. Drying Technol. & Equip. 2014, 11(1), 67–75.

[11]   Karaoglanoglou L et al, Innovative high value-added pine-oleoresin products. Proc. 4th National Conference of Chemical Engineering, Patras, May 2003; and Proc. 8th International Paint and Ink Symposium, Athens, May 2002.

[12]   Karaoglanoglou L. PhD Thesis, NTUA, Athens, 2017.

[13]   Kabel MA, Carvalheiro F, Garrote G, Avgerinos E, Koukios EG, Parajo JC, Girio FM, Schols HA, Voragen, AGJ. Hydrothermally treated xylan rich by-products yield different classes of xylo-oligosaccharides. Carbohydr. Polym. 2002, 50, (191–200).

[14]   Koukios E. Biorefinery, Biomass Course Notes, NTUA, Athens, Creative Commons, 2013, https://ocw.aoc.ntua.gr/modules/document/file.php/CHEMENG126/%CE%98%CE%B5%CE%BC%CE%B1%CF%84%CE%B9%CE%BA%CE%AE%20%CE%95%CE%BD%CF%8C%CF%84%CE%B7%CF%84%CE%B1%203/Section%203a.pdf

[15]   Claassen P, de Vrije T. Non-thermal production of pure hydrogen from biomass: HYVOLUTION. Int. J. Hydrogen Energy. 2006, 31(11), 1416–1423.

[16]   Muradov N, Veziroglu T. "Green" path from fossil-based to hydrogen economy: An overview of carbon-neutral technologies. Int. J. Hydrogen Energy. 2008, 33(23) 6804–6839.

[17]   Dunn S. Hydrogen futures: Toward a sustainable energy system. Int. J. Hydrog. Energy. 2002, 27, 235–264.

[18] Panagiotopoulos I, Bakker R, Budde M, de Vrije T, Claassen P, Koukios E. Fermentative hydrogen production from pretreated biomass: A comparative study. Biores. Technol. 2009, 100(24), 6331–6338.

[19] Panagiotopoulos I, Koukios E. Biohydrogen production from agricultural residues. In: Meyers, RA., Ed. Encyclopedia of sustainability science and technology. Springer-Verlag, New York, 2018, https://doi.org/10.1007/978-1-4939-2493-6

[20] Kongjan P, O-Thong S, Kotay M, Min B, Angelidaki I. Biohydrogen production from wheat straw hydrolysate by dark fermentation using extreme thermophilic mixed culture Biotechnol. Bioeng. 2010, 105(5), 899–908.

[21] Panagiotopoulos I, Bakker R, de Vrije T, Claassen P, Koukios E. (2013) Integration of first and second generation biofuels: Fermentative hydrogen production from wheat grain and straw. Biores. Technol. 2013, 128, 345–350.

[22] Özgür E, Peksel B. Biohydrogen production from barley straw hydrolysate through sequential dark and photofermentation. J. Cle. Pro. 2013, 52, 14–20.

[23] Hebborn E. The art forger's handbook. Cassell, London, 1998.

[24] Moutsatsou A, Terlixi A, Alexopoulou A, Koukios A. The contribution of visible and fluorescence light microscopy to a, non-invasive authenticity control of paper substrates of nineteenth century watercolour paintings. Cellul. Chem. Technol. 2017, 50(5–6), 486–495.

[25] Moutsatsou A. Doctor of Engineering Thesis, National Technical University of Athens, Athens, 2017.

[26] Turner S. The book of fine paper. Thames and Hudson, London, 1998.

[27] Kourakos N. Network and other digital tools in technology education, Doctoral Thesis, NTUA, Athens, 2018; http://thesis.ekt.gr/thesisBookReader/id/43134#page/1/mode/2up

[28] Davis F, Bagozzi R, Warshaw P. User acceptance of computer technology: a comparison of two theoretical models. Manage. Sci. 1989, 35, 982–1003.

Irina Volf, Iuliana Bejenari, and Valentin I. Popa

Chapter 6
Valuable biobased products through hydrothermal decomposition

6.1 Introduction

The biorefining concept offers an opportunity to revitalize the pulp and paper industry to produce not only pulp and paper but also high-value chemicals and biofuels, developing new technologies to penetrate new markets. Thus, pulp and paper industry may integrate along wood and annual plant used as raw materials, wastes resulted in the manufacturing processes (e.g., bark, saw dust, lignin, hemicelluloses, wastewaters, and cellulosic residues from paper recycling), and other external sources of biomass to obtain energy and products with added value.

From the point of view of sustainability, waste management, and development of energy-efficient and mitigating climate change, one of the major objectives in Europe is the conversion of waste into energy and value-added products. It was considered that the most important waste resources for conversion are agricultural waste and forest residues, municipal solid waste, and sludge from wastewater treatment [1].

Lignocellulosic biomass can be transformed into bioproducts and bioenergy through several processes, including biochemical, thermochemical, mechanical, and physical processes. Biological conversion processes such as alcoholic fermentation and anaerobic digestion, despite consuming low energy, require longer timeframes as compared to other processes [2].

Several thermochemical methods for biomass conversion such as torrefaction, hydrothermal carbonization (HTC), pyrolysis, and gasification are intensively studied [3]. Torrefaction aims at converting biomass into a coal-like material, which has better fuel characteristics than the original biomass. The process occurs around 290 °C in 10–60 min, and leads to hydrochar and gas yields of 80% and 20%, respectively. In the same temperature range (180–300 °C), HTC takes place. However, the process time is higher (1–16 h) and the main bioproducts result with different yields (50–80% for hydrochar, 5–20% for bio-oil, and 2–5% for gas). HTC mimes the natural process of coal formation that converted the major constituents of biomass (cellulose) into coal-like materials. This factitious coalification process was rediscovered and has been variously referred to as hot compressed water treatment, subcritical water treatment, wet torrefaction, and hydrothermal treatment [4].

Irina Volf, Iuliana Bejenari, Valentin I. Popa, Gheorghe Asachi Technical University of Iasi, Romania

https://doi.org/10.1515/9783110658842-006

Pyrolysis is a thermal decomposition at elevated temperatures in an inert atmosphere. It involves a change of chemical composition and is irreversible. To optimize the yield, the temperature profile is the most important factor [5]. Rapid heating and cooling rate (< 2 s at 500–1,000 °C) are recommended for minimizing the extent of secondary reactions. When the pyrolysis temperature reached up to 550 °C bio-oil yields are higher (75%), while the biochar and biogas yields are 12% and 13%, respectively. The slow pyrolysis uses slow heating rates (300–700 °C during hours or even days), which lead to higher char yields of 35% while the bio-oil and biogas yields are 30% and 35%, respectively.

Gasification converts the carbonaceous materials into carbon monoxide, hydrogen, and carbon dioxide. This is achieved by reacting the material at high temperatures (>700 °C) and very short residence time (10–20 s), without combustion, with a controlled amount of oxygen and/or steam. The yield of gas mixture called syngas (from synthesis gas) is 85%.

On the thermochemical processes, pyrolysis has high efficiency and flexibility, but is faced with a main obstacle that the high moisture content of biomass requires high heat for vaporization [6]. Subsequently, a lot of attention was given to hydrothermal conversion (HTC) processes that have been shown to be more cost-effective as compared to conventional thermal drying [7]. However, the hydrothermal process for biomass conversion has to deal with difficult collection of the products and high requirements for equipment.

6.2 Feedstock for hydrothermal conversion

When considering the potential of biologically renewable raw materials, the biomass-sustainable availability becomes decisive. However, the data for biomass are fragmented, and estimates vary significantly. In 2020, the EU-sustainable biomass supply is estimated at 375 million tons oil equivalent considering the following distribution: wastes 36, agricultural residues 106, rotational crops 0, perennial crops 52, landscape care wood 11, roundwood production 56, additional harvestable roundwood 35, primary forestry residues 19, secondary forestry residues 15, and tertiary forestry residues 45 [8].

As it is well known, biomass can be divided into two categories: nonlignocellulosic and lignocellulosic. Sewage sludge and animal manure fall into the first category (nonlignocellulosic). These mostly contain proteins, fatty acid, and a small quantity of hemicelluloses, cellulose, and lignin. The lignocellulosic category includes agricultural waste, biodegradable municipal wastes, and forest waste, and the main components of these are hemicelluloses, cellulose, and lignin. In a common assess, the biomass contains about 20–40% hemicelluloses, 40–60% cellulose, and 10–25% lignin [9]. Although the composition is highly dependent on the type of biomass, the maturity and the climate conditions are also decisive. For this reason, some feed-stocks from different categories of biomass have been characterized (Tab. 6.1) [10].

Tab. 6.1: The chemical composition of different types of biomass.

Type of biomass	Cellulose (D-glucose) Unit $(C_6H_{10}O_5)_n$	Hemicelluloses $C_5H_{10}O_5$	Lignin $C_9H_{10}O_2$, $C_{10}H_{12}O_3$, $C_{11}H_{14}O_4$	Extractives/ash
		%		
Woody biomass and residues from forestry and trees outside forests				
Beech wood	45.3	31.2	21.9	1.6
White poplar	49.0	25.6	23.1	0.2
European birch	48.5	25.1	19.4	0.3
White willow	49.6	26.7	22.7	0.3
Monterey pin	41.7	20.5	25.9	0.3
Douglas fir	42.0	23.5	27.8	0.4
Spruce wood	49.8	20.7	27	2.5
Wood bark	24.8	29.8	43.8	1.6
Ailanthus wood	46.7	26.6	26.2	0.5
Softwood (av.)	45.8	24.4	28	1.7
Hardwood (av.)	45.2	31.3	21.7	2.7
Japan cedar	35	24	33	8
Rubber tree	45.84	73.84	21.42	2.86
Bamboo	47	25	21	7
Energy crops and residues from agricultural and marginal land				
Oil palm shell	34.28	61.31	31.91	6.99
Sunflower shell	48.4	34.6	17	2.7
Almond shell	50.7	28.9	20.4	2.5
Hazelnut shell	26.8	30.4	42.9	3.3
Coconut shell	26.49	79.29	35.54	1.86
Walnut shell	25.6	22.7	52.3	2.8

Tab. 6.1 (continued)

Type of biomass	Cellulose (D-glucose) Unit $(C_6H_{10}O_5)_n$	Hemicelluloses $C_5H_{10}O_5$	Lignin $C_9H_{10}O_2$, $C_{10}H_{12}O_3$, $C_{11}H_{14}O_4$	Extractives/ash
		%		
Oil palm husk	34.28	61.31	31.91	6.99
Olive husk	24	23.6	48.4	9.4
Coconut husk	30.55	56.45	38.82	2.65
Rice husk	44.14	82.42	42.58	2.48
Hazelnut seed coat	29.6	15.7	53	1.4
Wheat straw	28.8	39.4	18.6	3.3
Rice straw	36.26	74.45	34.73	5.28
Corn stover	51.2	30.7	14.4	3.7
Corn stalk	42.43	68.18	21.73	3.27
Tobacco stalk	42.4	28.2	27	2.4
Tobacco leaf	36.3	34.4	12.1	17.2
Pineapple leaf	32.16	63.2	18.68	6.55
Coastal Bermuda grass	25	35.7	6.4	–
Switch grass	45	31.4	12	–
Organic waste (tertiary residues)				
Tea waste	30.2	19.9	40	9.9
Bagasse	49.2	74.98	19.54	1.28
Bagasse pith	45.18	72.16	22.13	1.24
Waste material	50.6	29.2	24.7	4.5
Newspaper	40–55	25–40	18–30	–

Source: Adapted from Nizamuddin et al. [10].

6.3 Inside the hydrothermal carbonization process

HTC was discovered in 1913 by Friedrich Bergius and is considered to be a relatively new process [11], which is generally used for the conversion of biomass with a high moisture content [12] into a high carbon-rich solid product, named hydrochar.

The use of wet biomass represents a huge benefit since the removal of the predrying process, before conversion, which leads to a significant decrease in the energy consumption [4]. At the same time, it is considered to be a cost-effective method that is carried out at relatively low temperatures 180–300 °C at autogenous pressure in the presence of water. Also, the process can be carried out under controlled pressure. HTC process is considered to be environmentally friendly, nontoxic, and easy to apply [13]. HTC technology uses discontinuous and semicontinuous systems, both of which make it less economically viable [14]. Table 6.2 reviewed the newest studies on HTC process by considering the types of biomass, conversion systems, and operating conditions.

Tab. 6.2: HTC process performed on different feedstock.

Type of biomass	Type of reactor	Working parameters		Solid yield	Reference
		Temperature	Time		
Woody biomass and residues from forestry and trees outside forests					
Eucalyptus bark	Stainless steel autoclave reactor	220–300 °C	2–10 h	40–46.4%	[15]
Wood chips	Teflon-lined stainless steel autoclaves	200 °C	6–48 h	51.8–60.6%	[16]
Loblolly pine	Parr reactor	240 °C	2–6 h	41,93–48.54%	[17]
Loblolly pine	Two-chamber reactor system	200–230 °C	1–5 h	62.44–79.09%	[18]
Lignin	Autoclave with a Pyrex-glass liner	200–350 °C	1 h	42–88%	[19]
Holocellulose	Stainless steel autoclave	220 °C	4–20 h	22–66.7%	[20]
Cellulose	Teflon-lined stainless steel autoclaves	200 °C	6–48 h	44.6–64.6%	[16]
Microcrystalline cellulose	Stainless steel autoclave	170–245 °C	4–11.8 h	42.7–99.5%	[21]
Energy crops and residues from agricultural and marginal land					
Corn stover	Stirred tank reactor	175–250 °C	240 min	35.82–62.92%	[22]
Tobacco stalks	Teflon-lined autoclave	180–260 °C	1–12 h	41 – 80%	[13]

Tab. 6.2 (continued)

Type of biomass	Type of reactor	Working parameters		Solid yield	Reference
		Temperature	Time		
Sunflower	Stainless steel autoclave	190–230 °C	20–45 h	28.3–39.7%	[23]
Walnut shell	Stainless steel autoclave	190–230 °C	20–45 h	23.2–29.3%	[23]
Watermelon peel waste	Stainless steel autoclave	190–260 °C	1–12 h	53.83–94.76%	[24]
Cassava rhizome	Stainless steel autoclave	160–200 °C	1–3 h	54–59%	[25]
Organic waste (tertiary residues)					
Lipid-extracted *Chlorella vulgaris*	Lab-scale reactor	180–240 °C	30 min	51.84–74.53%	[26]
Other sources					
Glucose	Teflon-lined stainless steel autoclave	200 °C	12– 48 h	43.1–48.4%	[16]
Chitin	Teflon-lined stainless steel autoclave	200 °C	6– 48 h	78.6–89.3%	[16]
Chitosan	Teflon-lined stainless steel autoclave	200 °C	6– 48 h	52.0–54.6%	[16]

The HTC is generally conducted between 160 °C and at most 300 °C and could be understood considering the diagram in Fig. 6.1. With an increase in the severity of reaction, that is, for the temperatures above 260 °C, the HTC further classified into two techniques: (i) hydrothermal liquefaction and (ii) hydrothermal vaporization or hydrothermal gasification, or supercritical water gasification [4].

The products obtained from HTC of lignocellulosic biomass are solid product (hydrochar), liquid (bio-oil mixed with water), and gas, mainly CO_2. The main quantity of biobased product resulted is hydrochar. Due to this and also to the many innovative value-added industrial and environment applications, the investigations on HTC are focused on the production of hydrochar. This has a higher carbon content and improved hydrophobic properties compared with the feedstock [25]. Hydrochar may have different properties depending on the different operating conditions of the HTC process and the characteristics of the raw material [27].

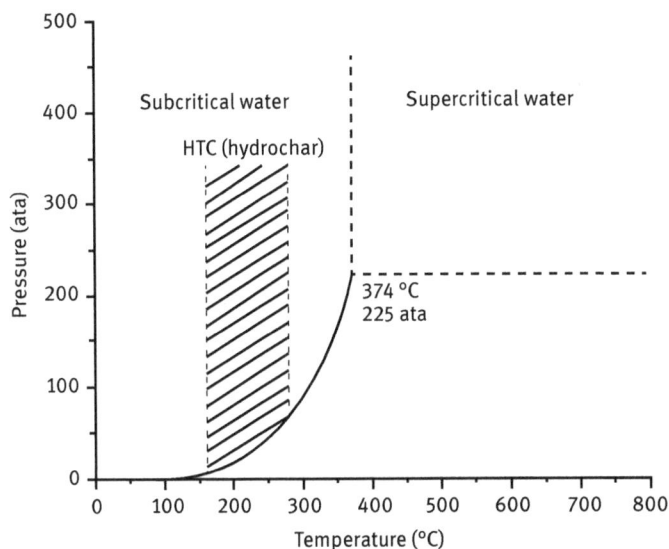

Fig. 6.1: Water phase diagram and the HTC insight field.

6.4 The reaction mechanism

The feedstock and the hydrochar resulted after HTC have different chemical composition and structure due to numerous reactions that occur during the process.

The complex reaction mechanism is not known in detail, and only a separate discussion of general reaction that has been identified can provide useful information on the possibilities of influencing the reaction.

The degradation of biomass in the HTC process is due to several chemical reactions that are extremely connected, concurrent and collaborative: hydrolysis, dehydration, decarboxylation, aromatization, and recondensation [28].

The first reaction in the process is hydrolysis. In this stage, the primary compounds are degraded to more simple molecules: hemicelluloses are converted to pentoses, cellulose to oligosaccharides or even to hexoses, and lignin to polyphenols with different molecular weight. The hydrolysis takes place during a high temperature and a short residence time and leads to breaking the chemical structure of biomass leading to the cleavage of mainly ester and ether bonds of the biomacromolecules by addition of water molecules. Cellulose is considered to be hydrolyzed under hydrothermal conditions above 200 °C, hemicelluloses around 180 °C, while lignin around 200 °C or more due to its high amount of ether bonds [29]. The dehydration reactions remove water from the biomass matrix, eliminating hydroxyl groups. Dehydration significantly carbonizes biomass by decreasing H/C and O/C ratios [29], and opens the way for biomolecules as furfurals, benzene, and organic acids that could be subjected to aromatization or decarboxylation reactions. Aromatization reaction appears due to dehydration and decarboxylation, which is characterized by replacing the double bonds such as C=O and C=C with simple bonds such as hydroxyl and carboxyl groups. After aromatization, liquid fragments of biomacromolecules' "biocrude" occur in the system, concurrently with CO_2, resulting in decarboxylation of the organic acids. All these reactions occurring so far are even more complicated by polymerization of fragments of biomolecules and by solid–solid reactions responsible for lignin conversion. The recondensation of degraded products during the carbonization process leads to hydrochar formation [30].

Three other possible reaction types have been reported to be competing reactions under hydrothermal conditions: pyrolytic reactions [31], Fischer–Tropsch-type reactions [32], and transformation reactions within the lignin molecule instead of a fragmentation with subsequent polymerization [33]. They might play a role in HTC, which has not been investigated in detail so far.

Finally, a heterogeneous mixture of solid, liquid, and gaseous phases is obtained.

6.5 The HTC process parameters

6.5.1 Temperature

Temperature is the key parameter in HTC because it controls all the reactions taking place in the process and improves the conversion breaking the intermolecular bonds [9]. The yield in solid product is higher in temperature range of 150–200 °C, while the yield in liquid phase is dominant in the temperature range of 250–350 °C. The HTC process can be conducted up to a maximum temperature of 350 °C. Generally, when the temperature rises more than 200–250 °C, the yield in solid product decreases while the yields of gaseous and liquid products increase [10]. In some cases, the yield variations are quite important as it was reported by Machado et al. [22]. The hydrochar yield decreases from 62.92% to 35.82%, while the yield of liquid and gaseous products increases from 35.43% to 54.59%, and from 1.49% to 9.58%, respectively. Considering the higher heating value (HHV), an increase with the increase of the temperature has been reported. The HHV of hydrochar obtained from conversion of food waste at 200, 250, and 300 °C increased from 20.81% to 28.98% and 31%, respectively [34]. The same influence was highlighted for the fixed carbon. Performing HTC from 180 °C to 250 °C, the fixed carbon has grown from 19.73% to 47.02% [35]. In terms of oxygen and hydrogen content, their concentration decreases with the increase of temperature. Converting grape pomace from 175 °C up to 300 °C, the content of oxygen decreases from 32.95% to 29.04%, and the hydrogen from 6.21% to 5.41% [36].

These behaviors could be explained by higher degradation of cellulose and lignin at higher temperatures, while at low temperatures hemicellulose degradation occurs [9].

6.5.2 Time

HTC of biomass is considered a slow reaction, and the residence time could be from several minutes to a few days [10]. However, a long residence time increases the reaction severity [29]. It is considered that the reaction time affects only hydrolysis reactions but up to a certain time beyond which it has no specific impact on the process. In general, higher amounts of solid products are obtained at higher reaction residence time [10].

Residence time has a similar influence on HTC process as temperature, but more mitigated. This means at a long residence time the yields in hydrochar are low, while at short residence time the yields are higher [37]. However, with the increase of time an improvement of the physical properties such as porosity, pore volume, and surface area of hydrochar has been observed [9].

The residence time has an insignificant influence on the HHV of hydrochar. An HTC on loblolly pine carried out at 240 °C during 2, 4, and 6 h revealed an increase of only 4% of HHV, more precisely from 26.52% to 30.58%. The same appearance in the case of the fixed carbon where a minimal increase (by 2%, from 50.43% to 52.18%) was reported [17].

6.5.3 Water-to-biomass ratio

In order to increase the efficiency of the hydrothermal reactions, it is important to consider the water amounts in order to ensure a complete dispersion of biomass. A good dispersion leads to occurrence of the hydrothermal reactions as well as the mass and heat transfer. The water-to-biomass ratio could be influenced by the density, structure and hydrophobicity of the feedstock. A high biomass density entails a large quantity of water to ensure the heat and mass transfer. Similarly, if the structure of the biomass is more porous, water can penetrate more easily to the pores and less water is required [9].

6.5.4 Pressure

The pressure has an important impact on the reaction network accordingly to the Le Chatelier principle. It is noteworthy that pressure is a parameter strongly determined by temperature. In a hydrothermal reaction, the pressure increases isotropically with the temperature [37]. However, the pressure has a small impact on the physical structure of hydrochar expressed by the increase of hydrogen and the reduction of oxygen content [9].

In some cases, at laboratory scale, the reactor is pressurized before starting the HTC procedure in order to ensure no vapor formation. The autogenous pressure generated during the process can maintain water in liquid phase, ensuring the absence of vapors [9].

6.5.5 Particle size

Smaller particles undergo better conversions resulting in lower hydrochar yields with higher calorific values. Reducing the particle size of the feedstock can help water penetrate more easily into the vegetal matrix, and can provide a larger surface area leading to improved heat and mass transfer and improving the reaction kinetics. At the end, a better conversion of the biomass is obtained with a lower yield in solid products. It is important to keep in mind that reducing particle size increases energy costs that may outweigh the benefits. Therefore, an optimal size must be chosen [9].

6.5.6 Catalyst

In the HTC process, the use of catalysts leads to a change of the products yield. In order to obtain a greater amount of coal an acid catalyst is advisable, while to increase the quantity of liquid products a basic catalyst is recommended to be used [9]. The type of acid used plays an important role; small amounts of Arrhenius acids, for example, generally catalyze dehydration [38].

Catalysts influence mostly the hydrolysis reaction that takes place in the process. Hydrolysis (side) products can, therefore, act as autocatalysts, but the effect is limited at higher temperatures due to the fact that the acidic dissociation constant decreases [29].

Another benefit of using the catalyst is the reduction of NO_x emissions, through catalytic chemical reactions that rapidly convert NO_x into nitrogen and water. Catalyst must be efficient, thermally stable, cost-effective, and to have a good selectivity. Some examples of catalysts used in the conversion process are K_2CO_3, $FeCl_2$ and $FeCl_3$, Na_2CO_3, NaOH, KOH, RbOH, CsOH, NH_4Cl, and C_2H_5OH [10]. The effect of CO_2 as acid catalyst (by ionization to carbonic acid) has been pointed out as well [39].

6.5.7 pH

Many studies on HTC reported a decrease of pH during process. It could be explained considering the variety of organic acids such as acetic, formic, lactic, and levulinic acid that have been identified as (intermediate) in the HTC of biomass. A neutral to weakly acidic environment appears to be necessary to achieve a simulation of natural coalification [40].

6.6 Industrial and environmental applications

In order to provide suitable applications, the **characterization of hydrochar** is a crucial step. Usual, the following aspects are considered: the proximate and coalification degree, surface functionality characterization, aromatic structure characteristics, and morphological and textural character.

The proximate and ultimate analysis of hydrochar is necessary to ensure the efficient utilization of hydrochar as a fuel. The proximate analysis includes volatiles, fixed carbon, and ash. The carbonization degree of the hydrochar could be revealed by the elemental composition special H/C and O/C ratios.

The different feedstock types and extremely different process conditions influence the structural characteristics of hydrochar. Fourier transform infrared (FTIR) techniques are employed in most studies to analyze the surface functional groups of raw biomass and hydrochar. The FTIR spectra can provide an investigation of hydrochar evolution, and many studies have used this technique to analyze the structural transformation [11].

Aromatic structure characteristics revealed by solid-state 13 °C magic angle spinning NMR has been employed and provides valuable complementary data to the surface functionality analysis.

Morphological and textural characters are generally studied by SEM, which provides micrographs showing the physical properties and surface morphology. Pore size distribution and specific surface area are generally used to analyze the textural changes of hydrochar.

However, to date, a systematic study on the textural characteristics and the application of hydrochar to soil are not fulfilled.

6.6.1 Cost-effective adsorbent

A global problem that threatens human health is the pollution with heavy metal ions, pesticides, drug residues, and other contaminants. In this view, efficient and cost-effective adsorbents are assiduously tested. The literature reports a lot of natural materials used as efficient adsorbent: peat moss [41, 42], green macroalgae [43, 44], brown seaweeds [45], invasive water plants [46], rapeseed waste [47], pine bark [48], and many others.

Among them, porous materials like hydrochar are of interest for the adsorption of organic or inorganic contaminants. A comparative analysis in terms of hydrochar and biochar adsorption capacity revealed different mechanisms for copper waste-water removal [49]. For hydrochar, the main mechanism was the exchange of ions (H^+ and/or metal ions) to and from the solution while for the biochar a physical adsorption on the surface is rather possible. A better adsorption was reported for hydrochar (4.46 mg/g), while the biochar has an adsorption capacity of 2.75 mg/g [49]. The higher adsorption rate could be explained due to the presence of oxygen-rich functional groups on the surface of hydrochars [4].

Generally, hydrochars have weak characteristics for sorption: its pore volume and low surface area provide fewer sorption sites for contaminants, and its negative surface charge is not suitable for negatively charged substances. However, some studies have shown that hydrochar has strong absorption capacities for both non-polar and polar organic contaminants due to its various surface functionalities [30]. The feedstock used for hydrochar and the operation conditions are the main factors influencing the adsorption capacity of the contaminants [50]. The improvement of the absorption capacity of the hydrochar can be achieved by activating it using different treatments. Table 6.3 summarized the results of several studies focused on the hydrochar adsorption of contaminants. The modified hydrochar has a better sorption capacity compared to the unmodified hydrochar [30].

The hydrochar resulted from different feedstock have a great ability to remove Pb and Cd ions, as well as organic compounds such as methylene blue or malachite green. These results represent a new opportunity to synthesize carbon-rich materials from biomass in order to generate cost-efficient adsorbent to remove hazardous pollutants from environment [37].

6.6.2 Carbon-rich activated materials

In order to improve the pore volume and the low surface of hydrochar, an activation treatment using chemical or physical procedures is recommended.

A **chemical activation** of hydrochar can be achieved in two ways. The first method consists of adding chemical agents to the reactor in which the conversion of biomass takes place, while the second method refers to washing the hydrochar after HTC with a chemical agent [30]. The most commonly used agents for chemical activation are phosphoric acid (H_3PO_4), zinc chloride ($ZnCl_2$), and potassium hydroxide (KOH) [50].

For example, using the phosphoric acid, the hydrochar obtained from rice husk shows a surface area of 2,610 m^2/g, while using KOH like chemical activating agent,

Tab. 6.3: Sorption rates of some contaminants using hydrochar from different feedstock.

Hydrochar from	Modification	Contaminant	Sorption rate (mg/g)	References
Woody biomass and residues from forestry and trees outside forests				
Oak wood	None	PO_4-P	26.6 ± 10.3	[51]
Oak wood	None	NH_4-N	109.7 ± 14.1	[51]
Hickory wood	None	Acetone	25.22	[50]
Hickory wood	KOH	Acetone	48.81	[50]
Hickory wood	H_3PO_4	Acetone	148.25	[50]
Hickory wood	None	Cyclohexane	15.52	[50]
Hickory wood	KOH	Cyclohexane	67.11	[50]
Hickory wood	H_3PO_4	Cyclohexane	158.99	[50]
Energy crops and residues from agricultural and marginal land				
Peanut hull	None	Acetone	17.95	[50]
Peanut hull	KOH	Acetone	58.75	[50]
Peanut hull	H_3PO_4	Acetone	110.47	[50]
Peanut hull	None	Cyclohexane	25.49	[50]
Peanut hull	KOH	Cyclohexane	99.77	[50]
Peanut hull	H_3PO_4	Cyclohexane	98.83	[50]
Peanut hull	H_2O_2	Pb	22.82	[52]
Peanut hull	None	Pb	0.88	[52]
Peanut hull	None	Methylene blue	38.55	[53]
Rice straw	Lanthanum	Phosphorus	61.57	[54]
Rice straw	None	Congo red	222.1	[55]
Rice straw	None	Berberine hydrochloride	174.0	[55]
Rice straw	None	2-Naphthol	48.7	[55]
Rice straw	None	Zn	112.8	[55]
Rice straw	None	Cu	114.9	[55]
Pinewood sawdust	None	Pb	2.20	[56]
Pinewood sawdust	H_2O_2	Pb	92.80	[56]

Tab. 6.3 (continued)

Hydrochar from	Modification	Contaminant	Sorption rate (mg/g)	References
Wheat straw	KOH	Cd	38.75	[57]
Sawdust	KOH	Cd	40.78	[57]
Corn stalk	KOH	Cd	30.40	[57]
Corn straw	None	Anthrazine	1.18	[58]
Corn straw	None	Cr	4.07	[58]
Corn straw	None	Cd	1.04	[58]
Corn stover	None	Pb	32.67	[59]
Corn stover	H_3PO_4	Pb	353.4	[59]
Corn cobs	None	Cr	7.239	[60]
Corn cobs	None	Ni	2.277	[60]
Acerola	None	Methylene blue	108.88	[61]
Organic waste (tertiary residues)				
Swine solid	None	Cd	27.18	[62]
Swine solid	None	Sb	3.98	[62]
Poultry litter	None	Cd	19.8	[62]
Poultry litter	None	Sb	2.24	[62]
Municipal waste	None	PO_4-P	5.1 ± 3.8	[51]
Municipal waste	None	NH_4-N	146.4 ± 5.8	[51]

the Brunauer-Emmett-Teller (BET) surface areas increase at 3,362 m^2/g [63]. The hydrochar activated with phosphoric acid has microporous characteristics, while the activated hydrochar with potassium hydroxide has several mesoporous characteristics. These structural differences are mainly due to different activation mechanisms [64]. The hydrochar's porosity obtained from glucose and sucrose was investigated by Jain et al. [3], considering the effect of various chemical activating agents. Using KOH, the BET surface area was around 3,100 m^2/g with a KOH to precursor ratio of 4:1 and 5:1. In addition, different total surface oxygen content was reported accordingly to the activation agent: NaOH > KOH > H_3PO_4 > CO_2.

The chemically activated hydrochar showed better adsorption properties compared to a nonactivated hydrochar. Adsorption tests performed for heavy metals and organic pollutants (pharmaceuticals, dyes, pesticides, etc.) removal argue the increase of adsorption capacity [30].

In the case of **physical activation**, the feedstock is decomposed in an inert atmosphere and activated with a light oxidizing gas such as carbon dioxide or steam. Although the energy consumption in this case is higher than for chemical activation, the physically activated hydrochar has a larger surface area and a higher volume of pores. Thus, the activated hydrocarbons are characterized by a better ability to remove contaminants from aqueous solutions [30].

The hydrochar obtained from rice husk and pine wood physically activated with CO_2 leads to the increase of BET surface area, which is 446 and 569 m^2/g for rice husk and pinewood-derived carbons, respectively [63].

6.6.3 Soil amendment

Most types of hydrochars have a low nutrient content; therefore, it is preferable to use them in combination with a fertilizer in order to reduce the amount of fertilizer and improve its effects. The pores on the surface of the hydrocarbon absorb the nutrients and then slowly remove them into the soil [30].

Hydrochar is a hydrophobic material and contains a few polar functional groups on the surface. However, when it is introduced into the soil, hydrochar comes into contact with the oxygen and water present in the soil and forms carboxylic and phenolic bonds. As a result, hydrochar becomes more hydrophilic in time [4] and could improve water retention capacity, soil properties, cation and anion exchange capacity, extractable nutrient, water-stable aggregation, pH, and so on.

A synergistic effect with fertilizers on the growth of plants was also encountered. For phaseolus beans and barley crops, an increase of the mass yield was reported, while for leek and sugar beet a decrease of the mass yield has been noticed [30]. However, the detrimental effect on crops was mitigated when the soil was incubated with a hydrochar suspension 3 months before sowing [4]. By adding too much hydrochar into the soil, an inhibition of *Taraxacum* plant growth was observed [3].

The properties of hydrochars including the stability against microbial degradation depend significantly on HTC process conditions (residence time and temperature) and soil properties. Hydrochar require pretreatments before being used as a soil amendment. The modified hydrochar showed better immobilization properties of heavy metals from contaminated soils [65].

In general, the main benefits of using hydrochars as soil amendment are nutrient release, mineralization, carbon sequestration, and as a consequence a better germination and growth of different crops. The hydrocarbon can improve soil fertility by absorbing nutrients from its surface and preventing their loss by sanding. The effect of several hydrochars on soil nitrogen dynamics and plant growth was studied [9]. One interesting result was the higher C:N ratio of hydrochar decreased the N content of the plant. Therefore, one of the key characteristics of a suitable hydrochars is low N content. Due to the release of nutrients, the presence of micronutrients such as Ca, Mg, Zn, Cu, Na, and Cl can help crop growth. However, a maximum amount of these should be taken into account. For example, a high concentration of Na and Cl will increase the electrical conductivity of the soil that can harm the growth of sensitive crops [9].

6.6.4 Carbon and greenhouse gases sequestration

The conversion of biomass into carbon-rich materials through the HTC process has the maximum efficiency for collecting and binding carbon dioxide from the air. Therefore, the use of the hydrothermal process for the conversion of biomass into coal-like materials could be the most effective tool for sequestration of carbon dioxide [37]. Large amounts of noxious gases are not generated during the HTC process, and hydrochar particles are not prone to autoignition due to the high concentration of surface oxygen groups.

However, there are few studies that have investigated the potential of hydrochar for carbon sequestration. Due to the low stability of the hydrochar in soil, some authors are not in favor of the hydrochars [4]. It was reported that hydrochar obtained from corn silage is decomposed rapidly and increased carbon dioxide and methane emissions from the atmosphere. This is most likely due to increase of microbial activity caused by the mild decomposition of carbon from the hydrochars [66]. However, a decrease in nitrogen dioxide emissions was observed for hydrochars obtained from beet chip and bark [30].

In conclusion, detailed research is needed to optimize the application of hydrochars in soil for carbon sequestration with low greenhouse gas emissions and better stability [67].

6.6.5 Solid fuel

Due to the low energy density, ash, and moisture content present in the feedstock, the biomass is considered to be a low-quality fuel. In order to reduce storage and transportation costs, biomass is typically processed into pellets. The conversion of biomass through HTC process offers the possibility of obtaining hydrochars with improved properties, such as: mass density and higher combustion [30].

Hydrochar pellets had a high mechanical strength compared to the pellet from crude biomass, being 1.8 times higher. Hydrochar pellets had improved combustion characteristics: high values of combustion temperature, decreases maximum weight loss rates, and reduced residues. High mechanical durability and significantly improved combustion characteristics suggest that hydrochar pellets are more suitable

as solid fuels compared to raw biomass pellets. Therefore, HTC combined with pallet-
ization offers an alternative for the production of biofuels from biomass [68].

6.6.6 Medical applications

The pharmaceutical or medical application of hydrochar is rarely investigated. An example is the fluorescent quantum dots produced from silk hydrochar that has a good
hemocompatibility, low toxicity, and a high cellular uptake being used in medical imaging for detecting diseases at the cellular level. These dots are safer and an effective
material to be used in living organisms due to a lower amount of heavy metals and a
better location than quantum dots synthesized from other materials. Furthermore, it
also displayed a strong blue fluorescence under certain wavelengths, making it a good
substitute for dyes that are traditionally used in medical imaging [30].

Other interesting point of view is the use of HTC as a tool to increase the antioxidant activity and level of nutraceutical in natural extracts. The hydrothermal treatment of different parts of watermelons (flesh, white rind, and green rind) releases
some phenolic compounds (catechol and derivatives) with important antioxidant
effects [30].

References

[1]   Nanda S, Azargohar R, Dalai AK, Kozinski JA. An assessment on the sustainability of
      lignocellulosic biomass for biorefining. Renew. Sustain. Energy Rev. 2015, 50, 925–941.
[2]   Demirbaş A. Estimating of structural composition of wood and non-wood biomass samples.
      Energy Sources. 2005, 27, 761–767.
[3]   Zhang Z, Zhu Z, Shen B, Liu L. Insights into biochar and hydrochar production and
      applications: A review. Energy. 2019, 171, 581–598.
[4]   Kambo HS, Dutta A. Comparative evaluation of torrefaction and hydrothermal carbonization
      of lignocellulosic biomass for the production of solid biofuel. Energy Convers. Manage. 2015,
      105, 746–755.
[5]   Zaman CZ, Pal K, Yehye W, Sagadevan S, Johan RB. Pyrolysis: A sustainable way to generate
      energy from waste. Pyrolyses Pyrolysis Samer M., Ed, IntertechOpen, 2017 DOI: 10.5772/
      intechopen.69036
[6]   Yip K, Tian F, Hayashi J, Wu H. Effect of alkali and alkaline earth metallic species on biochar
      reactivity and syngas compositions during steam gasification. Energy Fuels. 2009, 24, 173–81.
[7]   Saxena R, Adhikari D, Goyal H. Biomass-based energy fuel through biochemical routes: A
      review. Renew. Sustain. Energy Rev 2009, 13, 167–178.
[8]   www.biomassfutures.eu.
[9]   Heidari M, Dutta A, Acharya B, Mahmud S. A review of the current knowledge and challenges
      of hydrothermal carbonization for biomass conversion. J. Energy Inst 2019 doi:https://doi.
      org/10.1016/j.joei.2018.12.003.

[10] Nizamuddina S, Baloch HA, Griffin GJ, Mubarak NM, Bhutto AW, Abro R, Mazari SA, Ali BS. An overview of effect of process parameters on hydrothermal carbonization of biomass. Renew. Sustain. Energy Rev. 2017, 73, 1289–1299.

[11] Wang T, Zhai Y, Zhu Y, Li C, Zeng G. A review of the hydrothermal carbonization of biomass waste for hydrochar formation: Process conditions, fundamentals, and physicochemical properties. Renew. Sustain. Energy Rev. 2018, 90, 223–247.

[12] Diakité M, Paul A, Jäger C, Pielert J, Mumme J. Chemical and morphological changes in hydrochars derived from microcrystalline cellulose and investigated by chromatographic, spectroscopic and adsorption techniques. Bioresour. Technol. 2013, 150, 98–105.

[13] Cai J, Li B, Chen C, Wang J, Zhao M, Zhang K. Hydrothermal carbonization of tobacco stalk for fuel application. Bioresour. Technol. 2016, 220, 305–311.

[14] Kumar M, Oyedun AO, Kumar A. A review on the current status of various hydrothermal technologies on biomass feedstock. Renew. Sustain. Energy Rev. 2018, 81, 1742–1770.

[15] Gao P, Zhou Y, Meng F, Zhang Y, Liu Z, Zhang W, Xue G. Preparation and characterization of hydrochar from waste eucalyptus bark by hydrothermal carbonization. Energy. 2016, 97, 238–245.

[16] Simsir H, Eltugral N, Karagoz S. Hydrothermal carbonization for the preparation of hydrochars from glucose, cellulose, chitin, chitosan and wood chips via low-temperature and their characterization. Bioresour. Technol. 2017, 246, 82–87.

[17] Wu Q, Yu S, Hao N, Wells T Jr, Meng X, Li M, Pu Y, Liu S, Ragauskas A. Characterization of products from hydrothermal carbonization of pine. Bioresour. Technol. 2017, 244, 78–83.

[18] Yan W, Hoekman K, Broch A, Coronella CJ. Effect of hydrothermal carbonization reaction parameters on the properties of hydrochar and pellets. Environ. Prog. Sustain. Energy. 2014, 33(3). doi:10.1002/ep

[19] Atta-Obenga E, Dawson-Andoh B, Seehra M, Geddam U, Poston J, Leisen J. Physico-chemical characterization of carbons produced from technical lignin by sub-critical hydrothermal carbonization. Biomass Bioenergy. 2017, 107, 172–181.

[20] Liu F, Yu R, Ji X, Guo M. Hydrothermal carbonization of holocellulose into hydrochar: Structural, chemical characteristics, and combustion behavior. Bioresour. Technol. 2018, 263, 508–516.

[21] Alvarez-Murillo A, Sabio E, Ledesma B, Roman S, Gonzalez-García CM. Generation of biofuel from hydrothermal carbonization of cellulose. Kinetics modelling. Energy. 2016, 94, 600–608.

[22] Machado NT, de Castro DAR, Santos MC, Araújoc ME, Lüderf U, Herklotza L, Wernera M, Mummee J, Hoffmanna T. Process analysis of hydrothermal carbonization of corn stover with subcritical H2O. J. Supercrit. Fluids. 2018, 136, 110–122.

[23] Román S, Nabais JMV, Laginhas C, Ledesma B, González JF. Hydrothermal carbonization as an effective way of densifying the energy content of biomass. Fuel Process. Technol. 2012, 103, 78–83.

[24] Chen X, Lin Q, He R, Zhao X, Li G. Hydrochar production from watermelon peel by hydrothermal Carbonization. Bioresour. Technol. 2017, 241, 236–243.

[25] Nakason K, Panyapinyopol B, Kanokkantapong V, Viriya-Empikul N, Kraithong W, Pavasant P. Characteristics of hydrochar and liquid fraction from hydrothermal carbonization of cassava rhizome. J. Energy Inst. 2018, 91, 184–193.

[26] Lee J, Lee K, Sohn D, Kim YM, Park KY. Hydrothermal carbonization of lipid extracted algae for hydrochar production and feasibility of using hydrochar as a solid fuel. Energy. 2018;153:913–920.

[27] Gascó G, Paz-Ferreiro J, Álvarez ML, Saa A, Méndez A. Biochars and hydrochars prepared by pyrolysis and hydrothermal carbonisation of pig manure. Waste Manage. 2018, 79, 395–403.

[28] Kabakci SB, Baran SS. Hydrothermal carbonization of various lignocellulosics: Fuel characteristics of hydrochars and surface characteristics of activated hydrochars. Waste Manage. 2019, 100, 259–268.

[29] Funke A, Ziegler F. Hydrothermal carbonization of biomass: A summary and discussion of chemical mechanisms for process engineering. Biofuels, Bioprod. Biorefin. 2010, 4, 160–177.

[30] Fang J, Zhan L, Ok YS, Gao B. Minireview of potential applications of hydrochar derived from hydrothermal carbonization of biomass. J. Ind. Eng. Chem. 2018, 57, 15–21.

[31] Chornet E, Overend RP. Biomass liquefaction: An overview, fundamentals of thermochemical biomass conversion. London New York, 1985, 967–1002, Springer, Dordrecht.

[32] McCollom TM, Ritter G, Simoneit BRT. Lipid synthesis under hydrothermal conditions by Fischer-Tropsch-type reactions. Origins Life Evol. Biosphere. 1999, 29, 153–166.

[33] Buchanan AC, Britt PF, Struss JA. Investigation of reaction pathways involved in lignin maturation. Energy Fuels. 1997, 11, 247–248.

[34] Saqib NU, Baroutian S, Sarmah AK. Physicochemical, structural and combustion characterization of food waste hydrochar obtained by hydrothermal carbonization. Bioresour. Technol. 2018, 266, 357–363.

[35] Sermyagina E, Saari J, Kaikko J, Vakkilainen E. Hydrothermal carbonization of coniferous biomass: Effect of process parameters on mass and energy yields. J. Anal. Appl. Pyrolysis. 2015, 113, 551–556.

[36] Pala M, Kantarli IC, Buyukisik HB, Yanik J. Hydrothermal carbonization and torrefaction of grape pomace: A comparative evaluation. Bioresour. Technol. 2014, 161, 255–262.

[37] Khan TA, Saud AS, Jamari SS, Rahim MHA, Park IW, Kim HJ. Hydrothermal carbonization of lignocellulosic biomass for carbon rich material preparation: A review. Biomass Bioenergy. 2019, 130, 105384.

[38] Peterson AA, Vogel F, Lachance RP, Fröling M and Antal MJ. Thermochemical biofuel production in hydrothermal media: A review of sub and supercritical water technologies. Energy Environ. Sci. 2008, 1, 32–65.

[39] Watanabe M, Sato T, Inomata H, Smith RL, Arai K, Kruse A, et al. Chemical reactions of C1 compounds in near-critical and supercritical water. Chem. Rev. 2004, 104, 5803–5821.

[40] van Krevelen DW. *Coal: Typology – Physics – Chemistry – Constitution (3rd Edition)*. Elsevier, Amsterdam 837–846 (1993).

[41] Balan C, Volf I, Bulai P, Bilba D, Macoveanu M. Removal of Cr (VI) from aqueous environment using peat moss: Equilibrium study. Environ. Eng. Manage. J. 2012, 11 (1), 21–28.

[42] Ungureanu G, Balan CD, Volf I. Application of Sphagnum moss peat in ecological remediation of oxyanions contaminated aqueous solutions. Environ. Eng. Manage. J. 2018, 17(4), 915–925.

[43] Ungureanu G, Filote C, Santos S, Boaventura R, Volf I, Botelho C. Antimony oxyanion uptake by green marine macroalgae. J. Environ. Chem. Eng. 2016, 4 (3), 3441–3450.

[44] Filote C, Volf I, Santos SCR, Botelho CMS. Bioadsorptive removal of Pb(II) from aqueous solution by the biorefinery waste of *Fucus spiralis*. Sci. Total Environ. 2019, 648, 1201–1209.

[45] Ungureanu G, Santos SCR, Volf I, Boaventura RAR, Botelho CMS.Biosorption of antimony oxyanions by brown seaweeds. Batch and column studies. J. Environ. Chem. Eng. 2017, 5(4), 3463–3471.

[46] Volf I, Rakoto N, Bulgariu L. Valorisation of Pistia stratiotes biomass as biosorbent for lead(II) ions removal from aqueous media. Sep. Sci. Technol. 2015, 50, 1577–1586.

[47] Tofan L, Paduraru C, Volf I, Toma O. Waste of rapeseed from biodiesel production as a potential biosorbent for heavy metal ions. BioResources. 2011, 6(4), 3727–3741.

[48] Mihailescu Amalinei RL, Miron A, Volf I, Paduraru C, Tofan L. Investigations on the feasibility of Romanian pine bark wastes conversion into a value-added sorbent for Cu(II) and Zn(II) ions. BioResources. 2012, 7 (1), 148–160.

[49] Liu Z, Zhang FS, Wu J. Characterization and application of chars produced from pinewood pyrolysis and hydrothermal treatment. Fuel. 2010, 89, 510–514.

[50] Zhang X, Gao B, Fang J, Zou W, Dong L, Cao C, Zhang J, Li Y, Wang H. Chemically activated hydrochar as an effective adsorbent for volatile organic compounds (VOCs). Chemosphere. 2019, 218, 680–686.

[51] Takaya CA, Fletcher LA, Singh S, Anyikude KU, Ross AB. Phosphate and ammonium sorption capacity of biochar and hydrochar from different wastes. Chemosphere. 2016, 145, 518–527.

[52] Xue Y, Gao B, Yao Y, Inyang M, Zhang M, Zimmerman AR, Ro KS. Hydrogen peroxide modification enhances the ability of biochar (hydrochar) produced from hydrothermal carbonization of peanut hull to remove aqueous heavy metals: Batch and column tests. Chem. Eng. J. 2012, 200–202, 673–680.

[53] Fang J, Gao B, Chen J, Zimmerman AR. Hydrochars derived from plant biomass under various conditions: Characterization and potential applications and impacts. Chem. Eng. J. 2015, 267, 253–259.

[54] Dai L, Wu B, Tan F, He M, Wang W, Qin H, Tang X, Zhu Q, Pan K, Hu Q. Engineered hydrochar composites for phosphorus removal/recovery: Lanthanum doped hydrochar prepared by hydrothermal carbonization of lanthanum pretreated rice straw. Bioresour. Technol. 2014, 161, 327–332.

[55] Li Y, Tsend N, Li T, Liu H, Yang R, Gai X, Wang H, Shan S. Microwave assisted hydrothermal preparation of rice straw hydrochars for adsorption of organics and heavy metals. Bioresour. Technol. 2019, 273, 136–143.

[56] Xia Y, Yang T, Zhu N, Li D, Chen Z, Lang Q, Liu Z, Jiao W. Enhanced adsorption of Pb(II) onto modified hydrochar: Modeling and mechanism analysis. Bioresour. Technol. 2019, 288, 121593.

[57] Sun K, Tang J, Gong Y, Zhang H. Characterization of potassium hydroxide (KOH) modified hydrochars from different feedstocks for enhanced removal of heavy metals from water. Environ. Sci. Pollut. Res. 2015, 22, 16640–16651.

[58] Liu Y, Ma S, Chen J. A novel pyro-hydrochar via sequential carbonization of biomass waste: Preparation, characterization and adsorption capacity. J. Clean Prod. 2018, 176, 187–195.

[59] Jiang Q, Xie W, Han S, Wang Y, Zhang Y. Enhanced adsorption of Pb(II) onto modified hydrochar by polyethyleneimine or H3PO4: An analysis of surface property and interface mechanism. Colloids Surf. A. 2019, 583, 123962.

[60] Shi Y, Zhang T, Ren H, Kruse A, Cui R. Polyethylene imine modified hydrochar adsorption for chromium (VI) and nickel (II) removal from aqueous solution. Bioresour. Technol. 2019, 247, 370–379.

[61] Nogueira GDR, Duarte CR, Barrozo MAS. Hydrothermal carbonization of acerola (Malphigia emarginata D.C.) wastes and its application as an adsorbent. Waste Manage. 2019, 95, 466–475.

[62] Han L, Sun H, Ro KS, Sun K, Libra JA, Xing B. Removal of antimony (III) and cadmium (II) from aqueous solution using animal manure-derived hydrochars and pyrochars. Bioresour. Technol. 2017, 234, 77–85.

[63] Jain A, Balasubramanian R, Srinivasan MP. Hydrothermal conversion of biomass waste to activated carbon with high porosity: A review. Chem. Eng. J. 2016, 283, 789–805.

[64] Yorgun S, Yildiz D. Preparation and characterization of activated carbons from Paulownia wood by chemical activation with H3PO4. J. Taiwan Inst. Chem. Eng. 2015, 53, 122–131.

[65] Xia Y, Liu H, Guo Y, Liu Z, Jiao W. Immobilization of heavy metals in contaminated soils by modified hydrochar: Efficiency, risk assessment and potential mechanisms. Sci. Total Environ. 2019, 685, 1201–1208.

[66] Malghani S, Gleixner G, Trumbore SE. Chars produced by slow pyrolysis and hydrothermal carbonization vary in carbon sequestration potential and greenhouse gases emissions. Soil Biol. Biochem. 2013, 62, 137–146.

[67] Saqib NU, Sharma HB, Baroutioan S, Dubey B, Sarmah AK. Valorisation of food waste via hydrothermal carbonisation and techno-economic feasibility assessment. Sci. Total Environ. 2019, 690, 261–276.

[68] Liu Z, Quek A, Balasubramanian R. Preparation and characterization of fuel pellets from woody biomass, agro-residues and their corresponding hydrochars. Appl. Energy. 2014, 113, 1315–1322.

Yue Shen, Bo Wang, and Runcang Sun

Chapter 7
Catalytic conversion of hydroxymethylfurfural and levulinic acid to biomass-based chemicals

7.1 Introduction

The rapid depletion of fossil fuels has led to an international effort to increase the use of renewable energy. The replacement for fossil fuels in this aspect comes from biomass, with biorefineries being presented as the future substitutes for petroleum refineries. In the same way, petroleum refineries use certain chemicals as the building blocks for more complex molecules such as polymers, and biorefinery will use simple molecules that can be readily obtained from a vast number of feedstock as a base for the synthesis of biopolymers and other large molecules [1, 2]. Among the most promising building blocks are hydroxymethylfurfural (HMF) and levulinic acid (LA), which are the subject of this chapter.

HMF is considered to be one of the few petroleum-derived chemicals. It is held to be a bridge between carbohydrate chemistry and industrial mineral oil-based organic chemistry [3, 4]. Meanwhile, HMF has some potential markets on its own, such as in fuel cells and the treatment of sickle cell disease [5, 6]. A variety of chemicals can also be produced from HMF, and some of the most important ones are listed in Tab. 7.1. HMF is a key platform chemical, and this depends on its availability and cost [7–10]. Some researchers have explored to decrease the cost of HMF in order to be used as a substitute for certain target chemicals. Torres et al. estimated the cost of HMF production using a semibatch biphasic reactor and compared it with the price of p-xylene, for which HMF can act as an alternative in the production of polyethylene terephthalate [11]. A minimum HMF cost of 1967–2165 $/ton for a fructose cost of 550 $/ton was obtained, depending on the solvent used at the extraction stage. However, the cost was higher than that of p-xylene. Since the cost of fructose is the main factor in the HMF price, it can be concluded that lower fructose costs are necessary alongside the development of more efficient processes for the HMF price to be competitive. Lately, Liu et al. stated that HMF produced at

Yue Shen, Institute of Environment and Sustainable Development in Agriculture, Chinese Academy of Agriculture Sciences, Beijing, China
Bo Wang, Key Lab Lignocellulosic Chemistry, Beijing Forestry University, Beijing, China
Runcang Sun, Centre for Lignocellulose Science and Engineering, and Liaoning Key Laboratory Pulp and Paper Engineering, Dalian Polytechnic University, Dalian, China

https://doi.org/10.1515/9783110658842-007

Tab. 7.1: Chemicals produced from HMF or LA and their potential application.

Substrate	Chemical	Potential market	Ref.
HMF	Formic acid (FA)	Chemical, textiles, road salt, catalysts, fuel cells	[24–29]
	Ethoxymethylfurfural (EMF)	Biofuels	
	5-Hydroxymethylfuroic acid	Polymers	
	2,5-Furandicarboxylic acid (FDCA)	Polymers, pharmaceuticals	
	Dimethylfuran (2,5-DMF)	Biofuels	
	2-Methylfuran	Biofuels	
	2,5-Diformylfuran (DFF)	Pharmaceuticals, fungicide	
	2,5-Di(hydroxymethyl) tetrahydrofuran (DHM-THF)	Solvents	
	2,5-Furandicarboxyaldehyde (FDC)	Polymers, resins	
LA	Diphenolic acid	Epoxy resins, lubricants, adhesives, paints, polymers	[16–19, 27]
	Succinic acid	Polymers, solvents, pesticides	
	δ-Aminolevulinic acid (DALA)	Herbicides, insecticides, cancer treatment	
	Methyltetrahydrofuran	Fuel additive, solvents	
	Ethyl levulinate	Fuel additive, food flavoring	
	γ-Valerolactone (GVL)	Solvents, fuel additive, biofuels, polymers	
	Different esters of LA	Plasticizers, solvents	
	Sodium/calcium levulinate	Antifreeze	
	1,4-Butanediol	Polymers, solvents, fine chemicals	
	Valeric (pentanoic) acid	Fuel additive	
	5-Nonanone	Paints, resins	

1,210 \$/ton would be cost competitive with the petroleum-derived *para*-xylene–terephthalic acid selling at 1,440 \$/ton, and this HMF price was achievable for a fructose price of 460 \$/ton [12].

LA was listed among the top 12 most promising value-added chemicals from biomass by the Biomass Program of the US Department of Energy in 2004 [13]. It is a promising chemical intermediate that is derived from HMF, and continues to rank highly in more recent reviews of the most important biorefinery target products. The recognition of the potential of LA is not a recent phenomenon. It was first identified in the 1870s, and in 1956 a report outlining numerous derivatives from LA in detail and their potential applications were published [14]. Although like this, the commercial applications of LA have been slow to develop. Among the reasons cited for these are the expensive raw materials, low yields, high equipment cost, problematic handling, and recovery [15]. However, it is currently used in several industries, including lubricants, adsorbents, electronics, personal care products, photography, batteries, and drug delivery systems. The production of LA should greatly enhance its use as a chemical intermediate, given these chemicals having numerous applications that can be produced from it. Table 7.1 also summarizes some important chemicals that can be synthesized from LA and the markets they can be used [16–19]. The price of LA was about 8, 800–13,200 $/ton in 2,000, which meant that it was mostly used as a special chemical. The small market size of LA at this time, around 450 ton annually, meant that it was produced largely from maleic anhydride and other petrochemicals [20, 21]. The price of LA in 2010 was around 3,200 $/ton, which was substantially lower than the earlier figures, but still too high for chemicals produced from LA to compete with those derived directly from petroleum [22]. Although LA prices remaining in the range of 5,000–8,000 $/ton in 2013, global LA consumption rises to around 2,600 tons, and is expected to increase steadily in the future, reaching 3,800 tons in 2020 [23]. The biorefinery process has been projected to lower LA cost to as low as 90–220 $/ton, which would make a whole range of LA-derived chemicals economically attractive.

All in all, while both HMF and LA have tremendous market potential, the realization of this potential depends on the production of large quantities of these chemicals sustainably and at low prices. Thus, it is noticeable that these problems have attracted the interest of many researchers. The large amount of research conducted on the synthesis of these two chemicals and the numerous reaction systems devised make it necessary to analyze the advantages and disadvantages of the different processes, particularly with respect to the feedstock used, the yields obtained, and the accompanying environmental influence. The requirements for taking the laboratory schemes toward industrial production also need to be identified and addressed. This chapter, therefore, provides an overview of the production of HMF and LA from different carbohydrates and lignocelluloses using various solvent and catalyst systems, then surveys their principal derivatives and discusses the relative merits of each molecule in the future. Special attention has been paid to the reaction mechanism for each process to gain insight into the activation of C–O and C–C bonds in the presence of hydrogen or oxygen.

7.2 Conversion of lignocelluloses to HMF

7.2.1 Possible pathways for the formation of HMF

The direct conversion of cellulose into HMF involves three key steps, including hydrolysis of cellulose to glucose, isomerization of glucose to fructose, and dehydration of fructose into HMF (Fig. 7.1). Generally, Brønsted acids are efficient catalysts for the hydrolysis of cellulose. The hydrolysis reaction is believed to proceed via protonation of the glycosidic oxygen in cellulose, followed by the cleavage of glycosidic bond [30]. Since the existence of huge hydrogen bonds in cellulose makes cellulose insoluble in most solvents, it is difficult for the active acid sites to access the glycosidic bond. Thus, harsh reaction conditions (e.g., supercritical or subcritical, high concentration of acids) or pretreatments of cellulose are generally required for the cleavage of glycosidic bonds in aqueous system. Besides, the use of ionic liquid (IL), which could dissolve cellulose at atmospheric conditions, would be helpful for the hydrolysis reaction under mild conditions. Such a homogeneous system makes the glycosidic bond more available to catalytically active sites. Moreover, IL was also found to have positive effects on the activation of glucose during its further conversion [31]. Therefore, high HMF yield could be obtained from one-pot conversion of cellulose when IL was used as the solvent in the presence of an acid catalyst. Mutarotation and isomerization are two important steps for the conversion of glucose to HMF in IL with a Lewis acid catalyst (e.g., $CrCl_2$). First, α-glucopyranose was reversely transformed to β-glucopyranose in the presence of a catalytic amount of [EMIM]MCl$_x$ complex. After quickly approaching the equilibrium of α- and β-monomers, the β-glucopyranose was subsequently isomerized to fructose through an open enol glucose-[EMIM]MCl$_x$ intermediate. Then, the resulting fructose was converted into HMF via dehydration reaction. Pidko et al. also found that the facile reactions of glucose ring opening and closure process involve coordination with a single Cr center in the presence of $CrCl_2$ [32]. Meanwhile, they observed that the transient self-organization of mononuclear Cr

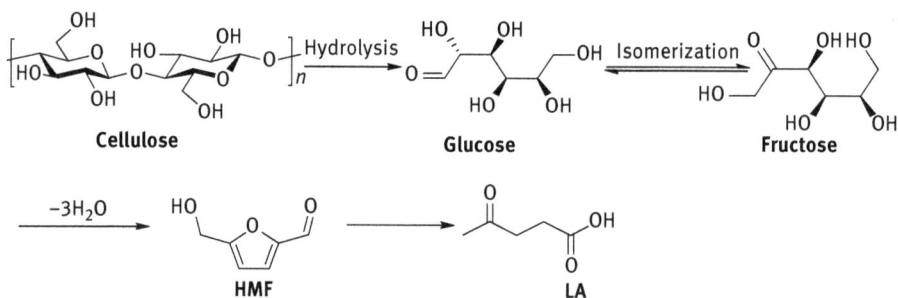

Fig. 7.1: Transformations of cellulose into value-added chemicals.

species into a binuclear complex with the open-form glucose could facilitate the H-shift reaction, which determines the rate of isomerization.

During the dehydration of fructose to HMF, there are two possible pathways. One is the cyclic route via the fructofuranosyl intermediate, and the other is an acyclic route via the enol intermediate (Fig. 7.2) [33]. Antal et al. explored the conversion of fructose and sucrose at different temperatures and pressures by monitoring the concentrations of D-fructose and the key products in the presence of H_2SO_4 [34]. They observed that 2,5-anhydromannose (a possible cyclic intermediate for HMF) can be readily converted to HMF. Additionally, when the reaction was carried out in D_2O solvent, they did not observe the formation of carbon-deuterium in HMF, which was expected to be formed by keto-enol tautomerism in the acyclic method. Based on the above results, they proposed a cyclic mechanism. Furthermore, by using ^1H and ^{13}C NMR technique, Amarasekara et al. identified a key cyclic intermediate (4R, 5R)-4-hydroxy-5-hydroxymethyl-4,5-dihydrofuran-2-carbaldehyde in DMSO-d_6 solvent, providing further evidence to the dehydration via cyclic route [35].

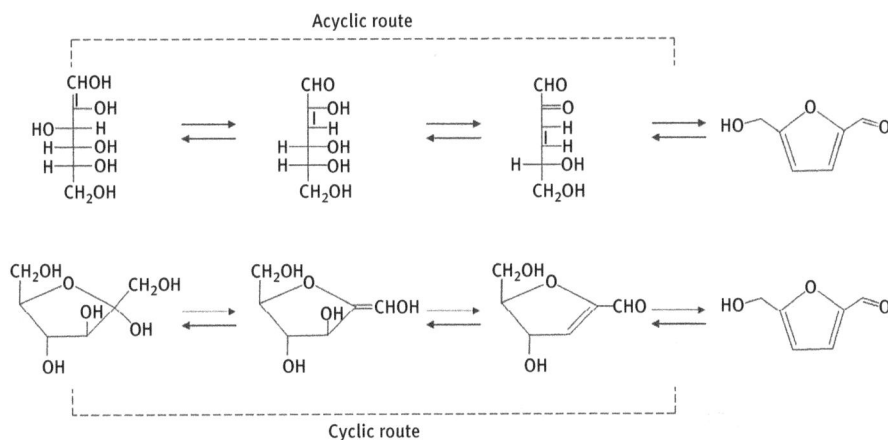

Fig. 7.2: Proposed mechanism for the dehydration of fructose to HMF.

7.2.2 Feedstocks

7.2.2.1 Monosaccharides
Since HMF is basically a molecule formed by the dehydration of a hexose molecule, and is itself used for LA production, the hexose has been the starting point for HMF and LA synthesis for many researchers. Among them, fructose is easier to convert

into HMF, but the wider availability of glucose means that it might be a better candidate as HMF feedstock [36]. One method of utilizing glucose as feedstock is to use a catalyst that can isomerize glucose to fructose. Solid catalysts (e.g., ZrO_2 and TiO_2) that can act as Lewis bases in addition to Lewis acids can be used for this purpose, with the basic sites on the catalyst isomerizing the glucose, and the acidic sites converting the resultant fructose to HMF and LA [37]. An integration of an immobilized glucose isomerase that converts glucose to fructose with an acid catalyst that dehydrates the fructose to HMF is another method that has been proposed [38].

7.2.2.2 Polysaccharides

In recent years, the direct conversion of polysaccharides to HMF and LA has been the subject of increasing research interest. A variety of polysaccharides have been identified as attractive candidates based on different parameters, such as starch being one of the cheapest and most abundant carbohydrates, chitin being the second-most abundant biopolymer on the earth, inulin being a carbohydrate that cannot be digested by humans, and cellulose being the major form of photosynthetically fixed carbon [39–42]. However, polysaccharides should be depolymerized via hydrolysis prior to dehydration of the monomer units to HMF. This results in the possibility of side reactions that would render HMF production from polysaccharides, a complex process [43].

7.2.2.3 Lignocelluloses

Sustainability concerns and commercial-scale applicability require that efforts should focus on the production of HMF from untreated lignocelluloses rather than edible crop-derived carbohydrates [44]. Lignocelluloses are a major type of biomass consisting primarily of cellulose, hemicelluloses, and lignin, with some amount of organic substances and inorganic ashes also present [45]. The cellulose and hemicellulose portions of the lignocelluloses are of primary importance since these are composed of hexoses such as glucose, mannose, and galactose [43]. Typically, the differences in the chemical and physical properties of cellulose and hemicelluloses necessitate separation of these two components before processing, and integrated conversion processes have also been reported [44, 46].

In conclusion, although monosaccharides are the easiest to convert to HMF and LA and are hence the preferred substrates tried by researchers testing new catalysts, solvents, or reaction schemes, polysaccharides and lignocelluloses are the feedstock that must eventually be used for any commercial unit to economically produce these products.

7.2.3 Catalyst and medium

A number of catalytic systems have been reported for the production of HMF. In addition to the catalyst, the reaction medium also plays a key role in the formation of HMF. Table 7.2 displays some typical results obtained by using different reaction media and catalysts. The following sections will highlight some efficient catalysts for the conversion of lignocelluloses in water, organic solvents, ILs, and biphasic systems.

Tab. 7.2: The production of HMF in different solvents.

Substrate	Solvent	Catalyst	Conditions		Yield (%)	Ref.
			T (°C)	t (h)		
Glucose	[EMIM]Cl	$CrCl_2$	120	3	70	[31]
	DMSO	$SnCl_4$–TBAB	100	2	69	[47]
	H_2O/DMSO MIBK/2-butanol	HCl	180	0.5	74	[48]
	H_2O–ChCl/MIBK	$AlCl_3$	180	0.25	70	[49]
Fructose	H_2O	H_3PO_4	240	0.03	65	[50]
	DMSO	Glu–TsOH	130	1.5	91	[51]
	DMSO	MIL–101(Cr)–SO_3H	120	1	90	[52]
	[BMIM]Cl	MI	175	0.02	97	[53]
	[HMIM]Cl	[HMIM]Cl	90	0.25	92	[54]
	H_2O/2-butanol	Ta_2O_5	160	1.7	90	[55]
Cellulose	H_2O	HZSM-5	190	4	46	[56]
	[BMIM]Cl	$CrCl_3$	150	0.17	54	[57]
	[EMIM]Cl	$CrCl_2$–$RuCl_3$	120	2	60	[58]
	[EMIM]Cl	$CrCl_3$–$CuCl_2$	120	8	57.5	[59]
	H_2O/THF	$NaHSO_4$–$ZnSO_4$	160	0.02	53	[60]
Wheat straw	DMA–LiCl/[EMIM]Cl	$CrCl_3$–HCl	140	2	48	[61]
Corn stalk	[EMIM]Cl	$CrCl_3$–HCl	160	0.25	61.4	[62]

7.2.3.1 Catalytic conversions in water

Because water is a green solvent, the catalytic transformation of carbohydrates in an aqueous solution is highly desirable. The production of HMF in water has been considered as an environmental-friendly process [63]. It has been reported that an HMF yield of 18.6% was obtained from fructose in hot compressed water under subcritical conditions of 240 °C and 3.35 MPa [64]. Despite no catalyst was present, the organic acids such as acetic and formic acids generated at high temperatures might act as genuine catalysts for the autocatalytic production of HMF. In addition to this, introduction of 1 wt% acetic acid into the reaction medium significantly enhanced the HMF yield under the same reaction conditions [65]. The addition of mineral acids (i.e., hydrochloric, sulfuric, and phosphoric acids) into the aqueous solution can also largely enhance the yield of HMF. For example, more than 64% yield of HMF was obtained from the conversion of fructose at 513 K for 10 s in the presence of HCl [66]. Asghari and Yoshida compared the catalytic performances of several mineral acids and organic acids such as oxalic, citric, maleic, or p-toluenesulfonic acid for the conversion of fructose at temperatures of 200–317 °C, pressures of 1.55–11.28 MPa, and reaction times of 75–180 s. It was demonstrated that H_3PO_4 with a pH of 2.0 was the most active catalyst and an HMF yield of 65.3% could be produced at 240 °C [50].

Although the mineral or organic acids played important roles in the improvement of HMF formation in homogenous catalytic solutions, these systems suffer from the demerits of the isolation of products and catalysts, the recovery of acids, and the corrosion of the reactor under harsh conditions. Besides, the production of large amount of acid wastes also leads to serious environmental problems. Hence, solid catalysts have received much attention [67]. A lot of effort has been devoted to develop stable and recyclable solid catalysts for the conversion of carbohydrates into HMF. TiO_2–ZrO_2, SO_4^{2-}/Ti-MCM-41, zirconium phosphates, and zeolites have been employed for HMF production in hot compressed water [62, 65, 68]. Nandiwale et al. found that mesoporous H-ZSM-5 was an efficient catalyst for the conversion of cellulose into HMF [56]. The reaction temperature had a significant influence on the conversion of cellulose. As the reaction temperature increased from 170 to 200 °C, the conversion increased from 37% to 77%, suggesting that higher temperature was favorable for the depolymerization of cellulose. The glucose dehydration was found to be accelerated at an elevated temperature of 190 °C, which afforded a 46% HMF yield. With a further increase in temperature from 190 to 200 °C, the yield of HMF rather decreased to 35%. It was attributed to an increase in the rate of rehydration and polymerization of HMF at higher temperature to LA and humin.

7.2.3.2 Catalytic conversions in ILs

The production of HMF in ILs has attracted great attention in recent years. Different substrates were used for HMF production in IL systems. Table 7.2 summarizes some of the results obtained to date. Lewis acid metal salts, in particular $CrCl_2$ and $CrCl_3$, were found to be efficient catalysts for the conversion of glucose and cellulose into furan derivatives in ILs [69]. In 2007, Zhao et al. first reported the conversion of glucose into HMF in IL, 1-ethyl-3-methyl-imidazolium chloride ([EMIM]Cl) [31]. Among various Lewis acids explored, $CrCl_2$ exerted the highest activity and offered a 70% HMF yield at 120 °C. Next, Qi et al. used $CrCl_3$ as a catalyst for the direct conversion of cellulose into HMF in [BMIM]Cl at 150 °C [57]. A 54% yield of HMF was obtained after a reaction time of 10 min. It was found that neutral ILs with Cl (such as [BMIM]Cl) as the counteranion were efficient for cellulose conversion. Binder et al. investigated a catalytic system that could directly convert untreated corn stalk into HMF with a 48% yield using $CrCl_3$-HCl as catalyst and DMA-LiCl/[EMIM]Cl as solvent [61].

The combination of $CrCl_2$ and $CrCl_3$ with other Lewis acid has been demonstrated to be a useful method to enhance the formation of HMF in ILs. Kim et al. have explored such combination for the direct conversion of cellulose in the [EMIM]Cl solvent [58]. They found that $CrCl_2$–$RuCl_3$ (4:1) was the most efficient and provided a HMF yield of 60% at 120 °C in 2 h. Su et al. explored the conversion of cellulose into HMF using a $CrCl_2$ and $CuCl_2$ combined catalyst in [EMIM]Cl [59]. An HMF yield of 57.5% was obtained after a reaction at 120 °C for 8 h. Wang et al. [62] obtained a 62% HMF yield from cellulose at 140 °C after 40 min by using the combination of $CrCl_3$ and LiCl (1:1) in [BMIM]Cl. With the $CrCl_3$–LiCl combination, a 61.4% yield of HMF could also be attained from wheat straw at 160 °C in [BMIM]Cl [62]. The coupling of $CrCl_2$ with an HY zeolite was also employed for the transformation of cellulose in [BMIM]Cl [70]. It was concluded that such a $CrCl_2$–zeolite catalytic system was very stable and insensitive to moisture and air. An HMF yield of 40% could be sustained after several repeated uses.

Despite $CrCl_2$ and $CrCl_3$ showed good performance in the conversion of glucose and cellulose to HMF in ILs, the toxicity of chromium limited the practical applications. A lower toxicity or nontoxic catalytic system for the production of HMF is desirable. Hu et al. found that $SnCl_4$ in 1-ethyl-3-methyl-imidazolium tetrafluoroborate ([EMIM]BF_4) could catalyze the conversion of glucose into HMF with a yield of 60% [71]. Stahlberg designed a catalytic system containing lanthanide metal salts such as $Yb(OTf)_3$ for the conversion of glucose in [BMIM]Cl, and a moderate HMF yield of 24% was attained [72]. Germanium chloride ($GeCl_4$) also exerted a good performance on the production of HMF from glucose in [BMIM]Cl [73]. The introduction of zeolite during the conversion of glucose led to an increase in HMF yield from 38.4% to 48.4%.

To conclude, ILs can act as a dual solvent and a catalyst for the conversion of carbohydrates into HMF with high selectivity and yield. However, the isolation of

HMF from IL is difficult and may need large amounts of extracting solvents. Furthermore, the high cost of IL and the difficulty in catalyst recovery need to be considered before its large-scale applications.

7.2.3.3 Catalytic conversions in biphasic systems

In 2006, a highly efficient water–organic biphasic system was developed by Dumesic and coworkers for the conversion of concentrated fructose into HMF [48]. In the aqueous phase, fructose dehydrated in the presence of HCl. The HMF formed was efficiently and rapidly extracted into the methyl isobutylketone (MIBK) organic phase to prevent the rehydration of HMF. By using DMSO and 2-butanol as phase modifier, 80% of fructose was converted, and an HMF selectivity of 75% was attained at 180 °C in 3 min. For the conversion of glucose, an HMF yield of 74% was obtained using the same biphasic system [74].

Liu et al. explored metal chloride-catalyzed conversion of glucose in a water/ MIBK biphasic system at 180 °C [49]. $AlCl_3$ afforded a 70% yield of HMF from glucose in the biphasic system containing choline chloride (ChCl) in the aqueous phase at 180 °C for 15 min. The optimized content of ChCl was 50 wt% in water. Using $FeCl_3$–$AlCl_3$ as the catalyst, cellulose could also be dehydrated to HMF with 49% yield at 200 °C in such a ChCl-assisted biphasic system. Recently, a mesoporous tin phosphate with aggregated nanoparticles of 10–15 nm in diameter has exerted an excellent catalytic activity for HMF synthesis [75]. The catalytic conversion of monosaccharides, disaccharides, and cellulose could give maximum yields of HMF of 77%, 51%, and 32%, respectively, under microwave-assisted heating at 150 °C. Large mesopores and nanoscale particle morphology were proposed to enable the access of the bulky carbohydrate molecules to the active sites. Water–acetone medium was also employed for the production of HMF under sub- and supercritical conditions [76]. The catalytic conversion of fructose was performed by using H_2SO_4 as a catalyst at reaction temperatures ranging from 180 to 300 °C. About 99% of fructose converted and HMF selectivity of 77% were obtained in acetone–water medium (9:1) at 180 °C for 2 min.

To sum up, the use of the biphasic system in combination with metal Lewis acid catalysts could catalyze the conversion of carbohydrates into HMF with high efficiencies. The main obstacle for the large-scale application of these biphasic systems is the complicated separation of inorganic and organic phases. The engineering aspects should be surveyed to make such complicated biphasic catalytic systems practical and cost-effective.

7.2.4 Derivatives

HMF has three chemical functionalities: the hydroxymethyl group, the aldehyde, and the furan ring itself. Together, these offer a diverse combination of chemistries for derivative synthesis. Herein, the hydroxymethyl group can be alkylated, acylated, substituted with nucleophiles, or oxidized to the aldehyde or carboxylic acid oxidation state, both of which have numerous derivatives. For example, the furan ring can undergo cycloaddition, electrophilic aromatic substitution, ring opening, or hydrogenation to the corresponding tetrahydrofuran. Chosen synthetic transformations can be applied to attain derivatives such as monomers for novel polymeric materials, special chemicals, or biofuels. Examples of practical applications of this chemistry are discussed below. In a different case, high-yielding and selected processes to specific derivatives will be highlighted.

7.2.4.1 Derivatization of the aldehyde or hydroxymethyl group

The OH group of HMF can be variously derivatized (Fig. 7.3). Acetylation of HMF **1** to **2** facilitates its separation from reaction media during its preparation from carbohydrates, as well as accelerating the deoxygenation of the OH group [77, 78]. Subsequently, intermolecular dehydration with the mesoporous MCM-41 gives the symmetric ether 5,5′(oxybis(methylene))bis(furan-2-carbaldehyde) **3** in high yield [79]. Substitution of the OH group for halogens (Cl or Br) can be carried out with the typical reagents (HX, SOX$_2$, PX$_n$, etc.) or affected in situ during the formation of HMF from various carbohydrates [79, 80]. Besides, the carbonyl group of HMF takes part in typical reactions of aromatic aldehydes, including the formation of various imines **5** and acetals **6**, the latter being of use both as novel surfactants and potential biodiesel fuels [81–83].

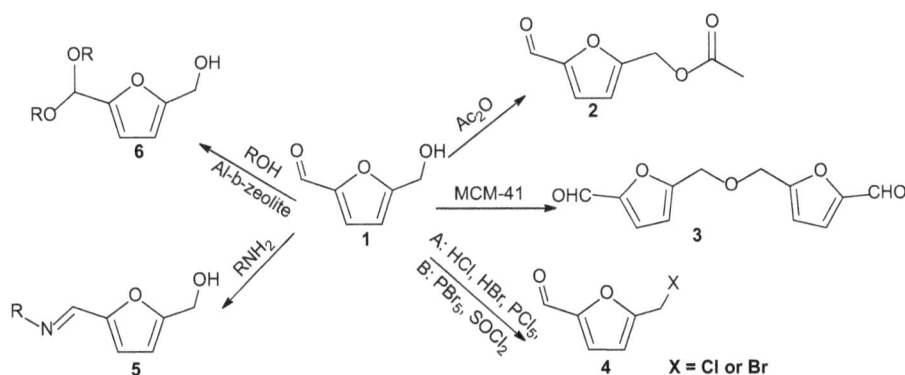

Fig. 7.3: Functional transformations of the aldehyde and hydroxymethyl group in HMF.

7.2.4.2 Oxidation reaction of the aldehyde or hydroxymethyl group

Oxidation of the OH group of HMF **1** to an aldehyde leads to DFF **7**, a monomer of considerable application in the polymer industry [84]. In 2012, Hu et al. reviewed the synthetic strategies of this molecule from HMF. Apart from various oxidants, such as Ba(MnO$_4$)$_2$, NaOCl, ceric ammonium nitrate, IBX, and dichromates, many promising catalytic approaches have recently been developed [85–90]. For example, using air as an oxidant and a silver-impregnated molecular sieve catalyst, a quantitative yield of DFF was obtained [91]. Additionally, comparable results have been seen with ruthenium-based catalysts (Fig. 7.4) [92]. Hence, there is much promise in establishing an industrially viable route to the HMF starting material.

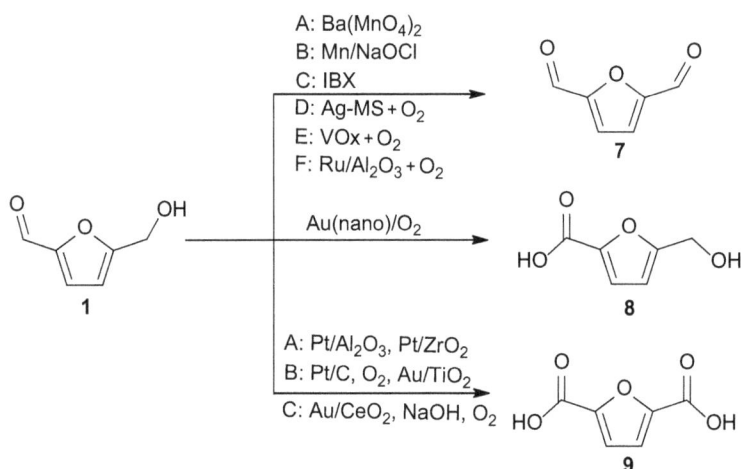

Fig. 7.4: Oxidation reaction of the aldehyde or hydroxymethyl group in HMF.

The oxidation rate of the HMF aldehyde group to the carboxylic acid is faster than that of the alcohol to the aldehyde substantially. Thus, the selective preparation of 5-hydroxymethyl-2-furancarboxylic acid **8** is possible. Casanova et al. investigated the synthesis of **8** using a gold nanoparticle catalyzed oxidation with molecular oxygen [93]. Similar results using Au/C, Au/TiO$_2$, and Au–Cu catalyst were reported [94, 95]. Moreover, oxidation of both OH and aldehyde groups to carboxylic acids gives FDCA **9**, a molecule acts as a renewable substitution for petroleum-derived terephthalic acid, which is used to produce PET, a dominant polymer that is widely used for the production of synthetic fibers and beverage containers [96]. Lew and Zope et al. reported the application of Au/C, Au/TiO$_2$, Pd/C, and Pt/C as catalysts in a basic reaction medium [97, 98]. Linga et al. investigated platinum catalysts on numerous supports (activated carbon, ZrO$_2$, Al$_2$O$_3$) in neutral, basic, and even acidic media and observed high yields of FDCA in all cases [99]. Casanova et al. used gold nanoparticle-based catalysts (e.g., Au/TiO_2 and Au/CeO_2) to achieve greater than 99% yield of FDCA under optimized reaction conditions (10 bar O_2, 130 °C, aq. NaOH) [100]. Due to the generation of a salt waste stream, the obstacle of this chemistry is the common application of basic media.

7.2.4.3 Reduction reaction of HMF

Reduction of the aldehyde of HMF **1** to a hydroxymethyl group gives 2,5-di(hydroxymethyl)furan **10**, a useful building block in the production of polyurethane foams [101]. Although $NaBH_4$ is the obvious reducing agent, catalytic hydrogenation is more industrially relevant to this process, and quantitative yields have been obtained with $Ir–ReO_x/SiO_2$, gold nanoparticles on Al_2O_3, and Pd/C with formic acid as the hydrogen source [102–104]. One of the most attractive furan derivatives to date is DMF **11**, the product that is attained from the reduction of both the OH and aldehyde groups of HMF to methyl groups [105]. In addition to being a high energy density and octane biofuel, DMF can also be converted into *p*-xylene [106]. A lot of effort toward DMF include hydrogenation of HMF using a CuRu/C catalyst with a 71% yield, a one-pot reaction of fructose with formic acid catalyzed by H_2SO_4 and Pd/C in 51% yield, and hydrogen transfer to HMF from supercritical methanol with a Cu-doped porous metal oxide catalyst (48% yield). Recently, Williams et al. reported an excellent 93% yield of DMF by hydrogenation of HMF using a novel Ru/Co_3O_4 catalyst (Fig. 7.5) [107].

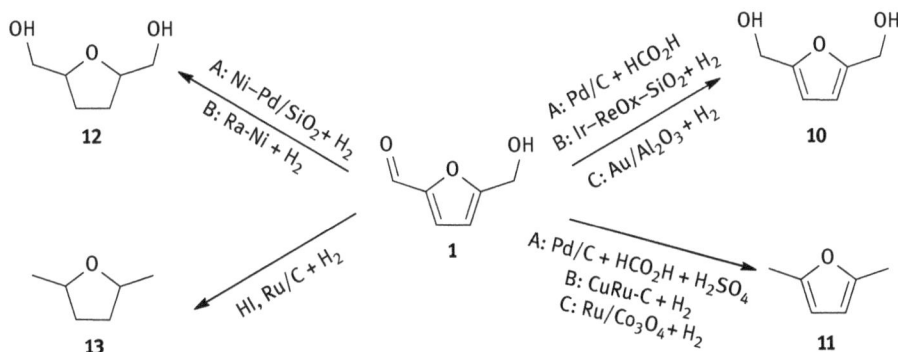

Fig. 7.5: Catalytic hydrogenations of HMF.

HMF to tetrahydrofuran through ring hydrogenation is also a highly useful strategy. This can be achieved by preserving oxygen functionality at the methyl positions, and giving 2,5-bis(hydroxymethyl)tetrahydrofuran (BHTHF) **12**. For example, hydrogenation of HMF on Ni–Pd/SiO$_2$ under mild conditions produced BHTHF in 95% yield, and the use of Ra–Ni or ceria-supported ruthenium catalysts gives similar results [108–110]. Sen and coworkers studied complete reduction of HMF to 2,5-dimethyltetrahydrofuran (DMTHF) **13**, where not only HMF but also fructose could be converted to DMTHF by hydrogenation using HI and a ruthenium catalyst [111, 112]. DMTHF has high energy content and great potential as a fuel; however, comparatively few studies have been devoted to its selective production from HMF and lignocelluloses.

7.2.4.4 Condensation reaction of HMF

Most recently, the production of simple hydrocarbons from biomass has attracted strong interest in biorefinery due to the fact that these products are considered to be substitutes for petroleum-derived alkanes, with vast markets to fuels and chemicals. HMF as a platform for extended carbon chain products has received much attention. The hydrodeoxygenation (HDO) gives products including diesel and aviation fuels, which depend on their hydrocarbon distribution. Aldol-type condensation reactions can take place in aqueous solution between HMF (or its derivatives) and biogenic ketones such as acetone [113]. The original work in this aspect was done by Dumesic group and involved various condensations of HMF or related molecules with acetone followed by HDO to give C$_1$ to C$_{15}$ alkanes (Fig. 7.6) [114]. Moreover, Liu and coworkers explored the benzoin condensation of HMF to give a dimer that could be submitted

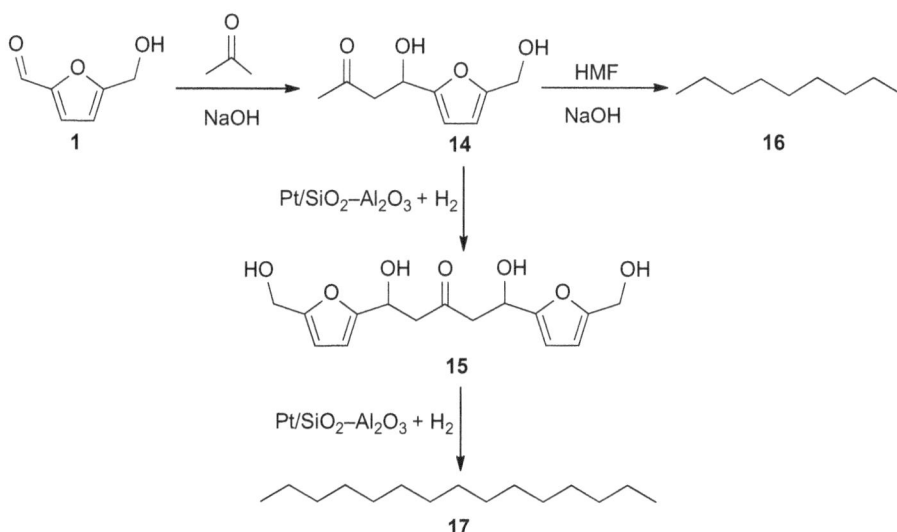

Fig. 7.6: Synthesis of hydrocarbons from HMF.

to HDO to give C_{10} to C_{12} alkanes [115]. Sutton and coworkers likewise employed simple aldol reaction between HMF and acetone to attain C_9 to C_{15} hydrocarbons [116]. The startup company Virent has piloted this sugar derivative to alkane process successfully [117].

7.2.4.5 Transformations involving cleavage of the furan ring

The HDO reaction involves first hydrogenation and finally hydrogenolytic cleavage of the ring of THF, and other useful derivatives also take advantage of ring-opening reactions. Rehydration of HMF in acidic media leads to hydrolytic ring opening to give equimolar quantities of LA and formic acid [118]. As noted in Section 7.1, LA is a platform chemical in its own right and its production and chemistry will be reviewed next. Adipic acid **18** is a high-volume commodity chemical used for making nylon polymers. The structural similarity between FDCA and adipic acid provides an obvious route by hydrogenating FDCA to 2,5-tetrahydrofurandicarboxylic acid followed by reductive cleavage (Fig. 7.7) [119]. Oxidative cleavage of the HMF ring can be achieved with oxygen. The reaction is carried out by irradiating an aerated solution of HMF containing a sensitizer. The reaction proceeds through the endoperoxide intermediate **19**, which undergoes ring opening to a butenolide in aqueous or alcoholic solvent, which subsequently cleaves to 4-oxopent-2-enoic acid or the corresponding ester **20** [120, 121].

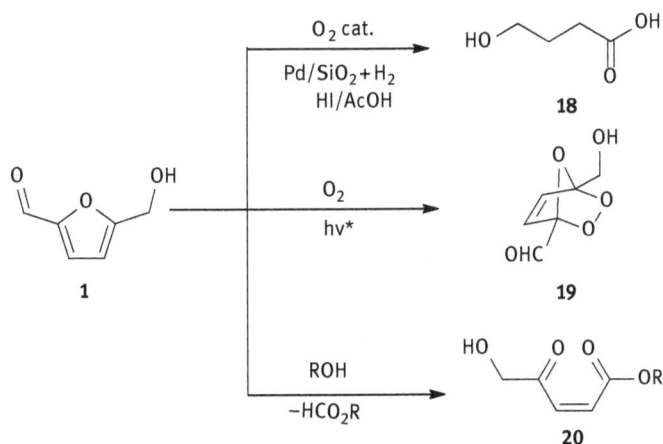

Fig. 7.7: Transformations of HMF involving furan ring cleavage.

7.3 Conversion of lignocelluloses into LA

7.3.1 Possible pathways for the formation of LA

The mineral acid-catalyzed conversion of lignocelluloses into LA has been widely studied in aqueous medium [122]. The formation of LA is proposed to proceed in several steps, including the dehydration of cellulose to HMF, and the subsequent rehydration of HMF to LA. Formic acid will be formed together with LA. Horvat and coworkers proposed a detailed mechanism for the formation of LA based on the analysis of ^{13}C NMR results [123]. As displayed in Fig. 7.8, an intermediate, 2,5-dioxo-3-hexenal was formed when HMF was transformed via several acid-catalyzed hydration and dehydration steps to open the furan rings. The C–C bond between C-1 and C-2 is highly unstable and can be easily cleaved to produce 5,5-dihydroxypent-3-en-2-one and formic acid. Then the former intermediate was converted to LA via several steps. The Brønsted acid catalyst plays dominant roles in the conversion of HMF into LA. Additionally, Zhang and coworkers investigated the conversions of unlabeled and ^{13}C-labeled fructose to identify the position of C–C bond cleavage via in situ ^{13}C and ^{1}H NMR results [124]. It was found that formic acid was generated from the aldehyde carbon C-1 but not from the hydroxymethyl carbon C-6. This observation was in good agreement with the mechanism proposed in Fig. 7.8. Table 7.3 depicts some typical catalytic systems for the conversions of carbohydrates into LA or levulinate in different media.

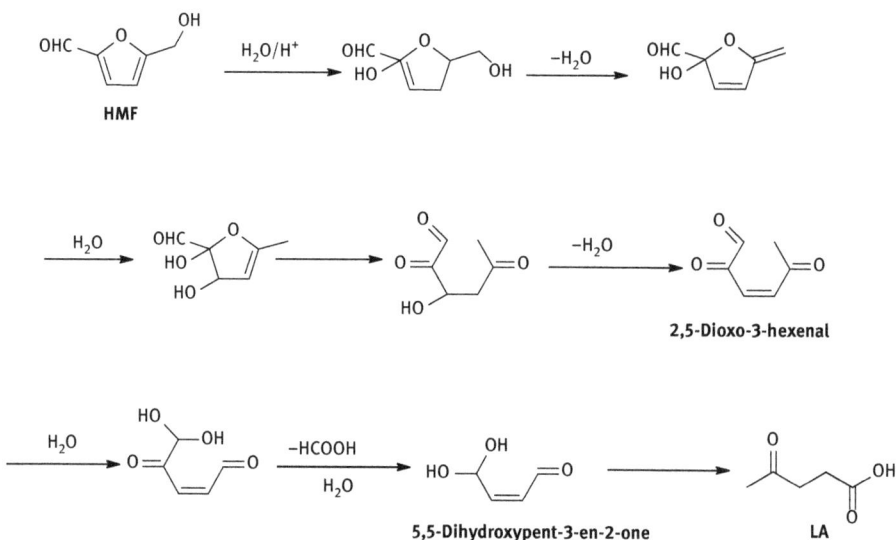

Fig. 7.8: Possible reaction pathways for the formation of LA.

Tab. 7.3: The production of LA in water or methyl levulinate in methanol.

Substrate	Solvent	Catalyst	Conditions		Yield (%)	Ref.
			T (°C)	t (h)		
Glucose	H_2O	Graphene–SO_3H	120	2	78	[125]
	H_2O	HCl	150	8	57	[126]
	CH_3OH	H_2SO_4	190	5	55	[127–130]
	CH_3OH	H-USY (Si/Al=6)			49	
	CH_3OH	Acidic TiO_2 nanoparticles	175	1	61	[131]
Fructose	CH_3OH	H-USY (Si/Al=6)			51	[130]
	CH_3OH	Acidic TiO_2 nanoparticles	175	1	80	[131]
Cellulose	H_2O	H_2SO_4	190	4	46	[132]
	H_2O	HCl	150	0.17	54	[133]
	H_2O	HCl	120	2	60	[134]
	H_2O	Amberlyst 70	120	8	57.5	[135]
	H_2O	Sulfated TiO_2	160	0.02	53	[136]
	CH_3OH	$Cs_{2.5}H_{0.5}PW_{12}O_{40}$	290	0.02	20	[129]
	CH_3OH	H-USY (Si/Al=6)	200		13	[129]
	CH_3OH	Acidic TiO_2 nanoparticles	175	20	42	[131]

7.3.2 Catalytic conversions in aqueous media

Mineral acids, in particular strong Brønsted acids, show excellent performances for the catalytic conversion of cellulose into LA [122]. Heeres and coworkers investigated H_2SO_4-catalyzed conversion of MCC in water at temperatures ranging from 150 to 200 °C [132]. They obtained an LA yield of 60% in the conversion of MCC at a relatively low temperature (150 °C). Detailed kinetic studies pointed out that the activation energy for LA formation was lower than that for glucose decomposition and humin formation. This suggests that a low temperature favors the formation of LA. Shen and Wyman carried out a comprehensive study on the conversion of cellulose into LA in the presence of HCl [133]. They obtained an LA yield of 60% with 1.5 wt% of cellulose and a 3.2 wt% concentration of acid at 180 °C. It was concluded that higher acid concentrations and higher temperatures accelerated the reaction rate and reduced the reaction time for attaining the maximum LA yield, but could not further increase the LA yield. Moreover, Wettstein and coworkers developed a water–GVL (1:1) biphasic system for the conversion of cellulose [134]. In this system, a high yield of LA (70%) was obtained by using 35 wt% HCl as catalyst at 155 °C. The formed LA in aqueous phase can be extracted into the GVL phase, suppressing the further conversion of LA.

However, the homogeneous acid-catalyzed processes have drawbacks including the difficulty in catalyst separation and repeated uses. To overcome these problems, several solid acids, in particular those containing SO_3H groups (e.g., commercial resin, Amberlyst 70), were used for the direct conversion of cellulose to LA [135]. A two-step process was designed for obtaining high yields of LA. First, cellulose was converted at moderate temperatures to produce water-soluble compounds (glucose and HMF) without any catalyst. Second, Amberlyst 70 was used for catalyzing the conversion of water-soluble compounds to LA at 160 °C. An LA yield of 28% was achieved for the conversion of cellulose with an initial loading of 29 wt%. Besides, the SO_3H-functionalized graphene oxide could one-pot convert glucose into LA with a high yield of LA (78%) at 200 °C [125]. It was suggested that both the oxygen-containing groups and layered structure of the graphene oxide played key roles in the adsorption and the transferring of the reactants. Wang et al. employed a sulfated TiO_2 as a catalyst for the conversion of cellulose into LA. An LA yield of 32% was achieved after a reaction at 240 °C for 15 min [136]. Lai et al. presented a recyclable catalyst (Fe_3O_4–SBA–SO_3H) by grafting the SO_3H groups onto a composite of mesoporous silica and magnetic iron oxide for the conversion of cellulose into LA [137]. Such a catalyst could not only be easily recycle by magnetic field, but also catalyze the conversion of cellulose with a loading of up to 6 wt%, and provide a high LA yield of 42%. Recently, Fu and coworkers developed a sulfonated chloromethyl polystyrene (CP) resin (CP-SO_3H) for the conversion of cellulose. This solid catalyst contained acid sites (–SO_3H) and cellulose-binding sites (–Cl). It afforded a high yield of LA up to 65.5% by converting MCC at 170 °C [138].

Although high yields of LA can be obtained by using solid catalysts with SO_3H groups, the large-scale application of such catalysts is still limited due to the leaching of acidic groups. Recently, ZrO_2 was used as a stable catalyst for the direct conversion of cellulose to LA with air and water via an aqueous phase partial oxidation process [139]. The superoxide species on the ZrO_2 surface was proposed to play a key role in breaking down the glycosidic bonds in cellulose, enhancing the formation of LA. In this process, ZrO_2 afforded an LA yield of 50% at 240 °C under a low O_2 partial pressure, and the formation of formic acid was largely suppressed. The catalyst could be recycled several times without significant deactivation.

7.3.3 Catalytic conversions in alcohol media

The conversion of lignocelluloses in alcohols may lead to the formation of alkyl levulinates. Wu et al. compared a variety of acid catalysts including inorganic and organic acids for the alcoholysis of cellulose in subcritical methanol [127]. Both H_2SO_4 and *para*-toluenesulfonic acid exhibited good performances than other acids like H_3PO_4, acetic, and formic acids. The highest yield of methyl levulinate was 55%, which was obtained using 0.02 mol/L H_2SO_4 as catalyst at 190 °C for 5 h. Furthermore, Tominaga et al. found that a mixture containing Brønsted and Lewis acids could efficiently catalyze the direct conversion of cellulose in methanol into methyl levulinate [128]. About 75% yield of levulinate was obtained using the combination of 2-naphthalenesulfonic acid as a Brønsted acid and $In(OTf)_3$ as a Lewis acid at 180 °C for 5 h. The conversion of cellulose involved two tandem steps: the alcoholysis of cellulose to methyl glucosides catalyzed by Brønsted acid in the first step, and the subsequent conversion of methyl glucosides to methyl levulinate catalyzed by the Lewis acid in the second step.

In addition, Demolis et al. studied the one-pot conversion of MCC in supercritical methanol or methanol–water mixtures catalyzed by solid acid sulfated zirconia and polyoxometalates [129]. A methyl levulinate yield of 20% was obtained at 190 °C using $Cs_{2.5}H_{0.5}PW_{12}O_{40}$ catalyst. H-USY zeolite with a Si/Al ratio of 6 could catalyze the conversions of several sugars in methanol and ethanol, providing methyl or ethyl levulinate with yields of 40–50% at 200 °C [130]. For the conversion of cellulose, the yield of methyl levulinate was 13% in methanol. The catalyst could be reused at least five times without significant deactivation. Recently, acidic TiO_2 nanoparticles (anatase) were also found to be efficient for the production of methyl levulinates from different biomass-derived carbohydrates in methanol [131]. The methyl levulinate yield obtained from fructose was 80% with the stable and recyclable TiO_2 nanoparticles at 175 °C. Remarkable catalytic activity of the catalyst is due to in situ sulfation on the surface and their excellent dispersion in reaction media.

7.3.4 Derivatives

Like HMF, LA is a platform chemical that can generate a series of derivatives with applications across a range of markets. There are two functional groups in LA: the ketone and the carboxylic acid. Some of the more attractive derivatives with the potential to unlock important industrial markets were discussed as follows.

7.3.4.1 Esters, amides, ketals, alcohols, and ethers

Alkyl levulinate esters **23** can be prepared by homogeneous acid-catalyzed esterification of LA, reaction of LA with olefins, or by reaction of alcohols with the cyclic ester angelica lactone **22** given form the dehydration of LA (Fig. 7.9) [140, 141]. They can also form acetals **29** in an acid-catalyzed reaction with alcohols. The products are variously useful as green solvents, plasticizers, and monomers for renewable polymers [142]. All kinds of levulinate esters were tested as blends with diesel fuel and biodiesel, where they not only acted as oxygenates but also improved the cold-flow properties of the fuel [143].

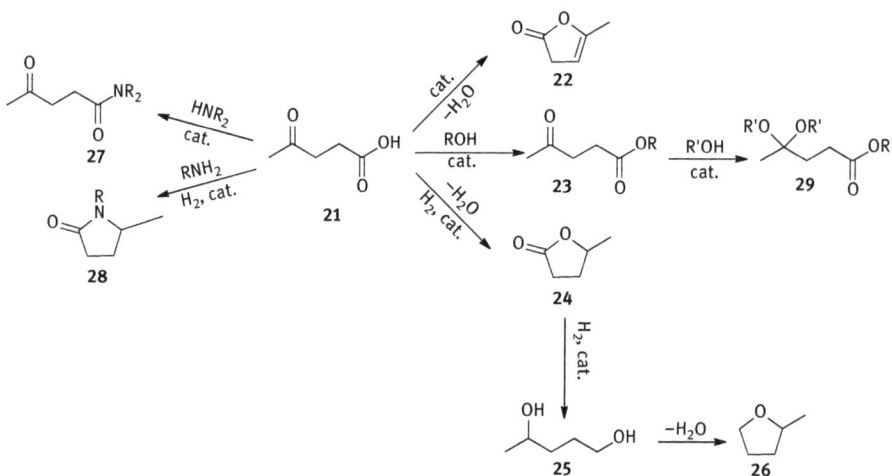

Fig. 7.9: Synthetic transformations of LA: esters, lactones, alcohols, ethers, amides, and lactams.

Reduction of the keto group of LA to the alcohol followed by cyclization generates γ-valerolactone (GVL) **24** [144]. It has found wide application as a green solvent and precursor to polymers, chemicals, and a series of biofuels [145]. For example, catalytic hydrogenation of GVL produces pentane-1,4-diol (1,4-PDO) **25** [146]. In addition, cyclodehydration of 1,4-PDO provides 2-methyltetrahydrofuran **26**, which is a component of Environmental Protection Agency (EPA)-approved P-series flex fuels [147].

Not only LA but also angelica lactone **22** can react with secondary amines to give the corresponding amides **27** [148]. Reductive amination of LA with primary amines gives 5-methyl-2-pyrrolidones (MPDs) **28**, which is an important solvent with a range of industrial applications. Recently, Wei et al. studied the reductive amination of LA with various primary alkyl and aromatic amines using a cyclometallated iridium complex catalyst and formic acid [149]. They also reported a catalyst-free synthesis of MPDs using a combination of triethylamine and formic acid for transfer hydrogenation [150].

7.3.4.2 Transformation into fuels

GVL 24 is a starting material for pentanoate esters that have been shown to have excellent fuel properties [151]. Pentanoic acid 31 derived from GVL can be catalytically upgraded to 5-nonanone 32 by decarboxylative ketonization (Fig. 7.10). 5-Nonanone can then be variously processed to nonane 33 or other hydrocarbons [152, 153]. Additionally, GVL can also be converted to butenes using a silica–alumina catalyst, which are oligomerized on solid acid to produce a mixture of C_8–C_{16} alkenes, which can be hydrogenated to drop-in fuels [154].

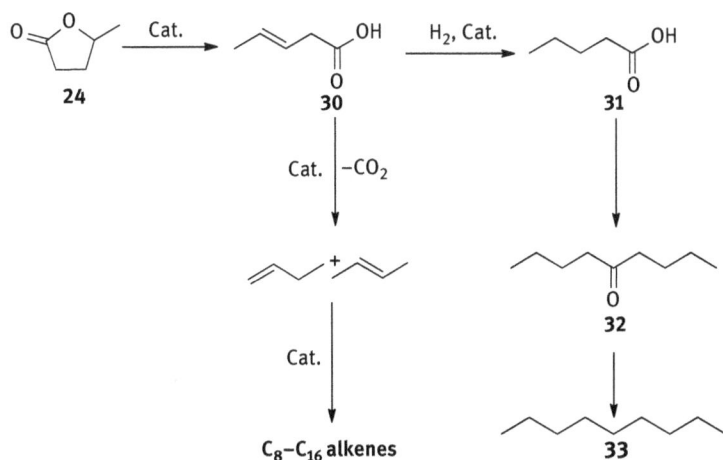

Fig. 7.10: The routes of GVL to hydrocarbon.

Mascal et al. investigated a synthetic approach to C_7 to C_{10} hydrocarbons using LA as the starting material [155]. First, LA was dehydrated to angelica lactone 22 using inexpensive and recyclable montmorillonite K10 catalyst. It was then dimerized over solid K_2CO_3 to give the angelica lactone dimer 34 in 94% yield. Angelica lactone dimer 34 has a C_{10} backbone and gives the branched C_{10} hydrocarbon 3-ethyl-4-methylheptane 35 as the major product when subjected to HDO, alongside other branched C_7–C_9 products (Fig. 7.11). Ir–ReO$_x$/SiO$_2$ and Pt–ReO$_x$/C catalysts performed best, both giving 88% yield of hydrocarbons from 34. Considering that LA is available in greater than 80% conversion from biomass, gasoline-range hydrocarbons of greater than 60% are achievable by this approach. Case et al. reported high yields of deoxygenated hydrocarbons through thermal decomposition of mixtures of LA and formic acid [156]. Contrasting with the pyrolysis oils attained from lignocelluloses, the product is nonviscous, of neutral pH, low in oxygen, and has high energy content.

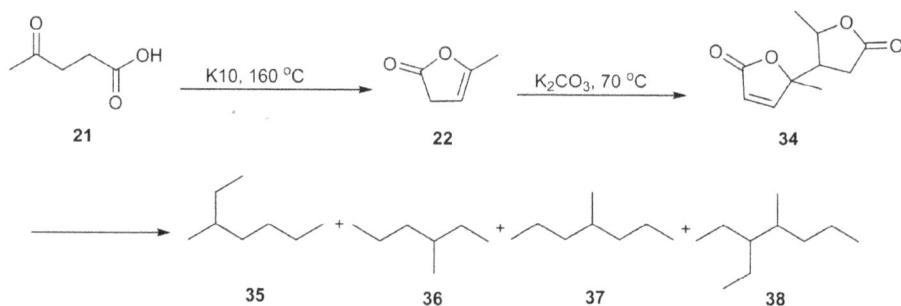

Fig. 7.11: The transformations of LA to hydrocarbons.

7.3.4.3 Transformations leading to renewable monomers, solvents, and special chemicals

Diphenolic acid (DPA) **39** is a condensation product of LA and two molecules of phenol. It is known as a renewable substitute of bisphenol A (BPA), a high volume chemical that has estrogenic activity used for making polycarbonate plastics [157]. Thus, DPA has the potential to substitute BPA across the range of polymer markets [158].

The oxidation of LA can also lead to useful derivatives. High-temperature oxidation of LA with O_2 over a V_2O_5 catalyst gives succinic acid **40** in high yield (Fig. 7.12) [159]. Nitric acid was also used as the oxidant under milder reaction conditions with 52% yield of succinic acid yields [160]. Succinic acid is a platform for γ-butyrolactone, 1,4-butanediol, and tetrahydrofuran, which have important applications in the solvents, special chemicals, and polymer markets.

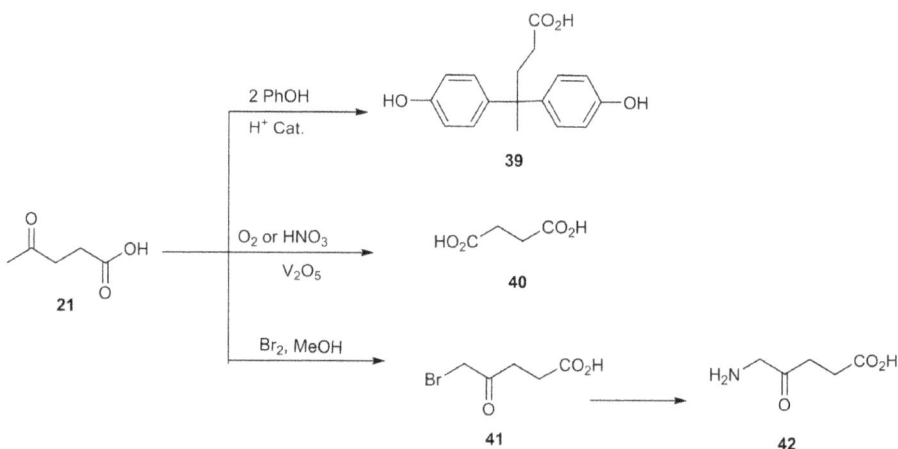

Fig. 7.12: Preparation of diphenolic acid, succinic acid, and 5-bromolevulinic acid **41**.

Besides, the synthesis of the δ-aminolevulinic acid **42** can be obtained directly from LA. This method relies on the halogenation of the methyl group of LA, followed by substitution with a nitrogen nucleophile, then transformed into the primary amine. The halogenation reaction can be approached by the direct bromination of LA in refluxing methanol to generate 5-bromolevulinic **41** [161, 162].

7.4 Conclusion and outlook

The catalytic transformation of carbohydrates into platform or building-block chemicals is promising for the utilization of the abundant lignocelluloses. This chapter contributes to highlight some recent advances in the developments of efficient catalysts and catalytic systems for the production of HMF and LA. The primary, secondary, tertiary, and other generations of their derivatives can be alternatives of fermentation products and petrochemicals. Although a mountain of work devoted to HMF's advancement as a cellulosic platform chemical, practical issues threat to hinder its further development. The biggest challenge still to overcome is the use of feedstock because of the unstable essence of HMF and the harsh reaction conditions required. The important thing of success of HMF is an economical production approach. Avantium Chemicals operated a pilot plant for the production of furan-based chemicals and plastics. HMF was in situ converted into its ether to help improve yield and purification. A similar strategy was followed by Mascal that produced 5-chloromethylfurfural (CMF). In effect, since HMF and CMF are interchangeable, CMF serves every derivative class and application that can be accessed from HMF. The biorefinery as such is not new, but the rapid development in the old days has brought the issue of commercialization to prominence inevitably. It will be interesting to wait and see which approach has the market in the years to come.

Acknowledgments: This work was financially supported by the Beijing Higher Education Young Elite Teacher Project (YETP0765), Fundamental Research Funds for the Central Universities (BLYJ201519), New Century Excellent Talents in University (NCET-13-0671), National Natural Science Foundation of China (31170556), and Program of International S &T Cooperation of China (2015DFG31860).

References

[1] Werpy T, Petersen G. Top value added chemicals from biomass. Pacific Northwest National Laboratory. National Renewable Energy Laboratory.

[2] Boisen A, Christensen TB, Fu W, Gorbanev YY, Hansen TS, Jensen JS. Process integration for the conversion of glucose to 2, 5-furandicarboxylic acid. Chem. Eng. Res. Des. 2009, 87, 1318–1327.

[3]   Bicker M, Kaiser D, Ott L, Vogel H. Dehydration of D-fructose to hydroxyl -methylfurfural in sub- and supercritical fluids. J. Supercrit. Fluids. 2005, 36, 118–126.
[4]   Lichtenthaler FW, Peters S. Carbohydrates as green raw materials for the chemical industry. Comp. Rend. Chim. 2004, 7, 65–90.
[5]   Breen JP. The effects of temperature and working electrode material on the power output of a hydroxymethylfurfural fuel cell. Chemical Engineering Oregon State University.
[6]   Abdulmalik O, Safo MK, Chen Q, et al. 5-Hydroxymethyl-2-furfural modifies intracellular sickle haemoglobin and inhibits sickling of red blood cells. Br. J. Haematol. 2005, 128, 55261.
[7]   Van Putten RJ, Van der Waal JC, de Jong E, Rasrendra CB, Heeres HJ, de Vries JG. Hydroxymethylfurfural, a versatile platform chemical made from renewable resources. Chem. Rev. 2013, 113, 1499–1597.
[8]   Kunz M. Hydroxymethylfurfural, a Possible Basic Chemical For Industrial Intermediates. In: Fuchs A., ed. Studies in plant science. Amsterdam, Elsevier, 1993, 149–160.
[9]   Lichtenthaler FW. Unsaturated O- and N-heterocycles from carbohydrate feedstocks. Acc. Chem. Res. 2002, 35, 728–737.
[10] Bicker M, Hirth J, Vogel H. Dehydration of fructose to 5-hydroxymethylfurfural in sub- and supercritical acetone. Green Chem. 2012, 5, 280–284.
[11] Torres AI, Daoutidis P, Tsapatsis M. Continuous production of 5-hydroxy methylfurfural from fructose: A design case study. Energy Environ. Sci. 2010, 3, 1560–1572.
[12] Liu W, Richard, Zheng F, Li J, Cooper A. An ionic liquid reaction and separation process for production of hydroxymethylfurfural from sugars. AIChE J. 2013, 60, 300–314.
[13] Bozell JJ, Petersen GR. Technology development for the production of biobased products from biorefinery carbohydrates. US Department of Energy's "Top 10" revisited. Green Chem. 2010, 12, 539–554.
[14] Leonard RH. Levulinic acid as a basic chemical raw material. Ind. Eng. Chem. 1956, 48, 1330–1341.
[15] Bozell JJ, Moens L, Elliott DC, et al. Production of levulinic acid and use as a platform chemical for derived products. Resour. Conserv. Recycl. 2000, 28, 227–239.
[16] Ghorpade V, Hanna MA. Industrial applications for levulinic acid. In: Campbell GM, Webb C, McKee SL., eds. Cereals: Novel uses and processes, Springer, Manchester, 1997, 49–55.
[17] Harmsen P, Hackmann M. Green building blocks for biobased plastics. 8, Wageningen UR Food& Biobased Research, Wageningen, 2013, 306–324.
[18] Xin L, Zhang Z, Qi J, et al. Electricity storage in biofuels: Selective electrocatalytic reduction of levulinic acid to valeric acid or gamma-valerolactone. ChemSusChem. 2013, 6, 674–686.
[19] Serrano Ruiz JC, Wang D, Dumesic JA. Catalytic upgrading of levulinic acid to 5-nonanone. Green Chem. 2010, 12, 574–577.
[20] Fitzpatrick, SW. Commercialization of the biofine technology for levulinic acid production from paper sludge. BioMetics, Inc.
[21] Moens L. Sugar Cane as a Renewable Feedstock for the Chemical Industry: Challenges and Opportunities. In: Sugar processing research conference. New Orleans, 2002, 26–41.
[22] Patel AD, Serrano Ruiz JC, Dumesic JA, Anex RP. Techno-economic analysis of 5-nonanone production from levulinic acid, Chem. Eng. J. 2010, 160, 311–321.
[23] James S. Global levulinic acid market expected to reach 3,820 Tons by 2020. Grand view research, 2014.
[24] Yang ZZ, Deng J, Pan T, Guo QX, Fu Y. A one-pot approach for conversion of fructose to 2, 5-diformylfuran by combination of Fe_3O_4-SBA-SO_3H and K-OMS-2. Green Chem. 2005, 14, 29869.

[25] Roman LY, Barrett CJ, Liu ZY, Dumesic JA. Production of dimethylfuran for liquid fuels from biomass derived carbohydrates. Nature. 2007, 447, 982–985.
[26] Li G, Li N, Li S, et al. Synthesis of renewable diesel with hydroxyacetone and 2-methyl-furan. Chem. Commun. 2013, 49, 5727–5729.
[27] Hayes DJ, Steve F, Hayes MHB, Ross JRH. The Biofine Process Production of Levulinic Acid, Furfural and Formic Acid from Lignocellulosic Feedstock. In: Kamm B, Gruber P, Kamm M., eds. Biorefineries industrial processes and products. New York, USA, Wiley-VCH, 2006, 139–164.
[28] Moreau C, Belgacem MN, Gandini A. Recent catalytic advances in the chemistry of substituted furans from carbohydrates and in the ensuing polymers. Top. Catal. 2007, 27, 11–30.
[29] Zakrzewska M, Bogel Łukasik E, Bogel Łukasik R. Ionic liquid-mediated formation of 5-hydroxymethylfurfural a promising biomass-derived building block. Chem. Rev. 2011, 111, 397–417.
[30] Rinaldi R, Schüth F. Acid hydrolysis of cellulose as the entry point into biorefinery schemes. ChemSusChem. 2009, 2, 1096–1107.
[31] Zhao HB, Holladay JE, Brown H, Zhang ZC. Metal chlorides in ionic liquid solvents convert sugars to 5-hydroxymethylfurfural Science. 2007, 316, 1597–1600.
[32] Pidko EA, Degirmenci V, van Santen RA, Hensen EJM. Glucose activation by transient Cr^{2+} dimers. Angew. Chem. Int. Ed. 2010, 49, 2530–2534.
[33] Tong XL, Ma Y, Li YD. Biomass into chemicals: Conversion of sugars to furan derivatives by catalytic processes. Appl. Catal. A. 2010, 385, 1–13.
[34] Antal MJ, Mok WSL, Richards GN. Kinetic-studies of the reactions of ketoses and aldoses in water at high-temperature 1. Mechanism of formation of 5-(hydroxymethyl)-2-furaldehyde from D-fructose and sucrose. Carbohydr. Res. 1990, 199, 91–109.
[35] Amarasekara AS, Williams LD, Ebede CC. Mechanism of the dehydration of D-fructose to 5-hydroxymethylfurfural in dimethyl sulfoxide at 150 degrees C: An NMR study. Carbohydr. Res. 2008, 343, 3021–3024.
[36] Qi X, Watanabe M, Aida TM, Smith RL. Synergistic conversion of glucose into 5-hydroxymethylfurfural in ionic liquid-water mixtures. Bioresour. Technol. 2012, 109, 224–228.
[37] Qi X, Watanabe M, Aida TM, Smith RL. Jr. Catalytical conversion of fructose and glucose into 5-hydroxymethylfurfural in hot compressed water by microwave heating. Catal. Commun. 2008, 9, 2244–2249.
[38] Huang R, Qi W, Su R, He Z. Integrating enzymatic and acid catalysis to convert glucose into 5-hydroxymethylfurfural. Chem. Commun. 2010, 46, 1115–1117.
[39] Yang Y, Xiang X, Tong D, Hu C, Abu Omar MM. One-pot synthesis of 5-hydroxymethylfurfural directly from starch over SO_4^{2-}/ZrO_2-Al_2O_3 solid catalyst. Bioresour. Technol. 2012, 116, 302–306.
[40] Omari KW, Besaw JE, Kerton FM. Hydrolysis of chitosan to yield levulinic acid and 5-hydroxymethylfurfural in water under microwave irradiation. Green Chem. 2012, 14, 1480–1487.
[41] Qi X, Watanabe M, Aida TM, Smith RL. Jr. Efficient one-pot production of 5-hydroxymethylfurfural from inulin in ionic liquids. Green Chem. 2010, 12, 1855–1860.
[42] Mascal M, Nikitin EB. Direct, high-yield conversion of cellulose into biofuel. Angew. Chem. Int. Ed. 2008, 120, 8042–8044.
[43] Rosatella AA, Simeonov SP, Frade RFM, Afonso, CAM. 5-Hydroxymethylfurfural (HMF) as a building block platform: Biological properties, synthesis and synthetic applications. Green Chem. 2011, 13, 754–793.

[44] Dutta S, De S, Alam MI, Abu Omar MM, Saha B. Direct conversion of cellulose and lignocellulosic biomass into chemicals and biofuel with metal chloride catalysts. J. Catal. 2012, 288, 8–15.

[45] Kamm B, Gerhardt M, Dautzenberg G. Catalytic Processes of Lignocellulosic Feedstock Conversion For Production Of Furfural, Levulinic Acid, and Formic Acid-Based Fuel Components. In: Suib SL., ed. New and future developments in catalysis: Catalytic biomass conversion. 2013, 91–113.

[46] Alonso DM, Wettstein SG, Mellmer MA, Gurbuz EI, Dumesic JA. Integrated conversion of hemicellulose and cellulose from lignocellulosic biomass. Energy Environ. Sci. 2013, 6, 76–80.

[47] Zhou L, Liang R, Ma Z, Wu T, Wu Y. Conversion of cellulose to HMF in ionic liquid catalyzed by bifunctional ionic liquids. Bioresour. Technol. 2013, 129, 450–455.

[48] Roman LY, Chheda JN, Dumesic JA. Phase modifiers promote efficient production of hydroxymethylfurfural from fructose. Science. 2006, 312, 1933–1937.

[49] Liu F, Audemar M, Vigier KDO, et al. Selectivity enhancement in the aqueous acid-catalyzed conversion of glucose to 5-hydro-xymethylfurfural induced by choline chloride. Green Chem. 2013, 15, 3205–3213.

[50] Asghari FS, Yoshida H. Acid-catalyzed production of 5-hydroxymethylfurfural from D-fructose in subcritical water. Ind. Eng. Chem. Res. 2009, 45, 2163–2173.

[51] Wang JJ, Xu WJ, Ren JW, Liu XH, Lu GZ, Wang YQ. Efficient catalytic conversion of fructose into hydroxymethylfurfural by a novel carbon-based solid acid. Green Chem. 2011, 13, 2678–2681.

[52] Chen J, Li K, Chen L, Liu R, Huang X, Ye D. Conversion of fructose into 5-hydroxymethylfurfural catalyzed by recyclable sulfonic acid-functionalized metal-organic frameworks. Green Chem. 2014, 16, 2490–2499.

[53] Li, CZ, Zhao ZB, Cai HL, Wang AQ, Zhang T. Microwave-promoted conversion of concentrated fructose into 5-hydroxymethylfurfural in ionic liquids in the absence of catalysts. Biomass Bioenerg. 2011, 35, 2013–2017.

[54] Moreau C, Finiels A, Vanoye L. Dehydration of fructose and sucrose into 5-hydroxymethylfurfural in the presence of 1-H-3-methyl imidazolium chloride acting both as solvent and catalyst. J. Mol. Catal. A: Chem. 2006, 253, 165–169.

[55] Yang F, Liu Q, Yue M, Bai X, Du, Y. Tantalum compounds as heterogeneous catalysts for saccharide dehydration to 5-hydroxymethyl-furfural. Chem. Commun. 2011, 47, 4469–4471.

[56] Nandiwale KY, Galande ND, Thakur P, Sawant SD, Zambre VP, Bokade VV. One-pot synthesis of 5-hydroxymethylfurfural by cellulose hydrolysis over highly active bimodal micro/mesoporous H-ZSM-5 catalyst. ACS Sustainable Chem. Eng. 2014, 2, 1928–1932.

[57] Qi X, Watanabe M, Aida TM, Smith RL. Jr. Fast transformation of glucose and di-/polysaccharides into 5-hydroxymethylfurfural by microwave heating in an ionic liquid/catalyst system. ChemSusChem. 2010, 3, 1071–1077.

[58] Kim Z, Jeong J, Lee D, et al. Direct transformation of cellulose into 5-hydroxymethyl-2-furfural using a combination of metal chlorides in imidazolium ionic liquid. Green Chem. 2011, 13, 1503–1506.

[59] Su Y, Brown HM, Huang X, Zhang ZC. Single-step conversion of cellulose to 5-hydroxymethylfurfural (HMF), A versatile platform chemical. Appl. Catal. A Gen. 2009, 361, 117–22.

[60] Shi N, Liu Q, Zhang Q, Wang T, Ma L. High yield production of 5-hydroxymethylfurfural from cellulose by high concentration of sulfates in biphasic system. Green Chem. 2013, 15, 1967–1974.

[61] Binder JB, Raines RT. Simple chemical transformation of lignocellulosic biomass into furans for fuels and chemicals. J. Am. Chem. Soc. 2009, 131, 1979–1985.

[62] Wang P, Yu H, Zhan S, Wang S. Catalytic hydrolysis of lignocellulosic biomass into 5-hydroxymethylfurfural in ionic liquid. Bioresour. Technol. 2011, 102, 4179–4183.

[63] Dashtban M, Gilbert A, Fatehi P. Recent advancements in the production of hydroxymethylfurfural. RSC Adv. 2014, 4, 2037–2050.

[64] Asghari FS, Yoshida H. Dehydration of fructose to 5-hydro-xymethylfurfural in sub-critical water over heterogeneous zirconium phosphate catalysts. Carbohydr. Res. 2006, 341, 2379–2387.

[65] Li Y, Lu X, Yuan L, Liu X. Fructose decomposition kinetics in organic acids-enriched high temperature liquid water. Biomass Bioenerg. 2009, 33, 1182–1187.

[66] Asghari FS, Yoshida H. Kinetics of the decomposition of fructose catalyzed by hydrochloric acid in subcritical water: Formation of 5-hydroxymethylfurfural, levulinic, and formic acids. Ind. Eng. Chem. Res. 2007, 46, 7703–7710.

[67] Rinaldi R, Schüth F. Design of solid catalysts for the conversion of biomass. Energy Environ. Sci. 2009, 2, 610–626.

[68] Jiang CW, Zhong X, Luo Z. An improved kinetic model for cellulose hydrolysis to 5-hydroxymethylfurfural using the solid SO42/Ti-MCM-41 catalyst. RSC Adv. 2014, 4, 15216–15224.

[69] Mascal M, Nikitin EB. Direct, high-yield conversion of cellulose into biofuel. Angew. Chem. Int. Ed. 2008, 47, 7924–7226.

[70] Tan M, Zhao L, Zhang Y. Production of 5-hydroxymethylfurfural from cellulose in $CrCl_2$/zeolite/BMIMCl system. Biomass Bioenerg. 2011, 35, 1367–1370.

[71] Hu SQ, Zhang ZF, Song JL, Zhou YX, Han BX. Efficient conversion of glucose into 5-hydroxymethylfurfural catalyzed by a common Lewis acid $SnCl_4$ in an ionic liquid. Green Chem. 2009, 11, 1746–1749.

[72] Stahlberg T, Sorensen MG, Riisager A. Direct conversion of glucose to 5-(hydroxymethyl) furfural in ionic liquids with lanthanide catalysts. Green Chem. 2010, 12, 321–325.

[73] Zhang ZH, Wang Q, Xie HB, Liu WJ, Zhao ZB. Catalytic conversion of carbohydrates into 5-hydroxymethylfurfural by germanium(IV) chloride in ionic liquids. ChemSusChem. 2011, 4, 131–138.

[74] Chheda JN, Roman LY, Dumesic JA. Production of 5-hydroxymethylfurfural and furfural by dehydration of bimass-derived mono- and polysaccharides. Green Chem. 2007, 9, 342–350.

[75] Dutta A, Gupta D, Patra AK, Saha B, Bhaumik A. Synthesis of 5-hydroxymethylfurural from carbohydrates using large-pore mesoporous tin phosphate. ChemSusChem. 2014, 7, 925–933.

[76] Bicker M, Hirth J, Vogel H. Dehydration of fructose to 5-hydroxymethylfurfural in sub-and supercritical acetone. Green Chem. 2003, 5, 280–284.

[77] Hu C, Yang Y, Yan H, et al. Preparation of 5-acetoxy methylfurfural from carbohydrates. CN 10163331 A, 2009.

[78] Rauchfuss TB, Thananatthanachon T. Efficient method for preparing 2,5-dimethyl -furan. US 201113092816, 2011.

[79] Sanda K, Rigal L, Gaset A. Synthesis of 5-(bromomethyl)- and of 5-(chloromethyl) -2-furancarboxaldehyde. Carbohydr. Res. 1989, 187, 15–23.

[80] Bredihhin A, Maeorg U, Vares L. Evaluation of carbohydrates and lignocellulosic biomass from different wood species as raw material for the synthesis of 5-bromomethyfurfural. Carbohydr. Res. 2012, 375, 63–67.

[81] Cukalovic A, Stevens CV. Production of biobased HMF derivatives by reductive amination. Green Chem. 2010, 12, 1201–1206.

[82] Arias KS, Al Resayes SI, Climent MJ, Corma A, Iborra S. From biomass to chemicals: Synthesis of precursors of biodegradable surfactants from 5-hydroxymethylfurfural. ChemSusChem. 2013, 6, 123–131.

[83] Balakrishnan M, Sacia ER, Bell AT. Etherification and reductive etherification of 5-(hydroxymethyl)furfural: 5-(alkoxymethyl)furfurals and 2,5-bis(alkoxymethyl) furans as potential bio-diesel candidates. Green Chem. 2012, 14, 1626–1634.

[84] Gandini A, Belgacem MN. Furans in polymer chemistry. Prog. Polym. Sci. 1997, 22, 1203–1379.

[85] Hu L, Zhao G, Hao W, Tang X, Sun Y, Lin L, Liu S. Catalytic conversion of biomass derived carbohydrates into fuels and chemicals via furanic aldehydes. RSC Adv. 2012, 2, 11184–11206.

[86] Elhajj T, Masroua A, Martin JC, Descotes G. Synthese de l'hydroxymethyl-5-furanne carboxaldehyde-2et de ses derives par traitement acide de sucres sur resines echangeuses d'ions. Bull. Soc. Chim. Fr. 1987, 5, 855–860.

[87] Amarasekara AS, Green D, McMillan E. Efficient oxidation of 5-hydroxymethylfurfural to 2, 5-diformylfuran using Mn(III)-salen catalysts. Catal. Commun. 2008, 9, 286–288.

[88] Mehdi H, Bodor A, Lantos D, Horvath IT, DeVos DE, Binnemans K. Imidazolium ionic liquids as solvents for cerium (IV)-mediated oxidation reactions. J. Org. Chem. 2007, 72, 517–524.

[89] Yoon HJ, Choi JW, Jang HS, et al. Selective oxidation of 5-hydroxymethylfurfural to 2, 5-diformylfuran by polymer-supported IBX amide. SynLett. 2011, 2, 165–168.

[90] Cotier L, Descotes G, Lewkowski J, Skowronski R. Ultrasonically accelerated syntheses of furan-2, 5-dicarbaldehyde from 5-hydroxymethyl-2-furfural. Org. Prep. Proc. Int. 1995, 27, 564–566.

[91] Yadav GD, Sharma RV. Biomass derived chemicals: Environmentally benign process for oxidation of 5-hydroxymethylfurfural to 2,5-diformylfuran by using nano-fibrous Ag-OMS-2-catalyst. Appl. Catal. B. Environ. 2014, 147, 293–301.

[92] Nie J, Xie J, Liu H. Activated carbon-supported ruthenium as an efficient catalyst for selective aerobic oxidation of 5-hydroxymethylfurfural to 2,5-diformylfuran. Chin. J. Catal. 2013, 34, 871–875.

[93] Casanova O, Iborra I, Corma A. Biomass into chemicals: Aerobic oxidation of 5-hydroxymethyl-2-furfural into 2,5-furandicarboxylic acid with gold nanoparticle catalysts. ChemSusChem. 2009, 2, 1138–1144.

[94] Davis SE, Houk LR, Tamargo EC, Datye AK, Davis RJ. Oxidation of 5-hydroxy methylfurfural over supported Pt, Pd and Au catalysts. Catal. Today. 2011, 160, 55–60.

[95] Pasini T, Piccinini M, Blosi M, et al. Selective oxidation of 5-hydroxymethyl-2-furfural using supported gold-copper nanoparticles. Green Chem. 2011, 13, 2091–2099.

[96] de Jong E, Dam MA, Sipos L, Gruter GJM. Furan Dicarboxylic Acid (FDCA), a Versatile Building Block for a Very Interesting Class of Polyesters. In: Smith PB, Gross RA., eds. Biobased monomers, polymers, and materials. ACS Symp Ser, 2012, 1, 1–13.

[97] Lew BW. Method of producing dehydromucic acid. US 35052264A, 1967.

[98] Zope BN, Davis SE, Davis RJ. Influence of reaction conditions on diacid formation during Au-catalyzed oxidation of glycerol and hydroxymethylfurfural. Top. Catal. 2012, 55, 24–32.

[99] Lilga MA, Hallen RT, Gray M. Production of oxidized derivatives of 5-hydroxymethylfurfural (HMF). Top. Catal. 2010, 53, 1264–1269.

[100] Casanova O, Iborra S, Corma A. Biomass into chemicals: One pot-base free oxidative esterification of 5-hydroxymethyl-2-furfural into 2,5-dimethylfuroate with gold on nanoparticulated ceria. J. Catal. 2009, 265, 109–116.

[101] Moreau C, Belgacem MN, Gandini A. Recent catalytic advances in the chemistry of substituted furans from carbohydrates and in the ensuing polymers. Top. Catal. 2004, 27, 11–30.

[102] Tamura M, Tokonami K, Nakagawa Y, Tomishige K. Rapid synthesis of unsaturated alcohols under mild conditions by highly selective hydrogenation. Chem. Commun. 2013, 49, 7034–7036.

[103] Ohyama J, Esaki A, Yamamoto Y, Arai S, Satsuma A. Selective hydrogenation of 2-hydroxymethyl-5-furfural to 2,5-bis(hydroxymethyl)furan over gold sub-nano clusters. RSC Adv. 2013, 3, 1033–1036.

[104] Thananatthanachon T, Rauchfuss TB. Efficient production of the liquid fuel 2, 5-dimethylfuran from fructose using formic acid as a reagent. Angew. Chem. Int. Ed. 2010, 49, 6616–6618.

[105] Hansen TS, Barta K, Anastas PT, Ford PC, Riisager A. One-pot reduction of 5-hydroxymethylfurfural via hydrogen transfer from supercritical methanol. Green Chem. 2012, 14, 2457–2461.

[106] Roman LY, Barrett CJ, Liu ZY, Dumesic JA. Production of dimethylfuran for liquid fuels from biomass-derived carbohydrates. Nature. 2007, 447, 982–985.

[107] Williams CL, Chang CC, Do P, et al. Cycloaddition of biomass-derived furans for catalytic production of renewable p-xylene. ACS Catal. 2010, 2, 935–939.

[108] Nakagawa Y, Tomishige K. Total hydrogenation of furan derivatives over silica supported Ni-Pd alloy catalyst. Catal. Commun. 2010, 12, 154–156.

[109] Yao S, Wang X, Jiang Y, Wu F, Chen X, Mu X. One-step conversion of biomass derived 5-hydroxymethylfurfural to 1,2,6-hexanetriol over Ni-Co-Al mixed oxide catalysts under mild conditions. ACS Sustain. Chem. Eng. 2014, 2, 173–180.

[110] Alamillo R, Tucker M, Chia M, Pagan Torres Y, Dumesic J. The selective hydrogenation of biomass-derived 5-hydroxymethylfurfural using heterogeneous catalysts. Green Chem. 2012, 14, 1413–1419.

[111] Grochowski MR, Yang W, Sen A. Mechanistic study of a one-step catalytic conversion of fructose to 2, 5-dimethyltetrahydrofuran. Chem. Eur. J. 2012, 18, 12363–12371.

[112] Yang W, Sen A. One-step catalytic transformation of carbohydrates and cellulosic biomass to 2, 5-dimethyltetrahydrofuran for liquid fuel. ChemSusChem. 2010, 3, 597–603.

[113] Chheda JN, Dumesic JA. An overview of dehydration, aldol-condensation and hydrogenation processes for production of liquid alkanes from biomass-derived carbohydrates. Catal. Today. 2007, 123, 59–70.

[114] Huber GW, Chheda JN, Barrett CJ, Dumesic JA. Production of liquid alkanes by aqueous-phase processing of biomass-derived carbohydrates. Science. 2005, 308, 1446–1450.

[115] Liu D, Chen EYX. Diesel and alkane fuels from biomass by organocatalysis and metal-acid tandem catalysis. ChemSusChem. 2013, 6, 2236–2239.

[116] Sutton AD, Waldie FD, Wu R, Schlaf M. 'Pete' Silks (III) LA, Gordon JC: The hydrodeoxygenation of bioderived furans into alkanes. Nature. Chem. 2013, 5, 428–432.

[117] Virent. Inc. http://www.virent.com. Accsessed Jan 2014, 14.

[118] Girisuta B, Janssen LPBM, Heeres HJ. A kinetic study on the decomposition of 5-hydroxymethylfurfural into levulinic acid. Green Chem. 2006, 8, 701–709.

[119] Boussie TR, Dias EL, Fresco ZM, Murphy VJ, Shoemaker J, Archer R, Jiang H Production of adipic acid and derivatives from carbohydrate-containing materials. US 20100317823 A1, 2010.

[120] Cottier L, Descotes G, Eymard L, Rapp K. Syntheses of γ-oxo acids or γ-oxo esters by photooxygenation of furanic compounds and reduction under ultrasound: Application to the synthesis of 5-aminolevulinic acid hydrochloride. Synthesis. 1995, 303–306.

[121] Marisa C, Ilaria D, Marotta R, Roberto A, Vincenzo C. Production of 5-hydroxy-4-keto-2-pentenoic acid by photo-oxidation of 5-hydroxymethylfurfural with singlet oxygen: A kinetic investigation. J. Photochem. Photobiol. A. 2010, 210, 69–76.

[122] Rackemann DW, Doherty WO. The conversion of lignocellulosics to levulinic acid. Biofuels. Bioprod. Biorefin. 2011, 5, 198–214.

[123] Horvat J, Klaic B, Metelko B, Sunjic V. Mechanism of levulinic acid formation. Tetrahedron Lett. 1985, 26, 2111–2114.

[124] Zhang J, Weitz E. An in situ NMR study of the mechanism for the catalytic conversion of fructose to 5-hydroxymethylfurfural and then to levulinic acid using ^{13}C labeled D-fructose. ACS Catal. 2012, 2, 1211–1218.

[125] Upare PP, Yoon J, Kim MY, Kang HY, Hwang DW, Hwang YK, Kung HH, Chang JS. Chemical conversion of biomass-derived hexose sugars to levulinic acid over sulfonic acid-functionalized graphene oxide catalysts. Green Chem. 2013, 15, 2935–2943.

[126] Weingarten R, Cho J, Xing R, Conner WC Jr, Huber GW. Kinetics and reaction engineering of levulinic acid production from aqueous glucose solutions. ChemSusChem. 2012, 5, 1280–1290.

[127] Wu X, Fu J, Lu X. One-pot preparation of methyl levulinate from catalytic alcoholysis of cellulose in near-critical methanol. Carbohydr. Res. 2012, 358, 37–39.

[128] Tominaga K, Mori A, Fukushima Y, Shimada S, Sato K. Mixed-acid systems for the catalytic synthesis of methyl levulinate from cellulose. Green Chem. 2011, 13, 810–812.

[129] Demolis A, Essayem N, Rataboul F. Synthesis and applications of alkyl levulinates. ACS Sustain. Chem. Eng. 2014, 2, 1338–1352.

[130] Saravanamurugan S, Riisager A. Zeolite catalyzed transformation of carbohydrates to alkyl levulinates. ChemCatChem. 2013, 5, 1754–1757.

[131] Kuo CH, Poyraz AS, Jin L, et al. Heterogeneous acidic TiO_2 nano- particles for efficient conversion of biomass derived carbohydrates. Green Chem. 2014, 16, 785–791.

[132] Girisuta B, Janssen LPBM, Heeres HJ. Kinetic study on the acid-catalyzed hydrolysis of cellulose to levulinic acid. Ind. Eng. Chem. Res. 2007, 46, 1696–1708.

[133] Shen J, Wyman CE. Hydrochloric acid-catalyzed levulinic acid formation from cellulose: Data and kinetic model to maximize yields. AIChE J. 2012, 58, 236–246.

[134] Wettstein SG, Alonso DM, Chong Y, Dumesic JA. Production of levulinic acid and gamma-valerolactone (GVL) from cellulose using GVL as a solvent in biphasic systems. Energy Environ. Sci. 2012, 5, 8199–8203.

[135] Weingarten R, Conner WC Jr, Huber GW. Production of levulinic acid from cellulose by hydrothermal decomposition combined with aqueous phase dehydration with a solid acid catalyst. Energy Environ. Sci. 2012, 5, 7559–7574.

[136] Wang P, Zhan S, Yu H. Production of levulinic acid from cellulose catalyzed by environmental-friendly catalyst. Adv. Mater. Res. 2010, 96, 183–187.

[137] Lai DM, Deng L, Guo QX, Fu Y. Hydrolysis of biomass by magnetic solid acid. Energy Environ. Sci. 2011, 4, 3552–3557.

[138] Zuo Y, Zhang Y, Fu Y. Catalytic conversion of cellulose into levulinic acid by a sulfonated chloromethyl polystyrene solid acid catalyst. ChemCatChem. 2014, 6, 753–757.

[139] Lin H, Strull J, Liu Y, et al. High yield production of levulinic acid by catalytic partial oxidation of cellulose in aqueous media. Energy Environ. Sci. 2012, 5, 9773–9777.

[140] Bart HJ, Reidetschlager J, Schatka K, Lehmann A. Kinetics of esterification of levulinic acid with n-butanol by homogeneous catalysis. Ind. Eng. Chem. Res. 1994, 33, 21–25.

[141] Manzer LE. Preparation of levulinic acid esters from alpha-angelica lactone and alcohols. WO 2005097724, 2005.

[142] Leibig C, Mullen B, Mullen T, Rieth L, Badarinarayana V. Cellulosic-derived levulinic ketal esters: A new building block. ACS Symp. Ser. 2011, 1063, 111–116.

[143] Windom BC, Lovestead TM, Mascal M, Nikitin EB, Bruno TJ. Advanced distillation curve analysis on ethyl levulinate as a diesel fuel oxygenate and a hybrid biodiesel fuel. Energy Fuels. 2011, 25, 1878–1890.

[144] Zhang J, Wu S, Li B, Zhang H. Advances in the catalytic production of valuable levulinic acid derivatives. ChemCatChem. 2012, 4, 1230–1237.

[145] Alonso DM, Wettstein SG, Dumesic JA. Gamma-valerolactone, a sustainable platform molecule derived from lignocellulosic biomass. Green Chem. 2013, 15, 584–595.

[146] Corbel DL, Ly BK, Minh DP, et al. Heterogeneous catalytic hydrogenation of biobased levulinic and succinic acids in aqueous solutions. ChemSusChem. 2013, 6, 2388–2395.

[147] Pace V, Hoyos P, Castoldi L, Dominguez de, Maria P, Alcantara AR. 2-Methyltetrahydrofuran (2-MeTHF): A biomass-derived solvent with broad application in organic chemistry. ChemSusChem. 2012, 5, 1369–1379.

[148] Haskelberg L. Some derivatives of levulinic acid. J. Am. Chem. Soc. 1948, 70, 2830–2831.

[149] Wei Y, Wang C, Jiang X, Xue D, Li J, Xiao J. Highly efficient transformation of levulinic acid into pyrrolidinones by iridium catalysed transfer hydrogenation. Chem. Commun. 2013, 49, 5408–5410.

[150] Wei Y, Wang C, Jiang X, Xue D, Liu ZT, Xiao, J. Catalyst-free transformation of levulinic acid into pyrrolidinones with formic acid. Green Chem. 2014, 16, 1093–1096.

[151] Lange JP, Price R, Ayoub PM, et al. Valeric biofuels: A platform of cellulosic transportation fuels. Angew. Chem. Int. Ed. 2010, 49, 4479–4483.

[152] Serrano, Ruiz JC, Wang D, Dumesic JA. Catalytic upgrading of levulinic acid to 5-nonanone. Green Chem. 2010, 12, 574–577.

[153] West RM, Liu ZY, Peter M, Dumesic JA. Liquid alkanes with targeted molecular weights from biomass-derived carbohydrates. ChemSusChem. 2008, 1, 417–424.

[154] Bond JQ, Alonso DM, Wang D, West RM, Dumesic JA. Integrated catalytic conversion of γ-valerolactone to liquid alkenes for transportation fuels. Science. 2010, 327, 1110–1114.

[155] Mascal M, Dutta S, Gandarias I. The angelica lactone dimer as a renewable feedstock for hydrodeoxygenation: Simple, high-yield synthesis of branched $C_7–C_{10}$ gasoline-like hydrocarbons. Angew. Chem. Int. Ed. 2014, 53, 1854–1857.

[156] Case PA, van Heiningen ARP, Wheeler MC. Liquid hydrocarbon fuels from cellulosic feedstocks via thermal deoxygenation of levulinic acid and formic acid salt mixtures. Green Chem. 2012, 14, 85–89.

[157] Vandenberg LN, Maffini MV, Sonnenschein C, Rubin BS, Soto AM. Bisphenol-A and the great divide: A review of controversies in the field of endocrine disruption. Endocr. Rev. 2009, 30, 75–95.

[158] Bozell JJ, Moens L, Elliott DC, et al. Production of levulinic acid and use as a platform chemical for derived products. Resour. Conserv. Recycl. 2000, 28, 227–239.

[159] Lester W, Vitcha JF. Preparation of succinic acid. US 2676186 A, 1954.

[160] Van DKF, Van EDS, Van HJ. Succinic acid from biomass. WO 2012044168 A1, 2012.

[161] Ha HJ, Lee SK, Ha YJ, Park JW. Selective bromination of ketones. A convenient synthesis of 5-aminolevulinic acid. Synth. Commun. 1994, 24, 2557–2562.

[162] Manny AJ, Kjelleberg S, Kumar N, de Nys R, Read RW, Steinberg P. Reinvestigation of the sulfuric acid-catalysed cyclisation of brominated 2-alkyllevulinic acid to 3-alkyl-5-methylene-2(5H)-furanones. Tetrahedron. 1997, 53, 15813–15826.

Milichovský Miloslav

Chapter 8
Chemistry and physics of cellulose and cellulose substance

8.1 Introduction

The unique properties and recent universal focus on natural material resources has put cellulose and cellulose derivatives into the sphere of intensive scientific effort and consequently to the attention of industrial companies. Nature provides wonderful examples of composite structures that involve cellulose. Cellulose is a fibrous, tough, water-insoluble substance, which is found in the protective cell walls of plants, particularly in stalks, stems, trunks, and all woody portions of plant tissues. The properties of wood, for instance, result from the unique interplay between Nanoscale domains of cellulose, hemicelluloses, and lignin [1]. The manner in which such elements are organized into larger structures is critical to the survival of trees and other plants. The hierarchical organization of wood is based on the natural composite paradigm of providing maximum strength with the minimum amount of material from the most efficient economy of biosynthesis [2]. Even some animal species make use of cellulosic nano-structures, such as some members of the tunicates ("sea squirts") family, the sea alga *Valoniaventricosa* [3], *Chaetomorphamelagonium* [4], and Bacterial cellulose, a polysaccharide synthesized in abundance, for example, by *Acetobacterxylinum* [5]. In all of these organisms, cellulose serves as a "scaffold" to evolve further mechanical support of highly organized living and growing matter. Knowledge of the supramolecular and hypermolecular structure of cellulose, accompanied by changes during its chemical or mechanical treatment is important not only for technical or biomedical applications, but also predominantly as a novel approach to better understand and control the aging of cellulose materials, for example, in paper and paper products and nanocomposites. During the last 150 years of intensive research, an abundance of experimental data and information has been collected concerning the ultrastructure and morphology of cellulose, native cellulosic substances, and cellulosic materials. Although the development and utilization of cellulose have a long history, the understanding of its chemistry and structure is relatively new, and many living polymer scientists have spent their entire working lives developing our present knowledge. Despite the fact that the polymer theory is established and the chemical structure of cellulose is accepted without dispute, the ultrastructure of cellulose remains controversial on several issues. Despite the degree to which

Milichovský Miloslav, University of Pardubice, Faculty of Chemical Technology, Institute of Chemistry and Technology of Macromolecular Substances, Pardubice, Czech Republic

https://doi.org/10.1515/9783110658842-008

cellulose has been investigated, its structural features have not been identified with absolute clarity and new information is constantly being discovered [6, 7]. With regard to the solid state structure of cellulose, progress has been made but a lack of information and theoretical ideas exist involving the behavior of cellulose in a wet state and water suspensions. Infrared spectroscopy, Raman spectroscopy, single-crystal X-ray studies, high-resolution nuclear magnetic resonance (NMR), dark-field electron microscopy, and electron diffraction, supported by the recent applications of the solid-state, cross-polarization/magic angle spinning (CP/MAS) NMR, ^{13}C-CP/MAS solid-state NMR, electron spectroscopy for chemical applications, photoacoustic Fourier transform infrared spectroscopy (FTIR), secondary ion mass spectrometry, and fast atom bombardment mass spectrometry have added considerable information to our knowledge of the solid-state structure of cellulose [8]. However, as we have learned from the past, no single physical method is sufficiently sensitive for all parameters of interest, and no single structural method can give answers to all the questions.

8.2 Basic chemistry of cellulose

Materials of a cellulosic nature are typically structured matter exhibiting a conventional molecular, supramolecular, and hypermolecular structure. The Supramolecular structure is introduced by molecular chain bundles of cellulose called *elementary fibrils* connected mutually into microfibrils. Elementary fibrils are cemented together by hemicelluloses and lignin. The Hypermolecular structures of cellulosic materials are then assembled through microfibrils accumulating into fibrils, followed by fibrillary bundles of fibrils forming the fibrous P, S1, and S2 parts of the cellular wall, which display different morphology [1]. The cellulose structure is based on 180° turn-screw D-1,4-glucopyranoside units connected with β-glycosidic bonds that form cellulose polymeric chains, giving rise to amorphous and various crystalline domain formations that are considered Allomorphs [9, 10]. It has been shown [11-14] that β-D-glucose exists in a pyranose form and that the latter is in the 4C1 chair conformation, which is the lowest energy structure for β-D-glucopyranose. Degrees of substitution, degree of polymerization (DP), and distribution of the molecular mass, as well as the distribution of substituents, strongly influence the properties of cellulose and cellulose derivatives, and are also significant factors in their characterization. According to Schultz and Marx [15, 16], the DP of cellulose depends on the species of wood and other plants, giving a DP of cellulose in the range of 6,200 ± 600 for linters and cotton cellulose, 8,000 for flax fiber, 3,300 for pinewood cellulose, and 500–3,000 for dissolving pulp.

Cellulose derivatives are obtained by several functionalization methods [17]. The acidolysis of carbohydrates is one of the oldest degradation methods for polymeric carbohydrates, but these reactions are very aggressive and lead to a varied concentration of undesirable degradation products of cellulose oligomers. A milder alternative has been developed using pivalic anhydride/$BF_3.Et_2O/CH_2Cl_2$ for the degradation of 2,3,6-tri-O-acetylcellulose (pivaloylysis). Glycosidation reactions allow access to linear products (monodisperse celluloses) as well as to new branched cellulose units [18].

8.3 Epimolecular characteristic of cellulose and lignocellulose materials

As well known, the environment components are entering the cellulosic and ligno-cellulosic materials as paper, plywood and so on, evoking their degradation being finished by those Agin, especially in coincidence with intensively electromagneti-cally radiation as it is typical for the photochemical controlled processes. Most important are relative humidity [19–22], environmental cleanliness (sulfur and nitrogen oxides, ozone) [23, 24], and predominantly presence of microorganisms (molds, bacteria) and insects [25, 26]. Obviously, environment forming components entry into every material especially a porous one and hit its molecules through it's an Epimolecular (i.e., above molecular) structure. For instance, micropores of po-rous matter of attoliter volume containing condensed vapor can then serve as zep-tomole-scale chemical reactors [27–29]. To well understand the behavior of those materials, it is consequently important to have a thorough knowledge not only of their chemical compositions defined by molecules forming pore walls, for example, cellulose, lignin, and hemicelluloses in paper, collagen in leather, parchment, and so on, but also to have considerable information about epimolecular, that is, supra-molecular and hypermolecular characters of these ones.

8.3.1 Supramolecular characteristics of cellulose and lignocellulosic materials

Factual Pore matter, that is, the matter of pore walls, has oriented, that is, crystal-line subunit and nonoriented, that is, amorphous structures of micro- and submi-croreticular character cemented together with the oriented subunits. Furthermore, the pore walls consist of the following:
-   structurated discontinuities as nanofibrils, elementary fibrils, microfibrils, fi-brils, and so on in controlled or stochastically controlled organization, that is, *heterogeneous reticular pore walls* as cell wall of plants and pulp;
-   consistently nonstructurated gel matter in xerogel or hydrogel form, that is, *ho-mogeneous reticular pore walls* as cell wall of leather.

Currently, the supramolecular structure of cellulose, in its many forms, remains one of the most studied areas of investigation in polymer science [6]. Cellulose molecules as a linear macromolecule consisting of anhydroglucose units chain-wise interlinked and of coil conformation are entangled into Nano-strands forming the so-called *elementary fibrils*. The following arrangement of these fibrillary bundles, called *microfibrils*, is sufficiently regular that shows cellulose exhibits a crystalline X-ray diffraction pattern. Although the chemical constitution of cellulose has been established, its morphology and Crystalline structure continue to be a source of interest and sometimes controversy. Diffraction studies show light and dark areas along cellulose microfibrils, which have been attributed to crystalline and amorphous cellulose, respectively. Another portion of the microfibrils consists of less highly ordered cellulose molecules, called the *amorphous* or *paracrystalline region*. In general, the "crystalline" regions alternate with less well-ordered "amorphous" regions, and there is no connection between the length of the crystalline regions and the molecular chain length. The molecules gradually transit from regions of high-lattice order to regions of low-lattice order. The high strength of the cellulose fibers may be due to the chain ends being held in the lattices by strong hydrogen bonding forces. The simplest conceivable model has "straight" cellulose chains isotropically distributed in the sample [30]. Cellulose from *Valonia* is a highly crystalline product and consequently it is frequently the object of studies. Verlhac and coworkers [31] suggested that the amorphous material consists mostly of surface chains, which in the large *Valonia* microfibrils makes up a low percentage of the total. The surface of crystalline cellulose or indeed areas of "amorphous" cellulose probably still possess a degree of order. Thus, "amorphous" cellulose cannot be considered truly amorphous as, by definition, an amorphous material is one that is formless or lacks a definite shape. Evidence has been put forward for the existence of more than one polymorph of cellulose in native samples [7]. Six Polymorphs of cellulose (I, II, III_I, III_{II}, IV_I, and IV_{II}) were described. Cellulose I, or native cellulose, is the form found in nature, but the crystal structures of native celluloses show structural differences that depend on the source of the cellulose [32]. Simon and coworkers [33] postulated that a form of crystalline cellulose existed near the surface of a crystal, which differed from the structure found at the center of the crystal. These two crystalline forms were termed *celluloses I_α* and *I_β* [34]. Both allomorphs possess symmetry, which is very close to twofold screw symmetry; this leads to the conclusion that the repeating unit along the chain is cellobiose. Celluloses produced by primitive organisms were said to have a dominant I_α component, while those produced by the higher plants have a dominant I_β form. I_α and I_β were found to have the same conformation of the heavy atom skeleton, but differed in their hydrogen bonding patterns. Horii and coworkers [35] suggested that the two ^{13}C-NMR spectra, obtained for polymorphs I_α and I_β, correspond to the resonances for the two-chain and eight-chain unit cell regions of cellulose. The polymorph I_α can be converted to the more stable I_β phase by annealing in various media. Annealing at a temperature of 270 °C converts most of the I_α cellulose to the I_β form. Bacterial cellulose has the highest percentage of I_α polymorph, which is 70%. The existence of I_α and I_β polymorphs in cellulose samples may affect the reactivity of native cellulose, as I_α is metastable and thus more reactive than I_β, according to Yamamoto [36]. The triclinic and the monoclinic structures correspond to I_α and I_β, according to the electron diffraction spectra. The triclinic phase is metastable and annealing it in dilute alkali at 260 °C converts it into the monoclinic form [37, 38]. Essentially, these structures are very similar and interconversion between them is achieved by the slipping of intersheet hydrogen bonds between cellulose sheets to give a slightly different pattern of cellobiose repeating [38].

In the case of cellulose biosynthesis, the groups or rosettes of particles, or terminal complexes (TC), are seen in the plasma membrane when viewed by freeze fracture techniques [39]. These groups of TC can be seen to be associated with the ends of microfibrils (collections of Cellulose chains), and are involved in the elongation of whole cellulose microfibrils. Thus, all the chains in one microfibril would have to be elongated by the complex at the same rate. This requirement means that the complex would need to be composed of many subunits, each elongating a single chain at a time. Crystalline Cellulose I is not the most stable form of cellulose. It is unlikely to be synthesized by the crystallization of preformed cellulose chains, as such a process carried out in vitro gives rise to the thermodynamically more favorable Cellulose II. Cellulose II, the second most extensively studied form, may be obtained from cellulose I by either of two processes:

- regeneration, which is the solubilization of cellulose I in a solvent followed by reprecipitation upon dilution in water to give cellulose II; or
- mercerization, which is the process of swelling native fibers in concentrated sodium hydroxide, to yield cellulose II on removal of the swelling agent. It should be noted that regeneration gives a higher level of conversion of cellulose I to cellulose II [40]. Kuga and coworkers [41] reported a mutant strain of *Acetobacterxylinum* as producing native folded-chain cellulose II. In the mercerization process (the treatment of cellulose I with alkali to achieve cellulose II) no solubilization occurs, which seems to imply that the fibrous structure of the cellulose would be maintained.

The mechanical properties of natural and regenerated cellulosic fibers have been found to be completely different, with natural fibers exhibiting near linear behavior, whereas the regenerated fibers show nonlinear behavior [8]. Differences in the hydrogen bonding patterns reported for models of cellulose I and II are not solely derived from deviations in hydroxymethyl conformation, but also from the fact that the polarity of the chains is popularly thought to differ; a parallel arrangement [3] is attributed to cellulose I and an antiparallel arrangement is proposed for cellulose II. Celluloses III_I and III_{II} may be obtained reversibly from celluloses I and II, respectively. Celluloses III_I and III_{II} [42] are formed, in a reversible process, from celluloses I and II, respectively, by treatment with liquid ammonia or some amines, and the subsequent evaporation of excess ammonia [43]. Polymorphs IV_I and IV_{II} [44] may be prepared by heating celluloses III_I and III_{II}, respectively, to 206 °C in glycerol. Cellulose IV is also produced by heating cellulose I or II in glycerol at 280 °C, or by boiling the cellulose–ethylene diamine complex in dimethyl formamide. Howsomon and Sisson [45] have suggested that cellulose III and IV can be regarded as disordered forms of celluloses II and I, respectively.

Based on stress invariance of the cellulose fiber, Eichhorn and coworkers [8] have concluded that the material structure of cellulose can be modeled using a modified series aggregate model structure, composed of parallel orientated and mutual regularly shifting elementary fibrils. In such a structure, the stress is uniform and equal within each element (composed of crystalline and amorphous regions). The surprising result is, however, that despite the different crystal structures of natural and regenerated cellulose (cellulose I and II), the rate of shift with respect to stress is equal for each type. Frey-Wysling and Mühlethaler [46, 47] indicated that Microfibrils are composed of Elementary fibrils having an average width of 3.5 nm. The elementary fibrils seem to be crystalline along their entire length. Pozgaj and coworkers [48] described elementary fibrils as having a width in the range of 3.5–10 nm and length of 30–80 nm. By using the ^{13}C-CP/MAS NMR spectroscopy technique, Heux et al. [49] established that the cellulose extracted from sugar beet pulp exhibited a crystallite size of 4 nm, which was in good agreement with the transmission electron microscopy (TEM) observations. High-resolution solid-state ^{13}C-NMR spectra have been taken using several native cellulosic materials, as well as regenerated low-DP cellulose I, implying that Native celluloses exhibit heterogeneous crystalline structures [50]. Therefore, the possibility that native celluloses are biosynthetically tailored composites certainly exists. From an excellent correlation of results obtained separately by the two methods of analysis, it was found highly probable that the chains of *Valonia* cellulose packed with parallel polarity maintained a twofold screw axis for the rigid components of the glucose ring, but showed some flexibility of the hydroxymethyl group rotations. No evidence for crystallite orientation other than parallel to the fiber axis was found [4]. Topographical images of the crystallites (highly crystalline regions of the original cellulose fiber) gained by using atomic force microscopy (AFM) have revealed corrugations across the surface of each crystal, with three spacing related to the 0.52 nm glucose interval, the 1.04 nm cellobiose repeat distance, and a ~ 0.6 nm repeat, matching the intermolecular spacing between chains [51]. A priori crystal structure prediction of native celluloses, by using a chain pairing molecular model, indicates [52] that only a few of all the low-energy three-chain models yield closely packed three-dimensional arrangements. The cellulose chains in the selected models form layers, stabilized by interchain hydrogen bonds. The emergence of several energy minima suggests that parallel chains of cellulose can be paired in a variety of stable orientations. The two best crystal lattice predictions were for a triclinic (P_1) and a monoclinic ($P2_1$) arrangement with the following unit cell dimensions: $a = 0.63$, $b = 0.69$, $c = 1.036$ nm, $\alpha = 113.0$ °, $\beta = 121.1°$, $\gamma = 76.0°$, and $a = 0.87$, $b = 0.75$, $c = 1.036$ nm, and $\gamma = 94.1$, respectively. They correspond closely to the respective lattice symmetry and unit-cell dimensions that have been reported for cellulose I_α and I_β allomorphs. At present, the internal structural features of cellulose microcrystals remain undetermined, although some NMR spectroscopic evidence indicates that the inner structure may exhibit [51] a high degree of organization. Obviously, Elementary fibrils, also known as *cellulose crystallites, cellulose nanorods, cellulose whiskers, nanocellulose*, and so on, are the smallest elements of all cellulose substances, which are characterized by the highly structured organization. With respect to the size of C, O, and H atoms, length and orientation of the chemical bonds, and assuming a 4C1 chair conformation of the anhydroglucopyranose unit it is possible to formulate a model of elementary fibrils. Hydrogen bonding within cellulose chains may determine the "straightness" of the chain. Interchain hydrogen bonds might introduce order or disorder into the system, depending on its regularity. When considering hydrogen bonding, it is essential to note the conformation of the C(6) hydroxymethyl group [52]. Due to inter- and intramolecular hydrogen bonds among the macromolecular cellulose chains, the nanostrands of elementary fibrils are formed by mutually self-crossing the cellulose macromolecules at an angle of 60° (Fig. 8.1) and regularly alternating the entanglement into a half part structure of opposite cellulose chains, resulting in extraordinary strength of the elementary fibril as similar as a *Czech Easter Monday* the whipping tool "pomlazka" (Fig. 8.2). The arrangement of both ends of every elementary fibril are a result of the dispersion of cellulose chain lengths, which are more disordered as they are formed after the common connection, resulting in a less compact Amorphous cellulose part of the cellulosic matter – a microfibril. Evidence exists

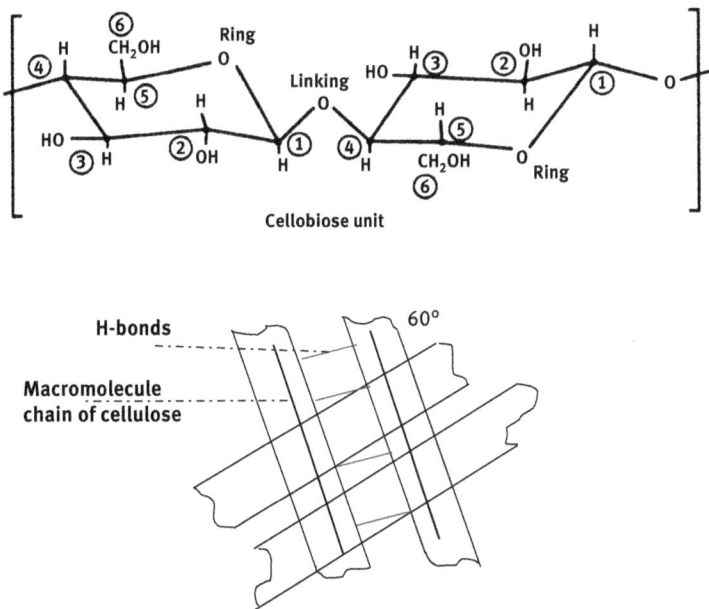

Fig. 8.1: Schematic representation of part cellulose chain nanostrands in two planes: one located above and the other twisted mutually and forming elementary fibrils.

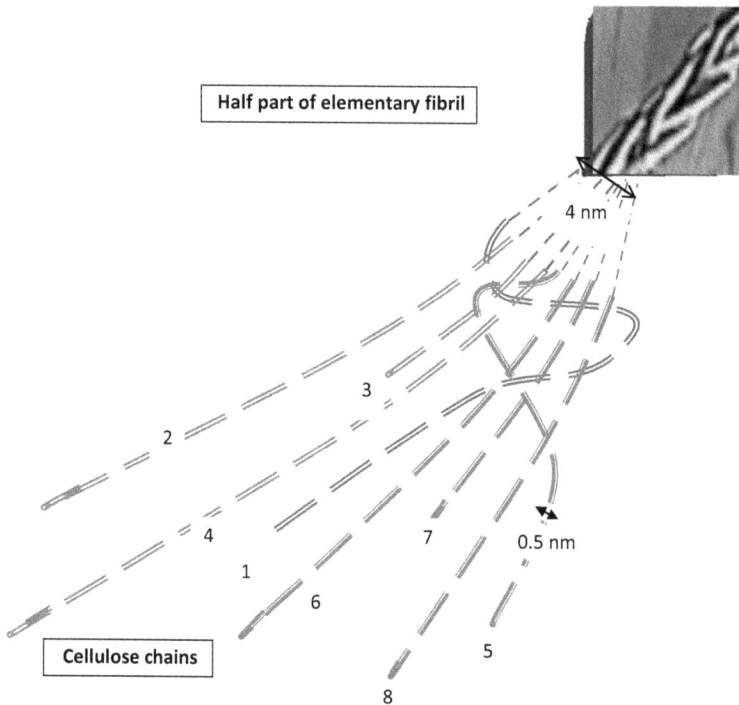

Fig. 8.2: Schematic representation of the half part of elementary fibrils formed by raveling the individual cellulosic chains into a nanoentanglement, similar to the whipping tool made of willow twigs.

detailing the varying extent of the compact matter of cellulose and is presented in Fig. 8.3. Histograms of grey values of pixels measured by synchrotron X-ray microtomography on ID19 multipurpose beamline, ESRF (European Synchrotron Radiation Facility), Grenoble, have revealed the existence of three and two types of matter in papers (bleached linters) with different compactness, which were prepared from only Defibrillation and beaten linter slurry, respectively. In the case of paper made from beaten linters, the disappearance of the high compactness of the third part was most likely due to homogenization during the Beating process.

Preston [53] proposed a model for the microfibril in which one central crystalline core is embedded in a paracrystalline cortex of molecular chains, which lie parallel along the microfibril length, but otherwise are not stacked in a crystalline array. The dimensions of the crystalline core vary with, but are smaller than, the microfibril width, which varies by 30 to 400 nm. Many studies, based on a variety of physical and chemical methods, have indicated that the microfibrils are not completely crystalline, but instead contain two distinctly different regions. The amorphous material in native fibers is composed partly of Noncellulosic constituents (e.g., hemicelluloses and

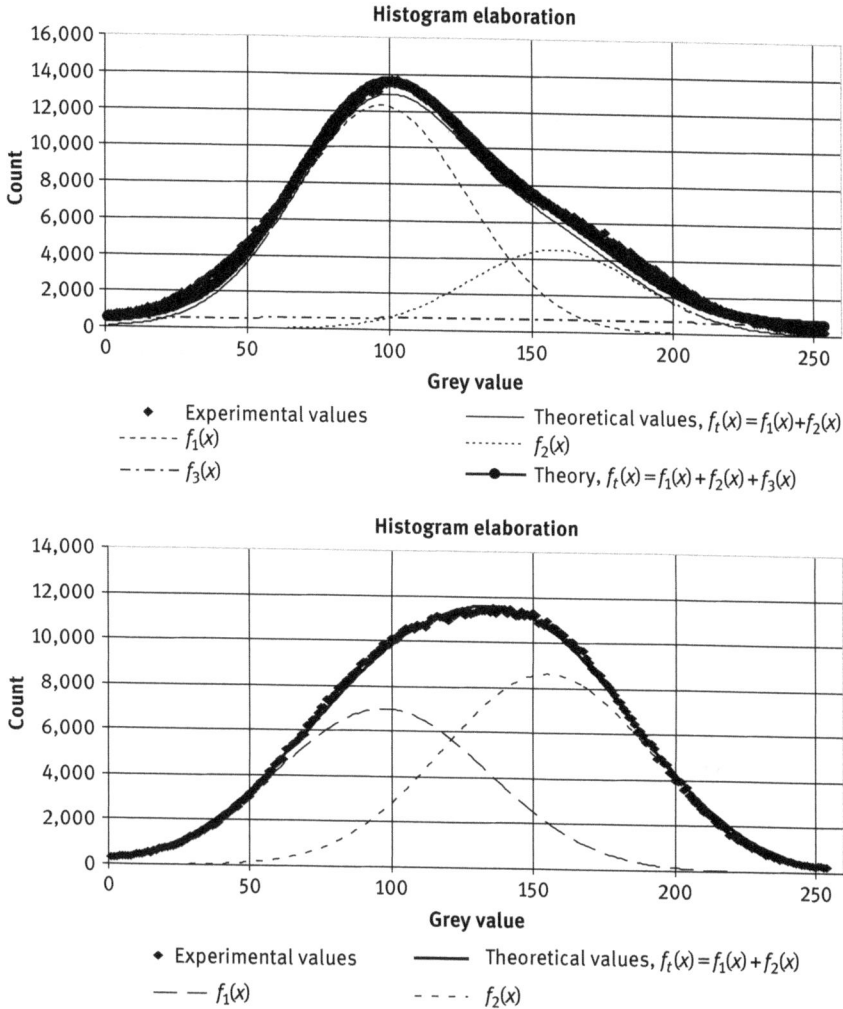

Histogram elaboration

Legend (first chart):
- ◆ Experimental values
- ----- $f_1(x)$
- ─·─·─ $f_3(x)$
- ——— Theoretical values, $f_t(x) = f_1(x) + f_2(x)$
- ·········· $f_2(x)$
- —●— Theory, $f_t(x) = f_1(x) + f_2(x) + f_3(x)$

Legend (second chart):
- ◆ Experimental values
- — — $f_1(x)$
- ——— Theoretical values, $f_t(x) = f_1(x) + f_2(x)$
- - - - - $f_2(x)$

Fig. 8.3: Histogram of grey values verified by using Gauss probability functions. Bleached linters, synchrotron X-ray microtomography on ID19 multipurpose beamline, ESRF, Grenoble, France. A paper was prepared by dewatering of pulp slurry composed only of *defibrillated linters*. Sample 100628/CZ/Linters/A. Slice 132/256. Std Dev = 46.29. Results of verification: $f_1(x)$: mean = 97; dispersion = 30; $f_2(x)$: mean = 158; dispersion = 30; background $f_3(x)$: mean = 152; dispersion = 160. $F_1/F_t = 0.655$; $F_2/F_t = 0.240$; $F_3/F_t = 0.105$ if $\int_1^{255} f_i(x)\,dx = F_i$ and $\sum^{F_i} = F_t$. A second paper was prepared by dewatering of pulp slurry composed only of *beaten linters*. Sample 100628/CZ/Linters /B. Slice 239/256. Std Dev = 45.77. Results of verification: $f_1(x)$: mean = 97; dispersion = 37; $f_2(x)$: mean = 155; dispersion = 37. $F_1/F_t = 0.448$; $F_2/F_t = 0.552$ if $\sum^{F_i} = F_t$.

lignin). There is also an appreciable amount of cellulose, which is in an amorphous state. It is believed that the long cellulose molecules protrude at the surface of the crystallites to give a fringe structure. These protruding chains permit association to take place between the cellulose and noncellulosic constituents of the fibers, thus forming an amorphous mixture of cellulose and noncellulosic substances. Noncellulosic polysaccharides (e.g., hemicelluloses such as xyloglycans, gluco-mannans, and glucuronoxylans) exhibit a strong interaction with the microfibril surface and lead to a further apparent disorder [49]. Owing to intramolecular and Intermolecular H-bonds, the cellulose molecules are mutually bound to form the fiber structure of an elementary fibril. The main types of H-bond forming a super-molecular cellulose structure are presented in Fig. 8.4. Intramolecular H-bond are formed either between different planes or on the same plane in which separate cellulose chains are found that occupy the most advantageous 4C1 conformation. Two or three types of qualitatively different H-bonds can be distinguished [54, 55], following the scheme presented in Fig. 8.4:

Intramolecular H-bonds	Strength order of Particular
These include:	H bonds
– C'(2)OH–C(6)OH	The strength order is:
– C'(3)OH–O(ring)	
	intra C'(6)OH– C'(2)OH >
	intra C'(3)OH– O(ring) ÷
Intermolecular H bonds	inter C'(6)OH– C(2)OH ÷
H-bonds linking up single cellulose	intra C'(6)OH– C(6)OH ÷
chains in the plane located one	intra C'(2)OH– O(ring) ÷
above the other:	intra C'(6)OH– O(linking) >
– C'(2)OH–O(ring)	intra C'(6)OH– C(2)OH in the plane >
– C'(3)OH–O(ring)	intra C'(3)OH– O(ring) >
– C'(6)OH–O(ring)	intra C'(2)OH– C(3)OH
– C'(2)OH–C(3)OH	
– C'(6)OH–C(2)OH	
– C'(6)OH–C(6)OH	
– C'(6)OH–O(lin king)	
H-bonds linking single cellulose	
chains located in the same plane:	
– C'(6)OH–C(6)OH	
– C'(6)OH–C(2)OH	
– C'(6)OH–O(ring)	

Fig. 8.4: The main types of H-bonds possibly participating in the formation of the supramolecular structure of cellulose. Compare with the cellobiose unit in Fig. 8.1.

1.   H-bonds of irreversible nature:
   –   glycosidic oxygen with hydrogen atoms of several hydroxyl groups [C'(6)
       OH–O(1);
   –   hemiacetal oxygen with hydrogen atoms of several hydroxyl groups [C'(3)
       OH–O(5), C'(2)OH–O(5), C'(6)OH–O(5)].

2. Reversible H-bonds between amphoteric primary and secondary hydroxyl groups (groups with the ability to form hydrogen bonds with water where both act as the H-donor group and the H-accepting group, depending on the structure and composition of molecules or macromolecules in which they are present):

    – $[C'(2)OH - C(3)OH, C'(6)OH - C(6)OH]$

3. Slightly irreversible type: between primary and secondary hydroxyl groups such as

    – $C'(6)OH - C(2)OH$

These qualitatively different H-bonds cause to occur in cellulose oriented (crystalline), less oriented, and nonoriented amorphous zones [49]. Transitions among these zones are not smooth, but the orientation and degree of coherence differ in various parts of the fiber. Different arrangements can exist even in oriented zones. The destruction of H-bonds with water and water solutions will depend not only on their strength but also on the nature of the H-bonds. The destruction can be reversible or irreversible. The H-bonds formed by H-amphoteric groups are usually reversible H-bonds, while the groups of only donor or acceptor usually form the Irreversible H-bonds [56, 57]. The occurrence of H-amphoteric groups in the system causes a decrease of the crystalline zones. Moreover, cellulose can easily accommodate hydrophobic appending chains to overcome adverse interactions with nonpolar composite matrices [58], and the high melting temperature of cellulose Nano-crystals cellulose makes it a very attractive feature for designing materials that need to perform at high temperatures.

Intramolecular hydrogen bonds have been found to play an important role in the determination of crystallite modulus and the chain deformation mechanism [59]. The intramolecular hydrogen bonds stiffen the chains along their axis, while the intermolecular hydrogen bonds are responsible for chain packing and aggregation [60]. Recently, increased attention was devoted to nanocellulose or cellulose nanowhiskers, because of their large modulus of elasticity, strength, and large aspect ratio. Cellulose whiskers, made from tunicates, are quasi-perfect crystals with 18×9 nm lateral dimensions [17] and an aspect ratio of around 100. They have a tensile modulus (150 GPa) much higher than the usual modulus of polymers (around 3 GPa below the glass transition temperature T_g) and can form hydrogen bonds with their neighbors, thanks to the numerous hydroxyl groups on their surface. A crystallite modulus of native cellulose along the chain axis was calculated by Tashiro and Kobayashi [61], based on the X-ray analyzed molecular conformation and the force constants used in the vibrational analysis. The calculated values are 172.9 and 70.8 GPa for the cases with and without the intramolecular hydrogen bonds taken into account, respectively. The smallest structural elements of cellulosic fiber, that is, elementary fibers and microfibers may still be fused to adjacent structures forming fibrils and fibrillary bundles of thread-like shape, which are then twirled into a hypermolecular structure of varyingly porous walls of fiber with well-defined morphology. These nano-sized fibrils stick out of the surface of the cellulose fiber and they are of great importance for their paper-forming ability due to their important bonding potential [62]. The presence of nanosized fibrils on the surface of intact cellulosic fibers was later demonstrated in a series of scanning electron micrographs published by Alince [63]. These domains possess very high strength, approximately in the order or greater than a comparable steel structure. The density of cellulose crystals, which may be determined by crystallography, is related to the substance structure. The density of crystalline cellulose, as found in a single crystal, is 1.59 g/cm³, whereas that of pure natural fiber cellulose reaches only 1.55 g/cm³ [64, 65]. The experimental densities are 1.582 and 1.599 g/cm³ for I_α and I_β phases, respectively. The density of the I_α allomorph was very well predicted by Mazeau and Heux [66], using molecular dynamics modeling, and the density of the I_β allomorph is slightly lower than the expected value, by less than 3%. As a general trend for the polymer, the density of the Amorphous phase is expected to be 10–20% lower than that of the crystalline phase. The density of the amorphous systems should then be in the range of 1.28 and 1.44 g/cm³. Generally, in the case of a multicomponent material, the overall density is

$$\rho_s = 1 \Big/ \left(\sum X_i / \rho_i \right), \quad \text{for} \quad \sum X_i = 1$$

where X_i and ρ_i are the absolute concentration (w/w) and density of the ith component of the solid part of the material, respectively; as the volume of the multicomponent material V_S is equal to $\Sigma V_i = \Sigma m_i/\rho_i$, all matter m_S equals Σm_i and $\rho_S = m_S/V_S$; $\rho_i = m_i/V_i$; $X_i = m_i/m_S$.

8.3.2 Hypermolecular characteristic of cellulose and lignocellulosic materials

Hypermolecular structure and its quality are crucial in entering of substances composing an environment into porous matters. The biomatter such as wood and paper depending upon mechanical, thermomechanical, chemomechanical, and chemical method of wood refinery have similar chemical (i.e., molecular composition and its character) and supramolecular (i.e., Skeletal cellulosic matter) structure [67], but the hypermolecular structure is more different. In contrast with the paper, the plywood is more orthotropic inclusive pore structure, although both the biomatter were created by similar type of nanoprocess, that is, a "bottom up," but according to different creation plan, the stochastically papermaking process from cell fibers and their fragments versus the effective sophisticated plan of plant growth from monosaccharides molecules controlled by its genetic appurtenances. However, surface-molecular properties of pore walls interfaces are similar to their inclusive behavior.

Evidently Pore matter exist in **oriented pore structure** forms, for example, plywood, and **nonoriented structure** forms, for example, paper. The pores are divided
according to their structure and shape.
Shape of pores:

- open hole pores of double-sided character, that is, through flow pores, for example, all filter materials;
- open hole pores of one-sided character, for example, microsacks, microgaps,
  and so on;
- closed holes, that is, hollow holes, for example, microhollow holes.

Pore structures:

- oriented pore structures, that is, structured pore materials, for example, leather;
- nonoriented haphazard pore structures, that is, nonstructurated pore materials, for example paper.

As is typical for all pore matters predominantly for microporous matters, their tendency
to Vapour absorption of volatile substances especially hygroscopicity behavior is main
property of these ones. Due to specific properties of pore matter, some porous matter
appears especial behavior during its moistening process, because these ones have oriented the pore structures and a swelling followed by viscous-elastic changes of the pore
walls. As a result, in contradiction with paper and pulp matter [68, 69], an amount of
condensed water during moistening process is at first increased followed by depression
of this one. An example of such behavior where the pore material with *oriented pore
structure* during water vapor absorption and swelling of its pore walls is leather, that is,
collagen matter. However, evidently the Hygroscopicity or hydrophobicity of microporous materials is given not only by pore properties but also with surface-molecular
properties of pore walls interface, that is, by both the quality and quantity of pore matters. Adhesion tension of water, σ_T, is a qualitative parameter that directly controls behavior of interface the pore walls of porous materials with high specific surface and
their moisture content. The adhesion tension is connected with the surface–molecular
properties by relation $\sigma_T = y_{s,g} - y_{s,l}$, where $y_{s,g}$ and $y_{s,l}$ are surface tensions of the wall
interface characterizing its affinity to air and to liquid (usually water), respectively.
With an increase in both the surface tensions the affinity to a relevant environment decreases and vice versa, but the adhesion tension is not sensitive to qualitative surface
changes if the both surface tensions $y_{s,g}$ and $y_{s,l}$ are changed in parallel.

No simple method or solution exists to evaluate hypermolecular characteristics
of pore matter. Cesek et al. [70–72] have developed a methodology based upon molecular groping of pores by using the vapor condensation method. The method is
based upon controlled humidification and vapors absorption of selected organic
substances (toluene and methyl ethyl ketone) in porous nonstructuralized flat material and enables us to evaluate also surface-molecular properties of pore interface.
The methodology consists of following steps:

- gravimetrical measuring a kinetic of humidification at 49%, 75%, and 97% of relative air humidity of porous plane figured sample with defined geometry especially its outer area (for more details see also Section 8.5.3);
- gravimetrical measuring a kinetic of vapor absorption of methyl–ethyl ketone and toluene by porous flat sample;
- fitting of received kinetic data by use of well-defined theoretical mathematical relationship; for more details see [68, 69];
- calculation or detection the density of solid-state matter and partial (apparent) density of sample;
- dependence an absorption rate of sample humidification in the beginning of sorption process, v^h_o versus the relative interval of pores ($r_{min} \le r \le r(\varphi_{rel})$), $\Delta\varepsilon_r$, filled in steady state with condensed liquid;
- at last by using proper software in xls format we can evaluate the following parameters.

8.3.2.1 Hypermolecular structure parameters

- $\varepsilon_i(48\%)$, $\varepsilon_j(75\%)$, $\varepsilon_k(97\%)$ – wet sample porosity (V/V), that is, the porosity fulfilled with condensed water at RH 48%, 75%, and 97%, respectively.
- (i, j, k) – Effective maximal pore diameter (µm) the smaller pore inclusive, fulfilled with condensed water at RH 48%, 75%, and 97%, can be evaluated by using eq. (8.1) [72]

$$r(\varphi_{rel}) \equiv i(\varphi_{irel}), j(\varphi_{jrel}), k(\varphi_{krel}) = \frac{4M\,\sigma_T(1+\sin\,\theta)}{\rho_l\,RT\,\ln\,(1/\varphi_{irel},\,\varphi_{jrel},\,\varphi_{krel})} \qquad (8.1)$$

where $i(\varphi_{irel})$, $j(\varphi_{jrel})$, $k(\varphi_{krel})$ are Effective maximal pore diameter (µm) smaller pores inclusive, fulfilled with condensed water at RH 48%, 75%, and 97%, respectively; M is molar mass of water (kg/mol); T is the temperature (K); σ_T is Adhesion tension of liquid to pore surface (N/m); in capillary $\sigma_T = \gamma_{l,g}(1-\sin\theta)/\cos\theta$; θ is the Contact angle between the surface of the capillary tube and water's internal meniscus (this angle is usually equal to 0°); $\gamma_{l,g}$ is the liquid surface tension (water, $72.2\ 10^{-3}$ N/m at $T = 295$ K); ρ_l is the liquid density (kg/m^3); R is the gas constant (8.314 J/(Kmol)).

- ε, $(\varepsilon-\varepsilon_k)/\varepsilon$, $\varepsilon_i/\varepsilon$, $(\varepsilon_j-\varepsilon_i)/\varepsilon$, $(\varepsilon_k-\varepsilon_i)/\varepsilon$ – the total porosity of the porous sample (V/V), relative extent of the big pores in the porous material, relative extent of the small pores in the porous material, extent of pore volume dispersion in the porous material, respectively.
- Effective mean pore diameter and pore size dispersiveness (µm) provided that pore size distribution function is controlled by theoretical Gauss distribution.

- v^h_o (48%), v^h_o (75%), and v^h_o (97%) – absorption rate of sample humidification in the beginning of sorption of porous sample (1/day) at 48%, 75%, and 97% relative air humidity, respectively.
- Y_e (48%), Y_e (75%), and Y_e (97%) – the steady-state moisture of porous sample (g water/g of o.d. sample) at 48%, 75%, and 97% relative air humidity, respectively.

8.3.2.2 Surface-molecular properties of pore interface
- σ_T (mN/m) – adhesion tension of pore interface to water,
- y_s (mN/m) – Surface tension of pore–wall with air,
- y_d, y_p, y^+, y^- (mN/m) – Dispersion and polar part of surface tension of pore interface with air, polar part of surface tension of pore interface with air, cation-active polar part of surface tension of pore interface with air, anion-active polar part of surface tension of pore interface with air, respectively,
- d, p, p^+, p^- – relative dispersion (nonpolar) part of surface tension, relative polar part of surface tension, relative cation-active and anion-active polar part of surface tension of pore interface with air, respectively. It was established that vapor absorption process is controlled by diffusion and vapor condensation, that is, the maximum adhesivity, that is, an affinity of liquid molecules to microsurface molecules of pores, is achieved if the following conditions are fulfilled:

1. $y_s = y_l$
2. $p_s = p_l$ and $d_s = d_l$
3. $p_s^+/p_l^+ = p_l^-/p_s^-$ or $y_s^+ . y_s^- = y_l^+ . y_l^-$

This relationship is can be better represented in the following schematic presentation

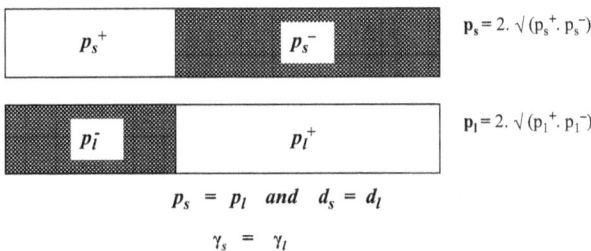

$p_s = 2. \sqrt{(p_s^+ . p_s^-)}$

$p_l = 2. \sqrt{(p_l^+ . p_l^-)}$

$p_s = p_l$ and $d_s = d_l$

$\gamma_s = \gamma_l$

However, for more detailed characterization of quality of pore-walls surface molecules during interactions with both environments (s–g or s–l type) are useable information about character of those surface molecules interactions as polarity (p) or dispersity parts (d) of these ones [70, 73], their **hydrophilicity** defined as p/p_{water}, their relative acidity or basicity y^+/y^- [73], and so on. Predominantly, the

relative basicity index y^+/y^- is important because it measures the basic orientation of water molecule to pore–wall surface molecule with hydrogen atoms in first basic nanolayer followed by nanolayers with more diffused character of this orientation and this one results generally in a character of forming hydration bonding system inside the micropore. The ratio y^+/y^- defines **the ability to form a hydration bonding system** [67, 74] among water molecules condensed inside of pores. With increasing value of this ratio, the ability to form Hydration bonding system increases as well, thereby achieving the maximal ability at $y^+/y^- = 1$ and vice versa. Reversely, nonbonding hydration system is prevailed in condensed water inside of pores. If the value of a relative basicity component is 1, then the basic water molecule orientation around whole micropore–wall surface is equilibrated with both the hydrogen atoms and oxygen atom of water orientation, that is, the surface of micropore contains 50% of nanolocalities groups of Proton-acceptor (PA) character and 50% nanolocalities with groups of Proton-donor (PD) character; thus, the basicity component (y^+ or PA monopole) is equal to acidity component (y^- or PD monopole) of $y_{s,g}$, which resulted in equilibrated influence of hydration forces inside the micropore. The hydration bonding system is formed. If the value of a relative basicity component is small or get closer zero, the basic water molecule orientation around whole micropore–wall surface is formed, with the oxygen atom of water orientation prevailing, that is, the surface of micropore contains only the nanolocalities with groups of PD characters; thus, the basicity component (i.e., PA monopole) is small or zero and in hydration system dominates only the acidity component (i.e., PD monopole of interface interaction with water molecules) of $y_{s,g}$, which resulted in prevailing influence of repulsive hydration forces inside the micropore. The hydration nonbonding system is formed. If value a relative basicity component is higher (predominantly if $y^+/y^- > 1$) the basic water molecule orientation around whole micropore–wall surface is formed, with the hydrogen atoms of water orientation prevailing, that is, the surface of micropore contains only the nanolocalities with groups of PA characters; thus, the basicity component (i.e., PA monopole) is high and in the hydration system dominates only the basicity component (i.e., PA monopole of interface interaction with water molecules) of $y_{s,g}$, which also resulted in prevailing influence of repulsive hydration forces inside the micropore. The hydration nonbonding system is formed.

Theoretically, at the same hydrophility, the water molecules are pulled into submicrospace of pores with principally prevailing hydration bonding system among those condense, that is, the ability of some micropores to form a hydration bonding system decreases the vapor tension of water condensed inside them. Otherwise, at comparable conditions and the same hydrophility of pore walls, the water molecules are better expelled from submicrospace of pores with principally prevailing hydration nonbonding system among those condensed, that is, the ability of some micropores to form a nonbonding hydration system increases the vapor tension of water condensed inside them. In other words, the hydration bonding system enables easer to occupy the pores, that is, at comparable pore–walls properties and conditions water condense in bigger pores, showing more hygroscopicity and vice versa.

Results of hypermolecular properties of paperboards prepared from typical mechanical and chemical Pulp are listed in Tab. 8.1 and 8.2. As a result of eq. (8.1), an effective maximal pore diameter fulfilled with condensed water at RH 48%, 75%, and 97% is strongly influenced by value of adhesion tension of water. The affinity of pore–wall molecules to water molecules increases with an increase in adhesion tension. With an increase in adhesion tension, the size of pores fulfilled with condensed water increases. Generally, all pulps, groundwood inclusive, are more Hydrophobicity than hydrophilic. As expected, adhesion tensions of water to bleached pulps achieve relatively higher values. However, the Groundwood is hydrophobic but its structure is formed of a large number of small pores. All pulps groundwood inclusive assert anion-active character of their polar intermolecular interactions. The porosity fulfilled with condensed water achieved higher values for small size of pores in case of groundwood paperboards compared with bleached hydrophilic pulps. Although more hydrophobic, the groundwood achieve maximal hygroscopicity because they have highest porosity formed by smaller pores – compare data in Tab. 8.2. Due to this fact, the maximal effective pore diameters fulfilled with water for groundwood achieved the lowest values. Differences among moistening rates of the paperboards of different composition are small because this rate is controlled by vapor diffusion, followed by those quick condensation in appropriate pores. Again, the groundwood have relatively or more hydrophobic character, but their pore structure is predominantly formed with a large number of small pores.

Tab. 8.1: Basic porosity properties of evaluated paperboards.

Paper parameters	DM (groundwood)	Sa J (bleached sulfate softwood pulp)	Sa L (bleached sulfate hardwood pulp)	Si (MgBi-sulfite spruce bleached pulp)
Apparent density, kg*m^{-3}	442.73	643.26	581.05	708.55
Porosity ε, %	71.6	59.29	63.2	55.2
ε_i (48%), %	2.76	3.34	3.07	3.67
ε_j (75%), %	4.80	5.21	5.27	5.88
ε_k (97%), %	12.78	13.73	15.19	16.57
i (µm)	0.2	0.6	0.5	0.5
j (µm)	0.5	1.5	1.2	1.2
k (µm)	5.1	14.3	11.7	11.0

Tab. 8.2: Basic porosity properties and surface-molecular properties of evaluated paperboards.

Paper parameters	DM (groundwood)	Sa J (bleached sulfate softwood pulp)	Sa L (bleached sulfate hardwood pulp)	Si (MgBi-sulfite spruce bleached pulp)
Apparent density, kg*m^{-3}	442.73	643.26	581.05	708.55
Porosity ε, %	71.6	59.29	63.2	55.2
Effective mean pore, μm	2.3	6.8	4.9	4.7
Dispersiveness of pore distribution, μm	3.4	26	10	12
σ_T, mN/m	2.6	7.5	6.1	5.7
γ_s, mN/m	33.2	33.7	33.5	33.5
γ_d, mN/m	30.1	30.1	30.1	30.1
γ^+, mN/m	0.28	0.28	0.28	0.28
γ^-, mN/m	8.6	11.6	10.7	10.5
γ_p, mN/m	3.1	3.6	3.5	3.4
d	0.91	0.89	0.90	0.90
p	0.09	0.11	0.10	0.10
p^+	0.01	0.01	0.01	0.01
p^-	0.26	0.34	0.32	0.31
Y_e (97%) (g/g)	0.29	0.22	0.26	0.24
Y_e (75%) (g/g)	0.11	0.08	0.09	0.08
Y_e (49%) (g/g)	0.06	0.05	0.05	0.05

8.4 Water and cellulose/cellulosic substances

Water is the key component in controlling the behavior of all biomaterials, living organisms, and so on. We have distinguished that liquid water in porous systems is as follows:

- Self-organised water, that is, a Bulk of water whose behavior is independent of the character and composition of the porous system; and
- Medium-enforced organised water that is, water organized in the porous system, particularly in the microreticular system, which is controlled by the

surface molecular properties of components forming this system. The states of liquid water are dependent on the specific surface area of the pores. The medium-enforced organized water is typical for microporous and nanoporous media, with the so-called natural nanostructured polymers being typically hygroscopic.

Undoubtedly, water plays an extraordinary role in the formation, manufacturing, aging, and influencing the properties of cellulose and cellulosic materials. The main focus of interest involves cellulose and cellulosic materials in a wet state, predominantly natural or in suspensions, during paper preparation. Formerly, a lot of attention was dedicated to the Electrokinetic behaviour of cellulosic fiber slurries during paper-making, that is, the pulp suspensions [75, 76]. The word "electrokinetic" implies that an electrical current or potential arises due to the relative movement between two phases, as in the case of anion-active cellulosic fibers or fines suspended in water. Broadly, all prevailing scientific works operate using the concept of Zeta-potential, cationic demand, polyelectrolytes, colloidal stability, and so on, because the charged nature of a cellulosic fiber surface is expected to play major roles in phenomena such as fiber dispersion, Flocculation, and the adhesion and adsorption of polyelectrolytes [77]. Using electrokinetic data, papermakers have used a variety of electrokinetic procedures to control and optimize the levels of additives. Electrokinetic data helped to predict the dosages of charged substances needed to achieve different balances of stability versus coagulation and flocculation of the suspensions.

8.4.1 Liquid crystalline cellulose suspensions

The era of nanotechnologies and nanomaterials evoked an increased interest of elementary cellulose components with a sophisticated structure, that is, cellulose crystallites [51]. The interest was concentrated upon stable colloidal suspensions of cellulose microcrystallites [78] prepared using optimized acid hydrolysis of native cellulose fibers, as above a critical concentration, the suspensions form a chiral nematic-ordered phase, or "colloid crystal" exhibiting chiral nematic (cholesteric) ordering – a liquid crystalline cellulose suspension. Revol and coworkers [79] reported that suspensions of acid-hydrolyzed cellulose crystallites can also form an ordered phase displaying chiral nematic orientation. Above critical concentrations, usually greater than ~1.5% (w/w), the suspensions separate into two phases, with the denser lower phase exhibiting birefringence [80] indicative of the chiral nematic phase. The rheology of rigid rod cellulose whisker suspensions has been broadly investigated [80, 81]. Preparation [82] conditions govern the properties of the individual cellulose microcrystallites, and hence the liquid crystalline phase separation of the cellulose suspensions. The typical dimensions of these crystallites are between 10 and 20 nm in diameter for a length of approximately 1 μm [83]. After hydrolysis conditions are optimized, the cellulose colloidal suspensions exhibit nematic liquid crystalline alignment. Nematic ordering is the alignment of fibrils in the same direction along an axis and has long been sought in the fiber industry with the promise of higher tensile strength. Liquid crystalline suspensions of cellulose microfibrils are particularly attractive because of their simplicity, consisting of charged microfibrils that "self-assemble" into large regions of cholesteric ordering. The microscopic study of algal cellulose [84], using TEM and AFM, supports the presence of such a chiral twist along the microfibrils. Without a magnetic field or a shear field to enhance the Liquid crystalline ordering of the cellulose microfibrils, the 2D small angle neutron spectroscopy (SANS) and X-ray scattering profiles are isotropic "powder" patterns. Magnetic and shear alignment of the cellulose microfibrils results in anisotropic scattering patterns. The magnetic field induces the orientation of the liquid crystalline phase. In aqueous suspension, these crystals are oriented so that their long axes become perpendicular to the field direction [85]. In addition, the observation that cellulose crystallites align perpendicular to the direction of a magnetic field is in direct contrast to existing alignment technologies, all of which exhibit positive diamagnetic susceptibility anisotropy and, therefore, align parallel to the field [51]. Observations [84] revealed that the "fingerprint" patterns, indicative of the chiral nematic phase, are deformed and disappear with increasing shear rate. SANS data obtained from an in situ shear cell placed in a neutron beam provided evidence that as the shear disrupts the chiral nematic phase, microfibrils exhibit nematic ordering with the microfibrils aligned parallel to the flow direction. Additionally, SANS data confirmed that the cholesteric axis of this phase aligns along the magnetic field, with implications that the distance between the microfibrils is shorter along the cholesteric axis than perpendicular to it. This is consistent with the hypothesis that cellulose microfibrils are helically twisted rods. It is hypothesized that the source of the chiral interaction is directly attributable to the packing of screwlike rods. As shown by Orts and coworkers [84], threaded rods can be packed more tightly when their main axes are offset, such that their "threads fit within the each other's grooves." Even a small twist in the cellulose microfibril would be sufficient to induce the observed chiral interaction. A more subtle explanation of this ordering is based on the anionic stabilization only via repulsion forces of the electrical double layers taking place, without any influence of the water molecules. Nevertheless, it was proved that the suspensions of cellulose crystallites were not stable at sufficiently high electrolyte concentrations. Apparently, a decrease in the double layer thickness increases the chiral interactions between the crystallites [86, 87]. The stability of the colloidal suspensions of the Rod-like shaped cellulose crystallites with negatively charged sulfate groups, that is, the isotropic-to-chiral nematic phase equilibrium, is sensitive to the nature of the counterions present in the suspension [88]. Although anisotropic phases did form under certain conditions, often either a gel or an isotropic phase was produced. Increasing the total concentration of colloidal suspensions resulted in an increase of the relative proportion of the chiral nematic phase with a higher density. For inorganic counterions, the critical concentration for ordered phase formation increases in the order $H^+ < Na^+ < K^+ < Cs^+$ [88]. Suspensions with H^+ counterions formed an ordered phase at the lowest concentrations of crystallites, that is, with the highest exclusion tendency. For the inorganic cation series, the hydration number and the hydrated ion size decreased with increasing atomic number in the order $Na^+ > K^+ > Cs^+$. When these hydrated cations bind to the negatively charged particle surface, a Hydration force will be generated between the particles. For hydrophilic colloidal suspensions (S), such as cellulose crystallites, the hydration force is repulsive and generally increases with the hydration number of the counterions, and so the repulsive hydration force is expected to be in the order S-Na > S-K > S-Cs. Obviously, the relative difference between existing attractive and repulsive forces among rodlike Cellulose crystallites in water was disturbed by the presence of additives, and due to this disturbance a new equilibrium is reached, introduced by the formation of the isotropic-to-chiral nematic phase. The phase separation of rodlike polyelectrolytes is among others that are determined by Interparticle forces, such as steric repulsion, electrostatic repulsion, and hydration and hydrophobic interactions [86]. The temperature influence connected with the action of the interparticle forces is well known. Dong and Gray [88] observed an anomalous abrupt decrease of the volume fraction of the anisotropic phase at temperatures of 40–50 °C, but only in the H-form of the cellulosic suspension; under the same conditions, the Na-form did not exhibit the same response. The assumed explanation is based on the desulfation reaction of the cellulose crystallites, because for the neutral salt-form suspensions, the desulfation reaction does not occur due to the absence of an acid catalyst; however, in other place of this work it is alleged that at 35 °C for 24 h, the desulfation of the acid form of the suspension is negligible and can be ignored. Unfortunately, a detailed description of the nematic anisotropic denser phase is not reported. Similar liquid crystalline characteristics, which are quite unusual for aqueous suspensions of cellulose microfibrils, have been observed by Dinand and coworkers [89] who noted that the microfibrils of parenchymal cell cellulose (PCC), prepared from the cell wall of delignified sugar beet cell ghosts, consisted of a loose network of cellulose microfibrils. The PCC suspensions did not flocculate or sediment as long as some hemicelluloses and pectin were maintained at the microfibril surface. However, flocculation occurred if these encrusts were removed by either strong alkali or a trifluoroacetic acid treatment; these observations indicate the formation of a structured microreticular system of cellulosic microfibrils, mutually connected due to the presence of water.

8.4.2 Wet-web strength and wet strength of cellulosic materials

The presence of surface functional groups that generate hydrolytically stable covalent bonds in the fiber joint will increase the wet strength. Furthermore, if such bonds can form in the wet end of the paper machine, they will contribute to the Wet-web strength. Wet tensile strength is a key functionality for tissue paper, paper towel, paper board for packaging, cellulosic filter sheets, and similar paper grades. The chemistry, mechanisms of action, and applications of commercial wet strength resins have been extensively studied [90, 91]. Wet strength can be increased by either surface treatment, that is, the surface modification of fibers, with small molecules, or by surface treatment with macromolecules, called *wet strength resins*. Conventional wet strength resins contain highly reactive chemical groups, which can crosslink the resin and graft them to the fiber surfaces. Mechanical entanglement, homocross, and cocrosslinking mechanisms are available [92–98]. By contrast, it is not obvious how polyamines, such as polyvinyl amine, polyethyleneimine, and so on, provide wet strength [99]. Laleg and Pikulik [100], working with never-dried webs (wet-web strength) and rewetted sheets (wet strength) made of mechanical pulps, proposed that three possibilities of the fibrous network crosslinking exist: chemical bond by amino-linkage between the carbonyl group and amine group, ionic attraction between the carboxyl and amine groups, and an irreversible H-bond between the –OH group of cellulosic materials and the $-NH_2$ group of the polyamine. It is known that the wet-web strength (initial strength of the paper) or its bond system, respectively, is determined by the mechanical entanglement of fibrous formations and by the mutual physical bonds between them. The extent of the mutual-bonding ability increases with an increase in the degree of beating, but this is detrimental to the mechanical entanglement. If a constant contribution to mechanical entanglement is assured (for instance, by a constant degree of beating), then the changes in the wet-web strength will indicate the changes in the wet state fiber bonds. It was observed [54, 101] that application of well-soluble nonionic urea, which strongly influenced the Hydration bond system, that is, makes it weaker; this bond system is effectively negatively influenced. The wet-web strength, measurable under comparable conditions with a relative hydration factor (for further details see Fig. 8.6), falls with the rise of repulsive hydration forces. DiFlavio and coworkers [102], in their comprehensive laboratory study of the mechanism of polyvinylamine (PVAm) wet-strengthening, have concluded that PVAm should not form many covalent bonds with untreated cellulose. However, PVAm gave low adhesion to untreated cellulose, whereas the 2,2,6,6,-tetramethylpiperidine-1-oxyl radical oxidized cellulose resulting in strong adhesion [103]. This treatment converts the C6 cellulose hydroxyl to carboxyl and aldehyde groups, which will react with the amine groups on PVAm. Amines will couple to aldehydes to form imines, but only a few terminal aldehyde groups, which are accessible to PVAm, will react with the amines in PVAm. Therefore, PVAm will not form many covalent bonds with untreated cellulose. They observed [102] that the occurrence of strong adhesion is practically unchanged from pH 3 to 9, which is beyond the range of imine formation. It suggests that physical bonding is also significant. But what bonds occur? DiFlavio and coworkers [102] proposed that PVAm strengthening is due to a combination of covalent bond formation and electrostatic bonding, but they also noted that it is impossible to decouple the effects of electrostatic and covalent bond formation. Or, does another sort of H-bond or physical bond exist?

8.4.3 Interactions in cellulosic fibrous slurries using enthalpiometric measurements

Previously, using the Microcalorimetric method [104, 105] for the investigation of amine interactions in cellulosic fibrous slurries, it was found that the Interactions of nitrogen compounds in aqueous pulp suspensions exhibit measurable exothermal or endothermal effects. These observations enable us to follow the character of firmness of the bonds being formed, as well as the mechanism of their formation. These types of interactions are due to the mechanism of releasing water molecules from active centers on the cellulose substrate and the formation of stronger H-bonds of the "nitrogen–hydrogen" type with amine groups, instead of the weaker " oxygen–hydrogen" type with water [105]. It has also been observed [88] that the intermolecular hydrogen bonding, generated from cellulose backbones of the H-form, is much stronger than that of the Na-form.

Two reasons for Heat effects in cellulosic fibrous slurries exist:
– formation of strong H-bonds and
– formation of strong hydration bonds due to oriented water molecules present at interacting interfaces.

The following conclusions have arisen from the results summarized in Tab. 8.3 [104]:
– Measurable heat changes are only exhibited by nitrogen compounds.
– Differential heat and corresponding entropy changes are constant throughout the whole concentrations series of dosed reaction components.
– The interactions lead (except diethylamine (DEtA)) to an increase in the entropy of the system.
– The ratio between the maximum concentrations of the separate adsorbed reaction components is practically a whole number.

These facts reveal that the mechanism of interactions of monomolecular compounds does not change with their doses. The increase of the system entropy can most likely be accounted for by the increased water mobility and, therefore, the state of water molecules being released from the active locations of the cellulose substrate formed by the pulp. The measured heat changes are therefore the result of the released heat coming from the formation of a new, stronger H-bond between the reaction component and the active locations of the cellulose, and the consumed heat during the release of water molecules from these active locations. The level of these resulting heat changes (differential heats, maximum released heat $-\Delta H_{max}$) can therefore be considered to be a relative criterion of the strength of the originating bonds. The results in Tab. 8.3

Tab. 8.3: Sorption data related to the interaction of 1% (w/v) water suspension of unbleached sulfite pulp with 0.5–4% (w/v) water solutions of monomolecular compounds [104]. T = 293 K.

Compound	Max. adsorbed amount, $n_{A\ max}$ (mol/g pulp)	$-\Delta H_{max}$ (J/g pulp)	ΔS (kJ/mol K)	Differential heat, $d\Delta H_r/dn_A$ (kJ/mol)
Methylethyl-ketone		Nonmeasurable heat		
Ethanol		Nonmeasurable heat		
Acetamide	12.5×10^{-4}	8.4	0.069	−6.7
Formamide	–	–	–	−3.3
Urea	24.8×10^{-4}	6.6	0.075	−2.8
Ammonia	25.5×10^{-4}	15.0	0.094	−5.9
Ethylamine	13.4×10^{-4}	23.0	0.107	−15.0
Diethylamine	3.9×10^{-4}	23.0	− 0.034	−50.0
Triethylamine	13.4×10^{-4}	13.0	0.053	−12.0
Ethylendiamine	6.7×10^{-4}	20.0	0.013	−30.8
Diethylentriamine	4.5×10^{-4}	10.0	0.023	−29.7

confirm that the strongest bonds are formed by DEtA or secondary amino nitrogen, respectively.

Supposing that, during an interaction, one ammonium molecule is substituted by one water molecule, the mechanism and its stoichiometry can be formulated. It follows that the originating H-bonds 'nitrogen vs. oxygen' are stronger than H-bonds, by which water molecules bond to the cellulose substrate, because the heat of this interaction is exothermal.

8.4.3.1 Interpretation of enthalpiometric observations

Heat changes will be greater if bonds other than H-bonds are formed; particularly if they are of different qualities. "Nitrogen–oxygen" bonds are qualitatively different from the usual "oxygen-hydration" bonds; H-bonds "nitrogen–oxygen" can be divided into two groups:

1)   the proton-donating group is on the nitrogen compound

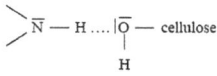

$$\diagdown \overline{N} - H \dots |\overline{O} - \text{cellulose}$$
$$\qquad\qquad\quad |$$
$$\qquad\qquad\quad H$$

(including ammonium, primary, and secondary amino groups);
2)   the proton-donating group is on the cellulose or on another compound

$$\diagdown N| \dots H - \overline{\overline{O}} - \text{cellulose}$$

(tertiary amino groups belong there).

These two types of H-bonds enable the formation of a wide Spectrum of H-bonds, depending particularly upon their polarity as in the H-bonds "hydrogen–oxygen." Continued polarization of the H-bonds of the second type leads to the formation of highly polarized bonds, which are formed by quaternary amino groups of an ionic type (third type). Breaking the Hydration layers around the interacting components will occur if the energy level of the originating bonds is the same or greater than the level of bonds within, which are fixed on the interphase of the water molecule or their clusters. The mechanism of this process, in the case of cellulosic material, can be schematically represented in the following way (Cell – representing the cellulosic material, K – the reactive component).

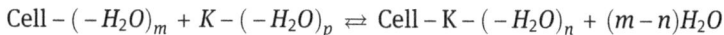

$$\text{Cell} - (-H_2O)_m + K - (-H_2O)_p \rightleftharpoons \text{Cell} - K - (-H_2O)_n + (m-n)H_2O$$

Due to the interaction of the hydrated molecules of the reactive component K with the surface of the cellulosic material, shifts or the release of water molecules and their clusters occur from the active locations on the cellulose substrate and within the hydration shell of the reactive component; thus, a more stable bond is formed. The $(m-n)$ value can be positive or negative. The process is naturally accompanied by corresponding energy and entropy changes. The formation of the bond between the cellulosic material and the reactive component K is accompanied by energy release, particularly if a qualitatively different bond is formed, from the original one, with water. The Release of water molecules and their clusters or shifts are linked to energy consumption; this is analogous within the ordered state region of separate water molecules and the K component, but in the reverse order. However, it can be assumed that greater entropy changes – increase in entropy – will appear in the water molecules than in the K component molecules – decrease in entropy. As heat effects only occur if energy changes happening within the system are at least on a molecular level (which are linked to the inner energy of the molecules, character and size of the bonds, etc.), heat effects will presumably only exhibit interactions leading to changes in hydration layers. This means that the interactions schematically represented above (1) will be accompanied by heat changes of exothermal and endothermal character. If $(m - n) = 0$ and the energy of the H-bond in the reactive component combined with water is approximately the same as in combination with the cellulose material, then the resulting heat effect will be zero. The amount of exothermal and endothermal heat will depend on the strength of the newly formed bonds and on the relative amount of water molecules, either released or displaced to another energy stage of water molecules and their clusters. The amount of exothermal heat will then relatively indicate the strength of the formed bond. The higher the released heat, the relatively stronger the bond and vice versa in the case of the endothermal character of the interaction: the lower the value of the consumed heat, the relatively firmer the bond [104]. The release mechanism of water molecules, as well as hydrogen bond formation, satisfactorily explains the wet-web strength and paper wet strength when using polyamine substances and other similar processes, such as the heat gelation of proteins.

8.4.4 Existence of water inclusions among cellulosic chains

Existence of small Water inclusions among cellulosic chains of the cellulose membrane with widths about 15 nm, that is, containing approximately 50 water molecules, was confirmed by the use of wide angle X-ray (WAX) and small angle X-ray (SAX) – scattering or diffraction methods [60]. The cellulose–water systems were modeled using a water-vapor-treated membrane, and soaked in liquid membranes employing light and heavy water. Using these methods, new results were obtained that were coherent for both the applied methods, WAX and SAX. They proved the existence of structural changes in the membrane, on a molecular level, caused by molecular water and that the changes depended on the type of penetrated water. It was also evidence for the existence of oriented water clusters in soaked samples. Simultaneously, the difference at a molecular level between light and D_2O was retained. As a result, Water clusters are formed in the soaked samples instead of in the water-vapor-treated samples, where no bulk water exists. Moreover, H_2O vapor modifies the cellulose structure to a greater extent than D_2O. The Weaker interaction of D_2O with cellulose corresponds to its lower permeation rate through the cellulose membrane. Inelastic neutron scattering experiments on various types of cellulose [106] revealed that amorphous cellulose is indeed practically 100% accessible to water, and that a nearly perfect exchange of OH groups into OD groups, after deuteration, is obtained in the accessible (disordered) regions by simple wetting. Hence, swelling and washing of amorphous cellulose in D_2O will lead to a full exchange of OH by OD. This means a 30% exchange of all hydrogen atoms in the cellulose molecule. On removing heavy water by drying the sample, OD groups are preserved as long as the sample is protected against atmospheric humidity or other protonated polar solvents. The explanation is based on additionally perturbing the hydrogen bonds between the cellulose molecules, due to the penetration of water molecules, and increased distance between cellulose molecules because of the swelling. It was also confirmed that all OH groups *inside* the crystalline parts (crystal cores) are unaffected by immersing the sample in D_2O or H_2O, that is, inaccessible. However, deuteration inside the crystals is possible using a relatively simple hydrothermal treatment technique. Samples were inserted into glass ampoules filled with 0.1 N NaOD/ D_2O. The sealed ampoules were kept at 210 °C for 30 min, and the morphology of the crystals was shown to be unaffected by this process [106].

8.4.5 Hydrogel structure of cellulose

All information and experiences have revealed that the smallest cellulosic particles are microcrystallites of rodlike shape (4–10 nm in width; 80–1,000 nm in length), enwrapped with anisotropic Hydrogel shells and exhibiting a gradual decrease in viscosity in the direction of bulk to water. These nanosized structured "submicrohydrogels" of fibrils stick out of the surface of the pulp fiber and have a great importance for their paper-forming ability, due to their secure bonding potential. The results of sorption – desorption hydrocolloid measurements, suggest a permanent evolution of "submicrohydrogels" these ones from pulps, which are dependent on the type and quality of the pulp (degree of beating), and water composition; the "submicrohydrogels" substances that predominantly influence the hydration forces. Soluble and colloidal substances, mostly of anionic character (polysaccharides, microcrystallites, monosaccharides, lignosulfonates in the case of sulfite pulp, and other low molecular compounds), are bonded by hydration forces to form a Vicinal water shell, of quasi-gel structure, with decreasing viscosity around the fibrous surface [107, 108]. These substances are released into the water environment during washing, at a pre-established temperature, in approximately the same amount [109], but depend on the quality and quantity of the acting hydration forces. The increased activity of repulsive hydration forces, that is, Antibonding activity, improved the evolution of colloidal substances into the bulk of the water and vice versa. It was found [109] that of the hydrocolloid substances released into the water environment during repeating pulps (bleached sulfite, sulfate, and unbleached sulfite pulp, respectively) extraction had a positive influence on the repulsive hydration forces, as a higher amount of substances were released into solution than that corresponding to the predicted value, except in the case of groundwood. More detailed measurements of hydration force activities, by use of the so-called hydration factor (for more details see below in chap. 8.5), confirmed an increase of the repulsive forces in systems with an increased concentration of released lignocellulosic substances. An increase of the hydration factor (g) [54, 101] also indicates a better release of soluble and colloidal substances from the pulp. However, the values for groundwood were in the $g < 1$ range, that is, in the range of prevailing attractive hydration forces in comparison with distilled water. In this case, the prevailing attractive hydration forces are weakened only upon an increased concentration of the released lignocellulosic substances. Apart from this one, there were other characterized auxiliary substances of ionic and non-ionic character effectively influencing the Hydration bonding/antibonding systems [54–56, 101, 104]. As well known, for example, an important role plays nonionic urea molecules in these systems because increased the activity of repulsive hydration forces. Recently, a method was published to enhance significantly the separation/removal of hemicelluloses from a softwood sulfite dissolving pulp by use of the urea/NaOH treatment [110]. Unfortunately, the author's explanation of this effect is irrational formally based only upon H-bond concept, although evidently the epimolecular interactions of this treatment are taking place in water medium.

8.5 H-bond ability and hydration bonding/ antibonding concept

A simple comparison of fiber size (approximate length 2 mm and width 20,000 nm), the length of H-bond (0.27 nm [111]), and water molecule (0.094 nm [112]) signals that it is practically impossible [57, 113, 114] to form a fibrous bonding system of paper without a supporting tool to draw the mutually interacting fiber surfaces near, during the H-bond formation. Without this tool, relatively huge fibers are not able to effectively approach with an accurate orientation of the PD group on one surface and the PA group on the other surface. Water molecules allow serviceable tool functioning, which enable the interacting fibers to form an optimal Fiber-network structure. The bonding system of paper has been formed due to this unique property of water [5, 113, 114]. Obviously, both the Hydration attractive and repulsive forces are responsible for the effective structural fibrous formation, that is, hydration bonding functions as a predecessor of H-bond formation (see Fig. 8.5). It is thought that water molecules enable the formation of natural cellulose crystals by favorable alignment of the chains through hydrogen-bonded bridging. This process is connected with the action of attractive hydration forces in the wet state. Part of the cellulose preparation is amorphous between these crystalline sections. The overall structure is formed of aggregated particles with extensive pores capable of holding relatively large amounts of water by micro- and nanocapillarity. The hydration bonding and antibonding concept (structure change in hydration layers (SCHL) theory) is based on the dipole nature of water molecules and on their two possible orientations in hydration spheres:

- orientation with the hydrogen atoms of water molecules in the direction of the active surface of the fiber and
- orientation with the oxygen atom of water molecules in the direction of the active surface of the fiber.

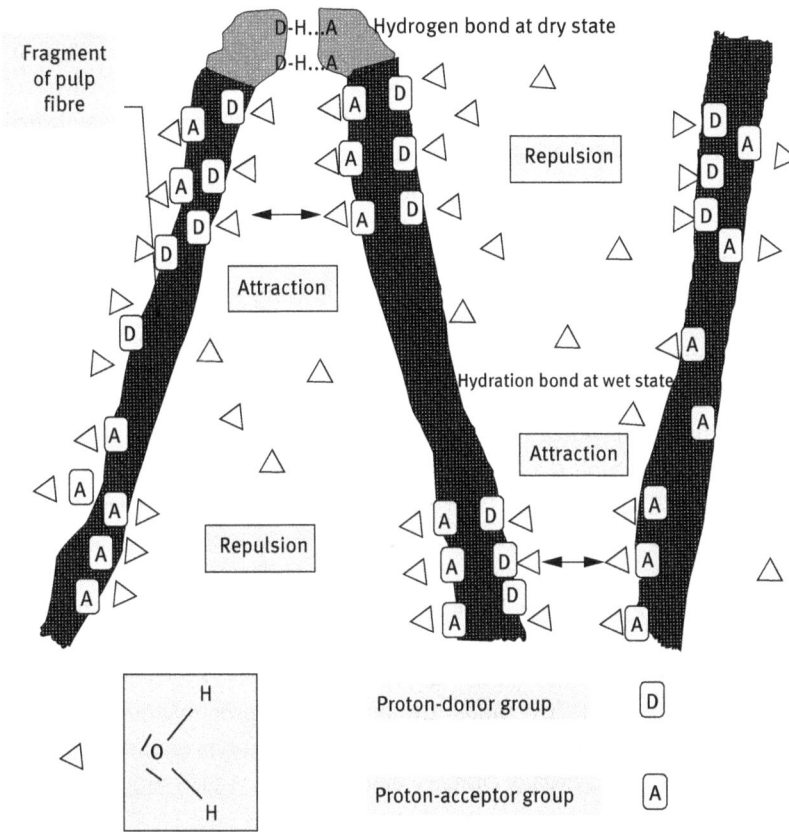

Fig. 8.5: Schematic representation of the origin and action of repulsive and attractive hydration forces during the formation of a hydrogen bond system among cellulosic fiber materials in water.

These two different Water molecule orientations are caused by the amphoteric character of water molecules with regard to accepting hydrogen bonds. A water molecule has the ability to join with other molecules or groups with hydrogen bonds, in which it can act as the hydrogen atom donor or acceptor. The Intermolecular force field, produced by the formed hydrogen bonds, will then spread by means of the other molecules, through the hydration sphere under the influence of this orientation of water molecules, becoming more and more diffuse until it equals a zero value within water. This effect gives rise to forces that cause action between the interacting surfaces – that is, Hydration forces [57, 113, 114].

If the water molecule orientation to each of the interacting surfaces is equal, then both surfaces will affect each other with repulsing hydration forces. On the other hand, if the orientation of water molecules is different, the two surfaces will attract. Theoretically, the repulsive forces are effective over a greater distance [57] but are smaller in nearest distance (approx. 2 nm) at comparable condition. Increased temperature depresses this behavior being accompanied with increasing its distance action but more deeply for action of repulsive than attractive hydration forces. This difference appears in the interactions of heterogeneous surfaces, in which repulsive and attractive hydration forces act simultaneously, as a kind of equilibrium is established in which the two surfaces reach a definite optimum distance from each other, which is also strongly dependent upon temperature. For example, between two interacted flat heterogeneous microsurfaces at 3 nm distance, in which repulsive and attractive hydration forces act simultaneously, the repulsion prevails at 20 °CC and the attraction at 80 °CC. During the interaction, a mutual diffusion of hydration spheres takes place, connected with a change of their structure. These structural changes are on the molecular level and should be accompanied by measurable temperature effects [57, 104, 115]. For hydration forces to originate, it is necessary for water molecules to reach a suitable orientation in the hydration spheres. This orientation is determined by H-donor and H-acceptor groups, their amount, and the strength of the hydrogen bond formed with water, or in other words, by the value of the interface tension $y_{l,s}$. All hydrated systems with a high specific surface area behave in this manner. These include various porous systems filled with water and its solutions, gels, quasi-gelatinous systems, and so on. According to this theory, the groups forming hydrogen bonds with water can be divided into three types:

- H-donor groups and molecules: such as primary alcoholic OH-groups, secondary amino groups in aliphatic compounds, and partially primary amino groups;
- amphoteric groups and molecules: such as H_2O, secondary alcoholic OH-groups in polysaccharides, and partially primary amino groups, amido groups, and so on; and
- H-acceptor groups: such as hemiacetal oxygen in saccharides, carbonyl groups, and tertiary amino groups.

The behavior of these groups considerably depends upon the surrounding groups, which are not necessarily linked to H-bonds. For example, surrounding groups having the −I effect will intensify the donor nature of oxygen in these groups. The mesomeric effect has the opposite influence: the hydroxyl groups will strengthen their acceptor nature and weaken the donor nature [116]. The relationship of this theory to electrokinetic behavior, following the electrical double layer theory, shows that the adjacent part of the electric double layer, around the hydrated hydrophilic phase interfaces, is formed by hydration layers [57, 116]. The other molecules of ionic and nonionic character are drawn in or expelled by electrostatic, van der Waals, or other forces into these layers. However, the thickness of the hydration layers is considerably greater than that of the inner layer, resulting from the classical Electrical double layer theory, and due to its diffuse character it is not limited to exact size. Its thickness depends highly on the effect of shear forces, that is, the velocity gradient. For this reason, the value of the Zeta-potential will be highly influenced by these factors.

Under comparable conditions, different Zeta-potential results can be obtained by different experimental methods, which differ both in absolute values and sign.

Coacervation of nanoreticular-hydrated systems, composed of hydrophilic anisometric oligomer molecules, during their dilution is a typical phenomenon where the hydration forces play the dominant role. It also seems reasonable to utilize these Quasi-hydrogel systems, that is, dilutable hydrogel, systems to measure the behavior of hydration forces using the so-called relative hydration factor [56, 101]. As represented in Fig. 8.6, under isothermal conditions, the volume of the original hydrated hydrophilic system V_o, for example, some polyesters or urea-formaldehyde (UF) precondensate, when gradually diluted with water or water solutions respectively, does not undergo any changes until the moment when the volume of diluted sample exceeds the critical value V_k. Consequently, the originally transparent liquid becomes turbid due to precipitated Microhydrogel particles; such behavior is due to the presence of hydrophilic anisometric oligomer molecules. The parts or ends of these molecules influence each other by attractive hydration forces, while other parts of the interacting molecules are influenced by repulsive hydration forces. The effect of the attractive forces among interacting molecules must, however, prevail over the repulsive forces. The uniform distribution of the oligomer molecule is thus reached throughout the volume of the water medium. When these systems are diluted, the molecules, their parts, and ends are drawn away from each other, and their mutual force of action is thus weakened. After exceeding the critical concentration or critical volume of added diluting water, coacervation takes place, owing to the fluctuation

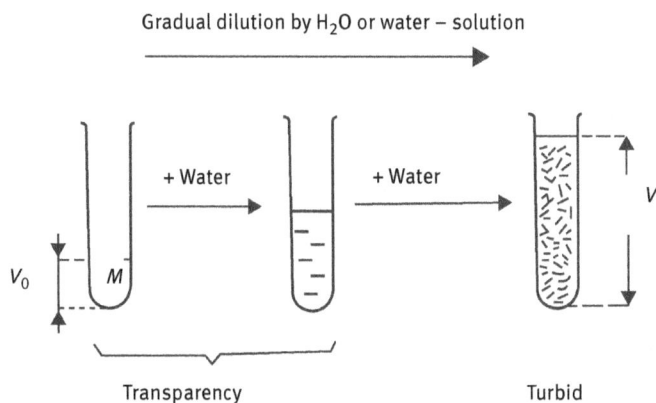

Fig. 8.6: Evaluation of the effect of hydration forces by means of a hydration hydrophilic modeling system.
M – Modeling system; CDD = V/V_0; $V = V_0 + V_k$; $CDD_0 = V_d/V_0$ – critical degree of dilution in distilled water; CDD = V_s/V_0 – critical degree of dilution in water solution; $g = CDD/CDC_0 = V_s/V_d$ – relative hydration factor.
If $g < 1$ – increasing of the attractive forces influence in the hydration bonding system; if $g > 1$ – increasing of the repulsive forces influence in the hydration bonding system.

of the affecting mutual forces and kinetic energy of the interacting molecules. In this case, the composition and properties of the separated coacervates correspond to the original concentrated system. The influence of the rate of added substances upon hydration forces can be evaluated by means of the so-called Hydration factor g = CDD/CDD_O, where CDD and CDD_O mean the critical degree of dilution in water solution and distilled water, respectively.

However, Kim and Yun [117] very interestingly described the discovery of a Smart cellulose material that can be used for biomimetic sensor/actuator devices and micro-electromechanical systems. This smart cellulose material is termed Electroactive paper (EAPap) and it can produce a large bending displacement with low actuation voltage and low power consumption. The authors proposed that electroactive paper is advantageous for many applications, such as microinsect robots, microflying objects, microelectromechanical systems, biosensors, and flexible electrical displays. Using this phenomenon it is also possible to explain and simulate muscle movement. EAPap is made from a cellulose film (cellophane) on which gold electrodes are deposited on both sides. An EAPap actuator was supported vertically in a controlled humidity and temperature environment chamber. By excitation of the voltage application to the actuator, a bending deformation is evoked. The authors believe that the actuation is due to a combination of two mechanisms: ion migration (diffusion of sodium ions to the anode?) and dipolar orientation. Again, in spite of the confusing and irrational explanation of the EAPap movement, the reported results have great inspiration for theoretical potential accompanied by practical challenges. The tip displacement of the EAPap actuator is dependent on the applied electric field, its frequency, EAPap sample thickness, and temperature, but predominantly on humidity. The humidity affects the displacement, where a high relative humidity leads to a large displacement; this behavior can be rationally explained by use of the SCHL theory [74]. The orientation of water molecules in immobilized layers around the cellulose macromolecules, in the layered structure of the EAPap actuator, is determined by the presence of PD groups or PA groups at their interacting surfaces. The overall film structure and shape are formed among structural cellulosic units due to hydrogen-bond bridging in the dry state and hydration-bond bridging in the wet state. The extent and intensity of this bonding system is determined by the size, concentration, and distribution of nanodomains with either an attractive or a repulsive force of action, that is, among opposite interacting nanosurfaces exhibiting the reverse or identical basic orientation of water molecules, respectively. The basic orientation of a water molecule is determined by the presence of surface PD groups or PA groups of cellulose. While hemiacetal and glycosidic oxygen in cellulose are typical PA groups, the hydroxyl groups can behave as PD and PA groups. Nevertheless, it is supposed that the behavior of most hydroxyl groups in cellulosic materials have more of a PD character. As a consequence, the domains of prevailing hydration-bond bridging are regularly distributed within the cellulosic material in a flat formation. Any disturbance of this distribution and paper strip curling is evoked because the inner tension equilibrium has been disrupted. Application of an oriented electric field to cellulosic material in a wet state results in the water molecules in the bonding nanodomains, contained nearest the electrodes, to be reoriented. However, reorientation at the cathode is different to reorientation at the anode at the anode; reorientation only occurs to the water molecules having been oriented to this pole with hydrogen atoms at the basic original state, and at the cathode it occurs only for those water molecules having been oriented to this pole with oxygen atoms at the basic original state. Moreover, the distribution of attractive forces formed around both the **A** and **D**, and the **D** and **A** nanocenters is not the same as prevailing **A** – **D** structure orientation in the bonding domains is presumed. In this situation, the application of a dc electric field evokes a weaker hydration bond system in layers near the anode and vice versa, a stronger hydration bond system in layers near the cathode. Due to this effect, the paper strip bends toward the anode. Logically, the effect is strongly dependent upon relative humidity, and the reorientation of water molecules is independent of the diffusion process and relatively quick. Obviously a similar effect, but at the microscale, explains the movement of muscles. The main presumption is the nonsymmetrical distribution of attractive forces formed around both the **A** and **D**, and the **D** and **A** nanocenters.

8.5.1 Rheosedimentation

The phenomenon, known as *rheosedimentation*, is typical of fibrous slurries with papermaking abilities [118, 119]. The macroreticular systems, comprised of a weak bonding system, are represented by papermaking pulp slurries composed of fibers of cellulosic or lignocellulosic character. It is a typical property of components, with marked Papermaking properties, that form fiber networks, which are only compressed during sedimentation, that is, this process behavior is called *rheosedimentation* [120, 121]. The basic condition of Rheosedimentation is an ability of the pulp fiber to form a network displaying special behavior, that is, due to a weak bonding system, the fiber network is compressed by gravity [118, 120]. The bonding system of the wet-web state is realized through the so-called Campbell effect followed by the origination of a H-bonding system, but only in water media, that is, among paper components with the so-called Hydro-cohesive properties. The ability of these components to form a comprehensive fiber space net in a water medium, that is, hydrated macroreticular system, has been well documented as typical of this behavior [74]. It is well known that paper matter exists in three forms:
-   the suspension form;
-   the wet-web paper form; and
-   the dry-web paper form.

Behavior of the first paper state corresponds to a general formula describing the continuum movement and it is typical for components with marked papermaking properties. Evidence supporting this description is the phenomenon called *rheosedimentation* as it is observable during sedimentation of diluted fiber slurries. The movement of the fiber space net during rheosedimentation is described by eq. (8.2), which follows from the solution of the general continuity equation:

$$\frac{t}{(h_0 - h)} = \alpha + \beta \cdot t \tag{8.2}$$

where $\alpha = 1/v_0$ is for initial fiber network concentration c_p; $v_s = 1/\alpha_s$ standardized rheosedimentation velocity for $c_p = 1$ kg/m^3; β, characterizing a final concentration of fiber network, c_k because $c_k = h_{00} \cdot x_1 / (h0 - 1/\beta)$.

Symbols x_1, h_0, and h are the concentration of paper slurry (kg/m^3), and the height of the boundary level of a fiber space net in water at time $t = 0$ and t, respectively. The parameters α and β characterize the velocity of sedimentation at high h_0 and a final concentration of the sedimented fiber space net, respectively. It was shown that:

- Rheosedimentation is definitive for every fiber component of papermaking ability and it is a useful indicator of this process.
- The sedimentation velocity and final concentration of the sedimented fiber components are dependent on the fiber morphology.
- The final concentration is sensitive to the interbonding ability of the fiber components that create the fiber suspension – it appears that with increasing this ability, the final concentration of the sediment decreases.
- Both of these parameters, that is, the sedimentation velocity and final concentration of the sediment are strongly dependent upon the Hydration ability of the fiber components – upon increasing this hydration ability, both parameters decrease.

The homogeneous pulp fiber network is formed at a suspension concentration higher than 1 kg/m^3. Dilution of the suspension to a concentration under 1 g/L is accompanied by Flocculation and rheosedimentation. However, the rheosedimented fiber network is not in a fully homogeneous state, due to the lack of shear forces (agitation) disturbing the rheosedimenting floccules.

It was shown [118] that for rheosedimenting fibers there are important three papermaking significant properties:

- their Bonding abilities,
- their flexibility, and
- their structure.

Regardless of mutual attractive forces approaching the microlocalities are important, the structure and geometry of interacting fibers are most important at start of rheosedimentation. On the other hand, the flexibility of individual fibers in connection with their Interbonding abilities predominantly determines compression degree of fiber network at final stage of rheosedimentation, that is, the Standard velocity of rheosedimentation v_s predominantly depends on geometry and structure of interacting fibers, the Final concentration of rheosediment, and c_k is mostly influenced by interbonding abilities and flexibility of fibers composing a fiber network [118]. At comparable condition – that is, at the same structure, geometry, and flexibility of interacting fibers – a relative decrease of c_k indicates fibers with improving bonding abilities and vice versa. Any measurement of the rheosedimentation indicates a close connectivity of rheosedimenting phase of the suspension form of a paper matter with its dry web paper form, especially with strength and mechanical properties of paper [118, 121]. Missing of pulp slurry rheosedimentation indicates the loss in ability of paper fiber components to form firm bonding system of dry paper web, that is, this pulp fiber has no paper-forming abilities. For this reason, chemical modification of cellulose being represented, for example, by nitrocellulose and oxycellulose with high degree of substitution leads to loss in its ability to rheosedimentate, although its water retention ability (WRV) is relatively high. The missing of pulp slurry rheosedimentation is well indicated by unrealistic values of standard velocity of rheosedimentation (v_s more than 20 mm/s). The lowest values of both v_s as well as partly c_k for sulfate softwood bleached pulp declare well-known best papermaking properties characterized by highest stiffness of their fibers and good bonding abilities of sulfate softwood pulps. In this content, the lower value of c_k of Mg bisulfite pulp ECF quality compared with TCF quality indicates better bonding ability of ECF versus TCF pulp, respectively, that is, the better strength properties of ECF pulp [117]. On the other hand, poor bonding abilities are shown by pulps as groundwood, fibers from waste paper [120], and Magnefite bleached hardwood pulp from beech [118].

Formerly it was also shown (Milichovský at al [122]) that rheosedimentation is strongly dependent on intensity of pulp beating, that is, with an increase in a degree of beating the rheosedimentation gets slow, and the standard rheosedimentation velocity can be used for the determination of character of the beating process. It was found theoretically and experimentally proved that with an increase in pulp beating, the standard velocity of rheosedimentation decreases markedly faster for fibrillation than for fiber cutting. This theory is in good agreement with experimental observations of pulp beating in dry state as well. Obviously, a pulp is during its milling, that is, dry beating, only fragmented into particles without any fiber fibrillation or structure changes. During milling process pulp fibers lose their bonding abilities, followed by loss of papermaking properties including the rheosedimentation ability [121]. Compared with the final concentration of rheosediment, c_k, the standard velocity of rheosedimentation v_s is much more dependent upon geometry and structure of interacting fibers. Linear dependency of v_s cf. c_k has been observed on the amount of milled pulp fibers in pulp slurry. This reality is proved by sheerer dependency of v_s cf. c_k on the amount of dry beaten fibers in pulp slurry as it is documented in [121]. Although the absolute values of v_s and c_k are approximately the same, the slope of v_s versus the amount of dry beaten fibers, x, is approximately one order bigger than c_k versus this one. However, the dependence c_k versus x is important because this one indicates a baneful influence of dry beating on inter-bonding abilities abilities of pulp. The values of c_k increases proportionally with amount of dry beaten fibers, x in pulp slurries.

8.5.2 Thermoresponsive hydrated macro-, micro-, and submicroreticular systems of cellulose

Hydrated reticular systems, that is, networks in water environments, characterize all bioentities and the products of their existence. We can identify these structures at the nano- (submicro-), micro-, and macroscale as submicro-, micro-, and macroreticular hydrated systems, respectively. Supramolecular and hypermolecular structures are typical, for example, hydrogels based on peptides, and fiber networks based on cellulose; a typical characteristic being the Phase transition temperature (lower critical solution temperature (LCST) or upper critical solution temperature (UCST)) of the Thermoresponsive hydrated reticular system (TRHRS), exhibiting a unique hydration–dehydration change. As illustrated in Fig. 8.7, the volume changes are typical in hydrogel TRHR systems. Below the LCST, the uncrosslinked polymer chains are soluble in water, whereas above the LCST the polymer chains form submicro- and microaggregates, which separate from solution. Above the LCST, the polymer starts a complex self-

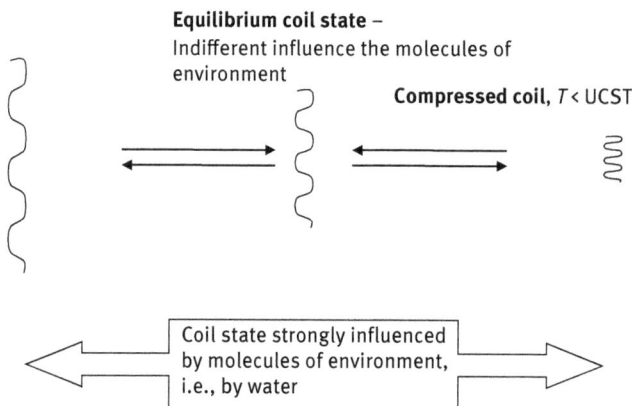

Expanded coil, $T < $ LCST

Equilibrium coil state –
Indifferent influence the molecules of environment

Compressed coil, $T < $ UCST

Coil state strongly influenced by molecules of environment, i.e., by water

Fig. 8.7: Schematic 2D illustration of linear macromolecule behavior in hydrated micro- and nanoreticular systems with LCST or UCST, that is, a lower or an upper critical solution temperature, respectively.

assembling process that leads to an aggregation of polymer chains, initially forming nano- and microparticles, which segregate from the solution [123]. Below the LCST, the crosslinked hydrogel is swollen and absorbs a significant amount of water, while above the LCST the hydrogel significantly releases free water and begins to shrink. Thermoresponsive hydrogels composed of crosslinked polymer chains undergo fast [124, 125], reversible structural changes from a swollen to a collapsed state by the expulsion of water. However, other types of thermoresponsive hydrogels exist, which are opposite to the LCST hydrogels, that is, hydrogels with an UCST. These hydrogels shrink at lower temperatures and swell at higher temperatures.

According to the behavior of the TRHRS during dilution [74], we can divide the gels into water dilutable and nondilutable, crosslinked 3D networks, or crosslinked 2D networks. Additionally, it is possible to divide the dilutable TRHRS into fully dilutable polymer solutions at $T <$ LCST (or $T >$ UCST), and Coacervated [54, 56, 126] submicro- or micro TRHRS, or flocculated macro-TRHRS. The dilutable [74] TRHRS coacervate or flocculate in the water environment due to weak bonds among the polymer chains, microparticles, and hydrogel particles or fibers and microfibers, respectively. This is typical for the submicro-, micro-, and macronetworks that are disrupted during the dilution process, that is, the quasi-hydrogels coacervate and the fiber networks flocculate, respectively. As the temperature changes, the transition of the sol–gel reversible hydrogels occurs due to nonchemical crosslinks being formed among the grafted and branched elements of the copolymers.

Poly (N-isopropylacrylamide (PNIPAAm) has been the most used polymer in thermoresponsive hydrogels with characteristic LCST. Above the LCST, at 32 °C in pure water, a reversible structural transition occurs from an expanded coil (soluble chains) to a compact globule (insoluble state) [127–132]. N,N'-methylenebisacrylamide (MBA) is utilized as a crosslinker [124, 133]. Such a macromolecular design widens the applicability of such systems to a variety of biomedical applications inclusive of the biomineralization of biodegradable substrates [134]. Such modifications are particularly important to tailor the LCST of PNIPAAm-based systems. Recently, a curiously self-oscillating system of hydrogel particles, rhythmically changing its volume, was described by Yoshida, which was based on PNIPAAm, MBA, and a covalently immobilized $Ru^{2+/3+}$ redox system using the so-called Belousov–Zhabotinsky oscillating reaction [135]. The UCST hydrogels are mainly composed of an Interpenetrating polymer network of polyacrylamide and poly(acrylic acid) or poly(acrylamide-co-butyl methacrylate) crosslinked with MBA [133]. The formation of helices (double or triple in polysaccharides such as agarose, amylose, cellulose derivatives, and carrageenans, or in gelatin, respectively) and the corresponding aggregation upon cooling, form physical junctions that are the basis of hydrogel formation [136]. Dilutable but coacervating quasi-hydrogels with UCST are represented by UF precondensates [54, 56]. We can conclude [69] that under normal conditions, that is, at room temperature and an inert environment, the PNIPAAm polymer has a preferred thermodynamically advantageous Coil conformation because of the Hydrophobic interaction among isopropyl pendant groups. However, due to a peculiar water activity, the coil conformation is stretched at temperatures below the LCST, in contradiction with the squeezed original coil conformation above the LCST, as the repulsive domain activity is weakened. The peculiar water activity is accompanied, below the LCST, by the origination of repulsive water action among the polymer chains and its segments, the submicro- and microcolloidal particles, and so on, arising from equal water molecule orientation at the interacting interface microdomains due to hetero- followed by homo-H-bonds, that is, a hydration antibonding system. Obviously, the width of vicinal immobilized water, within the interacting polymer interfaces, decreases with an increase in polymer concentration, because the improving disruptive action of the hydration repulsive forces is significantly weakened upon an increase in temperature. As a result, the LCST decreases with an increase in polymer concentration [137]. Under normal conditions, that is, at room temperature and an inert environment, the UCST-hydrated crosslinked and uncrosslinked polymers have a preferred thermodynamically advantageous long-chain structure, which is squeezed into a compressed coil conformation in a water environment due to the origination of a weak hydration bonding system. The hydration bonding system [54, 56, 74, 101] among polymer chains, its segments, submicro- and microcolloidal particles, and so on arises from the opposite water molecule orientation at the Interacting interface microdomains because of the hetero-H-bonds of the PA or PD groups of the polymer chains and homo-H-bonds among water molecules. As the temperature increases, the hydration bonding system is weakened because of the increase in the kinetic energy of the water molecules. Above the UCST, the hydration bonding system is weaker than the inner opposite stress of the compressed polymer chains of the TRHRS and the polymer chains expand. As a result, the crosslinked TRHRS swell and the polymer chains in the noncrosslinked TRHRS are dissolved. Below the UCST, the hydration bonding system is stronger than the inner opposite tension of the compressed flexible polymer chains and the long chains of the polymer compress. The process results in deswelling and coacervating of the crosslinked hydrogel and dissolved polymers, respectively. Obviously, the intrahydration bonds squeeze the long-chain uncrosslinked polymer structure into a compressed coil conformation and the interhydration bonds squeeze the long-chain crosslinked structures to a deswollen form at temperatures below the UCST. However, different characteristic behavior is observed if a hydrated reticular system is composed of relatively rigid rodlike particles, such as short polymer chains or fibers. The short polymer chains or fibers in hydrated submicro- or macroreticular systems, respectively, are formed through interhydration bonds and the interhydration repulsive domains, that is, mutually functioning hydration bonding and debonding sites. Typically, due to increased fluctuation of the bonding and debonding activities of the interacting microsites during dilution, the submicroreticular systems coacervate and the macroreticular systems flocculate. The submicro-weak bonding of the hydroreticular system, accompanied by coacervation during dilution, is well demonstrated by using a UF precondensate (Fig. 8.8) [54, 56].

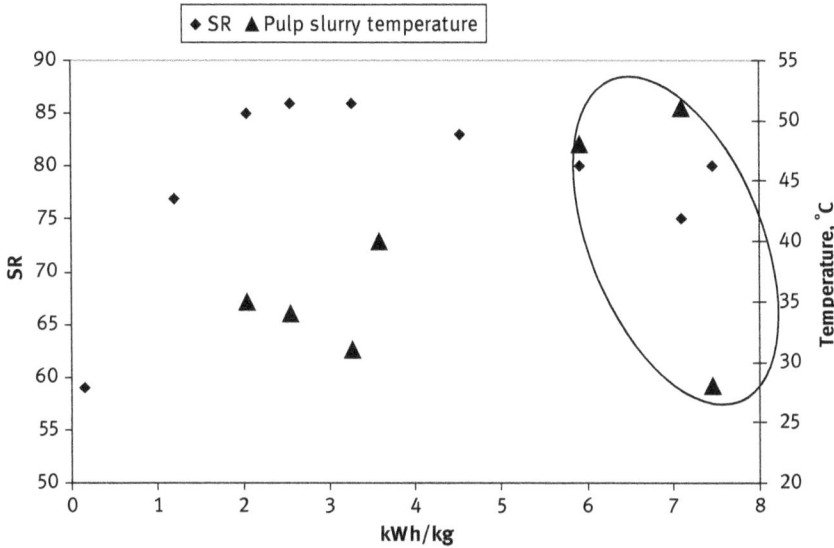

Fig. 8.8: Temperature of pulp slurry influence upon drainage ability of beaten hemp pulp. Note: SR – the degree of pulp beating according to Schopper–Riegler (ČSN EN ISO 5267-1) – the drainage ability of pulp slurry decreases with increasing of SR; effective beating energy in kWh/kg of oven-dried pulp fiber.
Beating conditions:
- laboratory toroidal beater;
- nonbleached hemp pulp prepared by alkaline cooking method; and
- pulp beated at 3% consistency during 49 min at approximately constant operating beater edge load.

It has been concluded that the water inside the pores of the lignocellulosic material is physically created by three sorts of water: the nonmoveable adjacent part of the immobilized water shell with the highest viscosity at a given temperature and flow conditions; transition part of the Immobilised water shell flowing with higher viscosity at a given temperature and flow conditions; and Bulk water in the middle part of the pores flowing with a normal viscosity coefficient at a given temperature.

8.5.3 Compete vapor activities in pores systems

As is well known, a paper as typical cellulosic or lignocellulosic material is appropriate substrate to Microbial growth substrate, but the bacterial growth requires a moist environment to grow well. Otherwise, a "false," or preventive, Antibacterial effect will be obtained. The experiments have confirmed [138] that the evolution of microbial growth occurs if relative humidity of air is about 70–75%. Well-known antimicrobial activity of essential oils (EO) molecules is mainly connected with water distribution restriction in reticular porous systems of bioactive matters, but the process is more complex. Accumulated experiences indicate noticeable Antimicrobial activity of essential oils effect of EO's vapors only if the air relative humidity is at least 70–75%. However, a noticeable Sporicidal effect of EO's vapours in environment with RH ≤ 15% has been observed. All the EOs have approximately the same influence but with different efficiency depending upon composition of EO relative to lignocellulosic pore's matter and its hypermolecular properties [139]. Generally, it seems that deepest influence have substances or groups of substances as linalyl-acetate, limonene, eucalyptol, citral, and ocimene. For this reason it is beneficial to use a special simple chromatographic device to separate the EO's vapors before their exposition [140].

The progress of the concentration of water and EO vapors condensed in lignocellulosic materials depends on the relative air humidity (RH) of environment in which they are located. Competitive absorption mechanisms for condensation of water and EO in the porous structure of the paper matter are described in detail by Česek et al. [69]. As indicated, an absorption capacity of water vapor for all pulp sorts is increased exponentially with an increase in relative air humidity that being influenced in small extent by pulp quality, but what is most important, this one is influenced in high extent by presence of EO molecules. For example, the presence of EO vapors caused a decrease in hygroscopic capacity of lignocellulosic materials at RH = 75% on a level as achieving at RH = 50%. This is probably the main reason of microbicide effect of all the EO vapors. Additionally, with depressing of environment humidity the living microorganism has tendency to transform into inactive spore state with minimum water content being stabile in micropores systems in relatively large extent of low relative air humidity's (RH ≤ 20 %). However as documented [141], an application of EO vapors suppress the water content in micropores practically to zero, which is probably the main reason of high sporicidal effect of EO's vapors being applied at low RH (RH ≤ 15 %). From a particular point of view, it is very important to better understand all the processes of mutual interaction and competition of water and chosen few EO with high antibacterial activities in hypermolecular matrix of cellulosic matter represented by paperboard of lignocellulosic character. In this context, crucial valuable basic information about these processes enables the measurements of kinetic of the absorption–desorption processes and their reversibility.

The process of vapor condensation, that is, vapor absorption, in porous web material takes place by diffusion of vapor from surrounding atmosphere into pores of porous flat medium followed by quick condensation in suitable pores. The suitable pores are all pores that are size characterized by the so-called equivalent cylindrical pore radius $r \leq r(\varphi_{rel})$. The value of maximal equivalent cylindrical pore radius, r (φ_{rel}), occupied by the condensed liquid, for example, water in the pore material at constant temperature and relative humidity, is expressed by the well-known Kelvin equation (see eq. (8.1)). It was demonstrated that the Kinetics of paper products moistening [71, 142] or one component vapor absorption of organic liquids [70] is satisfactorily described by the following originally empirical, later theoretically derived and experimentally verified eq. (8.3) [68]:

$$\Delta Y = \frac{at^d}{b + t^d} \tag{8.3}$$

where ΔY (=$Y - Y_i$) is increment of moisture content (mass of liquid per unit mass of dry solid), given as the difference between current moisture content, Y, and initial moisture content at the beginning of the experiment, Y_i, or by using moisture of sample y (w/w) according to recalculation $Y = y/(1 - y)$; t is the time (day) of sample storage at a given relative air humidity; d characterizes the rate of steady-state adjustment; ($a = \Delta Y_e$) is the hygroscopicity or absorption capacity of the porous material, and b (day d) is a parameter connected with a mobility of diffusing molecules by following equation: $b = \frac{x_{ra}^2 \varphi_{rel}}{dD_{ap}\Delta Y_e}$. The steady-state moisture or Steady-state absorption capacity of porous flat sample, Y_e, is given as $Y_e = a + Y_i$.

If $d > 1$ or $d < 1$, the porous web material of sample thickness, $2x_{ra}$, has stratified structure with worse uniformity where vapor diffusion takes place slowly or rapidly than that in an ideal web porous material, respectively, that is, the sample uniformity decreases. Additionally, it is possible to show that if $d \leq 1$, a humidification kinetic curve has a monotonic ascending character and if $d > 1$, a humidification kinetic curve has a typical S character with an inflexion point being moved to higher time values with increased d values.

Apparent or pseudo-diffusion coefficient: where $D_{ap}^1 = \frac{\varphi_{rel}x_{ra}^2}{}b\Delta Y_e$ is hence proportional to a given relative air humidity, φ_{rel}. For better comparison of Diffusivity results, these coefficients have been one relativized [141] with apparent diffusion coefficient of pure water at comparable conditions according to eq. (8.4) defined as $D_{rel} = D_{ap}/D_{ap}(H_2O$ at 49%), that is, a ratio of D_{ap} of water or EO vapors during a common absorption process at given relative air humidity (RH) to the apparent diffusion coefficient of pure water at RH of 49%, $D_{ap}(H_2O$ at 49%).

$$D_{rel} = (b \cdot a \cdot d)_{pure\ water\ vapour\ at\ RH\ 49\%} / (b \cdot a \cdot d)_{water\ or\ EO\ vapours\ at\ given\ RH} \tag{8.4}$$

If vapor concentration of absorbed molecules in surrounding atmosphere decreased under equilibrium, an inverse desorption process is evoked. Supposing that the reversion desorption process takes place according to similar mechanism as sorption process but only in reverse mode, a weight increment monitored gravimetrically during sorption experiments is then generally written in the form of two components: the vapor of water and EO absorption–desorption kineticeq. (8.5):

$$\Delta Y = \frac{a(H_2O)t^d}{b(H_2O) + t^d} - \frac{a(EO)t^d}{b(EO) + t^d} \qquad (8.5)$$

where $a_{(H2O)}$, $a_{(EO)}$ and $b_{(H2O)}$, $b_{(EO)}$ are the absorption capacities and mobility coefficients of water or EO component absorbed or desorbed in the porous materials, respectively. Kinetic of common absorption of water and EO vapors is described by similar equation but with a plus sign before second sum of eq. (8.5). The reversibility with respect to irreversibility of Reversibility of sorption processes is then calculated as a difference between absorption capacity $a_{(EO)absorption}$ achieved during common absorption process and $a_{(EO)desorption}$ of common desorption process at comparable conditions. If the difference is zero, then the sorption process is full reversible; otherwise, it is irreversible.

Experimental results confirmed that the absorption–desorption process itself is given by condensation of water vapor and evaporation of EO molecules in appropriate pores and its kinetic is very well described by theoretical eq. (8.5) [141]. An absorption capacity of water vapor for all pulp sorts is increased exponentially with an increase in relative humidity being influenced in small extent by pulp quality and the presence of EO molecules. It was shown that in comparison with pure water, an absorption capacity of EO in porous nonstructuralized material as paper is also inconstant at different RH and with increasing of RH increases as well as but relatively in more small extent. In contrast, the penetration of liquid water essentially never occurs for water to fulfil all the pores, but a considerable portion of the pores (~75%), in particular the largest, remains unfilled with condensed water. Furthermore, the absorption capacity of water vapor is markedly higher for absorption from clean humid atmosphere compared with polluted environment, for example, by EO. Moreover, it was also observed that most water absorption in clean water vapor environment exceeds a sum of absorbed components in composed atmosphere inclusive of water vapor. This is probably the main reason of microbicide effect of all the EO vapors.

By a close experimental and theoretical investigation, absorption–desorption phenomenon follows a simple rule [141]: the molecules of a substance slowly permeating porous mass crowd out those that permeate quickly. From the porous matter to escape, that is, the races to win, the molecules that are faster.

This behavior is then related to the reversibility of absorption–desorption processes. Due to water vapor, with the growth of the permeation velocity of EO vapors thorough porous environment of lignocellulosic materials relative to water vapor, the reversibility of their release increases. Also with the growth of the relative permeation velocity of water vapor, the displacement of water vapor from the porous environment is increased due to EO vapors, that is, the Reversibility of water displacement increases. This reality is evidenced also by irreversibility of these processes. It means that if the ratio of the apparent diffusion coefficient of water vapor and EO increases (i.e., the relative diffusivity of water vapor), the sorption irreversibility of EO vapors also increases, that is, the reversibility of EO absorption–desorption process increased with an increase in RH and is completely fulfilled at 97% of RH. On the other hand, at low RH, for example, at 49% of RH, due to rapid spread of noncondensed water molecules through a porous material, the EO molecules have lack of time to escape the pores of the lignocellulosic material and the desorption process in contrast with high RH is more irreversible. As a rule, with an increase in absorption, the diffusion is depressed and vice versa. Explanation of this behavior is established by the formation of Stratified structure of fulfilled nonuniform pores composed of water and EO condensed liquids and their vapors. During following Desorption process in contrast with or else slowly but frontal proceeding a cleaning process at 97% of RH, that is, cleaning of this porous material with quickly moving pure water vapor atmosphere at 49% of RH, the absorbed EO molecules also in hidden big pores are not released because relatively high mobile the water vapors are not allowed the both of condensed or vapor forms of EO to escape.

From practical point of view, in order to achieve maximal microbicide and sporicidal effects, it is reasonable to dose EO vapors at RH of 65–75% and lower to about 20%, respectively. Lot of devices for dosing EO vapors exist with different designs and control units, but only on common principle of dosing that is only mixture of vapors evolved from EO liquids. Recently, a Static EO vapors releaser was developed, which is provided with "chromatographic" unit that enables to separate the EO vapors [140]. For example, a simple static EO vapors releaser enables the separation of linalool and linalyl-acetate from vapor of lavender EO (*Lavandula angustifolia*).

8.5.4 Swelling and recycling

The supramolecular and hypermolecular structure of cellulosic materials is changed upon drying. The pores and lumens collapse and the hydrogel structures are transformed into xerogels. The opposite process – swelling – is not fully reversible with the characteristic where water flows into the porous cellulosic matter, followed by swelling, that is, a volume increase of the sample. The whole process is known as the *hornification phenomenon*. A few decades ago, Dobbins [143] described that pulp swelling is not evoked by the presence of soluble substances, but by water molecules with high polarity. The swelling is often connected with the charging of swollen cellulose predominantly with dissociation of the carboxylic groups within the cellulose. However, the diminished dissociation of carboxylic groups within the bulk of the cellulose and at its outermost surface was observed. The increased charge density and possibly the ordered water structure are assumed to cause the distinct differences in the behavior of the carboxylic groups inside the cellulosic layer and at the surface [144]. Hornification is known as an irreversible loss of fiber swelling during which most of the macropores and none of the micropores have irreversibly collapsed when the fibers are dried and rewetted [145]. This suggests that Hornification is caused by the formation of bonds, stable to water, between adjacent lamellae, that is, by hydrogen-hydration bonds. When the hornified fibers are refined, the macropores can be almost completely regenerated.

It was demonstrated theoretically and experimentally by Geffert et al. [146], the kinetics of the Swelling process and its kinetics, that is, the volume of microporous material increase, is not controlled by the water flow into the porous material taking place relatively most quickly but predominantly due to own hypermolecular and rheological (i.e. flexibility and elasticity) properties of wetted pore walls. It was shown that the all swelling process is possible to dismantle into two consecutive steps (see Fig. 8.9). First step is characterized by relatively quick water flow into empty shrunk pores connected with those filling by water, followed by their Proper swelling process evoked by osmotic pressure. Both steps are accompanied with an increase in own volume of pores measured as, for example, the paper sample thickness.

Provided that the volume filling by water, that is, the moisture content in pulp mass, is proportional to thickness of water-swellable fibers and that the kinetics describing this process has a consecutive character, then combining both steps results in eq. (8.6):

$$Y(t) = Y_1(t) + Y_2(t) \tag{8.6}$$

where Y_1 is described by the well-known GHM model (generalized hygroscopicity model) successively used for the description of kinetic of vapor absorption (condensation) in paper as shown by eq. (8.3) and the term Y_2 describes the slower binding of water molecules into swelling centers of the cellulosic material (see Fig. 8.9). The second step (i.e., proper swelling) is theoretically described by the following differential eq. (8.7):

$$\frac{dY2(t)}{dt} = k(c - Y_2(t)); Y_2(0) = 0 \tag{8.7}$$

that has a solution of the form (8.8),

$$Y_2(t) = c(1 - e^{-kt}) \tag{8.8}$$

where c is the equilibrium of the fluid content at $t \to \infty$ (%), t is the time of swelling (s), and k is the parameter describing the dynamics of water sorption of the material (s^{-1}), that is, a proper swelling. At equilibrium osmotic pressure, p_{osm} in the swelling centers is equal to the inner stress of stretched swollen cellulosic matter. Obviously, $p_{osm} \approx c RT$, where concentration, $c = n_e/V_e$ (mol/dm^3), that is, volume, V_e (dm^3) is given by the overall amount of sample water in the Swelling centers and n_e (in mol), is the amount of ion-active and nonion-active hydrophilic soluble and nonsoluble substances embedded at cellulosic interfaces; R is the gas constant; and T is the temperature in K. By connecting both eqs. (8.3) and (8.8) according to eq. (8.6) one finally gets eq. (8.9):

$$Y(t) = \frac{a \cdot t^d}{(b + t^d)} + c(1 - e^{-kt}) \tag{8.9}$$

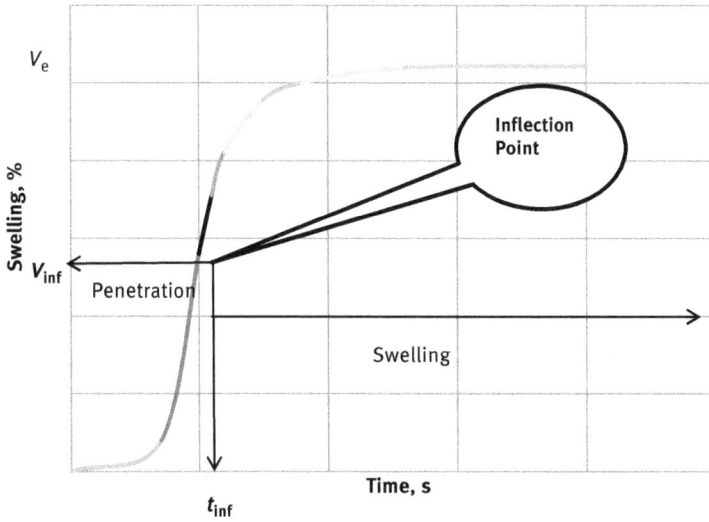

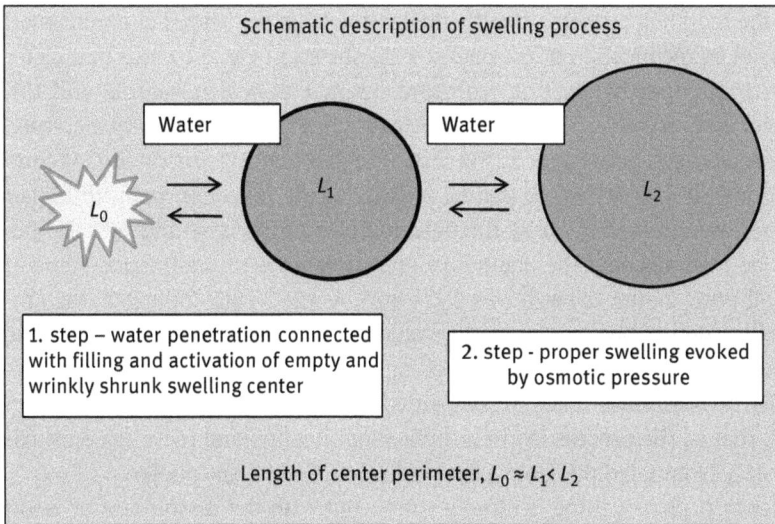

Fig. 8.9: Swelling kinetics of cellulosic material; relative volume sample increases (e.g., the paper sample thickness), % versus time.

The sum of $a + c$ gives the total Swelling capacity of material, and it was statistically derived based on empirical measurements [146].

Solár and coworkers [147] have described a simple method for the measurement of the Dimensional changes during swelling of wood when exposed to an environment composed of water. Geffert and coworkers [148] have studied this process by applying recycled cellulosic fibers. All obtained similar kinetic curves as shown in Fig. 8.9, except that of Recycled pulp fibers, but recently more detailed careful experiments have shown a similar character of these ones pushing only the inflection point at the beginning of the swelling process [146]. Paper swelling kinetics was investigated on the hand sheets made from bleached sulfate pulp composed of a blend of hardwood species (original pulp once industrial dried), subjected to an eightfold Recycling model at two temperatures of drying: 80 °C and 120 °C. The first treatment of fibers (0th recycling) represented a hand laceration of the pulp, defibering, beating to 29 °SR, paper hand-sheet preparation, and drying. During the process of the eightfold recycling model, pulp was returned repeatedly to the process of pulping, additional beating to 29 °SR, and drying [149]. Collected results in Fig. 8.10 demonstrate an influence of simple recycling and its condition upon hypermolecular character of cellulosic matter represented by the bleached sulfate pulp. As expected, at 80 °C, a slower reduction of the sorption capacity with an increasing number of recycling cycles was observed, even at each swelling function separately. A decrease in the change rate of the function Y_2 (at equilibrium, $y_2 = c/(a + c)$) was also observed, which could be explained by the more gradual closing of the pores and lumens, for example, by hornification of cellulose fibers. However, more information about the influence of the recycling process upon the hypermolecular properties of porous matter was received by rectification of parameter d. As shown in Fig. 8.10, this parameter was strongly influenced by beating (compare original unbeaten sample and 0th beaten sample) and recycling. The beating decreased d, while the recycling increased this one. These changes were larger in the case of higher temperature drying. At compared conditions, an increase of d, that is, shifting of the inflection point to higher values of time, was accompanied by the delaying of a proper swelling process, because it can be assumed that the amount of appropriate micro- and nanocenters of the porous cellulosic matter to swell was decreased. As the results indicated, the prolonged recycling noticeably increased the value of d, revealing a decrease in the amount of these appropriate potentially swelling centers. This fact was reflected in a higher rigidity of individual fibers accompanied by a depression of water sorption ability, WRV, that is, the process led to hornification. Beating and recycling approximately reversibly eliminated this unpleasant behavior of cellulosic matter.

The majority of the swelling is closely connected with the dissolution of wood cellulose fibers in NaOH–water. The fact that some cellulose chains did not dissolve while others did, although having the same molecular weight, indicates that these chains are less accessible and embedded in cell regions difficult to dissolve. It was shown [150] that the dissolution capacity of cellulose chains is strongly dependent on their localization within the cell wall structure and the cellulose/hemicelluloses complex. The presence of small amounts of hemicelluloses may prevent or decrease the solubility of cellulose, but it must also be kept in mind that what is called a "cellulose solution" is not a molecular solution in the thermodynamic sense, and that cellulose aggregates are mainly present [150]. Obviously, hemicelluloses connecting the elementary fibrils in the amorphous part of the microfibrils are the most important.

Influence of simple recycling and temperature upon paper swelling relative kinetic parameters

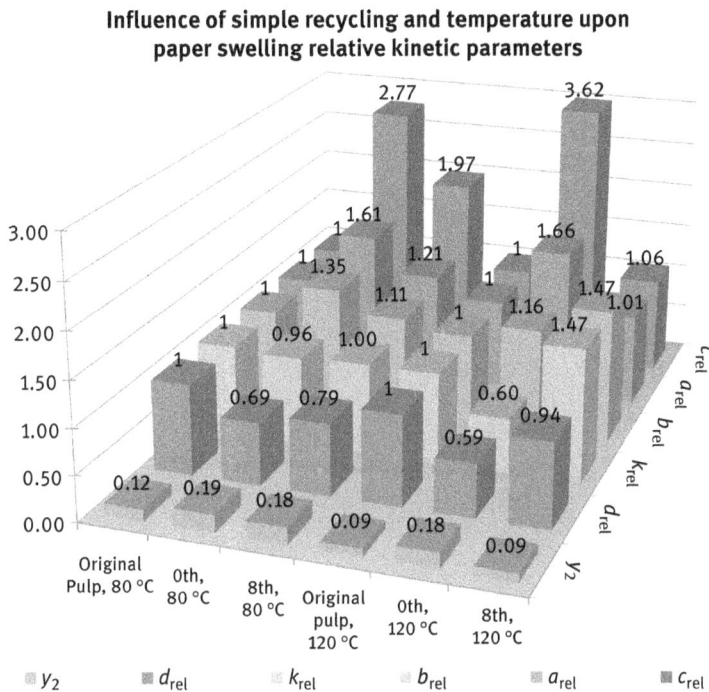

Fig. 8.10: Swelling relative kinetics parameters of paper prepared of bleached sulfate pulp and an influence the simple recycling and temperature [146]. $y_2 = c/(a + c)$.
Note: Relatives parameters a_{rel}, c_{rel}, k_{rel}, d_{rel}, b_{rel} are defined as a ratio of swelling parameters of the papers/swelling parameters of the original pulp. $y_2 = c/(a + c)$. Swelling kinetic parameters of the original pulp are presented in following table:

	a	c	k, s^{-1}	d	b
Original pulp, 80 °C	24.56	3.41	−0.027	2.46	0.47
Original pulp, 120 °C	23.38	2.28	−0.043	3.25	0.795

8.6 Conclusions

We have discussed about relatively simple macromolecules with a miraculous ability to form an infinite number of supramolecular and hypermolecular structures with fascinating shapes; cellulose represents the most abundant biopolymer of all plants (and also in some marine animals). Its existence in both an animate and inanimate nature is strongly connected with water. It is formed from water and carbon (carbon dioxide) during photosynthesis at the very beginning of its existence. It dies out during the Aging of the lignocellulosic matter of plants via degradation and destruction due to the chemical and biochemical influence of the environment;

the dehydration process is a typical example. Also typical is the Dehydration process, which occurs during the burning of cellulose; its liquefaction to hydrocarbons or condensation reactions occur during the carbonization of plant matter under appropriate conditions. The cycle of cellulose in nature is so brilliantly closed.

However, water plays the most important role during formation of the hydrogen bonds, that is, during the formation of supramolecular and hypermolecular cellulosic skeletal structures regardless of whether it is in nature or in industry; the simplest formation of this hypermolecular structure is paper.

Abbreviations

A	Proton-acceptor nanocenter
AFM	Atomic force microscopy
CDD and CDD_0	Critical degree of dilution in water-solution and distilled water
CP/MAS	Cross-polarization/magic angle spinning
D	Proton-donor nanocenter
DM	Groundwood never dried
DEtA	Diethyl amine
DP	Degree of polymerization
DS	Degrees of substitution
EAPap	Electroactive paper
ECF	Elementary chlorine free
EO	Essential oil
ESR	Electron spin resonance
ESCA	Electron spectroscopy for chemical applications
ESRF	European Synchrotron Radiation Facility
FTIR	Fourier-transformed infrared
g	Hydration factor
H-bonds	Hydrogen bonds
LCST	Low critical solution temperature
MBA	N,N'-methylenebisacrylamide
NMR	Nuclear magnetic resonance
PA, PD	Proton-donor, proton-acceptor
PCC	Parenchymal cell cellulose
PNIPAAm	Poly(N-isopropylacrylamide)
PVAm	Polyvinylamine
RH	Relative air humidity
S	Colloidal suspension
Sa J	Bleached sulfate softwood pulp
Sa L	Bleached sulfate hardwood pulp
Si	MgBi-sulfite spruce-bleached pulp
SANS	2D small angle neutron spectroscopy
SAX	Small angle X-ray
SCHL	Structure change in hydration layers
SEM	Scanning electron micrographs
SR	Degree of pulp beating according to Schopper–Riegler

TCF   Totally chlorine free
TC   Terminal complexes
TEM   Transmission electron microscopy
TEMPO   2,2,6,6,-Tetramethylpiperidine-1-oxyl radical
TRHRS   Thermoresponsive hydrated reticular system
UCST   Upper critical solution temperature
UF   Urea-formaldehyde
WAX   Wide angle X-ray
WRV   Water retention ability

References

[1]   Hon D N-S, Schiraischi N. Weathering and Photochemistry of Wood and cellulosic chemistry.
      2nd Marcel Dekker, New York, NY, USA, 2000, 513.
[2]   Wegner T H, Jones PE. Advancing cellulose-based nanotechnology Cellulose 2006, 13(2), 115.
[3]   Gardner K H, Blackwell J. The Structure of Native Cellulose Biopolymers 1974, 13, 1975.
[4]   Sarko A, Muggli R. Macromolecules, Packing Analysis of Carbohydrates and Polysaccharides.
      Valonia Cellulose and Cellulose II Macromolecules 1974, 7(4), 486.
[5]   Czaja W K, Young DJ, Kawecki M, Brown RM Jr.. The Future Prospects of Microbial Cellulose in
      Biomedical Applications Biomacromolecules 2007, 8(1), 1.
[6]   Hon D N-S. Cellulose: a random walk along its historical path Cellulose 1994, 1, 1.
[7]   O'Sullivan A C. Cellulose: the structure slowly unravels Cellulose 1997, 4, 173.
[8]   Eichhorn S J, Baillie CA, Zafeiropoulos N, Mwaikambo LY, Ansell MP, Dufresne A, Entwistle KM,
      Herrera-Franco PJ, Escamilla GC, Groom L, Hughes M, Hill C, Rials TG, Wild PM. Review. Current
      international research into cellulosic fibres and composites, J. Mater. Sci. 2001, 36, 2107.
[9]   Zugenmaier P. Crystalline cellulose and derivatives: characterization and structure. Springer
      Series in Wood Science, Eds., T.E. Timell and R. Sommer, Springer-Verlag, Berlin, Germany,
      2008, 106.
[10] Saxena I M, Kudlicka K, Okuda K, Brown RM Jr.. Characterization of genes in the cellulose-
      synthesizing operons (acs operon) of *Acterobacter xylinum*: Implications for cellulose
      crystallization, J. Bacteriol. 1994, 176(18), 5735.
[11] Michell A J. P.m.r. spectra and conformation of cellulose oligosaccharides in solution in
      methyl sulphoxide Carbohydr. Res. 1970, 12, 453.
[12] Ham J T, Williams DG. The crystal and molecular structure of methyl b-cellobioside-methanol
      Acta Crystallographica 1970, B26, 1373.
[13] Koch H J, Perlin AS. Synthesis and 13C NMR spectrum of D-glucose-3-d. bond-polarization
      differences between the anomers of D-glucose Carbohydr. Res. 1970, 15, 403.
[14] Rowland S O, Roberts EJ. The nature of accessible surfaces in the microstructure of native
      cellulose J. Polym. Sci., Part A: Polym. Chem. Edition 1972, 10, 2447.
[15] Schultz G V, Marx M. Über Molekulargewichte und Molekulargewichtsverteilungen nativen
      Cellulosen Makromol. Chem. 1954, 14(52). (In German)
[16] Marx M. Molekulargewichtsverteilungen von nativen Faser- und Holzcellulosen J. Polym. Sci.
      1958, 30, 119. (In German)
[17] Heinze T. Chemical Functionalization of Cellulose in Polysaccharide: structural diversity and
      functional versatility. Ed., S. Dumitriu, Marcel Dekker, New York, NY, USA and Basel,
      Hong Kong, China, 2004, 551.

[18] Arndt P, Bockholt K, Gerdes R, Huschens S, Pyplo J, Redlichl H, Samm K. Cellulose oligomers: preparation from cellulose triacetate, chemical transformations and reactions Cellulose 2003, 10, 75.

[19] Robertson D D. The evaluation of paper permanence and durability Tappi J. 1976, 59, 63.

[20] Zou X, Uesaka T, Gurnagul N. Prediction of paper permanence by accelerated aging I. Kinetic analysis of the aging process Cellulose 1996, 3, 243.

[21] Kato K L, Cameron RE. A rewiew of the relationship between thermally-accelerated ageing of paper and hornification Cellulose 1999, 6, 23.

[22] Piantanida G, Bicchieri M, Atomic force microscopy characterization of the ageing of pure cellulose paper Coluzza C. Polymer 2005, 46, 12313.

[23] Zervos S. Cellulose: structure and properties, derivatives and industrial uses. Eds., A. Lejeune, T. Deprez, Nova Science, New York, 2010, 155.

[24] Milichovský M, Češek B, Filipi M, Gojný J. Cellulose ageing as key process of paper destruction Przeglad Papierniczy 2013, 69, 291.

[25] Maková A. Deterioration of archive and library documents by microorganisms (in Slovakian), Knižnica 2005, 6(3), 4.

[26] Čabalová I, Fr KJ, Češek GB, Milichovský M, Mikala O, Tribulová T, Ďurkovič J. Changes in the chemical and physical properties of paper documents due to natural ageing BioResources 2017, 12(2), 2618.

[27] Anzenbacher P. Jr., M.A. Palacios, Polymer nanofibre junctions of attolitre volume serve as zeptomole-scale chemical reactors Nat. Chem. 2009, 1, 80.

[28] deMello A J, Wootton RCR. Miniaturization. Chemistry at the crossroads Nat. Chem. 2009, 1, 1.

[29] Fu K, Bohn PW. Nanopore Electrochemistry: A Nexus for Molecular Control of Electron Transfer ACS Cent. Sci. 2018, 2018(4), 20.

[30] Fink H-P, Philipp B, Paul D, Serimaa R, Paakkari T. The structure of amorphous cellulose as revealed by wide-angle X-ray scattering Polymer 1987, 28, 1265.

[31] Verlhac C, Dedier J, Chanzy H. Availability of surface hydroxyl groups in *Valonia* and bacterial cellulose J. Polym. Sci., Part A: Polym. Chem. Edition 1990, 28, 1171.

[32] Okano T, Koyanagi A. Structural Variation of Native Cellulose Related to Its Source Biopolymers 1986, 25, 851.

[33] Simon I, Scheraga HA, Manley RTJ. Structure of cellulose. 1. Low-energy conformations of single chains Macromolecules 1988, 21, 983.

[34] Atalla R H, Vanderhart DL. Studies on the structure of cellulose using Raman spectroscopy and solid state 13C NMR Proceedings of the 10th cellulose conference. Ed., C. Schuerch, John Wiley and Sons, New York, NY, USA, 1989, 169.

[35] Horii F, Hirai A, Kitamaru R. CP=MAS C-13 NMR spectra of the crystalline components of native cellulose Macromolecules 1987, 20, 2117.

[36] Yamamoto H, Horii F. In situ Crystallization of Bacterial Cellulose I. Influences of polymeric additives, stirring and temperature on the formation celluloses Ia and Ib as revealed by cross polarization-magic angle spinning (CP=MAS) 13C NMR spectroscopy Cellulose 1994, 1, 57.

[37] Sugiyama J, Vuong R, Hanzy H. Electron Diffraction Study on the Two Crystalline Phases Occurring in Native Cellulose from an Algal Cell Wall Macromolecules 1991, 24, 4168.

[38] Debzi E M, Chanzy H, Sugiyama J, Tekely P, Excoffier G. The Ia–Ib Transformation of Highly Crystalline Cellulose by Annealing in Various Mediums Macromolecules 1991, 24, 6816.

[39] Okuda K, Tsekos L, Brown RM Jr.. Cellulose microfibril assembly in *Erythrocladia subintegra Rosenv.*: an ideal system for understanding the relationship between synthesising complexes (TCs) and microfibril crystallization Protoplasma 1994, 180, 49.

[40] Kolpak F J, Blackwell J. Determination of the structure of cellulose II Macromolecules 1976, 9, 273.

[41] Kuga S, Takagi S, Brown RM Jr.. Native folded-chain cellulose II Polymer 1993, 34, 3293.

[42] Hayashi J, Sufoka A, Ohkita J, Watanabe S. The conformation of existence of cellulose III$_I$, III$_{II}$, IV$_I$ and IV$_{II}$ by X-ray method J. Polym. Sci: Lett. Edition 1975, 13, 23.

[43] Sarko A. Cellulose – How much do we know about its structure Wood and cellulosics: industrial utilization, biotechnology, structure and properties. Ed., J.F. Kennedy, Ellis Horwood, Chichester, UK, 1987, 55.

[44] Gardiner E S, Sarko A. Packing analysis of carbohydrates and polysaccharides. 16. The crystal structures of celluloses IV$_I$ and IV$_{II}$ Can. J. Chem. 1985, 63, 173.

[45] Howsomon J A, Sisson WA. Cellulose and cellulose derivatives. 2nd, Interscience, New York, NY, USA, 1954, 244.

[46] Frey-Wysling A, Mühlethaler M. Die Elementarfibrillen der Cellulose MakromolekulareChemie 1963, 62, 25. (In German)

[47] Mühlethaler K. Fine structure of natural polysaccharide systems J. Polym. Sci.: Part C 1969, 28, 305.

[48] Pozgaj A, Chovanec D, Kurjatko S, Babiak M. Struktura a Vlastnostidreva. Priroda, Bratislava, Slovakia, 1993, 57. (In Slovakian)

[49] Heux L, Dinand E, Vignon MR. Structural aspects in ultrathin cellulose microfibrils followed by 13C CP-MAS NMR Carbohydr. Polym. 1999, 40, 115.

[50] Vander Hart D L, Atalla RH. Studies of Microstructure in Native Celluloses Using Solid-state 13C NMR Macromolecules 1984, 17, 1465.

[51] Fleming K, Gray DG, Matthews S. Cellulose crystallites Chem. Eur. J. 2001, 7(9), 1831.

[52] Viëtor R J, Mazeau K, Lakin M, Pérez S. A Priori Crystal Structure Prediction of Native Celluloses Biopolymers 2000, 54, 342.

[53] Preston R D. The physical biology of plant cell walls. Chapman and Hall, London, UK, 1974, 139.

[54] Milichovsky M. A new concept of chemistry refining processes Tappi J. 1990, 74(10), 221.

[55] Milichovsky M. Behaviour of hydrophilic components in papermaking suspensions – part III. Practical aspects and behaviour of real highly concentrate papermaking systems Scientific papers of faculty of chemical technology, Pardubice 1994, 57, 157.

[56] Milichovsky M. Behaviour of hydrophilic components in papermaking suspensions . Part II. Experimental hydrated hydrophilic modelling system – its properties and behavior Scientific papers of faculty of chemical technology, Pardubice 1992/93, 56, 155.

[57] Milichovsky M. Behaviour of hydrophilic components in papermaking suspensions. Part I. Interactions among hydrated particles – Theory of Structural Changes in Hydrated Layers (SCHL) Scientific papers of faculty of chemical technology, Pardubice 1992/93, 56, 123.

[58] Hubbe M A, Rojas OJ, Lucia LA, Sain M. Cellulosic nanocomposites: A review BioResources 2008, 3, 929.

[59] Bréchet Y, Cavaillé J-Y-Y, Chabert E, Chazeau L, Dendievel R, Flandin L, Gauthier C. Polymer based nanocomposites: Effect of filler-filler and filler-matrix interactions Adv. Eng. Mater. 2001, 3(8), 571.

[60] Grigoriew H, Chmielewski AG. Capabilities of X-ray methods in studies of processes of permeation through dense membranes J. Memb. Sci. 1998, 142, 87.

[61] Tashiro K, Kobayashi M. Cellulose. Calculation of Crystallite Modulus of Native Cellulose Polym. Bull. 1985, 14, 213.

[62] Neuman R D, Berg JM, Claeson PM. Direct measurement of surface forces in papermaking and paper coating systems Nordic Pulp Pap. Res. J. 1993, 8(1), 96.

[63] Alince B. Porosity of swollen pulp revisited Nordic Pulp Pap. Res. J. 2002, 17(1), 71.

[64] Hermans P H. Physics and chemistry of cellulose fibres. Elsevier, New York, NY, USA, 1949, 13.

[65] Hermans P H, Weidinger A. X-ray studies on the crystallinity of cellulose J. Polym. Sci. 1949, 4, 135.

[66] Mazeau K, Heux L. Molecular Dynamics Simulations of Bulk Native Crystalline and Amorphous Structures of Cellulose J. Phys. Chem. B 2003, 107, 2394.

[67] Milichovský M. Pulp production and processing: from papermaking to high-tech products. Ed., V.I. Popa, Smithers Rapra Technology Ltd., Shawbury, UK, 2013, 155.

[68] Češek B, Milichovský M, Potůček F. Kinetics of vapour diffusion and condensation in natural porous cellulosic fibre web ISRN Mater. Sci. 2011, Article ID 794306, 2011, 1.

[69] Češek B, Milichovský M, Gojný J. Mutual competitive absorption of water and essential oils molecules by porose ligno-cellulosic materials J Biomater Nanobiotechnol. 2014, 5, 66.

[70] Češek B, Milichovský M, Potůček F. Evaluation of microstructural and hypermolecular properties of cellulosic materials in web form Cellul. Chem. Technol. 2006, 40(9–10), 705.

[71] Češek B, Milichovský M, Adámková G. Evaluation of hygroscopicity kinetics of paper, moulded fibre products and other porous materials Cellul. Chem. Technol. 2005, 39(3–4), 277.

[72] Milichovský M, Češek B, J. JG. Behaviour of lignocellulosic materials during wet aging in the presence of essential oils Cellul. Chem. Technol. 2015, 49(9–10), 853.

[73] van Oss C J, Chaudhury MK, Good, RJ. Monopolar surfaces Adv. Colloid. Interface. Sci. 1987, 28, 35.

[74] Milichovsky M. Water – A Key Substance to Comprehension of Stimuli-Responsive Hydrated Reticular Systems J Biomater Nanobiotechnol. 2010, 1, 17.

[75] Poppel E. Rheologie und elektrokinetischeVorgänge in der Papiertechnologie. VEB Fachbuchverlag Leipzig, Germany, 1977, 294. (In German)

[76] Milichovsky M. Mechanism of interaction within pulp suspension II. Electrokinetic properties of pulp mass Scientific Papers of Faculty of Chemical Technology, Pardubice. 1987, Vol. 50, 403. (In Czech)

[77] Hubbe M A. Electrokinetic potential of fibers BioResources 2006, 1(1), 116.

[78] Revol J-F, Bradford H, Giasson J, Marchessault RH, Gray DG. Helicoidal self-ordering of cellulose microfibrils in aqueous suspension Int. J. Biol. Macromol. 1992, 14, 170.

[79] Revol J-F, Godbout L, Dong X-M, Gray DG, Hanzy H, Maret G. Chiral nematic suspensions of cellulose crystallites; phase separation and magnetic field orientation Liq. Cryst. 1994, 16(1), 127.

[80] Marchessault R H, Morehead FF, Koch MJ. Some hydrodynamic properties of neutral suspensions of cellulose crystallites as related to size and shape J. Colloid. Sci. 1961, 16, 327.

[81] Bercea M, Navard P. Shear Dynamics of Aqueous Suspensions of Cellulose Whiskers Macromolecules 2000, 33, 6011.

[82] Dong X M, Revol J-F, Gray DG. Effect of microcrystallite preparation conditions on the formation of colloid crystals of cellulose Cellulose 1998, 5, 19.

[83] Terech P, Chazeau L, Cavaille JY. A smal-angle scattering study of cellulose whiskers in aqueous suspension Macromolecules 1999, 32, 1872.

[84] Orts W J, Godbout L, Marchessault RH, Revol J-F. Enhanced Ordering of Liquid Crystalline Suspensions of Cellulose Microfibrils: A Small Angle Neutron Scattering Study Macromolecules 1998, 31, 5717.

[85] Sugiyama J, Chanzy H, Maret G. Orientation of Cellulose Microcrystals by Strong Magnetic Fields Macromolecules 1992, 25, 4232.

[86] Stroobants A, Lekkerkerker HNW, Odijk T. Effect of Electrostatic Interaction on the Liquid Crystal Phase Transition in Solutions of Rodlike Polyelectrolytes Macromolecules 1986, 19, 2232.

[87] Dong X M, Kimura T, Revol J-F, Gray DG. Effects of ionic strength on phase separation of suspension of cellulose crystallites Langmuir 1996, 12, 2076.

[88] Dong X M, Gray DG. Effect of counterions on ordered phase formation in suspensions of charged rodlike cellulose crystallites Langmuir 1997, 13, 2404.

[89] Dinand E, Chanzy H, Vignon MR. Suspensions of cellulose microfibrils from sugar beet pulp Food Hydrocoll 1999, 13, 275.

[90]   Eklund D, Lindström T. Paper chemistry – an introduction. DT Paper Science Publications, Grankulla, Finland, 1991, 89.

[91]   Scott W E. Principles of wet end chemistry. TAPPI Press, Atlanta, GA, USA, 1996, 61.

[92]   Westfelt L. The chemistry of paper wet-strength I. A survey of mechanisms of wet-strength development Cellul. Chem. Technol. 1979, 13, 813.

[93]   Fredholm B, Samuelsson B, Westfelt A, Westfelt L. The chemistry of paper wet-strength II. Design and dynthesis of model polymers Cellul. Chem. Technol. 1981, 15, 247.

[94]   Haglind B, Samuelsson B, Westfelt A, Westfelt L. The chemistry of paper wet-strength III. Synthesis of grafts for model polymers Cellul. Chem. Technol. 1981, 15, 295.

[95]   Westfelt A, Westfelt L. The chemistry of paper wet-strength IV. Exploration of synthetic routes to a ketene dimer-containing model polymer Cellul. Chem. Technol. 1983, 17, 49.

[96]   Westfelt A, Westfelt L. The chemistry of paper wet-strength V. Reactivities of non- and co-cross-linking model polymers Cellul. Chem. Technol. 1983, 17, 165.

[97]   Samuelsson B, Westfelt L. The chemistry of paper wet-strength VI. Reactivities of homo-cross-linking model polymers Cellul. Chem. Technol. 1983, 17, 179.

[98]   Fredholm B, Samuelsson B, Westfelt A, Westfelt L. Chemistry of paper wet-strength VII. Effect of model polymers on wet-strength, water absorbency, and dry strength Cellul. Chem. Technol. 1983, 17, 279.

[99]   Espy H H. The mechanism of wet-strength development in paper: a r Tappi J. 1994, 78(4), 90.

[100]   Laleg M, Pikulík II. Wet-web strength increase by chitosan Nordic PulpPap. Res. J. 1991, 6(3), 99.

[101]   Milichovsky M. The role of hydration in papermaking suspension Cellul. Chem. Technol. 1992, 26(5), 607.

[102]   DiFlavio J-L, Bertoia R, Pelton R, Leduc M. Proceedings of the 13th fundamental research symposium, session 7: chemistry. Interscience Publisher, Cambridge, UK, September, 2005, 1293.

[103]   Saito T, Isogai A. A novel method to improve wet strength of paper Tappi J. 2005, 89(4), 3.

[104]   Milichovsky M, Velich V. A new method for investigation of interactions in paper slurries Cellul. Chem. Technol. 1989, 23, 743.

[105]   Sanders N D, Bashey AR. Proceedings of the TAPPI international paper physics conference. Hawaii, TAPPI Press, Technology Park, Atlanta, GA, USA, September, 1991, 473.

[106]   Müller M, Czihak C, Schober H, Nishiyama Y, Vogl G. All Disordered Regions of Native Cellulose Show Common Low-Frequency Dynamics Macromolecules 2000, 33, 1834.

[107]   Vodenicarova M, Milichovsky M, Cesek B. Influence of modified pulp during some adsorption processes Cellul. Chem. Technol. 2001, 35(1–2), 59.

[108]   Milichovsky M, Cesek B. Surface flocculation as a new tool for controlling adsorption processes Adsorpt. Sci. Technol. 2002, 20(9), 883.

[109]   Cesek B, Milichovsky M. Releasing of soluble and colloidal substances in pulp slurries Cellul. Chem. Technol. 1996, 30(3–4), 297.

[110]   Jianguo L, Liu X, Zheng Q, Chen L, Huang L, Ni Y, Ouyang X. Urea/NaOH system for enhancing the removal of hemicellulose from cellulosic fibers Cellulose 2019, 26, 6393–6400.

[111]   Nissan A H. Lectures on fiber science in paper. Ed., W.C. Walker, Joint Textbook from the Committee of Paper Industry, CPPA, Montreal, Canada and TAPPI, Atlanta, GA, USA, 1977,

[112]   Zeegers-Huyskens T, Huyskens P. Molecular interactions. Vol. 2, Ed., H. Ratajczak and W.J. Orville-Thomas, John Wiley & Sons, New York, NY, USA, 1981, 281.

[113]   Milichovsky M. O mechanizme vzaimodejstvij v bumagoobrazajustschich gidrofilnych sistemach Khimia drevesiny. 1990, Vol. 1, 69. (In Russian)

[114]   Milichovsky M. Theory of behavior of hydrophilic dispersion systems I. Theory of structural changes in hydration layers (SCHL) of interacting surfaces of dispersion systems Sci. Pap. Faculty Chem. Technol. Pardubice. 1988, 51, 71. (In Czech)

[115] Milichovsky M. Theory of behavior of hydrophilic dispersion systems III. Experimental evidence of SCHL theory Sci. Pap. Faculty Chem. Technol. Pardubice. 1988, 51, 149. (In Czech)

[116] Milichovsky M. Theory of behavior of hydrophilic dispersion systems II. Discussion of the knowledge following from the SCHL theory Sci. Pap. Faculty Chem. Technol. Pardubice. 1988, 51, 115. (In Czech)

[117] Kim J, Yun S. Discovery of cellulose as a smart material Macromolecules 2006, 39(12), 4202.

[118] Milichovsky M, Cesek B. Rheosedimentation – typical and characteristic phenomenon of paper matter Cellul. Chem. Technol. 2004, 38(5–6), 385.

[119] Fišerová M, Gigac J, Balberčák J. Sedimentation properties of hardwood kraft pulps suspensions Papír a Celulóza 2009, 64(11–12), 362.

[120] Milichovský M. Voda – klíčový fenomén při výrobě a užití papíru a papírenských výrobků Chemické Listy 2000, 94(9), 875. (In Czech)

[121] Češek B, Milichovský M. Rheosedimentation – a tool for evaluation of pulp fibre behaviour in wet state Papír a Celulóza 2005, 60(7–8), 224.

[122] Milichovský M, Šestauber K. Hodnocení mechanického opracování buničiny sedimentační metodou Papír a celulóza 1987, 42(9), V54. (In Czech)

[123] Mendes P M. Stimuli-responsive surfaces for bioapplication Chem. Soc. Rev. 2008, 37(9), 2512.

[124] Lutecki M, Strachotova B, Uchman M, Brus J, Plestil J, Slouf M, Strachota A, Matejka L. Thermosensitive PNIPA-Based organic-inorganic hydrogels Polym. J. 2006, 38(6), 527.

[125] Zhang X-Z, Wang F-J, Chu -C-C. Thermoresponsive hydrogel with rapid response dynamics J Mater. Sci.: Mater. Med. 2003, 14(3), 451.

[126] Edelmann K. in Lehrbuch der Kolloidchemie. Band, Vol. I, VEB, Deutscher Verlag der Wissenschaften, Berlin, Germany, 1962, 353. (In German)

[127] Wong J E, Gaharvar AK, Müller-Schulte D, Bahadur D, Richtering W. Dual-stimuli responsive PNi-PAM microgel achieved via layer-by layer assembly: magnetic and thermoresponsive J. Colloid. Interface. Sci. 2008, 324(5), 47.

[128] Xia F, Ge H, Hou Y, Sun TL, Chen L, Zhang GZ, Jiang L. Multiresponsive surfaces change between superhydrophilicity and superhydrophobicity Multiresponsive Adv. Mater. 2007, 19, 2520.

[129] Jochum F D, Theato P. Temperature and light-responsive polyacrylamides prepared by a double polymer analogous reaction of activated ester polymers Macromolecules 2009, 42(7), 5941.

[130] Jaber J A, Schlenoff JB. Polyelectrolyte multilayers with reversible thermal responsivity Macromolecules 2005, 38(7), 1300.

[131] Hou H, Kim W, Grunlan M, Han A. A thermore-sponsive hydrogel poly (N-isopropylacrylamide) micropatterning method using microfluidic techniques J. Micromechanical Microeng. 2009, 19(10), 1.

[132] da Silva R M P, Mano JF, Reis RL. Smart thermoresponsive coatings and surfaces for tissue engineering: switching cell-material boundaries Trends Biotechnol. 2006, 25(12), 577.

[133] Hatakeyma H, Kichuchi A, Yamato M, Okano T. Bio-functionalized thermoresponsive interfaces facilitating cell adhesion and proliferation Biomaterials 2006, 27(9), 5069.

[134] Xin-Cai X, Liang-Yin C, Sen-Mei C, Jia-Hua Z. Monodispersed thermoresponsive hydrogel microspheres with a volume phase transition driven by hydrogen bonding Polymer 2005, 46(3), 3199.

[135] Yoshida R. Self-Oscillating gels driven by the Belousov-Zhabotinsky reaction as novel smart materials Adv. Mater. 2010, 20, 1.

[136] Mano J F. Stimuli-responsive polymeric systems for biomedical applications Adv. Eng. Mater. 2008, 10(6), 515.

[137] Stabenfeldt S E, García AJ, LaPlaca MC. Thermoreversible laminin-functionalized hydrogel for tissue engineering J. Biomed. Mater. Res. Part A 2006, 3, 718.

[138] Hofbauer W, Krueger N, Breuer K, Sedlbauer K, Schoch Mould resistance assessment of building materials – Material specific isopleth-systems for practical applicaton in T. Proceedings of the 11th Int. Conf. on Indoor Air Quality and Climate. Indoor Air 2008. Copenhagen, Denmark, 2008, Paper ID 465.

[139] Křůmal K, Kubátková N, Večeřa Z, Mikuška P. P. Antimicrobial properties and chemical composition of liquid and gaseous phases of essential oils Chem. Pap. 2015, 69, 1084.

[140] Milichovský M, Češekand B, Gojný J. Zařízenínařízenéuvolňování par odpařovaných z kapalnéfázetvořenérůznětěkavýmisložkami (The device for controlled vapour releasing from liquid composed of volatile components). 2016. CZ 29 646 U1, Czech Republic.

[141] Milichovský M, Češek B, Mikalaand O, Gojný J. Water activity restriction by application of essentials oils Cellul. Chem. Technol. 2019, 53(3–4), 281.

[142] Češek B, Milichovskýin M Influence of porosity properties of paper products on theirs hygroscopicity behaviour the Proceedings of XV. Int. Papermaking Conference PROGRESS'05 – Efficiency of Papermaking and Paper Converting Processes, Wroclaw, Poland, 2005.

[143] Dobbins R J. The role of water in cellulose-solute interactions Tappi J. 1970, 53(12), 2284.

[144] Freudenberg U, Zimmermann R, Schmidt K, Behrens SH, Werner C. Charging and swelling of cellulose films J. Colloid. Interface. Sci. 2007, 309, 360.

[145] Maloney T C, Paulapuro H. The formation of pores in the cell wall J. Pulp Pap. Sci. 1999, 25(12), 430.

[146] Geffert A, Vacek O, Jankech A, Geffertová J, Milichovský M. Swelling of cellulosic porous materials – mathematical description and verification BioResources 2017, 12(3), 5017.

[147] Solár R, Mamoň M, Kurjatko S, Lang R, Vacek V. A simple method for determination of kinetics of radial, tangential and surface swelling of wood Drvna Industria 2006, 57(2), 75.

[148] Geffert A, Geffertová J, Vacek O Swelling of the recycled pulp fibers Proceedings of the 8th International Symposium on Selected Processes at the Wood Processing, Štúrovo, Publishing House of Technical University in Zvolen, Slovakia, 2009, p. 45.

[149] Geffertová J, Geffert A. Chapter 11, Recycling of the hardwood kraft pulp Material recycling – trends and perspectives. ed., D.Achilias, InTech, Rijeka, Croatia, 2012, 265.

[150] Le Moigne N, Navard P. Dissolution mechanisms of wood cellulose fibres in NaOH–water Cellulose 2010, 17, 31.

Florin Ciolacu

Chapter 9
Wood- and nonwood fibers in fibrous structures with common and high-tech applications

9.1 Paper and fibrous raw materials

The term "fibrous structure" is a relatively general one, defining structures whose structural components are filiform elements called fibers. Perhaps the most well-known *fibrous structure* with which we are permanently in contact every day is paper. Paper is one of the most valuable inventions of humans, with significant effects on its evolution and culture [1]. Paper was historically the main communication medium, and for a long time it was the only way to store and transmit information from one generation to another. Over time, there was a broad diversification of paper quality that is used, in addition to writing and printing, as a packaging material, and for various technical purposes (filtration of gases and liquids, electrical insulation, building of houses, and interior decoration). Some papers that are used for packaging or technical purposes require greater thicknesses than those for writing and printing, and are known as paperboard and cardboard. Through its domestic and technical applications, paper also played an important role in increasing the comfort of daily life. Moreover, paper is perceived as an art element too, not only as a medium for graphics or watercolor masterpieces, but as a result itself in the art of handmade paper manufacturing [2].

More than half of the paper used globally is for packaging (carton board and containerboard), and this area has been increasing consistently in recent years. In the last time, there have been substantial reductions in consumption of printing and writing paper, which globally represents about a quarter of paper use by volume. The most rapid growth is in sanitary paper (tissue), although it accounts for less than 10% of global volume at present.

Over time, from the famous Tsai Lun to the present day, a variety of fibrous raw materials have been used in paper manufacturing. In fact, identification of new potential raw materials and their processing technology development were the main challenges for papermakers. Until the mid-nineteenth century, rags (based on hemp, linen, and cotton) were the only source of raw fibers for papermaking [3]. Along with the development of mechanical and chemical pulping processes, wood became the main raw material for the pulp industry. The use of annual plants is a particularly

Florin Ciolacu, *"Gheorghe Asachi" Technical University of Iasi*, Romania

https://doi.org/10.1515/9783110658842-009

effective solution for low-resource wood regions, for large areas that are not used for agriculture, and for large surpluses of agricultural waste products (straw).

Today, paper is made from primary and/or secondary fibers, named in this way after the number of papermaking cycles completed. Primary or virgin fibers are separated from wood or annual plants by various methods – mechanical, chemical, or combined – which require weakening or breaking the bonds that provide structural cohesion of the vegetal tissue. Secondary fibers are produced from recovered paper and board, or from "paper and board for recycling," the new term that appears in the updated European Standard List of Grades of Paper and Board for Recycling (EN 643). In 2018, the European consumption of pulps was 32% for chemical wood pulp, 11.6% for mechanical and semichemical wood pulps, 0.3% for nonwood pulp, and 56% for pulp from paper for recycling [4]. Probably the most appropriate term to include both wood- and nonwood fibers would be that of pulp fibers and it will be used further.

9.2 Suitability of pulp fibers for papermaking

During pulping and papermaking process, fibers preserve a great part of their original vegetal tissue structure. This is also the main source of the explanations for the behavior of the fibers in the unit operations of papermaking. Obviously, the question arises: *what make lignocellulosic fibers suitable for papermaking?* The answer is relatively simple because there are many economical and technical reasons, which demonstrate that the pulp fibers are an ideal raw material for papermaking. The economic arguments are as follows:
- Fibers are the main constituent of plants, which are a renewable resource and abundantly available in nature.
- Papermakers are in apparent competition with other users of wood as a raw material, because they mainly use small diameter logs. Wood scrap, sawmill waste, agricultural residue, straw, grasses, and/or rags are acceptable sources of virgin fibers.
- Pulp fibers are reusable/recyclable.
- Pulp fibers are biodegradable.

From a technical point of view, pulp fibers are suitable for papermaking as follows [5]:
- Lignin, which "cements" individual fibers in the plant, is physically and chemically weaker than cellulose, making separation of pulp fibers possible by mechanical and/or chemical means.
- Pulp/wood fibers are made of multilayers of very small thread-like structures called fibrils. These fibrils can be exposed by beating/refining of the fibers and provide a very large area for bonding.

- Pulp fibers develop physical and chemical bonds with other fibers when changing from wet to dry conditions.
- Pulp fibers have a high tensile strength, but good flexibility and conformability.
- Pulp fibers are water-insoluble but are hydrophilic materials at the same time.
- Pulp fibers are chemically stable and relatively colorless (white).

Fibers used in papermaking can be classified by their origin as softwood fibers, hardwood fibers, and nonwood fibers. Depending on the original product composition, recycled pulp consists of some or all of the above-mentioned fiber types. Fiber properties affect the formation and consolidation of the paper structure during the papermaking process, and are responsible for the properties of paper.

The term "pulp" is attributed to the fibrous material resulting from complex manufacturing processes involving chemical and/or mechanical treatment of various types of plant materials. Pulp is used mainly as the major raw material for paper and board production, where the pulp fibers are mechanically modified to form a paper sheet, but also for chemical conversion to products, such as regenerated fibers and cellulose derivatives. Transformation of fibrous materials into paper as the final product is possible after completion of a technological process flow of varying complexity, depending on the characteristics of paper grade made (Fig. 9.1).

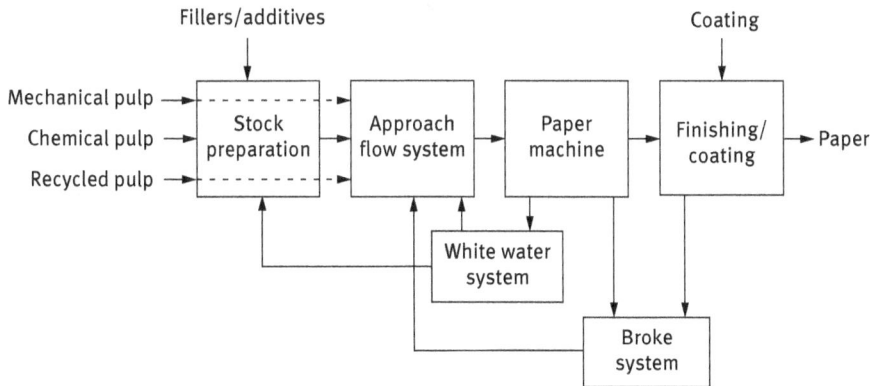

Fig. 9.1: Overview of the papermaking process.

Paper stock is a complex colloidal system in which the components (fibers, fine material, and fillers) interact with each other. The formation process of paper involves the distribution of the paper stock on the wire of a paper machine where the dewatering of suspension takes place and it formed a fibrous structure whose strength continuously increases. The strength of this unconsolidated structure is the result of adhesion forces and is not dependent on the fiber strength. During the formation process, paper passes successively through different phases of consolidation during which the moisture and the value of the bonding forces change. By passing the web from the wire section to wet presses and then to the drying stage, the paper structure gradually moves from a state of coagulation to a network, in which van der Waals forces and weak forces of friction develop between the fiber surfaces that are in contact. On drying, paper passes into a consolidated state where the fibers are strongly linked mainly by hydrogen bonds, and paper strength reaches the maximum value.

Unlike other engineering materials, paper has the following specific characteristics:

- *A heterogeneous composition* of component elements due to the presence of fibers with different lengths and origins, fillers, and sizing materials
- *Three-dimensional anisotropic distribution* of the structural elements due to the different orientation and size of fibers, and auxiliary materials
- The *porous nature of the structure* that controls the paper characteristics such as absorption capacity, permeability to air, hygroscopicity, hygro-expansiveness, and the irreversible modification of certain properties during drying
- *Action of bonding forces between the fibers* that determine the mechanical strength of the paper (friction forces, van der Waals forces, and hydrogen bonds)
- *The two-sidedness*, which is specific to most papers, is determined not only by structural anisotropy but also by the specificity of the manufacturing system. On the Fourdrinier paper machine, drainage takes place in one direction only and therefore the created web becomes asymmetric. Filler and fines are mainly situated on the top surface side of the paper, since drainage through the bottom surface "flushes" the bottom surface. Moreover, the paper web is in contact with the wire on one side and with the press felt on the other.

9.3 Are all types of fibers equally suitable for papermaking?

Currently, over 600 grades of paper and paperboard are known and manufactured, and if we take into account the variations on the basis weight, the number exceeds 10,000. For the manufacture of many types of paper and paperboard, only a relatively small number of fibrous raw materials are available, nearly 25 types of sulfite and kraft pulps, about 10 grades of mechanical pulp, several types of semichemical pulps, and a few pulps from rags and recovered paper. Almost all fibrous materials are suitable for the manufacture of most grades of paper if the right technology is applied, but only an optimum composition will allow achieving the desired paper properties at the lowest cost. To determine the optimal composition of pulp, it is necessary to know the papermaking properties of the fibrous raw materials.

9.4 Papermaking properties of pulp fibers

No physical or chemical fiber index can solely characterize papermaking properties of pulp fibers, as they depend on a complex number of characteristics reflecting the behavior of the fibers during refining, forming, and dewatering capacity of the wet paper web, and the effects on the final paper quality. During the refining process, the papermaking capability of the fibers results from their ability to fibrillate, from their shortening and thinning tendency, and from the speed of increasing the beating degree, while key features for paper sheet formation are the pulp drainability and wet strength of the web. Obviously, the papermaking properties of the fibers determine the structural, mechanical, rheological, optical, and dielectric characteristics of the paper. The papermaking properties of pulps are extensively described and reviewed in the relevant literature [6–10]. The papermaking properties of the fibers are themselves determined by the chemical composition of the fibers and of course by their morphological characteristics. The properties of fibers in a dry and wet state are also important.

9.5 Chemical composition

The chemical composition of the fibers is important because most of the physical properties of paper are strongly dependent on the chemical components. Of equal importance is the molecular structure and morphology of paper stock components. The chemical composition of cellulosic fibers can be distinguished by major chemical components (>1%), minor components (0.1–1%), trace components (0.01–0.1%), and microcomponents (<0.01%). Of course, the source material and pulping process have a decisive role on the chemical composition of the cellulose fibers. The major components of paper-grade pulps are cellulose, lignin, and hemicelluloses. Over time, many studies have tried to demonstrate the influence of the different chemical components of cellulosic fibers on their papermaking properties. Most of them focused on the influence of hemicelluloses and residual lignin content.

The content of *hemicelluloses* in chemical pulps is about half the original quantity existing in wood. Obviously, the chemical composition of the raw materials used to obtain the pulp is variable depending on the wood species and pulping process. Consequently, the content of hemicelluloses will differ. Thus, coniferous wood contains 24–33% lignin, 40–44% cellulose, and 25–30% hemicelluloses, to which are added small amounts of resin and extractive substances; hardwood has a lower lignin content (16–24%) and a corresponding higher hemicellulose content (30–35%) [11]. Due to their high reactivity and specific distribution in the cell wall structure (in the matrix between the cellulose fibrils in the cell wall, mainly in the outer layers of the fiber [12–14]), hemicelluloses are an important chemical component of paper-grade pulps. They are highly hydrophilic and have a high swelling capacity, promoting fibrillation and fiber hydration during the refining process. Generally, it was found that sulfite pulps with high hemicellulose content (about 14%) are easily refined and form papers with high rigidity, high tensile strength, and low greasy permeability. On the other hand, both kraft and sulfite pulps with only 8% hemicelluloses are beaten easier than those with 12% content. There is an optimum content of hemicelluloses that experience the highest indices of strength. A high content of hemicelluloses is in many cases equivalent to a lower content of alpha-cellulose, a component with long molecular chains that strongly influence fiber strength. The positive effect of the high content of hemicelluloses on the fiber-binding ability is diminished overall by the lower strength of the fibers. In conclusion, it is accepted that pulps have a good beating behavior and form papers with good resistance, if the alpha-cellulose content does not fall below 94–95% and the pentosans are at least 2.5–3%. It should be noted that not only the quantity but also the chemical composition and physical state of the hemicelluloses influence the papermaking properties of the chemical pulps. Hexosans, the main component of softwood hemicelluloses, have a stronger effect than pentosans, which predominate the hardwood hemicelluloses. In hardwoods, the predominant hemicelluloses are a partially acetylated (4-O-methylglucurono)xylan with a small proportion of glucomannan. In softwoods, the major hemicelluloses are partially acetylated galactoglucomannan. A smaller amount of an arabino-(4-O-methylglucurono)xylan is also present. This is the main chemical difference between softwoods and hardwoods [15–17]. When comparing two chemical pulps, sulfite and kraft, from the same wood with approximately equal yields and close hemicellulose content, the sulfite chemical pulp is beaten twice as fast, and its strength is between half and three quarters of the strength of the kraft pulp. This shows that the physical state of hemicelluloses also influences the papermaking properties of the fibers to a great extent.

Lignin, in the eyes of the papermaker, is an unwanted chemical component of chemical pulps because it prevents fiber softening, limits swelling, and reduces the beating rate and the ability of fibers to fibrillate. At high contents of lignin, fibers are rigid, brittle, and unable to establish strong links in the paper sheet. Upon wood delignification, the papermaking properties of the chemical pulp continuously improve as the lignin content decreases to a certain value, and the phenomenon is then reversed. This is explained by the fact that in the final stages of cooking, a strong degradation of the polysaccharides occurs, which has an adverse effect on pulp quality. In terms of mechanical resistance, an optimal content of lignin is about 9% for kraft pulps, but differs substantially for sulfite pulps depending on the type of strength test used. This makes it difficult to reach a general conclusion in this case. Unbleached sulfite pulps show different properties when compared with kraft pulps. The residual lignin is sulfonated and therefore rather hydrophilic and is extracted rather easily with effective washing. Because of the acidic pulping conditions, the level of hemicelluloses is very high; therefore, the unbleached pulp yield is high. This yield's advantage is easily sacrificed with careless bleaching. Alkaline extraction removes more materials from the sulfite pulp in comparison to the kraft pulp. The resin fraction in unbleached pulp can be quite high, because sulfite pulping does not saponify the resins. In contrast to kraft pulp, the strength of the sulfite fibers depends on the content of hemicelluloses. Removal of hemicelluloses results in a decrease of pulp strength [18].

The strength of the sulfite pulp is dependent on the pulping and bleaching conditions. Acidic pulping causes an initial degradation of the fiber structure. This degradation can be severe, despite the relatively high degree of polymerization of the cellulose chains. Upon beating, the breaking length increases due to the presence of the hemicelluloses, which produces good fiber-to-fiber bonding. In parallel, the tear strength decreases because of the shortening of the fibers. The extraction of the hemicelluloses, therefore, decreases the strength properties. In fact, due to distinct disadvantages of the sulfite cooking process (including all its modifications) over the kraft pulping technology (see above), the share of sulfite pulps in total fiber production steadily decreased from 60% in 1925 to only 3.7% in 2000 [8].

9.6 Morphological features of fibers

Depending on their origin, papermaking fibers differ mainly by morphological characteristics: fiber length, "diameter" or fiber width, wall thickness, density of the cell wall, and the shape of fiber tips, pits, and fenestrations. The morphological properties of the fibers do not essentially change during chemical pulping, but some dimensional changes occur during mechanical pulping.

Fiber properties vary significantly with wood species, growth condition, age, and the pulping and papermaking treatments. Due of the stochastic nature of these variables, the properties of the pulp may have a wide distribution. Each property derives from the contributions of each fiber, and is described by statistical numbers: mean, shape, and width of distribution. In order to determine the property of single fibers, many fibers must be measured and the results calculated using a statistical approach.

9.6.1 Fiber length and fines

There is no doubt that fiber length is one of the most important papermaking characteristics of fibers. A long fiber is able to establish a large number of bonds to other fibers and as such will be anchored more strongly in a fiber network compared with a short fiber. On the other hand, fibers that are too long form papers with a structure of great unevenness (poor formation). The papermaking fiber length is in the range of 1–6 mm. Hardwood fibers have an average length of around 1 mm, whereas the average length of a softwood fiber is 3 mm. Typically, the length of a softwood fiber is 100 times its width, while the width of the open (not collapsed) fiber is approximately 10 times the wall thickness. Tensile strength of the wet paper web increases rapidly with increasing fiber length. The tensile, burst, and tear strengths, and elongation at break of the dry paper also improve with increasing fiber length. The influence of the fiber length on the dynamic strength properties, such as tear strength, is much higher than for the equally important interfiber bonds. This is also proved by the well-known relationships established by Clark for the main strength indices [19 –24]. So, if the breaking length of the paper is proportional to the square root of the fiber length, the exponent of the fiber length term from the burst index equation increases to 1.0 and to 1.5 for the tear index equation. Fiber length can be assessed either directly or indirectly; the indirect methods split the pulp fibers into different size fractions and give an indication of the fiber length distribution of a pulp and its fines content. This is performed by repeated stepwise screening through screens of decreasing slot size. Fibers are retained predominantly according to their length, but flexibility of the fibers also affects the fractionation result (unsuitable for chemical pulps). Typical screen slots used for mechanical pulp on a Bauer–McNett apparatus are 30, 50, 100, and 200 mesh. The fines content is the material that passes through the screen with the smallest slot (200 mesh). A dynamic drainage jar with a 200-mesh (76 μm diameter holes) wire screen can also be used to measure the fines content [25]. Usually the sample is 500 mL of 0.5% consistency stock.

Determination of the average fiber length can be achieved by different microscopic counting methods [26], image analysis methods [27, 28], or by using various optical devices for measuring dilute fiber suspension movement [29–31]. Counting the fibers and determining their length allow calculation of the arithmetic mean fiber length, but this mean is not always the most useful. For different properties of paper, different averages are more suitable from the papermaker's point of view. In the case of fiber length, the length-weighted average is the most appropriate. The recent development of optical analyzers has made it possible to measure a large amount of fibers in a short time (10,000), including the measurement of fiber morphology parameters such as fiber width, fiber wall thickness, and external fibrillation (Fig. 9.2).

Today, commercial equipment for pulp fibers analyzing provides information regarding the fiber length distribution, average fiber length, and the fiber coarseness of a sample. More modern instruments (e.g., *MorFi Neo* from TECHPAP *Valmet Fibre Image Analyzer* known as *the Valmet FS-5*) also provide additional information about fiber width and fiber wall thickness. In addition, shape parameters such as kinks and fiber curl may be quantified by proper image analysis [29–31]. Even, the fines content, optical and gravimetric coarseness, hardwood/softwood ratio, and fibrillation can be determined; the measurements were made using two cameras that measure the fibers in a 50 μm wide chamber and then deliver the data to the software.

(a)

Contour length (L)

(c)

Fiber width

(b)

Projected length (l)

(d)

Wall thickness

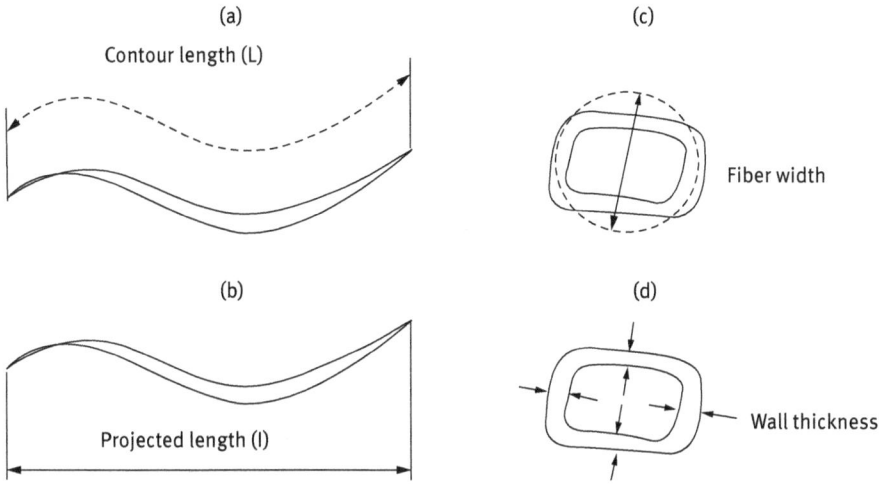

Fig. 9.2: Dimensional characteristics of fibers: (a) length of strained fiber; (b) length of curled fiber; (c) dimension of cross section; and (d) thickness of fiber wall.

The definition of curl index is the result of the difference between the true contour length (L) and the projected length (l) of the fiber divided by the projected length as follows:

$$\text{Curl index} = \frac{L - l}{l} \tag{9.1}$$

The curl index is calculated for each individual fiber. A curl index of zero indicates that no curl is present.

The kink is the abrupt change in fiber curvature. The most widely used definition for kink is Kibblewhite's equation. Kibblewhite established that large kinks in fibers had more of an impact on paper properties (i.e., tensile strength, tear, and so on) than small kinks. Therefore, his equation places greater importance on larger angled kinks as follows:

$$\text{Kink index} = \frac{2N_{21-45} + 3N_{46-90} + 4N_{91-180}}{L_{\text{tot}}} \tag{9.2}$$

where N is the number of kinks with kink angle values within the three intervals of 21–45°, 46–90°, and 91–180°, respectively, and L_{tot} is the sum of the fiber lengths [32].

The fiber length and coarseness have a large impact on paper properties such as tensile index [33–35], folding endurance [19], tear index at low bonding levels [19, 36], and improved formation [37, 38]. On the other hand, fiber curl and kink have been shown to affect the tensile index [39–41], tensile stiffness [42], tear index [39, 40], porosity, bulk, absorbency [41], and burst [42].

9.6.2 Cell wall thickness and fiber coarseness

In geographical areas where the seasons are clearly differentiated, softwood and some hardwood species show seasonal variations in growth and consequently in wood density and fiber size. During the spring season, thin-walled fibers are formed with large lumen called early wood fibers or springwood. With the change of season, there is a gradual change in fiber size due to the thickening of the walls and decreasing fiber lumen. Thick-walled fibers form sections of latewood or summerwood growth rings (annual). Wood fiber coarseness variations between early wood and latewood can be larger than those of softwood species [6]. The early wood fibers collapse more easily, they are more flexible and conform more easily to other fibers.

Fiber coarseness is defined as the weight per fiber length and is normally expressed in units of mg/m or g/m. Coarseness can be calculated from the number of fibers, their average length, and the total dry weight of the fibers in a sample. Fiber coarseness can also be obtained by multiplying the cross-sectional area by the density of the fiber cell wall. The fiber coarseness is somewhere around 0.1–0.3 mg/m; the thicker the fiber wall, the higher the coarseness value. Advanced fiber analyzers provide direct information on coarseness, the prerequisite being that the total dry weight of the fibers in the sample is known exactly [43]. Fiber length for a given species correlates with their coarseness, as long fibers are appreciated as being coarser than short fibers.

Sometimes, the longest fibers that appear to have the greatest coarseness lead to the strongest papers. In order to more accurately assess the effects of the dimensional characteristics of the fibers, various indices were calculated [44–49]:

$$\text{Slenderness ratio} = \frac{L}{D} \tag{9.3}$$

$$\text{Rünkel ratio} = \frac{2w}{d} \tag{9.4}$$

$$\text{Flexibility ratio} = \frac{d}{D}100 \tag{9.5}$$

$$\text{Mühlsteph ratio} = \frac{D^2 - d^2}{D^2}100 \tag{9.6}$$

$$\text{Rigidity coefficient} = \frac{w}{D}100 \tag{9.7}$$

$$F-\text{ratio} = \frac{L}{w}100 \tag{9.8}$$

where w is the cell wall thickness, d is the diameter of the lumen, D is the diameter of the fiber, and L is the length of the fiber.

The best papermaking pulp fibers are derived from wood species, whose fibers do not exceed the Rünkel ratio value of 1 and a value of 80 for the Mühlsteph ratio.

9.7 Wet fiber properties

The main papermaking characteristics of fibers in the slurry are swelling and specific properties of the swollen fibers: collapsibility, flexibility, and conformability. These features are defined schematically in Fig. 9.3.

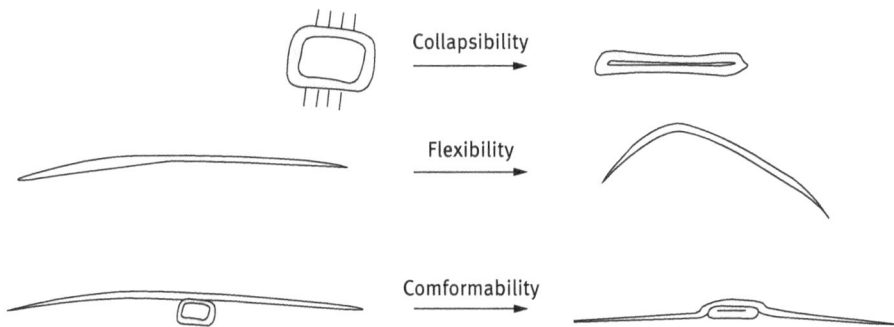

Fig. 9.3: Wet fiber properties.

Conformability is a combined property that is not directly measurable. Indirect information regarding conformability can be deduced from a combination of swelling (water retention), flexibility, and collapsibility measurements. *Conformability* of fibers is a general term that characterizes the behavior of the pulp fibers during the consolidation of the paper structure. Fibers with high conformability bend easily taking the shape of one another and form a dense strongly connected structure. Conformability depends on the cross-sectional dimensions of the fibers, the internal fibrillation, and the chemical composition and morphology of the cell wall.

Fibers have circular or rectangular sections in wood, but may flatten or collapse during the papermaking process. Under lateral pressure (wet pressing, drying, or calendaring), the pulp fibers collapse and change the tube structure into a double-layered strip. When collapsed fibers change, the lumen section alters, whereas the value of its perimeter does not. The collapse phenomenon is more specific for chemical pulp fibers rather than those of mechanical pulps. The compression force required to collapse the fiber laterally is in the range of 80–500 N/m for a normal pulp and decreases upon increasing the degree of pulping [50]. The main factor of influence on the *collapsibility* is the cell wall thickness. Numerous studies have indicated that the springwood fibers are more flexible [51, 52] and collapsible [53–57] than those of summerwood, the latter being stiffer and less collapsible. Furthermore, sulfite fibers collapse more easily than kraft fibers. The noncollapsed lumen of fibers produces an increase in light scattering. Collapsed fibers have a negative effect on the optical properties, but a positive effect on the binding ability of the fibers and thus on the strength properties of the paper.

The flexibility of wet fiber is a measure of its ability to deform and change shape, to bend or curve, under the action of external forces. Measurement of wet fiber flexibility is often limited to long fibers. Flexibility or its inverse, called wet fiber stiffness, can be measured by different methods but all are difficult to use and display a rather wide dispersion of results [58, 59]. The wet fiber flexibility decreases rapidly with increasing cell wall thickness, but increases with decreasing pulp yield. Wet stiffness is higher for kraft pulps than for sulfite pulps [60]. Generally, springwood fibers are approximately twice as flexible as summerwood fibers, and the flexibility increases linearly with the degree of beating [61]. Fibers of mechanical pulps have less flexibility than fibers of chemical pulps. The effects of the degree of beating and pulp yield on fiber flexibility are explained by increasing cell wall porosity and wall delaminating. Mohlin and coworkers [62] reported that fully bleached commercial pulps are more deformed than unbleached pulps. Deformation such as kinks, angular folds, and twists, which are prone to change the direction of the fiber axis, is found to have a negative effect on the tensile index, compressive strength, and so on [63].

9.8 Dry fiber properties

Despite the difficulties recorded when using measurement techniques, the mechanical properties of single fibers have enjoyed considerable attention [64–69]. By clamping a fiber on a microtensiometer, the tensile strength of single fibers may be measured directly. However, this technique is not easy and may damage the fiber wall such that a predetermined breaking zone is introduced. To determine a significant average value, a large number of fibers must be measured. The large distribution of fiber strength values is a result of the effects of nonuniformity of both the biological raw material and of the pulping processes. Moreover, due to the unfavorable statistics of single fiber measurements, the value of the zero-span strength of paper is commonly accepted as an indirect measure of fiber strength. A sheet is clamped in the tensile apparatus so that the clamps at rest are in contact, and most of the fibers bridging the gap are clamped from both sides. When the clamps separate, these fibers will be under stress. The tensile stress–strain behavior of various cellulose fibers is similar. Differences are due to the origin and type of fiber, as well as defects relating to the fiber's history, especially pulping and refining. Such weak points may include pores, cracks, and dislocations. The elastic modulus of elongation is determined by the initial linear section of the load–elongation curve; beyond the linear section, the fiber offers less resistance toward further stress of elongation. The load–elongation properties of wood fibers are heavily influenced by the wrapping angle of the cellulose fibrils within the secondary S2 layer. For small fibril angles (<10°), the fibers are highly elastic while for high angles, the behavior is more plastic. Moreover, when the angle of fibrils in the S2 layer is small, the load–elongation behavior of a single fiber is qualitatively similar to that of paper. Bergander [70] demonstrates, by measurements with confocal laser scanning microscopy, that the fibril angle in the springwood varies considerably, while in the latewood fibers it is rather constant and smaller.

At the structural level, the tensile behavior of fibers is more complex. When straining a fiber to a high degree, it will tend to twist due to its helical fibrillar structure. This phenomenon cannot happen in paper because of the surrounding network structure. The nonlinearity of the stress–strain dependence of single fibers comes from curling and various defects such as crimps, kinks, and microcompressions, which are corrected during fiber elongation. On the other hand, the absence of these defects determines an almost linearly elastic behavior of the fibers [71, 72].

The cell wall of cellulose fibers can have a large number of dislocations and other unevenness arising from wood chipping, pulping, and beating, or other operations from stock preparation (pumping, mixing, and cleaning). Also, there may be natural defects such as pits [73]. All these factors reduce the modulus, tensile strength, and breaking elongation of the fiber. Only substantial changes in the fiber axis direction (kinks, angular folds, twists) effectively reduce the zero-span strength, the tensile strength, and elasticity modulus of paper, as less significant deformations such as fiber curl just diminish their module and increase the tensile deformation of the fibers [74]. Typical values of the tensile strength of wood fibers are 100–200 mN, springwood fibers require lower values, whereas summerwood fibers require higher values. The breaking force of the summerwood fiber is higher than that required to break springwood fiber because the former has a larger cross-sectional area. The summerwood fibers also have higher breaking stress (540 MPa compared with only 390 MPa) [6, 75].

Pulping experiments have shown that the gentle removal of lignin during the cooking process does not alter the force required to break the fibers. Pulp may show a comparable fiber strength to wood; this leads to the conclusion that noncellulosic components – lignin and hemicelluloses – do not contribute to the tensile strength of the fibers. The pulping and bleaching reagents can degrade cellulose, reduce the degree of polymerization, and decrease the stiffness and strength of the fibers. This happens especially when it is used in more severe conditions or treatments of extended duration. Kraft pulping provides fibers that are more resistant than sulfite pulping [76, 77].

9.8.1 Effects of drying stress (Jentzen effect)

We cannot discuss the properties of dry fibers without mentioning the Jentzen effect. Jentzen [6, 72] showed that drying of never-dried virgin fibers under an axial tension increases the tensile strength and decreases the breaking strain. The opposite occurs when fibers are dried under axial compression. The mechanism of this phenomenon can be explained by the fact that the drying tension reduces fibril angles, aligns molecules parallel to the external load, straightens the fiber, and pulls out dislocations and other defects. A swollen state of the hemicelluloses matrix is necessary for the structure and molecular rearrangement of fiber to occur, and the high hemicellulose content is beneficial for the Jentzen effect. But what is more important is that dried and rewetted kraft fibers, which were dried again under load, show no Jentzen effect. This effect demonstrates that drying causes irreversible changes in the fiber wall structure and it also proves the generation of internal stresses during the drying process.

9.9 What happens to cellulosic fibers during papermaking?

The process, in which fibrous cellulosic materials are converted into stock that will be used on the paper machine, is called stock preparation. The paper stock may be composed of one grade of fiber or a mixture of fibrous raw materials. The nature and extent of the fibrous raw materials used for the paper stock composition depend on the quality characteristics and properties that are required by the final product. Stock preparation consists of several main technological operations: slushing, deflaking, and beating or refining by which pulp properties are tailored to fit the end-product requirements. The proper choice of technology for stock preparation allows manufacturing a wide range of paper grades using a relatively small number of fibrous raw materials. Balancing the efficiency and reliability of the papermaking process against end-product quality is a crucial factor in stock preparation optimization.

9.10 Pulp fibers in the slushing and deflaking processes

Paper mills can be an "integrated mill" consisting of several pulp and paper manufacturing lines, or a "nonintegrated mill" with solitary lines of paper or pulp production. Paper mills are integrated with (chemical) pulp mills, mechanical pulp mills, and/or deinked pulp mills.

In the cases where a pulp mill is integrated with a paper mill, pulp is pumped through a pipeline from the pulp mill into the paper mill. This eliminates the need for treatment and a pulping system for pulp bales. Moreover, pulp changes due to the drying process known as *hornification* are eliminated. In contrast, dissolved materials coming with the pulp slurry may disturb the wet-end processes of the paper machine [78].

In a nonintegrated paper mill, chemical pulp is supplied in the form of dried pulp sheets or bales, which must be subject to slushing. If necessary, pulp is further disintegrated in a deflaker. The main objective of these operations is to disintegrate cellulose sheets (bale) and turn them into a pumpable fibrous suspension by breaking the bonds created during the pulp dewatering and drying processes. Disintegration of the fibrous suspension ensures there are no visible flakes or bundles of fibers, ensuring a high degree of individualization of fibrous material which is sent for refining. During the wet disintegration stage, hydrating and swelling of the fibers also occurs, which improves efficiency and the effects of the subsequent refining operation.

Disintegration of fibrous materials must be performed without shortening the fibers, because their length influences the rate of stock drainage on the machine wire and the mechanical properties of the paper sheet; this condition is thoroughly satisfied by the pulpers. If the paper surface and strength properties are developed during refining, the disintegration and individualization of the fibers are key factors for optimal refining. Pulp is always very easy to disintegrate, but some pulps are more difficult to disintegrate into individual fibers than others. The main factors affecting the difficulty of disintegration are the type of pulp, pulp drying method, and the time and temperature of storage.

Softwood pulps dried in an air flotation dryer or on dryer cylinders are classified as pulps that are easy to disintegrate. On the other hand, the flash-dried pulps and in particular nonwood pulps that tend to form clumps (bagasse, reed, and straw) are considered to be difficult to disintegrate [79]. Without any connection to the pulp origin or method of drying, the frozen pulp bales require more intensive use of energy for slushing in comparison to normal pulps. As a general rule, the pulp types that are difficult to slush require deflaking prior to refining. The results of slushing and deflaking, and the degree of defibering are also affected by a wide variety of factors related to stock preparation history, slushing time, and energy and slushing conditions. Undoubtedly, the slushing time and energy have a strong effect on the slushing result. The operating mode of the pulper is also important. The pulpers can operate continuously at low consistency (4–5%) or in a batchwise system at medium or high consistency (6–8% or over 15%). A batch-type pulper yields more homogenous pulp slush than that obtained from a continuously operating pulper since, for instance, variations in the feed flow can be eliminated and all fibers receive the same amount of slushing energy. The specific slushing energy consumption depends on the type of pulp and operating conditions of the pulper;

typical values are 10 kWh/t for easily disintegrated pulps and 20 kWh/t for pulps that are difficult to disintegrate. The specific energy consumption is 50% higher in a batchwise operating system. Slushing energy can be described by two parameters: the first is a quantitative one – the amount of energy for slushing; and the second is a qualitative one – the intensity of slushing, which is expressed by installed power per cubic meter capacity of the pulper vat (kW/m^3). Increasing the slushing intensity improves the process efficiency, but the specific energy consumption is also raised. A typical value for the intensity of slushing in a medium-consistency pulper is 5–6 kW/m^3. The slushing result is affected by the slushing conditions: water temperature and pH value as well as consistency. Generally, a higher water temperature results in an acceleration of defibering and reduces the amount of energy needed. For example, an increase in water temperature from 10 to 40 °C reduces the time required for disintegration and the specific energy consumption by about 50%, while increasing the temperature from 40 to 60 °C has only a minor effect on the slushing rate [80]. The slushing temperature normally depends on the circulation water temperature. Pulp disintegration under alkaline conditions (pH > 7) is easier than under acidic conditions. The penetration of water into the fibers is more difficult in acidic conditions or if the concentration of cations in process water is high. Sometimes, the slushing pH is raised with sodium hydroxide.

Increasing the slushing consistency will improve the disintegration effect up to a certain limit (optimum consistency). The required time for disintegration will also increase with increased consistency. However, at high consistency it can be processed in one batch a larger amount of pulp, and the specific energy consumption is greatly reduced. Increased consistency from 4% to 7–8% reduces the energy consumption by 40–50%, but the use of an appropriate rotor is required. The optimum consistency for pulp slushing depends on the type of pulp as well as the type of pulper. For example, in vertical pulpers running at medium consistency, the optimum slushing consistency for bleached softwood pulp is 6–7%, while for bleached hardwood pulp it is 7–8%. At consistencies of more than 8%, pulp movement starts to slow down markedly, which will reduce the disintegration efficiency. Generally, consistency must be high enough to allow pulp movement and vortex formation. Flotation of pulp sheets or pieces of material on the surface of the slurry indicates a consistency that is too high. For the discharge of a medium consistency pulper, the pulp consistency is reduced to 4–5% by adding diluted water at the bottom of the vat, above the screen plate, and in the suction of the discharge pump. Higher consistencies (over 15%) require the use of a specially designed rotor.

In many cases, pulp fibers cannot be sufficiently separated from each other only by slushing. To complete the individualization of the fibers, a special equipment called deflaker is used. Defibering is more efficient in a deflaker than in a pulper, which saves time and slushing energy. Moreover, deflaking will yield more flexible fibers, which decreases the shortening effect in the refining stage. On the other hand, due to the breakup of the fiber bundles, pulp cleaning will be easier and the fiber loss is reduced. The operation of the deflaker is based on generating turbulent flows and a pressure pulsation of high frequency by changes in the pulp flow direction using a rotor with a high peripheral velocity (about 40 m/s) and a wide gap clearance of approximately 0.5 mm. Velocity gradients between the liquid layers generate forces that efficiently disintegrate fiber bundles. The energy requirement for deflaking depends on fiber types and their "drying history." A typical energy requirement varies from 20 to 40 kWh/bdmt. An operating consistency for deflaking is from 4% to 5%; a higher consistency value is worth using as it improves the defibering result. Typically, pumps are the factors that set limits for the consistency.

The result of pulp disintegration can be assessed by means of sample sheets by sight, optical methods from sample sheets, and stock quality degree (SQD). SQD is a measure of the fiber bundle disintegration efficiency. SQD equals the tensile strength of the laboratory sheets made from mill stock, expressed as a percentage of the tensile strength of lab sheets from the same pulp when completely disintegrated in standard laboratory disintegration equipment (eq. (9.9)). By "completely," it is meant that there are no visible fiber bundles and that the tensile strength increases less than 5% during the next 10,000 rotations of the standard lab disintegrator. Easily disintegrated pulps require an SQD of about 75% after slushing, and the rest of disintegration takes place in the first moments of refining. Difficult pulps require a higher SQD, which sometimes goes up to 100%, so that they are essentially disintegrated before refining:

$$SQD = \frac{\text{Sample sheet tensile strength}}{\text{Lab sheet tensile strength}} 100 \, [\%] \tag{9.9}$$

Energy savings are possible by optimizing the extent of both slushing and refining of the market pulp. Initial trials at mill level found 2 min to be a sufficient time for slushing the softwood pulp. However, for the hardwood pulp, the tensile strength development continued for an optimum slushing time of 10–12 min, reducing its subsequent refining requirement [81]. Tensile index proved a good measure, although other properties may better satisfy the properties sought from the pulp in question. Correct slushing, and in some cases extra disintegration in deflaker, creates the conditions for optimum refining with favorable effects in carrying out the required paper properties. The slushing is the moment when the processes of wetting and swelling the fibers (free swelling) start, making them flexible and releasing the fiber bonds.

9.11 Swelling of fibers

Swelling of fibers is a stage of great importance in papermaking process because it has a great impact on refining results. As a fiber becomes more swollen, the links between the units of the superstructure become weaker, the fiber is more flexible, and the cutting actions during refining are reduced. On the other hand, the swelling of fibers favors fibrillation, a transformation that influences decisively the paper strength.

The water–fiber interaction absorption refers to the process where water is taken up by the fibers from the external liquid phase. Various mechanisms have been suggested for the absorption process. Three types of water have been defined for water present inside a fiber: water of constitution, imbibed water (or bonded water), and free water (nonbonded water) [82]:

Water of constitution is that which remains at zero relative vapor pressure. It holds water by fairly strong valence bonds, probably hydrogen bonds. This water forms a monolayer on the cellulose surface.

Imbibed water is all the additional water held by the fiber wall when the relative water vapor is increased from zero to 100%. This water accumulates as further layers on top of the basic monolayer.

Free water is that held by the fiber, after fiber saturation (at 100% relative water vapor pressure) has occurred. This is held by capillary forces and is not bonded in any way.

Cohesion between the supramolecular structural units of the fibers is ensured by hydrogen bonds established between the hydroxyl groups of the cellulose macromolecular chains. The penetration of water molecules between these units determines their spacing and effective swelling. Fiber swelling under aqueous conditions can be understood in terms of the theories of gel swelling derived at the beginning of the twentieth century. From this starting point, Grignon and Scallan [83] considered that the swelling of the cellulose gel is caused by an osmotic pressure differential, resulting from a difference in concentration of mobile ions between the interior of the gel and the exterior solution. They also showed that a highly charged fiber swells more than the fiber with a lower charge. This was, however, later found to be true only for bulk-charged fibers, since Laine and coworkers [84] and Torgnysdotter and Wågberg [85, 86] found that an increase in the surface charge of the fibers has no significant influence on the swelling of the fiber wall.

Pulp fibers are a multiphase body. When immersed in water, the molecules of the water will try to penetrate into the fiber wall because the osmotic pressure of the water considerably exceeds that of the fiber. The accessibility of the specific domain decreases in the following series: surface of the fiber > surface of the fibrils > open pores and crevices with decreasing size > amorphous domain > crystalline domain. The degree of fiber swelling is determined by the balance between the osmotic pressure and the elastic restraining force of the fiber wall [87, 88]. The partial solubility of hemicelluloses in the cell wall is also an important mechanism of fiber swelling [89].

Many investigations have been focused on studying the swelling properties of the fiber wall, following the introduction of the polyelectrolyte gel concept, as a model for describing the swelling of pulp fibers. Generally, it has been found that the concept works, that is, the swelling shows changes with pH, fiber charge, and salt concentration [90]. The water affinity of the cellulose and fibers depends on the ability of the structural units of glucose to form hydrogen bonds, resulting in the hydrophilicity of these materials. Water can be directly bonded by hydrophilic groups and also indirectly through hydrogen bonds established between water molecules. The degree of fiber swelling is known to be related to the chemical environment and fiber properties [91–93]. Swelling is influenced primarily by the nature of the fibers; in this case, by the content of hemicelluloses and their distribution in the cell wall. Swelling capacity also depends on the fiber content of the lignin. Pulps with high lignin content swell less because some of the hydroxyl groups of the polysaccharides are engaged in links with the lignin polymer that exhibit high hydrophobicity. Kraft pulps swell more than the sulfite pulps and unbleached pulps swell more than the bleached pulps (lower hemicellulose content).

Simple immersion in water without any mechanical treatment results in the degree of swelling (% by volume) of 32% for cotton pulp and 30–40% for wood pulps. Depending on the nature of the fibers, the following hierarchy for swelling can be established: cotton linters pulp < bleached sulfite pulp < unbleached sulfite < bleached kraft pulp < unbleached kraft pulp. Swelling depends on drying, pulp consistency, the severity of drying, and the fiber hornification. Increasing the degree of hornification results in the swelling occurring to a smaller extent; therefore, papers that are obtained from previously dried celluloses have lower strength properties than those obtained from never-dried pulps. Second, the swelling capacity depends not only on the properties of cellulose but also on the nature of the liquid phase. The swelling of the fibers is strongly influenced by pH [94]; as the pH increases to values above 10, the degree of swelling increases significantly, and a maximum has been recorded between 10.4 and 11.3.

Swelling of the fibers depends on the features of the water or the presence of different cations that favor swelling. The degree of swelling in distilled water is lower than when using industrial process water and the swelling capacity depends on the nature of the cations in the following order: $Al^{3+} < Fe^{3+} < Ca^{2+} < Mg^{2+} < NH^{4+} < Na^+$ [82].

The strong connection between swelling and the tensile strength of papers from kraft and sulfite fibers has been shown by Scallan and Grignon [91]. The result of refining is also influenced by fiber swelling as indicated in the two recent studies by Hammar and coworkers [95] and Laivins and coworkers [96]. In their studies, the fiber swelling was altered by changing the counterions, sodium giving a higher swelling than calcium or hydrogen. Both studies reported improved refining efficiency when using sodium.

Well-slushed fibers are flexible, entirely wetted, and free of fiber flocs. A uniform and excellent refining result is ensured by these properties [97]. Refining the poorly slushed "dry fibers" will notably increase fiber shortening and impair the bonding ability of the pulp fines.

9.12 Pulp fibers in the refining process

Paper manufacture is based on the ability of fibers to form a network with a desired strength. The strength of the fiber network is first derived from the surface tension forces and after the solids content has increased (over 65%), it is derived from inter-fiber hydrogen bonds. Unbeaten fibers that have been chemically separated from wood exhibit poor bonding ability, since they are stiff and have a smooth surface. In this form, they are unsuitable ingredients for almost any kind of paper-grade production. Papers obtained from unbeaten fibers are characterized by cloudy transparency, high bulk, and unsatisfactory strength indices.

During refining, the bonding ability of the cellulose fibers is developed by the mechanical processing (beating) of fibers between the refiner bars. After the collapse of their structure and removal of the poorly soluble lignin-containing primary wall, the cellulosic fibers become more flexible. The breaking of the internal fiber bonds and continued swelling process of the fibers lead to an increase of their fiber bonding area. Refining is considered one of the most important operations of the papermaking process. During refining, the operator can change the fiber properties, and thus, almost all of the properties of the finished paper. The effects of refining on the paper properties are quite different as some of the paper properties are improved (e.g., tensile strength), while others are sacrificed (e.g., opacity) [98, 99]. This is the reason why the refining process is conducted as a compromise between gains and losses, regarding the end product, its conversion, and end user needs. The main aspects pursued by refining are closely related to the type of pulp and its role for that paper grade [100]. Thus, the role of the softwood pulp is to improve the strength properties of the paper. The softwood pulp fibers are long, strong, and flexible, the ideal properties for reinforcement fibers. The purpose of refining is to improve the reinforcement abilities of softwood pulp without shortening or damaging the fibers to any great extent. In the case of refining softwood pulp, the optimum treatment conditions occur when a specific edge load (SEL) of 2.0 W·s/m is used for an energy level of 80–120 kWh/ton [101]. The main reason for the use of hardwood pulp as a raw material is to improve the paper printability properties. Moreover, short fibers improve the paper formation and optical properties. In this case, the function of refining hardwood fibers is to develop bonding abilities without impairing the optical and printability properties of the fibers. In the case of refining hardwood pulp, the optimum treatment in terms of strength occurs at an SEL of 0.2 W·s/m over an energy range of 100–150 kWh/ton [101].

Recycled fiber pulp is used to an increasing share in the papermaking process. The properties of the recycled fiber pulp differ from those of the virgin pulp due to the irreversible changes that occur during drying and calendaring [102, 103]. The target of refining the recycled fiber pulp is to restore the fiber properties weakened during the recycling and the recycled fiber pulp production process to a level required by the end product. Generally, recycled fiber pulp is refined with a lower intensity than the virgin pulp [104].

9.12.1 Refining effects on fibers

The beating or refining process modifies fiber structures in many different ways. For a long time, the effects of refining on fibers were classified into two groups: physicochemical changes (swelling, removal of the primary wall, hydration, internal fibrillation, and external fibrillation) and mechanical changes (shortening and thinning of fibers) [105]. Lately, the literature includes a modern approach to the refining effects on fibers, which are simpler and more scientifically correct. Thus, the basic refining effects on individual fibers can be divided into primary and secondary effects [100]. It should be noted that dividing the refining effects into primary and secondary is based more on conditioning and succession criteria rather than the importance of the criterion. The primary effects of refining on fibers are as follows: (1) breaking and peeling off of external fiber layers, the creation of fines; (2) external fibrillation; (3) breakup of hydrogen bonds between inner fiber layers, with fiber wall delamination and local collapse; (4) partial dissolution of fiber wall components and the creation of hemicelluloses gel on the fiber surface; and (5) creation of invisible damage and weak points with fiber shortening. As a consequence of the primary effects, the following secondary effects are recorded during refining: (1) water penetration into the fiber wall and fiber swelling by the removal of external plies; (2) creation of new surfaces and fines; (3) elongation and/or compaction of fibers; (4) increase of fiber flexibility; and (5) straightening or curling of fibers (based on whether the refining consistency is low or high) [106].

9.12.1.1 Removal of the primary fiber wall

The occurrence of swelling is differentiated across the thickness of the cell wall [107]. The primary cell wall is permeable to water, but swells slightly and acts as a rigid mesh preventing swelling of the secondary wall. The primary wall is more or less resistant to the action of pulping and bleaching reagents; hence, paper-grade pulp fibers still contain the primary cell wall. On the other hand, treating the dissolving-grade pulps with alkali generally leads to the loss of the primary cell wall destroyed by the action of concentrated alkali. The primary cell wall is very sensitive to mechanical action, and under the action of the refiner bars it is immediately removed at the initial refining stage. This creates the possibility of the secondary wall swelling, but it swells in a different manner [108, 109]. The secondary wall, called the S2 layer, swells more than the S1 layer, and therefore, due to the swelling pressure that can reach 10^{11} N/m^2, the S1 layer breaks at the weaker areas and allows the S2 layer to swell at a greater extent (ballooning effect). As a consequence, the S1 layer detaches.

9.12.1.2 External fibrillation

After peeling off and delamination of the primary fiber wall and S1 layer, the external surface of the S2 layer is exposed to friction with other fibers, thus resulting in the creation of a fibrillated surface structure. Due to the microfibrillar structure of the S2 layer and the predominant orientation of the microfibrils in the direction of the fiber axis, this leads to strong external fibrillation, which means that a lot of fibrils are partially detached from the fiber surface and results in the buildup of a "hairy" fiber structure. Upon external fibrillation, the external fiber surface increases significantly and this facilitates interfiber bonding [110–112].

9.12.1.3 Internal fibrillation

The action of water is not limited to the fiber surface; it penetrates deep into the amorphous areas of cellulose, acting as a lubricant between the internal fiber layers. It can move under the action of the refiner bars and get into the most convenient position by delamination of the fiber cell wall. Separation of the cell wall structure into fibril layers has the immediate effect of increasing the fiber flexibility and plasticity and is called internal fibrillation. Due to internal fibrillation, the external fiber area becomes larger, and favorable conditions are created for interfiber bonding [113].

9.12.1.4 Fiber shortening

Under the action of the cutting edges of the refiner bars, fibers are sheared transversally or longitudinally. As a result of the shear action, thinning and shortening of the fiber occurs. The average length of the fibers decreases continuously throughout the refining process. Fiber shortening usually adversely affects the paper characteristics (especially strength properties since the fiber length is an important influential factor), which should be limited by the judicious choice of refining parameters. On the other hand, shortening of the fibers may also have a positive effect on sheet formation [114]. The shortening of the fibers strongly depends on the refining conditions. In principle, fiber breakage or shortening always occurs when momentary tensile or shear stresses, which develop during refining, exceed the strength of the fibers. As a general rule, a high refining intensity will enhance fiber shortening. There is no weaker or preferential area for fracturing along the fiber length, which means that the shortening of fibers is random and can take place at any point.

9.12.1.5 Creation of new surfaces and fines

During refining, new surfaces and fines are created in the pulp suspension due to the peeling off of the outer fiber surfaces, partial delamination of the fiber wall, cutting off of fiber wall parts (lamellar segments and microfibrils), and sections detached during the process of fiber shortening [115]. The creation of fine material does not necessarily exert a negative effect. On the contrary, the creation of fines will improve fiber bonding and paper strength properties. Fine material is composed of particles that have a thickness of 0.1–0.5 μm, a length of several tens of micrometers, and can pass through a 200 mesh wire. The fine material generated during refining is called secondary fine material. In addition, two other types of fine material are known: primary and tertiary. Primary fine material is found naturally in the pulp and is represented by parenchymal cells, radial cells, and vessel elements. In contrast, tertiary fine material is formed as a result of the action of the pumps and agitator rotor blades of the tanks, and it accumulates into a white water system as a result of "fractionation" of the paper stock on the machine wire.

9.12.1.6 Partial dissolution of the fiber wall

With the development of refining and exposing the internal structure of the cell wall, new amounts of hemicelluloses and lignin are subjected to the dissolving action of water and refining forces. The amount of dissolution ranges between 0.3% and 1% on weight for paper-grade pulp, but can increase by several points for high-yield pulps. The cyclic action of the compressive forces during refining also contributes to the redistribution of the cell wall components. Since the hemicelluloses "migrate" to the fiber surface and act as a kind of interfiber binder, this type of redistribution has a remarkable effect on the strength of the paper produced from the refined pulp. Numerous studies have demonstrated a strong correlation of the paper strength properties not only with the amount of hemicelluloses present in the pulp, but also with its composition [116, 117].

9.12.1.7 Increase in fiber conformability

Increasing fiber conformability is gained both from the already mentioned refining effects and from the different structural changes and/or damage produced due to mechanical fiber stress.

9.12.2 Effects on pulp properties

The effects of refining on individual fibers are reflected by changes of the pulp properties and subsequently on the paper properties. The effect of refining on pulp properties is immediately noticed by the papermaker as changes in the drainability and dewatering properties. As refining progresses, the pulp drainability is reduced, which is translated into an increase of the beating degree. A large number of tests are used to control the refining process; some determine the overall transformation of fibers, while others are able to assess the separate effects. The Schöpper–Riegler and Canadian Standard Freeness are the main tests that allow the assessment of the overall refining effects based on the dewatering properties of the pulps. Freeness of pulp is the ease with which water drains from the pulp through a wire mesh.

The main influences on pulp freeness are physically, in that the short fibers, vessel elements, and fibrils, produced during refining, block drainage more than the long fibers and chemically, in that their affinity for water decreases in the order: hemicelluloses > cellulose ≫ lignin. Therefore, the unrefined pulps are freer than the refined pulps, the long fiber component of a stock is freer than the fine fibers (fines), and the chemical pulps are freer than the mechanical pulps. Softwood pulps are freer than the hardwood pulps, which in turn are freer than the high hemicellulose nonwood pulps. Since the freeness measurement is an overall assessment of the effects of refining, it is possible to have two pulps of identical freeness, but with other properties that are very different.

For assessing certain types of fiber transformations, specific tests are used: pulp fractionation to assess mechanical changes, measurement of the external surface area to assess external fibrillation, and the water retention value (WRV) as a measure of physical and chemical transformations, and in particular of internal fibrillation. The effect of refining can additionally be detected as a decreasing pulp viscosity value.

During the processes of short circulation (blending, dosage, dilution, cleaning, screening, and deaeration), the cellulose fibers do not undergo significant changes. However, the effects of different additives or chemicals on the charging and fiber surface chemistry should not be overlooked. Furthermore, it has been demonstrated that highly charged wet-end additives will continue to contribute to the charge of papermaking fibers, even after they have been recycled [118, 119]. Likewise, the effects of hydrophobic sizing treatments may be passed down to subsequent generations of the recycled paper [119].

9.13 Cellulose fibers in the dewatering and drying processes

Papermaking is essentially a dewatering process of a suspension of cellulosic fibers. During the papermaking process, water is removed by drainage and vacuum in the forming section, by mechanical forces in the press section, and by evaporation and vapor transport in the drying section [120]. Parker [121] believes that paper forming is the result of three basic hydrodynamic processes: drainage, oriented shear, and turbulent flow. The main effects of drainage, which can be described as a flow through the wire, are increasing the consistency of the fibrous suspension and felting of the fibers to form a paper web. When the fibers are free to move independently of one another, draining begins with a filtering mechanism and the fibers are deposited on the wire in discrete layers. Filtering is the dominant mechanism of dewatering during paper formation using the Fourdrinier-type systems, and the paper web is characterized by a layered structure. When there is no longer the possibility that the fibers can move independently, they flocculate and form coherent networks, and dewatering occurs by thickening, resulting in a cloudy sheet structure. During drainage, fibrous suspensions spontaneously form networks, except when dilution is very high (formation in a laboratory) or when additional mixing energy is provided (which generates turbulence) [122, 123]. Dilution is a great solution for achieving dispersion, but in economic terms cannot be used to control the flocculation into the formation of a paper web on the machine wire. Additional dispersion of the paper stock should be achieved by generating turbulence and/or by initiating oriented shear (shear forces) during drainage. Both shear and turbulence are the result of variations in the flow rate of a fluid. The main difference between these phenomena is that for shear, the velocity vectors are oriented in the same direction, whereas for turbulence they have different directions. Oriented shear can be defined as a flow in layers that has a distinctive profile in the fibrous suspension. This type of flow can occur due to the difference in speed between the wire and the jet, and speed differences between layers of paper stock, determined by contact with a dandy roll or a top wire ("top-former"). Turbulence, defined as an irregular motion, destroys flocks and prevents flocculation, thereby improving formation. Turbulence has selective effects since it only disperses flocks with a fiber length of the same order of magnitude as the scale of the turbulence, while oriented shear affects all flocks. Turbulence can be positive at an early stage of forming (the jet exit from the headbox nozzle), but has an adverse effect on paper formation if it persists in the next stages (forming and dewatering). The hydrodynamic processes of formation, drainage, turbulence, and oriented shear occur simultaneously and are not totally independent, being present in both Fourdrinier-forming systems and twin-wire systems.

Although of great importance, the processes of formation and dewatering during the wire section have a negligible effect on the structure and properties of the pulp fibers. The effects of these processes are confined only to the fiber position within the network under consolidation. The change in the composition of the fibrous suspension, by the lower retention of fines and fillers, can be considered a secondary effect.

Wet pressing is the papermaking operation of expelling water from the paper web by mechanical compression in the nip, formed by two rolls or recently by a roll and a special part named the "shoe." The wet pressing process affects the structure and fiber properties, and thus, both the paper properties and papermaking potential of these fibers [124, 125]. In addition to water removal, the fibers are flattened and new bonding conditions develop, which contribute to a permanent densification of the paper sheet. Different studies have demonstrated decreases in the water-holding capacities of fibers that have been pressed under normal or severe conditions [126–128].

Decreasing the WRV seems to be an effect of the irreversible formation of new semicrystalline regions, as a result of establishing new hydrogen bonds between cellulosic surfaces within parts of the cell wall that come in contact due to the applied force [129]. The use of laser confocal microscopy and swelling measurements led to the conclusion that irreversible hornification (loss of swelling capacity when the fibers are rewetted) began above solid contents of 30–35% [130].

Drying is the papermaking operation in which the dry content of the paper is brought to an optimum, which largely determines the final characteristics of the paper. In the paper wet web, water is in the form of free water and bound water. Wet pressing removes more than 90% of the free water. The remaining free water and most of the bound water can only be removed by evaporation or by heating. There are three methods normally applied to paper and board drying, which differ in principle from each other by the method of energy supply: contact or cylinder drying, air drying, and radiation drying. The drying temperature and duration of the process vary widely, depending on the paper basis weight, the speed of the paper machine, and other design and operation parameters. In conventional paper machines, drying takes place under atmospheric pressure. For this reason, the temperature of the paper will tend to remain at the boiling point or lower, as long as the liquid water remains within the pore spaces adjacent to the sheet's surface [131]. When drying freely, the web will shrink. The degree of shrinkage at a paper machine depends, among other things, on the fiber orientation, the length of draws, and the tension of the dryer fabric. The shrinking of the paper web during the drying process can result in fibers moving closer to each other and shrinkage of the fibers themselves. Thus, the shrinking process can be divided into interfiber and intrafiber shrinking. The shrinkage of fiber is 1–2% in its longitudinal direction and 20–30% in its cross-direction. When the dry content is below the saturation point of the fibers, interfiber shrinkage will occur. Neither shrinkage of the fibers nor the formation of interfiber bonds are possible in the presence of free water, whereas the surface tension forces tend to draw each of the fibers closer as well as the fibrils detaching partly or entirely from the fibers.

9.14 Paper-based microfluidics

Microfluidics is the science and technology of systems that process or manipulate small amounts of fluids, usually in the range of microliters (10^{-6}) to picoliters (10^{-12}), using networks of channels with dimensions of tens to hundreds of micrometers.

The development of the microfluidics field was influenced by the achievements in the domain of molecular analysis in the capillary format, by the biodefense research efforts (after the Cold War), by the explosion of genomics in the 1980s, followed by the development of microanalysis in molecular biology and not ultimately by photolithography and associated technologies from microelectronics. Microfluidics has the potential to influence subject areas from chemical synthesis and biological analysis to optics and information technology [132].

The behavior of fluids at the microscale differs essentially from that of macroscale, in that factors such as surface tension, energy dissipation, and fluidic resistance start to dominate the system. The flow of fluids in narrow channels with size of around 100 nm to 500 μm is laminar rather than turbulent (very low values of the Reynolds number). A key consequence is co-flowing fluids do not necessarily mix in the traditional sense, as flow becomes laminar; molecular transport between them must often be through diffusion [133].

The first applications of microfluidic technologies have been in analysis, for which they offer a number of advantages: the ability to decrease samples and reagent consumptions and to carry out separations and detections with high resolution and sensitivity; low cost; run multiple analyses simultaneously; short times for analysis; and small footprints for the analytical devices [134].

Microfluidics have seen a rapid development as regards materials, methods of fabrication, and the components – microchannels that serve as pipes and other structures that form valves, mixers, microreactors, and pumps – which are essential elements of microchemical "factories" or "lab on a chip."

Even if initially the materials of the microfluidic devices consist only of silicon and glass substrates, as the field advanced new materials such as low-temperature cofired ceramics, elastomers, and thermoplastic polymers and not ultimately paper were evaluated [135]. Paper microfluidics is an emerging technology and substantially different from that of devices made of polymeric or inorganic materials.

Paper as a matrix for microfluidic devices offers many advantages compared to other materials, particularly due to its capability to transport liquid by capillary forces. Unlike conventional microfluidic platforms, no external pumps are required for fluid transport inside the paper because of capillary forces.

In addition, the paper consists of cellulose, a renewable and very cheap material, with low fabrication cost, lightweight, easy to stack, store, and transport, disposable, and biodegradable. The surface of the cellulose fibers can be modified with different chemical functions for the definition of hydrophobic barriers (i.e., small channels inside the paper sheets), or conjugated with sensing elements (i.e., immobilization) [136].

Paper-based microfluidic devices provide interesting applications, including clinical "point-of-care" diagnostics, food quality control, and environmental monitoring.

Microfluidic paper-based analytical devices are fabricated by micropatterning hydrophobic regions on paper. These regions define paths for liquids that spontaneously wick through the paper. There are several techniques and processes involving chemical modification and/or physical deposition that could be used: photolithography, plotting with an analogue plotter, ink jet etching, plasma treatment, paper cutting, wax printing, ink jet printing, flexography printing, screen printing, and laser treatment [137].

The photolithography method uses an epoxy-based negative photoresist (SU-8). Paper is impregnated with photoresist, dried, and exposed to UV light through a transparency mask, which can be printed. The unexposed photoresist can be washed out of the paper to form the hydrophobic microfluidic channels.

Wax printing is a simple and inexpensive method for patterning microfluidic structures in paper using a commercially available printer and hot plate. Patterns of solid wax are printed on the surface of paper, and a hot plate melts the wax so that it permeates through the paper.

Current research on paper-based microfluidic devices focuses only on commercially available filter or chromatography papers. Although filter paper is widely used, it does not always possess the desired physical characteristics, so other types of paper or paper modifications must be explored [137].

9.15 Paper-based electronics–"papertronics"

The development during the last decades of electronic and optoelectronic applications has renewed the interest in electronics manufactured on low-cost flexible plastic and paper substrates [138]. The main advantages with paper over flexible plastic substrates lie in the low cost, recyclability, and low environmental footprint. By printing electronic circuits directly on paper substrates, the use of paper could be extended to applications such as electronic displays and sensors. However, truly low-cost flexible electronics not only require a cheap substrate, such as paper, but also require a fast and low-cost manufacturing (printing) process and the use of inexpensive materials [139, 140].

The major obstacles when fabricating electronics on paper substrates are the large surface roughness, porosity, chemical impurities, and hygroscopicity of paper. But, the change in the dielectric constant and resistivity of paper substrates on exposure to humidity can be utilized for making humidity sensors. A real advantage with paper substrates over plastic substrates is, nevertheless, that paper is less dilative than low-cost plastics upon heating, which extends the annealing and sintering possibilities of, for example, nanoparticle inks. However, the requirements of the substrate depend very much on the precise application. Other types of low-cost electronics with relaxed requirements on dimensions, smoothness, and registration still have a great potential for being fabricated by printing directly onto paper substrates. Furthermore, the porosity (and roughness) of paper is actually advantageous in obtaining good adhesion of the printed materials and improving the sensitivity of sensors and is essential for paper-based energy storage devices and microfluidic applications.

Among the different transistors that have been fabricated directly onto paper substrates are IM-OFETs (Ion Modulated-Organic Field-Effect Transistors), ECTs (Electrochemical Transistors), and inorganic FETs (Field-Effect Transistors) with the paper serving as the dielectric. The ion-modulated transistors are especially suitable for paper substrates due to the insensitivity to the surface roughness and the thickness of the dielectric. The robustness of the ion-modulated transistors allows large-scale manufacturing directly onto recyclable paper substrates with promise for systems operating at low voltages but at lower speed [139].

A fabulous progress has been recorded in the recent years toward better substrates suitable for printed functionality, which will allow intelligent packages and applications. However, further progress is required in the electronic materials (stability, performance, and processability), printing techniques, and approaches for making electronic device applications feasible.

9.16 Conclusions

Over time, the raw materials used in the papermaking process progressed from simple rags and pasta to bleached chemical pulps, resulting from pulping processes conducted to meet the most stringent quality requirements of the different types of paper and paperboard.

Undoubtedly, pulp fibers are the most suitable raw material for papermaking. To support this claim, economic and technical arguments were presented in this chapter. It is true that not all pulp grades are equally suitable for making any sort of paper or paperboard; hence, a review of papermaking properties was shown to be necessary. Furthermore, proceeding sequentially through the operations of the papermaking process provides an overview, but also a detailed image of the modifications of the cellulose fibers. This review also revealed many phenomena and transformations still not well understood that represent a rich topic of research in the field.

In the near future, there are unlikely to be major changes in the raw materials used for papermaking, but use of secondary fibers will further increase achieving a sustainable recycling limit. In addition, new efforts investigating the recovery papermaking potential of secondary fibers will be engaged.

Even if papermaking is one of the most common uses of cellulosic fibers, to-day's applications such as paper-based microfluidic devices and paper-based electronics are a particularly attractive topic for research on high-tech applications of fibrous structures.

References

[1] Tsien T-H. Paper and Printing. In: Needham J., ed. Science and civilization in China. 5, Third printing ed. Cambridge University Press, Cambridge, UK, 1985.
[2] Hubbe MA, Bowden C. Handmade paper: A review of its history, craft, and science. BioResources. 2009, 4, 1736–1792.
[3] Barrett T. European papermaking techniques 1300–1800. The University of Iowa, Ames, Iowa, USA, 2012, Accessed December 1, 2019 at http://paper.lib.uiowa.edu/european.php
[4] CEPI, KEY STATISTICS 2018 European pulp & paper industry (Accessed December 1, 2019 at http://www.cepi.org/system/files/public/documents/publications/Final%20Key% 20Statistics%202018.pdf)
[5] *** Cellulose the 'DNA' of paper, Vol.18 / Issue-2 Feb.2014, (Accessed December 1, 2019, at , https://paperonweb.com/Articles/CelluloseDNAofPaper.pdf)
[6] Retulainen E, Niskanen K, Nilsen N. Fibres and Bonds. In: Gullichsen J, Paulapuro H., eds. Paper physics: Papermaking science and technology. 16, Fapet Oy, Helsinki, Finland, 1998, 55–87.
[7] Hiltunen E. Papermaking Properties of Pulp. In: Levlin J-E, Söderhjelm L., eds. Pulp and paper testing: Papermaking science and technology. Fapet Oy, Helsinki, Finland, 1999, 39–64.
[8] Gruber E. Fibre Properties. In: Sixta H., ed. Handbook of pulp. Wiley-VCH, Weinheim, Germany, 2006, 1269–1280.
[9] Brännvall E, Annergren G. Pulp Characterisation. In: Ek M, Gellerstedt G, Henriksson G., eds. Pulping chemistry and technology: Pulp and paper chemistry and technology. 2, Walter de Gruyter, Berlin, Germany, 2009, 429–459.
[10] Rydholm SA. Pulping processes. Malabar, Robert E. Krieger Publishing, Florida, FL, USA, 1965, 1140–1146.
[11] Klemm D, Schmauder H-P, Heinze T. Cellulose. In: Vandamme E, De Baets S, Steinbüchel A., eds. Biopolymers: Biology, chemistry, biotechnology, applications. 6, Part. Polysaccharides, Wiley-VCH, Weinheim, Germany, 2005, 277–312.
[12] Kerr AJ, Goring DAI. Ultrastructural arrangement of the wood cell wall. Cell. Chem. Technol. 1975, 9, 563–573.
[13] Fengel D, Wegener G. Wood chemistry, ultrastructure, reactions. Walter de Gruyter, Berlin, Germany, 1984.
[14] Salmén L, Olsson A-M. Interaction between hemicelluloses. lignin and cellulose: Structure-property relationships. J. Pulp Pap. Sci. 1998, 24, 98–103.
[15] Teleman A. Hemicelluloses and Pectins. In: Ek M, Gellerstedt G, Henriksson G., eds. Wood chemistry and wood biotechnology: Pulp and paper chemistry and technology. 1, Walter de Gruyter, Berlin, Germany, 2009, 101–120.
[16] Meier H. Localization of Polysaccharides in Wood Cell Walls. In: Higuchi T., ed. Biosynthesis and biodegradation of wood components. Academic Press, New York, NY, USA, 1985, 43–50.
[17] Popa VI, Spiridon I. Hemicelluloses: Structure and Properties. Dumitriu S., ed. Polysaccharides: Structural diversity and functional versatility. Marcel Dekker Inc, New York, NY, USA, 1998, 297–311.

[18]   Süss HU, Kronis JD, Taylor R. Selection of the best available technology for TCF bleaching of sulphite pulp, In Proceedings of Tappi Pulping Conference, Montreal, Canada, 1998, 851.

[19]   Clark J d'A. In pulp technology and treatment for paper. 2nd. Miller Freeman, San Francisco, CA, USA, 1985, 878.

[20]   Clark J d'A. Factors influencing apparent. Density and its effect on paper properties. Tappi-Tech. Association Pap. 1943, 26, 499.

[21]   Clark J d'A. Effects of fiber coarseness and length I. Bulk, burst, tear, fold, tensile test. Tappi. 1962, 45, 628–634.

[22]   Clark J d'A. Effects of fiber coarseness and length II. Improved means of measuring intrinsic strength and cohesiveness (Zero-Span). Tappi. 1965, 48, 180–184.

[23]   Clark J d'A. Freeness fallacies and facts. Sven. papp. 1970, 3, 54–62.

[24]   Clark J d'A. Components of the strength qualities of pulps. Tappi. 1973, 56, 122–125.

[25]   *** Fines Fraction of Paper Stock by Wet Screening, Tappi pm T 261, 1990.

[26]   *** Fibre Length of Pulp by Projection, Tappi cm T 232, 1985.

[27]   *** ISO 16065-1: Pulps; Determination of fibre length by automated optical analysis. Part 1: Polarized Light Method, 2001.

[28]   *** Fibre length of pulp and paper by automated optical analyzer using polarized light, TAPPI om T 271, 1998.

[29]   Robertson G, Olsen J, Allen P, Chan B, Seth R. Measurement of fiber length, coarseness, and shape with the fiber quality analyser. Tappi J. 1999, 82, 93–98.

[30]   Trepanier RJ. Automatic fiber length and shape measurement by image analysis. Tappi J. 1998, 81, 152–154.

[31]   Olson JA, Robertson AG, Finnigan TD, Turner RRH. An analyzer for fiber shape and length. J. Pulp Pap. Sci. 1995, 21, J367–J373.

[32]   Sutton P, Joss C, Crossely B Factors affecting fiber characteristics in pulp. In: TAPPI 2000 Pulping/Process & Product Quality Conference Proceedings, TAPPI Press, Boston, MA and Atlanta, GA, USA, 2000, Paper no.133, p.5.

[33]   Seth RS Fibre quality factors in papermaking 1: The importance of fibre length and strength. In: Proceedings Materials Research Society Symposium, San Francisco, CA, USA, 1990, 197, 125–141.

[34]   Retulainen E. Fibre properties as control variables in papermaking? Part 1: Fibre properties of key importance in the network. Pap. Timber. 1996, 78, 187–194.

[35]   Seth RS Fibre Quality Factors in Papermaking Proceedings: The Importance of Fibre Coarseness in the Proceedings Materials Research Society Symposium, San Francisco, CA, USA, 1990, 197, 143–161.

[36]   Seth RS, Page DH Fiber properties and tearing resistance. In: Proceedings of International Paper Physics Conference, Auberge Mont Gabriel, Mont-Rolland, Quebec, Canada, 1987, 9–16.

[37]   Kerekes RJ, Schell CJ. Effect of fibre length and coarseness on pulp flocculation. Tappi J. 1995, 78, 133–139.

[38]   Kerekes RJ, Soszynski RM, Tam Doo PA The flocculation of pulp fibres, In: Punton V ed. Papermaking Raw Materials – Transactions of the 8th Fundamental Research Symposium, FRC of Pulp and Paper Fundamental Research Society, London, UK, 1985, 265–310.

[39]   Hakanen A, Hartler N. Fibre deformation and strength potential of kraft pulp. Pap. Timber. 1995, 77, 339–344.

[40]   Mohlin U-B, Alfredsson C. Fibre deformation and its implications in pulp characterization. Nord. Pulp Pap. Res. J. 1990, 5, 172–179.

[41]   Page DH, Seth RS, Barbe M, Jordan B Curls, crimps, kinks and microcompressions in pulp fibres – their origin, measurement and significanc. In: Punton V ed. Papermaking Raw

Materials – Transactions of the 8th Fundamental Research Symposium, FRC of Pulp and
Paper Fundamental Research Society, London, UK, 1985, 183–227.

[42] Mohlin U-B, Dahlborn J. Fiber deformation and sheet strength. Tappi J. 1996, 79, 105–111.

[43] *** ISO 9184-6: Paper, board and pulps – fibre furnish analysis. Determination of Fibre Coarseness, 1990.

[44] Tutus A, Comlekcioglu N, Karaman S, Alma MH. Chemical composition and fiber properties of Crambe orientalis and Crambe tataria. Int. J. Agric. Biol. 2010, 12, 286–290.

[45] Tavasoli A, Dehghani M, Mahdavi S. Properties of kenaf (*Hibiscus cannabinus L.*) bast fiber reinforced bagasse soda pulp in comparison to long fiber. World Appl. Sci. J. 2011, 14, 906–909.

[46] Watson AA, Dadswell HE Influence of fibre morphology on paper properties. 1. Fibre length. Appita 1961, 14, 168–176.

[47] Dadswell HE, Watson AJ Influence of the morphology of wood pulp fibres on paper properties. In: Bolam F ed. Formation and Structure of Paper – Transactions of the 2nd Fundamental Research Symposium, FRC of Pulp and Paper Fundamental Research Society, London, UK, 1962, 537–572.

[48] Watson AA, Dadswell HE. Influence of fibre morphology on paper properties. 3. Length: diameter (L/O) ratio. 4. Micellar. angle. Appita. 1964, 17, 146–156.

[49] Page DH. The origin of the differences between sulphite and kraft pulps. J. Pulp Pap. Sci. 1983, 9, TR15–TR20.

[50] Hartler N, Nyrén J. Transverse compressibility of pulp fibers II: Influence of cooking, yield, beating, and drying. Tappi J. 1970, 53, 820–823.

[51] Luner P Wet fiber flexibility as an index of pulp and paper properties, new technologies in refining. In: Proceedings PIRA International Conference 'Advances in Refining Technologies', Birmingham, UK, 1986, Paper No.3.

[52] Hattula T, Niemi H. Sulfate pulp fiber flexibility and its effect on sheet strength. Pap. ja Puu. 1988, 70, 356–361.

[53] Uhmeier A, Salmén L. Repeated large radial compression of heated spruce. Nord. Pulp Pap. Res. J. 1996, 11, 171–176.

[54] Mohlin UB. Cellulose fiber bonding. (5) Conformability of pulp fibers. Sven. Papp. 1975, 78, 412–416.

[56] Smith WE, Byrd VL Fiber bonding and tensile stress-strain properties of earlywood and latewood handsheets. U.S. forest service, Research Paper FPL 193, 1972.

[57] Huang F, Lanouette R, Law K-N. Morphological changes of jack pine latewood and earlywood fibers in thermomechanical pulping. BioResources. 2012, 7, 1697–1712.

[58] Steadman R, Luner P The effect of wet fiber flexibility on sheet apparent density In: Punton V ed. Papermaking Raw Materials – Transactions of the 8th Fundamental Research Symposium, FRC of Pulp and Paper Fundamental Research Society, London, UK, 1985, 311–337.

[59] Tam Doo PA, Kerekes RJ. A method to measure wet fiber flexibility. Tappi J. 1981, 64, 113–116.

[60] Kerekes RJ, Tam Doo PA. Wet fibre flexibility of some major softwood species pulped by various processes. J. Pulp Pap. Sci. 1985, 11, J60–J61.

[61] Hattula T, Niemi H. Sulphate pulp fibre flexibility and its effect on sheet strength. Pap. ja Puu. 1988, 70, 356–361.

[62] Mohlin UB, Dahlbom J, Hornatowska J. Fibre deformation and sheet strength. Tappi J. 1996, 79, 105–111.

[63] Paavilainen L. Conformability–flexibility and collapsibility–of sulphate pulp fibres. Pap. ja Puu. 1993, 75, 689–702.

[64] Hardacker KW. Effects of Loading Rate, Span, and Beating on Individual Wood Fiber Tensile Properties. In: Page DH., ed. The physics and chemistry of wood pulp fibres. 8, TAPPI Special Technical Association Publication, Appleton, WI, USA, 1970, 201–216.

[65] Mott L, Shaler SM, Groom LH, Liang BH. The tensile testing of individual wood fibres using ESEM and video image analysis. Tappi J. 1995, 78, 143–148.

[66] Page DH, El-Hosseiny F, Winkler K. Behaviour of single wood fibres under axial tensile strain. Nature. 1971, 229, 252–253.

[67] Page DH, El- Hosseiny F. The mechanical properties of single wood pulp fibres. Part 6. Fibril angle and the shape of the stress-strain curve. J. Pulp Pap. Sci. 1983, 84, TR99–TR100.

[68] Bergander A, Salmén L. Cell wall properties and their effects on mechanical properties of fibers. J. Mater. Sci. 2002, 37(1), 151–156.

[69] De Magistris F, Salmén L. Finite element modelling of wood cells deformation transverse to the fibre axis. Nord. Pulp Pap. Res. J. 2008, 23, 240–246.

[70] Bergander A, Salmén L. Variations in transverse fiber wall properties: Relations between elastic properties and structure. Holzforschung. 2000, 54, 654–660.

[71] Jayne BA. Mechanical properties of wood fibers. Tappi. 1962, 42, 461–467.

[72] Jentzen CA. The effects of stress applied during drying on some of the properties of individual fibres. Tappi. 1964, 47, 412–418.

[73] Forgacs OL. Structural weaknesses in softwood pulp tracheids. Tappi. 1961, 44, 112–119.

[74] Mohlin U-B, Alfredsson C. Fibre deformation and its implications in pulp characterization. Nord. Pulp Pap. Res. J. 1990, 5(4), 172–179.

[75] Duncker B, Nordman L. Determination of strength of single fibres. Pap. ja Puu. 1965, 47, 539–552.

[76] Leopold B. Effect of pulp processing on individual fiber strength. Tappi. 1966, 49, 315–318.

[77] Leopold B, Thorpe J. Effect of pulping on strength properties of dry and wet pulp fibers from Norway spruce. Tappi. 1968, 51, 304–308.

[78] Sjöström L, Jacobs A, Rådeström R, Nordlund M. Effects of released organic substances on sizing efficiency–Influence of origin, composition and molecular properties of the organic material. Nord. Pulp Pap. Res. J. 2006, 21, 575–585.

[79] Lumiainen J. Refining of ECF, TCF and chlorine-bleached pulps in equal conditions. Pap. Technol. 1996, 37(6), 22–29.

[80] Savolainen A, Jussila T, Nikula S. Defibering and specific energy consumption in bale pulpers. Tappi J. 1991, 74, 147–153.

[81] Nuttall GH, Mott L. Optimisation of pulping and refining energy use–development of an in-house technique. Pap. Technol. 1998, 39, 39–49.

[82] Hoyland RW, Field R. A review of the transudation of water into paper–in five parts. Part 2 The cellulose-water relationship. wetting & cutting angles. Pap. Technol. 1976, 17, 216–219.

[83] Grignon J, Scallan AM. Effect of pH and neutral salts upon the swelling of cellulose gels. J. Appl. Polym. Sci. 1980, 25, 2829–2843.

[84] Laine J, Lindström T, Bremberg C, Glad-Nordmark G. Studies on topochemical modification of cellulosic fibers. Part 5. Comparison of the effects of surface and bulk chemical modification and beating of pulp on paper properties. Nord. Pulp Pap. Res. J. 2003, 18, 325–332.

[85] Torgnysdotter A, Wågberg L. Study of the joint strength between regenerated cellulose fibers and its influence on the sheet strength. Nord. Pulp Pap. Res. J. 2003, 18, 455–459.

[86] Torgnysdotter A, Wågberg L. Influence of electrostatic interactions on fibre/fibre joint and paper strength. Nord. Pulp Pap. Res. J. 2004, 19, 440–447.

[87] Lindström T. Chemical factors affecting the behavior of fibres during papermaking. Nord. Pulp Pap. Res. J. 1992, 7, 181–192.

[88] Wågberg L, Annergren G Physico-chemical characterisation of papermaking fibres. In: Baker
       CF ed. Transactions of the 11th Fundamental Research Symposium Volume 1, Pira
       International, Leatherhead, UK, 1997,1–82.
[89] Luukko K, Maloney TC. Swelling of mechanical pulp fines. Cellulose. 1999, 6, 123–135.
[90] Wågberg L. Interactions Between Fibres and Water and the Influence of Water on the Pore
       Structure of Wood Fibres. In: Ek M, Gellerstedt G, Henriksson G., eds. Pulp and paper
       chemistry and technology, paper chemistry and technology. 3, W. de Gruyter, Berlin,
       Germany, 2009, 39–64.
[91] Scallan AM, Grignon J. The effect of cations on pulp and paper properties. Sven. Pap. 1979,
       82, 40–47.
[92] Lindström T, Carlsson G. The effect of chemical environment on fiber swelling. Sven. Pap.
       1982, 85, R14–R20.
[93] Scallan AM, Tigerström AC. Swelling and elasticity of the cell walls of pulp fibres. J. Pulp
       Pap. Sci. 1992, 18, J188–J193.
[94] Lindström T, Kolman M. The effect of pH and electrolyte concentration during beating and
       sheet forming on paper strength. Sven. Pap. 1982, 85, 140–145.
[95] Hammar L-Å, Bäckström M, Htun M. Effect of the counterion on the beatability of unbleached
       kraft pulp.s. Nord. Pulp Pap. Res. J. 2000, 15, 189–193.
[96] Laivins G, Scallan T. Acidic versus alkaline beating of pulp. J. Pulp Pap. Sci. 2000, 26,
       228–233.
[97] Mohlin U-B Industrial refining of unbleached kraft pulps-the effect of pH and refining
       intensity. In: Proceedings of the 2002 TAPPI Technology Summit Conference, TAPPI, Atlanta,
       GA, USA, 2002, paper no. 26, 11.
[98] Ebeling K Critical review of current theories for the refining of chemical pulp. In: Proceedings
       of the International Symposium on Fundamental Concepts of Refining, IPST, Appleton, WI,
       USA, 1980.
[99] Page DH The beating of chemical pulps – The action and the effects. In: Baker CF, Punton VW
       eds. Fundamentals of Papermaking – Transactions of 9th Fundamental Research Symposium,
       FRC of Pulp and Paper Fundamental Research Society, London, UK, 1989, 1–38.
[100] Lumiainen J Refining of chemical pulp. In: Paulapuro H ed. Papermaking Part 1: Stock
       Preparation and Wet End. Papermaking Science and Technology, Fapet Oy Publication,
       Helsinki, Finland, 2000, 87–121.
[101] Baker CF. Good practice for refining the types of fibre found in modern paper furnishes. Tappi
       J. 1994, 78, 147–153.
[102] Howard RC, Bichard WJ. The basic effect of recycling on pulp properties. J. Pulp Pap. Sci.
       1992, 18, J151–159.
[103] Gurnagul N, Ju S, Page DH. Fibre-fibre bond strength of once-dried pulps. J. Pulp Pap. Sci.
       2001, 27, 88–91.
[104] Bergfeld D Low intensity refining of hardwood and deinked pulps with a new generation
       of filling in a double disc refiner, In: Proceedings of 53rd APPITA International Refining
       Conference, APPITA, Rotorua, 1999, 2, 787–793.
[105] Kibblewhite RP. Structural modifications to pulp fibers: definitions and role in papermaking.
       Tappi J. 1977, 60, 141–143.
[106] Seth RS The importance of fibre straightness for pulp strength, Pulp and Paper Canada 2006,
       107, 34–42.
[107] Olejnik K. Effect of the free swelling of refined cellulose fibres on the mechanical properties
       of paper. Fibres Text. East Eur. 2012, 20, 113–116.
[108] Maloney TC, Paulapuro H, Stenius P. Hydratation and swelling of pulp fibres measured with
       differential scanning calorimetry. Nord. Pulp Pap. Res. J. 1998, 13, 31–36.

[109] Maloney TC, Laine JE, Paulapuro H. Comments on the measurement of cell wall water. Tappi J. 1999, 82(9), 125–127.

[110] Kang T, Paulapuro H. Effect of external fibrillation on paper strength. Pulp Pap. Canada. 2006, 107, 51–54.

[111] Kang T, Somboon P, Paulapuro H. Fibrillation of mechanical pulp fibers. Pap. ja Puu. 2006, 88, 409–411.

[112] Joutsimo O, Robertsén L. The effect of mechanical treatment on softwood kraft pulp fibers. Fiber surf. layer, Pap. ja Puu. 2004, 86, 508–513.

[113] Joutsimo O, Robertsén L. The effect of mechanical treatment on softwood kraft pulp fibers. Pulp fiber prop. Pap. ja Puu. 2004, 86, 359–364.

[114] Koskenhely K, Nieminen K, Hiltunen E, Paulapuro H. Comparison of plate and conical fillings in refining of bleached softwood and hardwood pulps. Pap. ja Puu. 2005, 87, 458–463.

[115] Bhardwaj NK, Hoang V, Nguyen KL. Effect of refining on pulp surface charge accessible to polydadmac and FTIR characteristic bands of high yield kraft fibres. Bioresour. Technol. 2007, 98, 1647–1654.

[116] Lindstrom T. Chemical factors affecting the behaviour of fibres during papermaking. Nord. Pulp Pap. Res. J. 1992, 7, 181–192.

[117] Fardim P, Duran N. Modification of fibre surfaces during pulping and refining as analysed by SEM, XPS and ToF-SIMS. Coll. Surf A: Physicochem. Eng. Aspects. 2003, 223(1), 263–276.

[118] Grau U, Schuhmacher R, Kleemann S. Effect of recycling and the performance of dry-strength agents. Wochenbl. fur Pap. fabrik. 1996, 124, 729–735.

[119] Sjostrom L, Odberg L. Influence of wet-end chemicals on the recyclability of paper. Papier. 1997, 51, V69–V73.

[120] Ahrens FW, Xu H Effect of pulp drying history on pressing and drying. In: Proceedings of the Tappi Engineering/Process and Product Quality Conference, Tappi Press, Atlanta, GA, USA, 1999, 1009–1015.

[121] Parker J, Hergert OM. Simultaneous convergence–a new concept of headbox design. Tappi. 1968, 51, 425–432.

[122] Hubbe MA. Flocculation and redispersion of cellulosic fiber suspensions: a review of effects of hydrodynamic shear and polyelectrolytes. BioResources. 2007, 2, 296–331.

[123] Celzard A, Fierro V, Pizzi A. Flocculation of cellulose fibre suspensions: the contribution of percolation and effective-medium theories. Cellulose. 2008, 15, 803–814.

[124] Page DH. The collapse behaviour of pulp fibres. Tappi. 1967, 50, 449–455.

[125] Maloney TC, Li T-Q, Weise U, Paulapuro H. Intra- and inter-fibre pore closure in wet pressing. App. J. 1997, 50, 301–306.

[126] Robertson AA Some observations on the effects of drying papermaking fibers. Pulp and Paper Magazine of Canada, 1964, 65, T161–T168.

[127] Carlsson G, Lindstrom T. Hornification of cellulosic fibers during wet pressing. Sven. Pap. 1984, 87, R119–R125.

[128] Luo XL, Zhu JY, Gleisner R, Zhan HY. Effects of wet-pressing-induced fiber hornification on enzymatic saccharification of lignocelluloses. Cellulose. 2011, 18, 1055–1062.

[129] Scallan AM. Mechanisms of Hornification. In: Blanco A, Negro C, Gaspar I, Monte MC, Tijero J., eds. Proceedings of COST Action E1- Improvement of recyclability and the recycling paper industry of the future. Complutense University of Madrid, La Palmas de Gran Canaria, Spain, 1998, 312–314.

[130] Weise U, Paulapuro H. Relation between fiber shrinkage and hornification. Papier. 1996, 50, 328–335.

[131] Garvin SP, Pantaleo PF Measurements and evaluation of dryer section performance. In: Proceedings of Tappi Engineering Conference, Houston, Texas, 1976, Tappi Press, Atlanta, GA, USA, 1976, 125–132.
[132] Whitesides GM. The origins and the future of microfluidics. Nature. 2006, 442, 368–373.
[133] Tabeling P. Introduction to microfluidics. Oxford University Press, New York, 2005, 70–129.
[134] Manz A, Harrison DJ, Verpoorte EMJ, Fettinger JC, Paulu A, Ludi H, Widmer HM. Planar chips technology for miniaturization and integration of separation techniques into monitoring systems. Capillary electrophoresis on a chip. J. Chromatogr. A. 1992, 593, 253–258.
[135] Nge PN, Rogers CI, Woolley AT. Advances in microfluidic materials, functions, integration, and applications. Chem. Rev. 2013, 113, 2550–2583.
[136] Böhm A, Carstens F, Trieb C, Schabel S, Biesalski M. Engineering microfluidic papers: effect of fiber source and paper sheet properties on capillary-driven fluid flow. Microfluid. Nanofluid. 2014, 16, 789–799.
[137] Ciolacu F. Paper-based microfluidic devices on fibrous platforms with designed structure. Cell. Chem. Technol. 2018, 52, 863–871.
[138] Forrest SR. The path to ubiquitous and low-cost organic electronic appliances on plastic. Nature. 2004, 428, 911–918.
[139] Tobjork D, Osterbacka R. Paper electronics. Adv. Mater. 2011, 23, 1935–1961.
[140] Irimia-Vladu M. "Green" electronics: biodegradable and biocompatible materials and devices for sustainable future. Chem. Soc. Rev. 2014, 43, 588–610.

Daniela Rusu and Diana Elena Ciolacu

Chapter 10
Cellulose-based hydrogels: design, structure-related properties, and medical applications

10.1 Introduction

A well-known axiom in the broad area of hydrogels defines them as "gel-like materials which are easier to recognize than to define" [1, 2]. Such a material consists of cross-linked polymer chains that form a three-dimensional porous matrix, in which the empty space is filled with water molecules that act as a plasticizer. The multitude of cross-links between macromolecular backbones provides physical integrity, enhanced mechanical properties, and the water-insoluble feature, while the nature of the macromolecules and their environment (temperature, pH, and ionic strength) determine the hydrophilic characteristic.

Due to this peculiar combination, the contact between a hydrogel and water or aqueous fluids results in the swelling of the material due to the absorption of water without the dissolution of the network. The extent of the swelling process depends on the polymer–water interactions and comes to an end when these are opposed by the interactions between the polymeric chains, and a swelling equilibrium is reached.

There is a wide range of elements that influence this process, from the ones belonging to the (macro)molecular building blocks (like structure, conformation, chain branching and packing, hydrophilic/hydrophobic nature, hydrodynamic volume, and cross-link dimensions) to thermodynamics, kinetics, or the nature of the aqueous environment.

A hydrogel is, therefore, a "solid-like" aqueous solution, in which polymeric chains and the connection points between them are saturated with absorbed water under a thermodynamic equilibrium [3, 4]. Such a material can be observed in Fig. 10.1, which displays fully transparent, macroporous, swelled hydrogels based on cellulose [5].

The high-water uptake and overall hydrated nature, the swelling behavior, and the special chemical and mechanical traits provide a unique combination that is broadly used in a plethora of applications, which span agriculture, industry, cosmetics, and the biomedical field [1, 6, 7].

Daniela Rusu, Diana Elena Ciolacu, "Petru Poni" Institute of Macromolecular Chemistry, Iasi, Romania

https://doi.org/10.1515/9783110658842-010

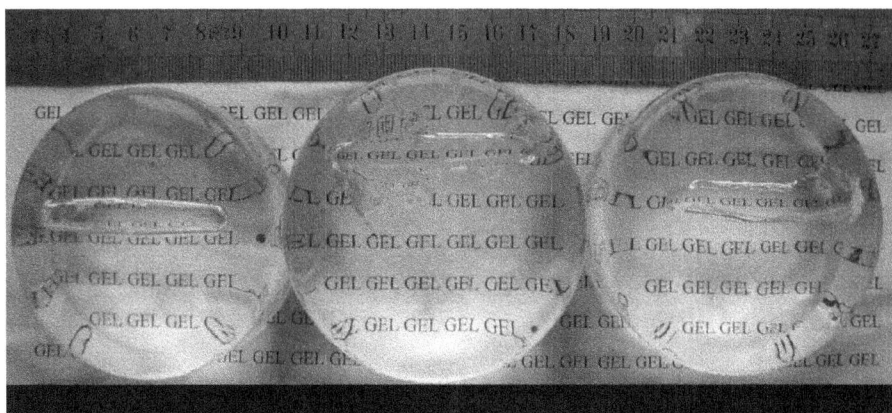

Fig. 10.1: Images of swollen cellulose hydrogels. Reproduced from [5], with permission from © 2007 John Wiley and Sons.

There is a wide range of natural and synthetic polymers and combinations therefrom that can be used to prepare hydrogels. The state of the art in the field is to employ natural macromolecules [such as cellulose, alginate, starch, chitin, gelatin, and collagen (COL)] in the construction of such networks, since these polymers provide many advantages over their synthetic counterparts: complex chemical structures, satisfying mechanical characteristics, reduced toxicity, biodegradability (and nontoxic degradation products), body-tissue mirroring, and improved biocompatibility [1, 8]. The latter is the pivotal feature for biomedical applications and has a twofold meaning. The first refers to biosafety, which means an appropriate host response in the absence of cytotoxicity, while the second implies bioactivity, namely the capacity to accomplish a specific task.

However, the innate biocompatibility of the natural polymers is diminished by the usage of the same synthetic cross-linkers, initiators, organic solvents, and emulsifiers as in the case of hydrogels based on purely synthetic macromolecules. Several purification methods (repeated washing, dialysis) were developed over time to surpass these toxicity-related issues [9, 10].

Cellulose is a key player in the field of hydrogels based on natural polymers, since it represents the most abundant natural macromolecule and a major constituent of plants and natural fibers. Moreover, microbial synthesis by some bacteria represents a secondary natural source of cellulose, besides cell walls of plants. Thus, due to the increasing demand for environmentally friendly and biocompatible products, various new functional materials from cellulose have been developed covering a wide range of applications.

In this chapter, the design, the preparation methods, and the main characteristics of cellulose-based hydrogels are highlighted. Furthermore, considering their extensive possibilities of applications, this study is focused only on the current and the most impactful contributions to biomedical science.

10.2 Cellulose–based hydrogels

Cellulose is a carbohydrate homopolymer with a linear backbone formed by β-D-glucopyranose units connected through β-1,4-glycosidic bonds (Fig. 10.2) [4]. The linear chain can encompass up to 1,500 β-glucose subunits in one backbone, each one of them being decorated with three hydroxyl groups in positions 2, 3, and 6. The adjacent ones behave as secondary alcohols, while the third one has a primary alcohol nature. These hydroxyl moieties have multiple functions within the macromolecular

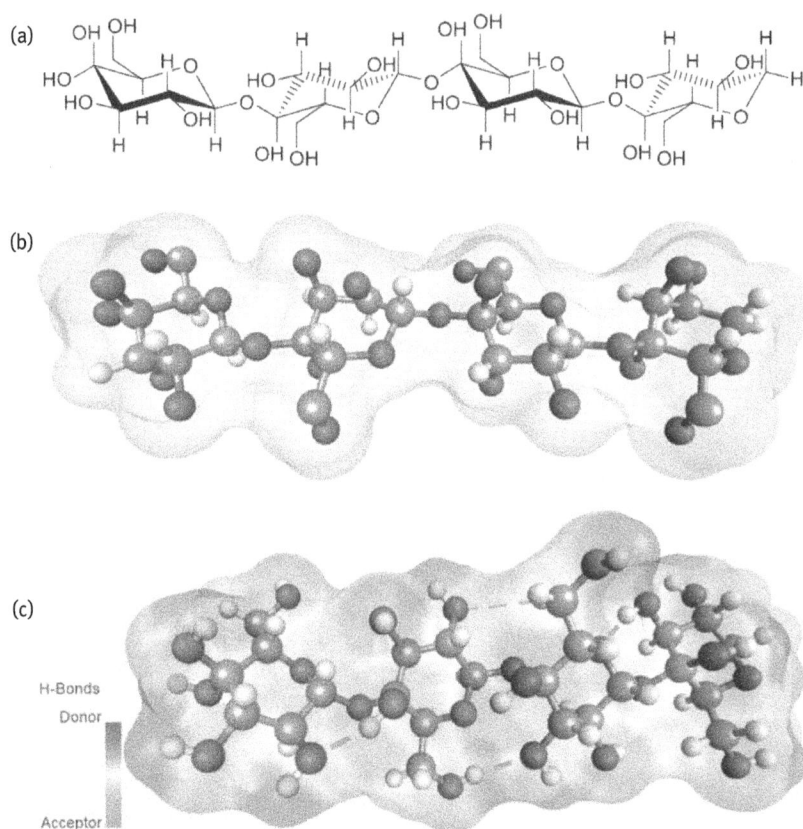

Fig. 10.2: The molecular structure of cellulose: (a) chemical structure drawing of a cellulose tetramer, (b) spatial molecular model, and (c) intermolecular hydrogen bonding. Reproduced from [4], with permission from © 2018 Springer Nature.

construct. On the one hand, they enable chemically driven reactions and modifications of the cellulose macromolecule. On the other hand, they provide a surface rich in donors and acceptors of hydrogen bonds, which are readily available for inter- and intramolecular interactions with other cellulose chains and with any absorbed water [7, 11]. The former leads to the tight, iterative packing of the linear macromolecules into planar sheets and further microfibrils. This is translated into a partially crystalline structure, low solubility in common solvents, poor processability, and inferior hydrophilicity as compared to other natural polymers.

The same hydroxyl units are the key to resolving this issue. Their reactivity enables various modifications, for example, their replacement with methyl groups that are not available for hydrogen bonding. This decreases the polymer–polymer interactions through hydrogen bonding and thus disturbs the structural regularity available for tight, crystalline packing, while it improves its solubility in water. As a result, the methylcellulose obtained in this manner is soluble in water and displays a modified version of the above features.

A similar strategy is used for a variety of cellulose modifications resulting in a plethora of derivatives with improved solubility and tailored characteristics. Some of the most well-known and widely employed cellulose derivatives are the ones obtained by alkylation and their charged salts, as presented in Fig. 10.3 [12].

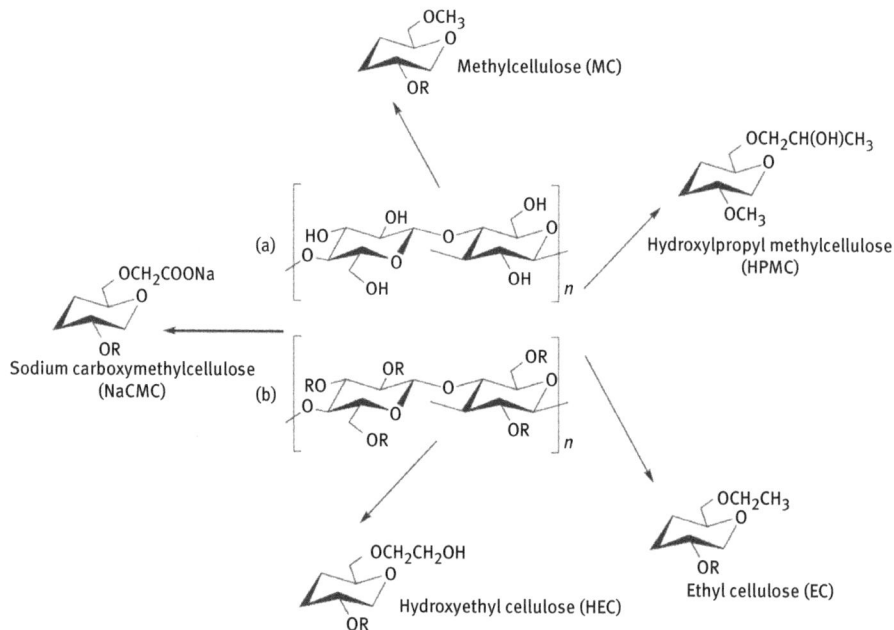

Fig. 10.3: General chemical structure of cellulose and its derivatives. Reproduced from [12], with permission from © 2018 Springer Nature.

The introduction of other functional moieties in the cellulosic backbone unlocks new interesting characteristics and processing options while maintaining most of the appealing innate features of the original cellulose. As a result, a wide range of derivatives are used nowadays in a broad spectrum of applications, as given in Tab. 10.1 [13].

Tab. 10.1: Main application areas of cellulose derivatives. Adapted from [13], under Creative Commons Attribution 3.0 License, from © 2015 IntechOpen.

Cellulose derivatives	Application areas
Ethers	
MC (methylcellulose) EC (Ethyl cellulose)	Foods, dyes, drugs, detergents, cosmetics, textiles, lacquers, finishing, inks, glues, electrical insulators, fire extinguishers, electrical appliances, water retainers, stabilizers, borehole liquids, etc.
CMC (carboxymethylcellulose) NaCMC (sodiumcarboxymethylcellulose) (E-466)	Foods, cosmetics, textiles, drugs, paper, detergents, glues, etc.
HEC (Hydroxyethyl cellulose)	Dyes, chemicals, liquid detergents, rubber, oil wells, polymerization emulsions, etc.
HPC (Hydroxypropyl cellulose) HPMC (Hydroxylpropyl methylcellulose)	Foods, drugs, papers, plastics, ceramics, gels (as water retainers), eye lenses, stabilizers, etc.
Esters	
Acetates (organic esters): cellulose acetate, cellulose triacetate, cellulose propionate, cellulose acetate propionate, propionate, cellulose acetate butyrate, etc.	Textiles, films, plates, sheaths, coils, isolation, and nonconductive parts
Filaments	Garments, threads, furniture, packaging, etc.
Plastics	Various
Tows (linen and cannabis fibers)	Textiles, yarns, cigarette filters, etc.
Nitrates (nitrocellulose) (inorganic ester)	Automotive, textiles, dyes, lacquers, polishes, films, explosives, cement, plates, etc.
Nitriles	Food and other packaging

Hydrogels represent the main form of use for these derivatives, which deliver a set of appealing features: biocompatibility and biodegradability, hydrophilicity, high mechanical strength and durability, thermal stability, low density, tailorable structure, low cost, and renewability. Moreover, the composition of such a material can be customized for a specific application by a proper selection of the partnering macromolecules and processing techniques. Consequently, they hold significant applicative potential in a broad range of areas like the food industry, pharmaceutics, agriculture, environmental protection, personal care products, and electronics.

Cellulose and cellulose derivative hydrogels are usually obtained by chemically cross-linking at low temperatures in aqueous solutions adjusted to solubilize and stabilize the starting macromolecules (e.g., water–urea mixtures and NaOH solutions), as it can be observed in Fig. 10.4 [14]. Several cross-linking agents are employed in this regard; despite its relatively toxic nature (which can be eliminated by the aforementioned method of purification), epichlorohydrin (ECH) is the most used cross-linking agent for cellulose hydrogels.

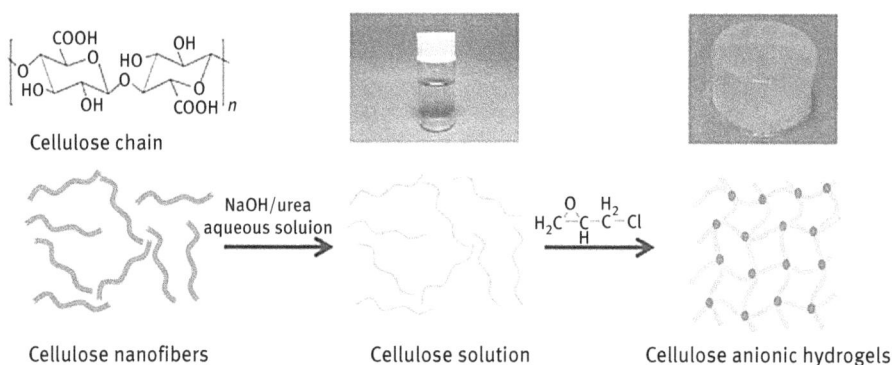

Fig. 10.4: Preparation of cellulose hydrogels by chemical cross-linking with ECH. Adapted from [14], with permission from © 2017 American Chemical Society.

The cellulose networks can also be prepared by physical cross-linking, based on the associations of polymeric chains at low temperatures, in order to lose hydration water. However, these reversible networks display several drawbacks, especially of physico-mechanical nature, which hamper their application.

Due to the diversity and complexity of the criteria and requirements imposed by the envisaged applications, cellulose is rarely the sole component of hydrogels. Additional low- and macromolecular reaction partners of natural or synthetic origin are chemically combined with cellulose to form composite hydrogels with customized features [15–17]. Figure 10.5 presents such an example by referring to the chemical cross-linking of cellulose with another natural polymer, chitosan (CS), and the addition of a small-sized reaction partner, rectorite [18].

A final comment must be made concerning the morphology of cellulose-based hydrogels since it represents a key feature in terms of structure–property relationship and potential application. Likewise, swelling ratio, water uptake, chemical and mechanical features, and the morphology of cellulose networks can be finely tailored and controlled by adjusting the nature and concentration of the starting

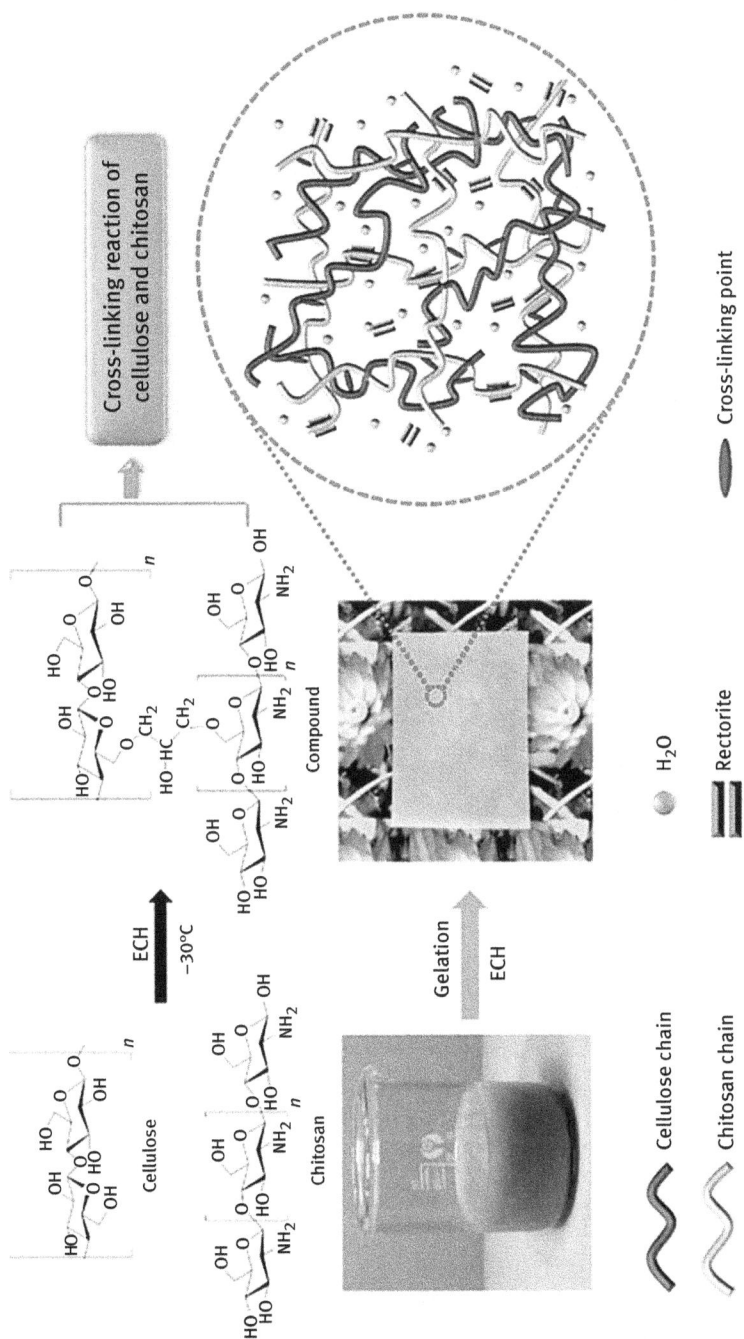

Fig. 10.5: Synthesis of a composite cellulose hydrogel. Reproduced from [18], with permission from © 2017 Royal Society of Chemistry.

gelling agents or by adapting the preparation method and subsequent treatments. Consequently, an abundance of porous architectures can be obtained, from macroporous interconnected networks to isolated fiber-like ensembles (Fig. 10.6) [19].

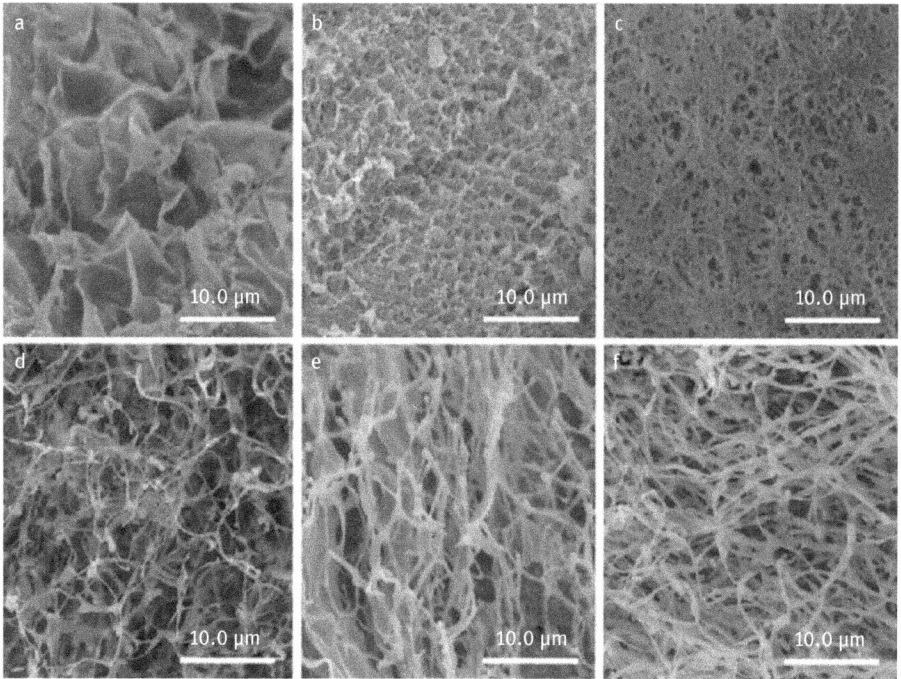

Fig. 10.6: Scanning electron microscopy micrographs of various cellulose hydrogels (from left to right: 2, 3, 4 wt% cellulose) synthesized by chemical cross-linking with ECH and various thermal treatments (a-c: heating at 50 °C for 20 h; d-f: freezing at −20 °C for 20 h). Reproduced from [19], with permission from © 2010 Elsevier.

The following sections of this chapter concentrate some of the most recent and solid research regarding the fundamental aspects, design, fabrication approaches, and main characteristics of cellulose-based hydrogels conceived for applications in the biomedical area, namely wound healing and drug delivery.

10.3 Biomedical applications

10.3.1 Wound dressing

Wound healing represents a dynamic process based on different intra- and extracellular mechanisms, which are activated with the purpose to replace damaged or injured tissues. It consists of separate, yet overlapping phases generally described as inflammation, development of new tissue (by migration and proliferation of cells), and remodeling (which represents the longest phase of the process). These skin damages occur because of multiple factors and are classified as acute and chronic by considering the nature of the restoration activity. For acute wounds, generally produced by mechanical or chemical accidents, the healing process is complete within 8–12 weeks, while the chronic-type ones imply a longer time to repair [20–23].

A large variety of wound dressings is commercially available and used depending on the nature of the injury as well as on the healing stage of the wound [24]. However, the topic is far from reaching the desired level of maturity, and more academic and industrial research is needed to achieve this goal.

An ideal wound dressing material should ensure biocompatibility, biodegradability, proper attachment, growth and proliferation of cells, and suitable physical, mechanical, and antibacterial properties. Nevertheless, the key requirement for such a material is to deliver an optimal level of moisture to the tissue of interest. Hydrogels qualify as perfect candidates in this regard since they provide an ideal hydration environment to sustain the healing process [25–28]. Different medical studies proved the connections between an adequate level of hydration and reduced local pain or increased healing rate. Of course, the nature of the wound plays a crucial role in selecting the most proper wound dressing [24, 29].

In this context, hydrogels represent valid materials for a broad range of wounds since they offer the possibility to finely tune and even control the quantity of water absorbed by their hydrophilic three-dimensional network [30, 31].

Hydrogels can provide a moist environment requested by a specific wound due to their extremely hydrated character. Polymers derived from natural sources are preferred candidates in this regard, by bringing several advantages over their synthetic counterparts: biodegradability, high biocompatibility, nontoxicity, and excellent similarity with the extracellular matrix [32, 33].

Cellulose and its derivatives are among the first choices in wound dressing because of their availability, biocompatibility, and complex physicomechanical features. Extensive research has been directed toward the manufacture of cellulose-based materials able to provide the right hydration level of specific skin injuries [34]. They are mostly developed by combining cellulose and various low- or/and macromolecular building blocks through various synthetic pathways in the same hydrogel matrix as to encapsulate the specific benefits of each component in the same material. The pharmaceutical effect is obtained either by loading the hydrogel with temporary bioactive compounds or by the irreversible functionalization of the matrix.

As a first example, thermoresponsive hydrogels composed of poly(N-isopropylacrylamide) and cellulose nanocrystals (CNC) were obtained by free-radical polymerization, without any additional cross-linker. Hydrogels with a higher content of CNC presented stronger mechanical properties and lower thermal stability. Moreover, the increased CNC amount revealed higher structural integrity upon hydrogel injection. After loading metronidazole, a frequently used antibiotic for skin infections, these hydrogels proved able to act as injectable materials for healing infected wounds [35].

Nanofibrillated cellulose (NFC) is another form of cellulose severely tested for biomedical usage. With fiber diameters below 20 nm and several micrometers in length, it is the smallest component of cellulose fibers and is employed in many wound dressing formulations to modulate the physical properties of hydrogels [36]. NFC displays a large surface area, which translates into enhanced water retention capacity and superior mechanical properties [37–39].

Basu et al. investigated the possibility of exploring NFC-based hydrogels as a potential platform for wound healing dressings. Ca^{2+}-cross-linked NFC hydrogels of different compositions revealed the capacity to maintain a proper moist environment for distinct types of injuries, based on water retention tests. The biocompatibility of these materials, tested against human dermal fibroblasts and monocyte cells, was confirmed by the high viability level, above 70%. Moreover, the NFC hydrogels did not induce any inflammatory response concerning cytokine secretion of blood-derived mononuclear cells. The monolayered cultures remained intact after these materials were removed from the wound, thus confirming their significant potential in wound dressing applications [40].

Other studies validated the noncytotoxicity of several scaffolds based on NFC hydrogels [41, 42], and underlined their capacity to reduce inflammation [43], and to support the adhesion, growth, and proliferation of cells [44, 45], which are pivotal criteria in wound healing applications.

Different chemical cross-linkers were used to further manipulate the mechanical features of hydrogel matrices, with the downside of promoting cytotoxicity or diminishing the cellular response [46]. Therefore, environmentally friendly alternatives are still needed to tune the structural properties of NFC hydrogel scaffolds for wound healing applications.

A possibility to address this issue is to appeal to hemicelluloses with distinct biological activities also developed for wound treatment usage [47]. The incorporation of hemicelluloses or their derivatives into the NFC porous structure would result in materials with enhanced biomedical and mechanical performances in the field, and various approaches were followed in this direction [48–50]. It has been observed that the pore architecture, cytocompatibility, and mechanical features of composite NFC–hydrogels can be controlled by the amount and type of incorporated hemicelluloses and also by the charge density of NFC, with the latter playing a key role concerning mechanical strength. At the same time, hemicelluloses containing NFC–hydrogels display high values of cell viability during NIH 3T3 fibroblasts culture. These all-polysaccharide hydrogels can be used in wound healing as support networks able to promote adhesion, growth, and proliferation of cells [51].

The combination of other cellulose derivatives with several polymers and low molecular reticulating agents was also deeply investigated in manifold hydrogel formulations to broaden the range of properties available for wound healing applications. Carboxymethyl cellulose (CMC), a hydrophilic polysaccharide with excellent water swelling, is another starred player in the field due to its advantageous biocompatibility, biodegradability, and low toxicity [52–54]. However, the preparation process of CMC-based hydrogels depends on finding the proper, biomedical-friendly cross-linking agent to obtain mechanically stable matrices with improved physical properties [55].

A very interesting example on the topic refers to sericin, a glue-like natural material found in the silkworms' cocoons. This protein imparts an abundant biological activity, being known as an antioxidant, antibacterial, anticoagulant, or anticancer agent. Also, due to its hydrophilic character, sericin maintains a proper level of moisture along with a minimal inflammatory response [56–58]. Therefore, many studies regarding the incorporation of sericin in scaffolds designed for wound dressing were reported. Engaging results showed that blending CMC and sericin without the involvement of a cross-linker leads to stable three-dimensional networks with tailorable properties, which are considered promising candidates for wound dressing applications [59–62].

Nayak and Kundu prepared porous blended matrices based on CMC and silk sericin (SS) with no cytotoxicity and low inflammatory response. The scaffolds support human keratinocytes attachment and proliferation, higher cellular viability being observed for the CMC-SS matrices, as compared to pure CMC. A CMC:SS ratio of 1:1 was found to be optimal in terms of biocompatibility and mechanical strength of the porous networks, a higher SS content leading to an unfavorable modification of the physicomechanical attributes [63].

Another study investigated the effect of CMC's molecular weight on the final characteristics of the CMC–SS hydrogels. While all tested molecular weights led to porous structures with good swelling capacity, the incorporation of sericin enhanced the mechanical stability, resistance to enzymatic degradation, and hydrolysis of these materials. The molecular weight conditioned the biodegradability and sericin release behavior of the matrices, with the low and middle molecular weight CMC showing the highest rate of enzymatic degradation and sericin release. In addition, the high sericin concentration allowed COL production from L929 cells, thus increasing the applicative potential of the CMC–SS materials as wound healing scaffolds [64].

Bacterial cellulose (BC) is another version of the versatile polysaccharide that has gained increased attention from the biomedical field in the last decade. This biopolymer comes close to a perfect candidate for skin regeneration and wound healing purposes, taking into account its physicomechanical properties, water retention ability, and high biocompatibility [65–69]. BC-based dressings offer excellent biorelated features for the treatment of burn wounds, acting like an artificial skin that allows a fast healing process and good protection against infections [70–73].

On the other hand, COL is a usual material in tissue restoration, which provides integrity, strength, and structure. Moreover, this biomaterial represents a part of the extracellular matrix, and is therefore biocompatible, nontoxic, nonimmunogenic, and plays a major role in each phase of the wound healing process [74].

Moraes et al. developed and evaluated wound dressings based on BC–COL hydrogels, and compared them with a commercial collagenase-based ointment. The in vivo tests in rats showed a faster rate of re-epithelialization and improved adhesive properties for the BC–COL formulations, especially in the first days of the wound healing. On the seventh postsurgery day, the BC–COL hydrogels exhibited superior results in the treatment of rat's skin wounds than collagenase, proving its efficiency as a wound dressing/skin regeneration agent [75].

In another direction, Mohamad et al. investigated the effect of electron beam irradiation and acrylic acid (AA) incorporation on biocompatibility and physicomechanical properties of BC hydrogels for wound healing. It was noticed that the mechanical stability, elasticity, flexibility, and overall cross-linking density of the hydrogels are enhanced by increasing the content of AA and beam irradiation doses. On the other side, the surface adhesiveness and swelling capacity were reduced in the same conditions. Skin sensitization and irritation examinations on rabbits revealed a nonallergic and nonirritant behavior of the BC–AA hydrogel, while cytotoxicity tests on human dermal fibroblasts showed an above 88% high viability, thus opening new directions in the area [76].

Based on these promising results, the same group evaluated the ability of these BC–AA hydrogels to act as human epidermal keratinocytes and dermal fibroblast carriers in wound dressings for large skin lesions. In vitro studies determined that these materials exhibit excellent cell attachment, viability, migration, and proliferation and permit also cell transfer. In vivo assays indicated that the cell-loaded scaffolds accelerate the wound healing process [77].

Other studies employed the SS to form BC composite porous films for wound healing applications. The BC–SS composites displayed a highly porous surface with uniform interconnected pores, and a swelling capacity depending on the SS concentration. The biocompatibility tests showed no cytotoxic effect for all tested composites and an enhanced fibroblast and keratinocyte attachment, growth, and proliferation. Furthermore, due to additional desirable structure-directed properties, these BC–SS biomaterials can be considered as prime candidates for wound healing applications [78].

Substantial research has been carried out on the development of antibacterial dressings for wound healing applications. Different studies relate to the modification of BC with the aim to introduce antibacterial features in hydrogel formulations in order to control the infection and to provide an optimum environment for wound healing. Three-dimensional networks involving the use of cellulose and chloramphenicol or 2,3-dialdehyde cellulose were developed and tested for their antibacterial efficacy against *Staphylococcus aureus*, *Streptococcus pneumoniae*, and *Escherichia coli.* In vitro antibacterial tests revealed the ability of these membranes to inhibit bacterial growth for an extended duration. Also, MTT assays (a colorimetric assay for evaluating cell metabolic activity, which measures the reduction of yellow 3-(4,5-dimethythiazol2-yl)-2,5-diphenyl tetrazolium bromide, MTT) indicated an increased capacity of promoting the adhesion and proliferation of L929 fibroblast cells, empowering their further study as materials for specific wound healing applications [79].

In another study, BC membranes loaded with Ag^+ coming from two different sources, $AgNO_3$ or Ag-containing zeolites (AgZ), were investigated for their antimicrobial effect. Release profile assays highlighted a constant and controlled release of a larger amount of Ag^+ coming from the BC hydrogels loaded with AgZ when compared with the $AgNO_3$ homologs. Antibacterial activity against *S. aureus* and *Pseudomonas aeruginosa* evaluated by inhibition tests confirmed that both $AgNO_3$- and AgZ-loaded BC hydrogels have a good antimicrobial character, and observed a significant long-term efficiency for the AgZ-BC materials [80].

Another study combined the excellent wound healing properties of BC with the antimicrobial activity of ZnO nanoparticles (ZnO-NPs) and prepared ZnO-NP-reinforced BC hydrogels to be used as wound dressings in burn treatment applications. The formulations exhibited excellent antibacterial activity against common burn wound pathogens, including *S. aureus* and *P. aeruginosa*. The histological analysis revealed good tissue regeneration and re-epithelialization triggered by the ZnO-NP-BC matrix and evidenced a significant healing activity (66%) [81].

10.3.2 Drug delivery

The highly porous structure of hydrogels and the capacity to control the swelling degree and cross-linking density in aqueous media render them the ability to incorporate and release different small bioactive entities into/from the gel matrix [82–84]. Therefore, these three-dimensional networks are deeply investigated in pharmaceutical applications as vehicles for the delivery of drugs and other bioactive molecules. Nowadays, the main interest in this regard is to prepare hydrogel formulations that provide the possibility to finely customize and overall control the release process within the matrix.

The release timeframe of drugs from these porous, watery materials, spans over a period of several hours up to several days [85, 86]. Generally, the swelling capacity of the formulations can be adjusted depending on the cross-linking degree, the chemical nature, and concentration of polymers involved in the matrix formation. At the same time, hydrogels developed as drug delivery vehicles should have optimum (bio) chemical and mechanical traits as to unlock the capacity to maintain stable polymeric networks until the envisaged biomolecules are released at a preset rate and for a preestablished time. Of course, the hydrogels used for this purpose must display biocompatibility, controlled biodegradability, and a nontoxic response during in vivo cytotoxicity tests [87–89].

Therefore, it comes simply naturally that the biological abundance, biocompatibility, biodegradability, and nontoxicity of hydrogels based on natural polymers render them extremely important for drug delivery applications [90].

The drugs can be loaded in hydrogel matrices by one of two strategies: (i) the drug slowly penetrates the polymeric matrix through a diffusion process when the hydrogel is placed in a solution containing the bioactive species; and (ii) a drug–polymer conjugate is formed through chemical interactions during hydrogel preparation [91]. The drug release mechanism depends on the pore's architecture, density and distribution, cross-linking degree, and swelling capacity of the hydrogels [92].

From the pharmacodynamics point of view, there are two crucial features concerning hydrogels used as base formulations: mechanical characteristics and drug loading capacity. The success of a hydrogel carrier is based on an optimum mechanical strength, which sustains the physical texture and integrity of the hydrogel until the bioactive agent is liberated at the envisaged rate and time. This strength is determined by an appropriate cross-linking degree, which is also responsible for the networks' porosity and subsequent loading capacity. Its optimization is a complex endeavor that needs to ensure a strong, yet elastic porous network capable to provide a tailorable cargo function.

Hydrogels based on cellulose and cellulose derivatives are broadly used in various forms as excipients in the pharmaceutical industry due to already-certified efficient characteristics for controlled drug release. These biomaterials can offer an optimal level of viscosity of the gel formulations, a crucial parameter related to polymer erosion and the diffusion of the pharmaceutical principle from the matrix [93]. The following section presents some of the most recent research in the field, organizing it according to the envisaged administration route: enteral (oral, rectal), parenteral (ocular, nasal, transdermal), and topical.

10.3.3 Hydrogels for oral drug delivery

The most convenient and common strategy for pharmaceutical delivery is the enteral or gastrointestinal oral administration, which provides a system-wide effect, but suffers frequently from the well-known hepatic adverse effect, weak targeting, and short residence time [94]. Consequently, this route requires repeated dosage administration, which many times lowers its success rate and determines severe side effects [95, 96]. In order to overcome these issues, hydrogels are used as smart drug delivery systems that can control the amount and release profile of the drug, thus improving efficacy and lowering toxicity. The production of such base formulation systems needs to be modified according to the pharmacokinetics of the drug, dosage form, and route or mechanism of administration. The latter represents the key factor of influence on the system's porosity, viscosity, and physicochemical features [97].

Hydroxypropyl methylcellulose (HPMC), CMC, methyl cellulose (MC), ethyl cellulose (EC), hydroxyethyl cellulose (HEC), hydroxypropyl cellulose (HPC), and alike are the most used cellulose derivatives in hydrogel formulations designed as carriers for (controlled) oral drug delivery due to their bioavailability, biocompatibility, excellent swelling behavior, stability, proper physicochemical characteristics, and cost-efficiency [98–104].

Intensive research efforts are dedicated to design, produce, and evaluate new drug carriers for systemic or targeted oral administration, which should be able to surpass current commercial products, to efficiently deliver new bioactive species, or to allow specificity or new drug release timeframes.

For example, Ciolacu et al. obtained highly porous cellulose-based hydrogels through physical and chemical gelation with ECH. It was observed that the cellulose content, ECH concentration, and gelation type are the main determinants of the formulations' swelling capacity and of the release behavior of procaine hydrochloride. The drug release behavior can be further adjusted while considering that the release and swelling are simultaneous phenomena [105].

In another study, Chang et al. prepared superabsorbent hydrogels based on cellulose and CMC, by using ECH as the cross-linking agent in a NaOH/urea aqueous system, and investigated them as drug carriers with a high swelling capacity. Cellulose acted as a solid backbone of the hydrogel matrix, while CMC improved pore size and distribution. The formulations displayed a high swelling degree in water (1,000%). This enabled subsequent large loading capacity, the release profiles of bovine serum albumin (BSA) being controlled by the CMC content. While encouraging, these results need further development to certify the potential controlled drug release in physiological conditions [106].

An interesting hydrogel system based on various amounts of CMC and AA was prepared as a colon-targeted drug carrier. A strong dependence between the swelling capacity indices and two pH values of the swelling medium (1 and 7) was observed for these formulations. The release profiles of theophylline showed the ability of these porous carriers to be used in colon-specific drug release [107].

In a recent study, Treesuppharata et al. developed composite hydrogels based on BC and gelatin for controlled drug delivery systems by using glutaraldehyde as the cross-linking agent. The presence of BC in gelatin-based matrices enhanced the mechanical properties and dimensional stability of the formulations and enabled high swelling capacities of these biocompatible porous networks in water (400–600%). However, further investigations are necessary to determine the carrier potential of these hydrogels in drug release [108].

Pandey et al. evaluated the release behavior of theophylline from biocompatible, oral drug delivery vehicles composed of BC–acrylamide hydrogels. Release assays revealed a higher amount of drug delivered at pH 7.4 when compared with tests performed at pH 1.5. In vivo cytotoxicity tests on mice indicated the noncytotoxicity of the formulations, as no histopathological changes or toxic responses were noticed in contrast to the control mice. The study concluded that these pH-sensitive hydrogel matrices qualify as potential theophylline carriers for oral drug delivery [109].

Nanocrystalline cellulose (CNC) has evolved as a very interesting base formulation in the fabrication of new biomaterials in targeted drug delivery due to its excellent physicochemical properties, biodegradability, biocompatibility, low toxicity, and abundance [110].

As an example, a very recent study describes composite hydrogels based on various ratios of CNC and CS as new carriers for the controlled-release of theophylline. It was observed that the swelling degree of the hydrogels increased with the CS concentration while enabling the release of a high bioagent amount at pH 1.5 in a gastric-simulating fluid. Without any doubt, these CNC–CS hydrogels have a high drug carrier potential and must be further investigated as efficient theophylline carriers [111].

The combinations of cellulose and cyclodextrins (CDs) are considered intelligent drug delivery systems due to a fascinating, tailorable mixture of features. The high swelling degree of cellulose merges with CD capacity to form inclusion complexes with a vast range of bioactive guest molecules [112–122].

For example, Dufresne et al. reported a series of CNC hydrogels containing CDs as potential drug carriers for doxorubicin (DOX). In the first step, β-CD was grafted on the CNC surface, followed by the host–guest in situ inclusion of pluronic polymers and α-CD. The external polyethylene glycol chains enhance the loading degree of CNC by facilitating the dispersion of nanocrystals in the hydrogel matrix. In vitro release studies showed that DOX is gradually liberated from the hydrogel system, which displays an extended drug release character [123].

Malik et al. developed and investigated different hydrogel formulations based on CMC and β-CD as carrier systems for acyclovir, an antiviral drug used in the occurrence of the herpes simplex virus. Drug release studies in simulated gastrointestinal medium revealed a controlled release pattern of acyclovir from these porous stable polymeric networks. The β-CD induced improved mechanical strength and enhanced release capability of these hydrogels and empowers them as promising acyclovir delivery systems [124].

Cellulose-β-CD hydrogels have been explored by Zhang et al. as carriers for the controlled delivery of 5-fluorouracil (5-FU), aniline blue (AnB), and BSA. It was established that an increase in the β-CD amount generates a reduction in the swelling capacity of hydrogels. Different drug release profiles and weak interactions with the β-CD molecules were noticed for 5-FU and BSA. AnB was the only bioactive agent to form inclusion complexes with β-CD, the drug being fully absorbed by the hydrogel matrix. The results confirm the adequacy of these formulations as suitable vehicles for the release of various hydrophobic drugs [125].

10.3.4 Hydrogels for rectal drug delivery

The rectal administration route is a highly effective systemic enteral pathway for a large variety of bioactive species, which are absorbed by the highly vascularized rectum's walls and mucosa in an operative and expressive manner. This route is a good alternative to the intravenous administration and provides a partial bypass of the unfavorable hepatic effect. This enables a larger availability for pharmaceutical compounds as compared to the other administration routes. Nevertheless, rectal drug delivery implies specific selection criteria for the base formulations in terms of viscosity, physicomechanical features, and liberation of the cargo molecule, as it can be noticed in the following examples.

Considering the benefits obtained by the rectal, enteral administration of non-steroidal anti-inflammatory drugs, a research group followed the formulation and evaluation of HPMC hydrogels for the delivery of diazepam. The results showed that the hydrogels incorporating different drug amounts presented good physico-chemical properties, proper viscosity, and efficient antimicrobial qualities. In vitro release studies demonstrated a sustained and controlled delivery of diazepam over a 3 h period. Moreover, the investigated formulations proved stable for a prolonged period of 4 months at 26 °C [126].

Koffia et al. obtained and characterized thermosensitive gels as to be used for rectal delivery of quinine. The rheological features of poloxamer 407 and the bioad-hesivity of HPMC were combined in order to obtain formulations with suitable characteristics for this type of administration: in vivo tests performed on rabbits indicated a good, tightly correlated viscoelasticity and bioadhesivity of the carrier and evidenced the advantages of the envisaged delivery method as compared to the injectable pathway [127].

Another group prepared mucoadhesive hydrogels based on HPMC and Carbopol (a cross-linked polymer of acrylic nature), loaded them with microspheres composed of sodium CS and diclofenac (DFS-CS) and studied their potential as rectal drug delivery systems. The consistency of the hydrogels was directly correlated to the HPMC amount used in the formulations. In vitro pharmacokinetic assays showed a controlled release of the drug over 6 h. The mucoadhesivity of the matrices allowed a good attachment of the DFS-CS microspheres at the surface of the rectal mucosa. Furthermore, histopathological inspections indicated the lowest irritant response of these formulations when compared to other delivery forms of the bioactive species [128].

10.3.5 Hydrogels for ocular drug delivery

The ocular administration route is a parenteral pathway used for the delivery of bioactive agents with topical (local) or systemic effects. The route bypasses the skin and mucous membranes but implies highly demanding and complex prerequisites for the base formulations.

Cappello et al. investigated some hydrogels based on HP-β-CD for the ocular delivery of 0.3% rufloxacin in rabbits. A supplemental 0.25% concentration of HPMC was enough to improve the solubility of the drug and thus reduce the β-CD quantity used with the same purpose. Preliminary data regarding the pharmacokinetic behavior in rabbits showed that HP-β-CD formulations with added HPMC offer a good bioavailability and absorption of rufloxacin in ocular, parenteral administration [129].

10.3.6 Hydrogels for intranasal drug delivery

Intranasal administration, either systemic or topical, provides the advantage of a highly vascularized, thin mucosa that allows for the rapid passage of the bioactive species through a single layer, directly into the systemic circulation. Its downsides are related to the low volume available for administration and the high sensitivity and irritability of the nasal epithelium.

Several soluble and insoluble cellulose derivatives such as HPMC, MC, CMC, HPC, EC, and microcrystalline cellulose have been demonstrated to be successful in enhancing the intranasal absorption of drugs [130]. Based on their proper mucoadhesive properties and high viscosity, these cellulose derivatives can stabilize various drug formulations and boost their delivery or extend the release timeframe through this administration route, thus enhancing its bioavailability toward new bioactive agents [131–133].

Another study shows that the nasal release of dopamine had the highest success regarding bioavailability in beagle dogs. Hydrogels containing 2% HPC and 5% azone were capable to increase the nasal absorption bioavailability with an extended retaining duration of dopamine in the nasal cavity. Moreover, high plasma levels were registered during the use of these systems up to 7 h [134].

El-Gizawy et al. prepared several nasal drug delivery systems based on a mixture of mucoadhesive polymers, such as HPC, MC, and Carbopol, for the controlled delivery of oxybutynin chloride. The addition of small amounts of EDTA and β-CD (0.5% and 1.8%, respectively) to these formulations significantly enhanced the absorption and bioavailability of the pharmaceutical agent, as confirmed by in vivo and in situ evaluations [135].

Bioadhesive gel formulations containing HPMC, HEC, and MC were investigated for the nasal administration of ciprofloxacin in rabbits. The HPMC hydrogels revealed the highest bioavailability, quite similar to the values obtained by oral administration. Meanwhile, addition of Tween 80 nonionic surfactant to the formulations containing MC and HEC led to a 25% increase in the bioavailability of these materials.

The results showed that the parenteral nasal administration route of ciprofloxacin through HPMC carriers represents a viable alternative to the oral, systemic pathway [136].

10.3.7 Hydrogels for transdermal drug delivery

In the transdermal route of administration, the bioactive agents are delivered through the skin following a systemic pharmacodynamic effect. The biggest issue in this regard is the overprotective function of the skin. The drug is forced to permeate the two layers of the skin, epidermis and dermis, and their various sublayers to enter the systemic circulation and fulfill its primary function. Therefore, delivery systems like gels and patches must be tailored to overcome this obstacle and many times encompass permeation enhancers while maintaining biocompatibility and nontoxicity. Moreover, they are obliged to present specific physicomechanical characteristics to adapt to the skin nature or to be delivered by injection.

Kouchak and Handali studied in vitro the permeation of aminophylline carried out by HPMC-based hydrogels through shed snakeskin. The presence of various drug absorption enhancers such as sodium tauroglycocholate, lauric acid, and ethanol boosted the permeation effect. The study concluded that the type and content of penetration enhancers greatly impact the transdermal absorption of aminophylline [137].

In another direction, Nandy et al. prepared various systems based on NaCMC or MC for the controlled delivery of atenolol by the same iontophoretic mechanism. L-Menthol was also used in these formulations to enhance the drug's penetration capacity through excised abdominal rat skin. Preliminary results indicated good incorporation and stability of atenolol in cellulose derivative-based hydrogels. The gel formulations containing 2% L-menthol and 1.5% atenolol were afterward attested as promising, electrically assisted, transdermal vehicles to achieve the desired atenolol release level [138].

Another study reports the transdermal delivery of celecoxib from different gel formulations by iontophoresis. HPMC, NaCMC, sodium alginate, and Carbopol 934P were used as gelling agents to obtain formulations with proper physicochemical properties. The formulation containing 4% HPMC showed the highest spreadability on the rat skin and the highest celecoxib (41.5%) release rate in a 5 h timeframe. It was concluded that celecoxib can be optimally delivered by the iontophoretic transdermal way through HPMC-based hydrogels [139].

Bertsch et al. investigated injectable CNC hydrogels as pH-responsive delivery systems for different bioactive principles. Release studies of protein BSA, tetracycline (TC), and DOX were performed in two mediums: physiological saline and simulated gastric juice. Different release patterns were observed: a burst release of TC within 2 days and a prolonged 2-week release for BSA and DOX. DOX was the single cargo molecule entirely released from the hydrogel matrix. A superior release mechanism at pH 2 was observed for TC as compared to BSA due to electrostatic complexation and charge inversion. The study certifies that CNC-based hydrogels can be used as pH-responsive delivery systems for various drugs [140].

10.3.8 Hydrogels for topical drug delivery

Topical administration implies the application of the drug carrier to a specific area or, in a few cases, in the human body. It usually refers to the epicutaneous administration, meaning the direct application of the pharmaceutical formula on the skin. The pharmacodynamic effect of this administration route is generally local, although a systemic transdermal pathway is sometimes considered. The base formulations are extremely diverse, encompassing creams, gels, ointments, pastes, powders, and others. Naturally, hydrogel-based formulations are among the main players in the field, and the cellulose-containing ones being heavily studied for the manipulation of the drug's pharmacokinetics.

Different antifungal bioactive agents like fluconazole, bifonazole, and clotrimazole are integrated into cellulose hydrogels through various emulsion systems in order to improve the emulsion viscosity necessary for topical-based applications. For example, Shahin et al. prepared jojoba oil-based emulgel materials for the controlled delivery of clotrimazole using HPMC and Carbopol934 P as gelling agents and further carriers. The efficacy and stability of these formulations were compared with two commercially available creams, Candistan® and Canesten®. The emulgel formulations based on the combination of HPMC and Carbopol 934 showed excellent stability and superior drug release profiles and antimycotic activity compared to the commercially available creams. The study evidenced the major impact of the gelling agent on the viscosity, consistency, and release mechanism of clotrimazole [141].

Sabale et al. developed HPMC-based hydrogels containing oleic acid microemulsion for the topical administration of bifonazole. Different concentrations of HPMC were used to increase the viscosity, solubility, skin permeability, and release profile of the bioactive product. Rat skin absorption assays underlined an 80% permeability of bifonazole in less than 10 h, the sustained delivery of drugs being influenced by the HPMC content. A comparative study with a similar commercial cream evidenced that the obtained formulations display 3-month stability and good antifungal activity, both in the absence of skin irritation [142].

Another group studied the in vitro release behavior of fluconazole from different gel-based vehicles, with the aim to fully deliver the active principle and maintain it within the skin. Microemulsions based on NaCMC as the gelling agent, and propylene glycol and diethylene glycol monoethyl ether used as fluconazole solvents and as permeation enhancers were investigated in vitro on pig's ear skin. The assays

revealed that the NaCMC emulsions containing the latter glycolic derivative can release the entire drug dose and improve its skin absorption, regardless of the viscosity of formulations. Moreover, this system showed superior antifungal activity.
Supplementary clinical evaluations are being conducted to confirm the potential of
these microemulsions in the topical administration of fluconazole [143].

Marcos et al. developed several gels based on HEC, α-CD, and poloxamer aiming
to improve the solubilization and, afterward, the topical release of griseofulvin. The
bioavailability of the gels, in the presence and absence of griseofulvin, was easily adjusted using various ratios of polymers and α-CD at different temperatures. The study
revealed that the addition of HEC increased the viscosity of the formulations and allowed a sustained release of the drug at 37 °C. These polymeric networks open new
opportunities in the development of novel carriers for the prolonged topical delivery
of specific drugs [144].

10.4 Conclusions

Cellulose-based hydrogels represent tailorable three-dimensional polymeric networks
with a unique set of features: biocompatibility and biodegradability, hydrophilicity,
high mechanical strength and durability, thermal stability, low density, tailorable
structure, low cost, and renewability. Moreover, the composition of such a material
can be customized for a specific application by a proper selection of the partnering
macromolecules and processing techniques. Therefore, they hold significant applicative potential in a broad range of areas such as food industry, pharmaceutics, agriculture, environmental protection, personal care products, and electronics.

This chapter presents an overview of the state of the art in the field of cellulose-
based hydrogels for biomedical usage. It incorporates some of the most recent and
solid research regarding the fundamental aspects, design, fabrication approaches,
and main characteristics of cellulose-based hydrogels, with a focus on some of their
most important applications in the biomedical area.

Cellulose and its derivatives are among the first choices in wound dressing
because of their availability, biocompatibility, and complex physicomechanical
features. Extensive research has been directed toward the manufacture of cellulose-based materials able to provide the right hydration level of specific skin injuries
and display antibacterial features. They are mostly developed by combining cellulose
and various low- or/and macromolecular building blocks in the same hydrogel matrix
as to encapsulate the specific benefits of each component in the same material. The
pharmaceutical effect is obtained either by loading the hydrogel with temporary bioactive compounds or by the irreversible functionalization of the matrix. The performance of these materials can be further improved to face new challenges in the field
and to expand their application spectrum.

These three-dimensional networks are deeply investigated in pharmaceutical applications as vehicles for the delivery of drugs and other bioactive molecules. The main interest in this regard is to prepare hydrogel formulations that provide the possibility to finely customize and overall control the delivery process within the matrix in terms of release concentration and period. The release timeframe from these porous of drugs from these porous watery materials spans over a period of several hours up to several days. At the same time, hydrogels developed as drug delivery vehicles should have optimum (bio)chemical and mechanical traits as to unlock the capacity to maintain stable polymeric networks until the envisaged biomolecules are released at a present rate and for a pre-established time.

From the pharmacodynamics point of view, there are two crucial features concerning hydrogels used as base formulations: mechanical characteristics and drug loading capacity. The success of a hydrogel carrier is based on an optimum mechanical strength that sustains the physical texture and integrity of the hydrogel until the bioactive agent is liberated at the envisaged rate and time.

Despite the huge number of drug delivery systems based on cellulose hydrogels reported by the mainstream article and patent literature, only a limited number of materials reached the necessary level of maturity as to be included in the last evaluation stages, and even fewer made it into the real market. At the same time, new, complex challenges, requisites, and obstacles appear in the field, and the solid body of knowledge acquired till now must be used and enriched in an interdisciplinary manner as to properly approach them.

Acknowledgments: This work was supported by a grant of the Romanian Ministry of Research and Innovation, CCCDI – UEFISCDI, project number PN-III-P1-1.2-PCCDI-2017-0697/13PCCDI/2018 within PNCDI III.

References

[1]   Mohite BV, Koli SH, Patil SV. Bacterial Cellulose-Based Hydrogels: Synthesis, Properties, and Applications, In: Mondal M., eds. Cellulose-based superabsorbent hydrogels. Polymers and polymeric composites: A reference series, Springer, Heidelberg, 2018, 1–22.
[2]   Jordan LD. The problem of gel structure. Colloid Chem. 1926, 1, 767–782.
[3]   Tanaka T. Gels. Sci. Am. 1981, 244, 124–136.
[4]   Shanks RA, Pardo IRM. Cellulose Solubility, Gelation, and Absorbency Compared with Designed Synthetic Polymers, In: Mondal M., eds. Cellulose-based superabsorbent hydrogels. Polymers and polymeric Composites: A reference series, Springer, Heidelberg, 2018, 1–26.
[5]   Zhou J, Chang C, Zhang R, Zhang L. Hydrogels prepared from unsubstituted cellulose in NaOH/Urea aqueous solution. Macromol. Biosci. 2007, 7, 804–809.
[6]   Ghosh T, Katiyar V. Cellulose-Based Hydrogel Films for Food Packaging, In: Mondal M., eds. Cellulose-based superabsorbent hydrogels. Polymers and polymeric composites: A reference series, Springer, Heidelberg, 2018, 1–25.

[7]   Ciolacu DE. Structure-Property Relationships in Cellulose-Based Hydrogels, In: Mondal M. eds. Cellulose-based superabsorbent hydrogels. Polymers and polymeric composites: A reference series, Springer, Heidelberg, 2018, 1–32.

[8]   Johari NS, Ahmad I, Halib N. Comparison study of hydrogels properties synthesized with micro- and nano- size bacterial cellulose particles extracted from nata de coco. Chem. Biochem. Eng. Q. 2012, 26, 399–404.

[9]   Mallik AK, et al.. Benefits of Renewable Hydrogels over Acrylate- and Acrylamide-Based Hydrogels, In: Mondal M. eds. Cellulose-based superabsorbent hydrogels. Polymers and Polymeric composites: A reference series, Springer, Heidelberg, 2018, 1–47.

[10]   Peppas N, Huang Y, Torres-Lugo M, Ward J, Zhang J. Physicochemical foundations and structural design of hydrogels in medicine and biology. Annu. Rev. Biomed. Eng. 2000, 2, 9–29.

[11]   Lee HV, Hamid SBA, Zain SK. Conversion of lignocellulosic biomass to nanocellulose: Structure and chemical process. Sci. World J. 2014, 2014, 1–20.

[12]   Mondal MIH, Haque MO. Cellulosic Hydrogels: A Greener Solution of Sustainability, In: Mondal M., eds. Cellulose-based superabsorbent hydrogels. Polymers and polymeric composites: A reference series, Springer, Heidelberg, 2018, 1–33.

[13]   Çankaya N. Cellulose Grafting by Atom Transfer Radical Polymerization Method, In: Poletto M., eds. Cellulose – Fundamental Aspects and Current Trends, IntechOpen, 2015. Doi: 10.5772/61707.

[14]   Zhang H, Yang M, Luan Q, Tang H, Huang F, Xiang X, Yang C, Bao Y. Cellulose anionic hydrogels based on cellulose nanofibers as natural stimulants for seed germination and seedling growth. J. Agric. Food Chem. 2017, 65, 3785–3791.

[15]   Hasan AMA, Abdel-Raouf MES. Cellulose-Based Superabsorbent Hydrogels, In: Mondal M., eds. Cellulose-based superabsorbent hydrogels. Polymers and polymeric composites: A reference series, Springer, Heidelberg, 2018, 1–23.

[16]   Gomez-Dıaz D, Navaza JM. Rheological characterization of aqueous solutions of the food additive carboxymethyl cellulose. Elec. J. Env. Agricult. Food Chem. 2002, 1, 1579–1587.

[17]   Sannino A, Esposito A, Nicolais L, Del Nobile MA, Giovane A, Balestrieri C, Esposito R, Agresti M. Cellulose-based hydrogels as body water retainers. J. Mater. Sci. Mater. Med. 2000, 11, 247–253.

[18]   Tu H, Yu Y, Chen J, Shi X, Zhou J, Deng H, Du Y. Highly cost-effective and high-strength hydrogels as dye adsorbents from natural polymers: chitosan and cellulose. Polym. Chem. 2017, 8, 2913–2921.

[19]   Chang C, Zhang L, Zhoua J, Zhang L, Kennedy JF. Structure and properties of hydrogels prepared from cellulose in NaOH/urea aqueous solutions. Carbohydr. Polym. 2010, 82, 122–127.

[20]   Boateng JS, Matthews KH, Stevens HN, Eccleston GM. Wound healing dressings and drug delivery systems: A review. J. Pharm. Sci. 2008, 97, 2892–2923.

[21]   Gurtner GC, Werner S, Barrandon Y, Longaker MT. Wound repair and regeneration. Nature. 2008, 453, 314–321.

[22]   Guo SA, Di Pietro LA. Factors affecting wound healing. J. Dent. Res. 2010, 89, 219–229.

[23]   Han G, Ceilley R. Chronic wound healing: A review of current management and treatments. Adv, Ther. 2017, 34, 599–610.

[24]   Abdelrahman T, Newton H. Wound dressings: Principles and practice. Surgery (Oxford). 2011, 29, 491–495.

[25]   Boateng SJ, Matthews HK, Stevens NEH, Eccleston MG. Wound healing dressing and drug delivery system: A review. J. Pharm. Sci. 2008, 97, 2892–2923.

[26]   Lloyd LL, Kennedy JF, Methacanon P, Paterson M, Knill CJ. Carbohydrate polymers as wound management aids. Carbohydr. Polym. 1998, 37, 315–322.

[27] Wang J, Wei J. Interpenetrating network hydrogels with high strength and transparency for potential use as external dressings. Mater. Sci. Eng. C Mater. Biol. Appl. 2017, 80, 460–467.

[28] Ogawa A, Nakayama S, Uehara M, Mori Y, Takahashi M, Aiba T, Kurosaki Y. Pharmaceutical properties of a low-substituted hydroxypropyl cellulose (L-HPC) hydrogel as a novel external dressing. Int. J. Pharm. 2014, 477, 546–552.

[29] Ovington LG. Advances in wound dressings. Clin. Dermatol. 2007, 25, 33–38.

[30] Ng VW, Chan JM, Sardon H, Ono RJ, García JM, Yang YY, Hedrick JL. Antimicrobial hydrogels: a new weapon in the arsenal against multidrug-resistant infections. Adv. Drug Deliv. Rev. 2014, 78, 46–62.

[31] Hoffman AS. Hydrogels for biomedical applications. Adv. Drug Deliv. Rev. 2012, 64, 18–23.

[32] Augustine R, Rajendran R, Cvelbar U, Mozetič M, George A. Biopolymers for Health Food and Cosmetic Applications, In: Thomas S, Durand D, Chassenieux C, Jyotishkumar P. eds. Handbook of biopolymer-based materials: from blends and composites to gels and complex networks, Wiley, Singapore, 2013, 801–849.

[33] Steinbüchel A. Perspectives for biotechnological production and utilization of biopolymers: metabolic engineering of polyhydroxyalkanoate biosynthesis pathways as a successful example. Macromol. Biosci. 2001, 1, 1–24.

[34] Mogoşanu GD, Grymezescu MA. Natural and synthetic polymers for wounds and burns dressing. Int. J. Pharm. 2014, 463, 127–136.

[35] Zubik K, Singhsa P, Wang Y, Manuspiya H, Narain R. Thermo-responsive poly (N-isopropylacrylamide)-cellulose nanocrystals hybrid hydrogels for wound dressing. Polymers. 2017, 9(4), 119–136.

[36] Stelte W, Sanadi AR. Preparation and characterization of cellulose nanofibers from two commercial hardwood and softwood pulps. Ind. Eng. Chem. Res. 2009, 48, 11211–219.

[37] Abraham E, Deepa B, Pothan LA, Jacob M, Thomas S, Cvelbar U, Anandjiwala R. Extraction of nanocellulose fibrils from lignocellulosic fibers: A novel approach. Carbohydr. Polym. 2011, 86, 1468–1475.

[38] Eyholzer C, Borges de Couraca A, Duc F, Bourbon PE, Tingaut P, Zimmermann T, Manson JAE, Oskman K. Biocomposite hydrogels with carboxymethylated, nanofibrillated cellulose powder for replacement of the nucleus pulposus. Biomacromolecules. 2011, 12, 1419–1427.

[39] Nair S, Zhu JY, Deng Y, Ragauskas AJ. Hydrogels prepared from cross-linked nanofibrillated cellulose sandeep. ACS Sustain. Chem. Eng. 2014, 2, 772–780.

[40] Basu A, Lindh J, Ålander E, Strømme M, Ferraz N. On the use of ion-crosslinked nanocellulose hydrogels for wound healing solutions: physicochemical properties and application-oriented biocompatibility studies. Carbohydr. Polym. 2017, 174, 299–308.

[41] Alexandrescu L, Syverud K, Gatti A, Chinga-Carrasco G. Cytotoxicity tests of cellulose nanofibril-based structures. Cellulose 2013, 20, 1765–1775.

[42] Chinga-Carrasco G, Syverud K. Pretreatment-dependent surface chemistry of wood nanocellulose for pH-sensitive hydrogels. J. Biomater. Appl. 2014, 29, 423–432.

[43] Mertaniemi H, Escobedo-Lucea C, Sanz-Garcia A, Gandía C, Mäkitie A, Partanen J, Ikkala O, Yliperttula M. Human stem cell decorated nanocellulose threads for biomedical applications. Biomaterials. 2016, 82, 208–220.

[44] Lin N, Dufresne A. Nanocellulose in biomedicine: Current status and future prospect. Eur. Polym. J. 2014, 59, 302–325.

[45] Bacakova L, Pajorova J, Bacakova M, Skogberg A, Kallio P, Kolarova K, Svorcik V. Versatile Application of Nanocellulose: From Industry to Skin Tissue Engineering and Wound Healing. Nanomaterials (Basel). 2019, 9, 164.

[46] Borges AC, Eyholzer C, Duc F, Bourban PE, Tingaut P, Zimmermann T, Pioletti DP, Månson JA. Nanofibrillated cellulose composite hydrogel for the replacement of the nucleus pulposus. Acta Biomater. 2011, 7, 3412–3421.
[47] Liu J, Willför S, Xu C. A review of bioactive plant polysaccharides: biological activities, functionalization, and biomedical applications. Bioact. Carbohydr. Diet Fibre. 2015, 5, 31–61.
[48] Lucenius, et al.. Nanocomposite films based on cellulose nanofibrils and water-soluble polysaccharides. React. Funct. Polym. 2014, 85, 167–174.
[49] Missoum K, Belgacem MN, Bras J. Nanofibrillated Cellulose Surface Modification: A Review. Materials. 2013, 6, 1745–1766.
[50] Zimmermann T, Bordeanu N, Strub E. Properties of nanofibrillated cellulose from different raw materials and its reinforcement potential. Carbohydr. Polym. 2010, 79, 1086–1093.
[51] Liu J, Chinga-Carrasco G, Cheng F, Xu W, Willför S, Syverud K, Xu C. Hemicellulose-reinforced nanocellulose hydrogels for wound healing application. Cellulose 2016, 23, 3129–3143.
[52] Sannino A, Demitri C, Madaghiele M. Biodegradable Cellulose-based Hydrogels: Design and Applications. Materials (Basel). 2009, 2, 353–373.
[53] Pasqui D, Torricelli P, Cagna MD, Fini M, Barbucci R. Carboxymethyl cellulose-hydroxyapatite hybrid hydrogel as a composite material for bone tissue engineering applications. J. Biomed. Mater. Res. A. 2014, 102A, 1568–1579.
[54] Bajpai AK, Giri A. Water sorption behaviour of highly swelling (carboxy methylcellulose-g-polyacrylamide) hydrogels and release of potassium nitrate as agrochemical. Carbohydr. Polym. 2003, 53, 271–279.
[55] Nerurkar NL, Elliott DM, Mauck RL. Mechanical design criteria for intervertebral disc tissue engineering. J. Biomech. 2010, 43, 1017–1030.
[56] Nagai N, Murao T, Ito Y, Okamoto N, Sasaki M. Enhancing effects of sericin on corneal wound healing in rat debrided corneal epithelium. Biol. Pharm. Bull. 2009, 32, 933–936.
[57] Aramwit P, Kanokpanont S, De-Eknamkul W, Srichana T. Monitoring of inflammatory mediators induced by silk sericin. J. Biosci. Bioeng. 2009, 107, 556–561.
[58] Hasatsri S, Yamdech R, Chanvorachote P, Aramwit P. Physical and biological assessments of the innovative bilayered wound dressing made of silk and gelatin for clinical applications. J. Biomater. Appl. 2015, 29, 1304–1313.
[59] Aramwit P, Siritienthong T, Srichana T, Ratanavaraporn J. Accelerated healing of full-thickness wounds by genipin-crosslinked silk sericin/PVA scaffolds. Cells Tissues Organs (Print). 2013, 197, 224–238.
[60] Siritienthong T, Ratanavaraporn J, Aramwit P. Development of ethyl alcohol-precipitated silk sericin/polyvinyl alcohol scaffolds for accelerated healing of full-thickness wounds. Int. J. Pharm. 2012, 439, 175–186.
[61] Aramwit P, Ratanavaraporn J, Ekgasit S, Tongsakul D, Bang N. A green salt leaching technique to produce sericin/PVA/glycerin scaffolds with distinguished characteristics for wound-dressing applications. J. Biomed. Mater. Res Part B: Appl. Biomater. 2015, 103, 915–924.
[62] Siritientong T, Angspatt A, Ratanavaraporn J, Aramwit P. Clinical potential of a silk sericin-releasing bioactive wound dressing for the treatment of split-thickness skin graft donor sites. Pharm. Res. 2014, 31, 104–116.
[63] Nayak S, Kundu SC. Sericin-carboxymethyl cellulose porous matrices as cellular wound dressing material. J. Biomed. Mater. Res. A. 2014, 102, 1928–1940.
[64] Siritientong T, Aramwit P. Characteristics of carboxymethyl cellulose/sericin hydrogels and the influence of molecular weight of carboxymethyl cellulose. Macromol. Res. 2015, 23, 861–866.
[65] Römling U. Molecular biology of cellulose production in bacteria. Res. Microbiol. 2002, 153, 205–212.

[66] Helenius G, Bäckdahl H, Bodin A, Nannmark U, Gatenholm P, Risberg B. In vivo biocompatibility of bacterial cellulose. J. Biomed. Mater. Res. A. 2006, 76, 431–438.

[67] Huang Y, Zhu C, Yang J, Nie Y, Chen C, Sun D. Recent advances in bacterial cellulose. Cellulose 2014, 21, 1–30.

[68] Kucińska-Lipka J, Gubanska I, Janik H. Bacterial cellulose in the field of wound healing and regenerative medicine of skin: recent trends and future prospectives. Polym. Bull. 2015, 72, 2399–2419.

[69] Fu L, Zhang J, Yang G. Present status and applications of bacterial cellulose-based materials for skin tissue repair. Carbohydr. Polym. 2013, 92, 1432–1442.

[70] Klemm D, Heublein B, Fink HP, Bohn A. Cellulose: fascinating biopolymer and sustainable raw material. Angew. Chem. Int. Ed. Engl. 2005, 44, 3358–3393.

[71] Czaja WK, Young DJ, Kawecki M, Brown RM. The future prospects of microbial cellulose in biomedical applications. Biomacromolecules. 2007, 8, 1–12.

[72] Jonas R, Farah LF. Production and application of microbial cellulose. Polym. Degrad. Stab. 1998, 59, 101–106.

[73] Andrade FK, Costa R, Domingues L, Soares R, Gama MF. Improving bacterial cellulose for blood vessel replacement: Functionalization with a chimeric protein containing a cellulose-binding module and an adhesion peptide. Acta Biomater. 2010, 6, 4034–4041.

[74] Boyar V. Collagen: Providing a Key to the Wound Healing Kingdom. Wound management and prevention. 2019, 65(8), 1–21.

[75] Moraes PRFDS, Saska S, Barud H, Lima LRD, Martins VDCA, Plepis AMDG, Ribeiro SJL, Gaspar AMM. Bacterial cellulose/collagen hydrogel for wound healing. Mater. Res. 2016, 19, 106–116.

[76] Mohamad N, et al. Characterization and biocompatibility evaluation of bacterial cellulose-based wound dressing hydrogel: effect of electron beam irradiation doses and concentration of acrylic acid. J. Biomed. Mater. Res. Part B Appl. Biomater. 2017, 105, 2553–2564.

[77] Xi Loh EY, Mohamad N, Fauzi MB, Ng MH, Ng SF, Mohd Amin MCI. Development of a bacterial cellulose-based hydrogel cell carrier containing keratinocytes and fibroblasts for full-thickness wound healing. Sci. Rep. 2018, 8, 2875.

[78] Lamboni L, Li Y, Liu J, Yang G. Silk sericin-functionalized bacterial cellulose as a potential wound-healing biomaterial. Biomacromolecules. 2016, 17, 3076–3084.

[79] Laçin NT. Development of biodegradable antibacterial cellulose based hydrogel membranes for wound healing. Int. J. Biol. Macromol. 2014, 67, 22–27.

[80] Gupta A, Low WL, Radecka I, Britland ST, Mohd Amin MCI, Martin C. Characterisation and in vitro antimicrobial activity of biosynthetic silver-loaded bacterial cellulose hydrogels. J. Microencaps. 2016, 33, 725–734.

[81] Khalid A, Khan R, Ul-Islam M, Khan T, Wahid F. Bacterial cellulose-zinc oxide nanocomposites as a novel dressing system for burn wounds. Carbohydr. Polym. 2017, 164, 214–221.

[82] Nokhodchi A, Raja S, Patel P, Asare-Addo K. The role of oral controlled release matrix tablets in drug delivery systems. Bioimpacts. 2012, 2, 175–187.

[83] Sharpe LA, Daily AM, Horava SD, Peppas NA. Therapeutic applications of hydrogels in oral drug delivery. Expert Opin. Drug. Deliv. 2014, 11, 901–915.

[84] Buwalda SJ, Vermonden T, Hennink WE. Hydrogels for therapeutic delivery: current developments and future directions. Biomacromolecules. 2017, 18, 316–330.

[85] Lee SC, Kwon K, Park K. Hydrogels for delivery of bioactive agents: a historical perspective. Adv. Drug Deliv. Rev. 2013, 65, 17–20.

[86] Simões S, Figueiras A, Veiga F. Modular hydrogels for drug delivery. J. Biomater. Nanobiotechnol. 2012, 3, 185–199.

[87] Das N, Bera T, Mukherjee A. Biomaterial hydrogels for different biomedical applications. Int. J. Pharm. Bio. Sci. 2012, 3, 586–595.

[88] De SK, Aluru N, Johnson B, Crone W, Beebe DJ, Moore J. Equilibrium swelling and kinetics of pH-responsive hydrogels: models, experiments, and simulations. J. Microelectromech. Syst. 2002, 11, 544–555.

[89] Grassi M, Sandolo C, Perin D, Coviello T, Lapasin R, Grassi G. Structural characterization of calcium alginate matrices by means of mechanical and release tests. Molecules. 2009, 14, 3003–3017.

[90] Lin CC, Metters AT. Hydrogels in controlled release formulations: network design and mathematical modeling. Adv. Drug Deliv. Rev. 2006, 58, 1379–1408.

[91] Prashant PK, Vivek BR, Deepashree ND, Pranav PP. Hydrogels as a drug delivery system and applications: a review. Int. J. Pharm. Sci. 2012, 4, 1–7.

[92] Nguyen KT, West JL. Photopolymerizable hydrogels for tissue engineering applications. Biomaterials. 2002, 23, 4307–4314.

[93] Das N. Biodegradable Hydrogels for Controlled Drug Delivery, Mondal MIH, ed., Cellulose-based superabsorbent hydrogels, polymers and polymeric composites: A reference series, Springer, Heidelberg, 2018, 1–41.

[94] Florence AT, Jani PU. Novel oral drug formulations. Drug Saf. 1994, 10, 233–266.

[95] Langer R. Drug delivery and targeting. Nature. 1998, 392, 5–10.

[96] Liechty WB, Kryscio DR, Slaughter BV, Peppas NA. Polymers for drug delivery systems. Ann. Rev. Chem. Biomol. Eng. 2010, 1, 149–173.

[97] Li J, Mooney DJ. Designing hydrogels for controlled drug delivery. Nat. Rev. Mater. 2016, 1, 16071.

[98] Ferrero C, Massuelle D, Doelker E. Towards elucidation of the drug release mechanism from compressed hydrophilic matrices made of cellulose ethers. II. Evaluation of a possible swelling-controlled drug release mechanism using dimensionless analysis. J. Control Release. 2010, 141, 223–233.

[99] Ferrero C, Massuelle D, Jeannerat D, Doelker E. Towards elucidation of the drug release mechanism from compressed hydrophilic matrices made of cellulose ethers. I. Pulse-field-gradient spin-echo NMR study of sodium salicylate diffusivity in swollen hydrogels with respect to polymer matrix physical structure. J. Control Release. 2008, 128, 71–79.

[100] Nerurkar J, Jun HW, Price JC, Park MO. Controlled-release matrix tablets of ibuprofen using cellulose ethers and carrageenans: effect of formulation factors on dissolution rates. Eur. J. Pharm. Biopharm. 2005, 61, 56–68.

[101] Saša B, Odon P, Stane S, Julijana K. Analysis of surface properties of cellulose ethers and drug release from their matrix tablets. Eur. J. Pharm. Sci. 2006, 27, 375–383.

[102] Barzegar-Jalali M, Valizadeha H, Siahi Shadbad MR, Adibkia K, Mohammadi G, Farahani A, Arash Z, Nokhodchi A. Cogrinding as an approach to enhance dissolution rate of a poorly water-soluble drug (gliclazide). Powder Technol. 2010, 197, 150–158.

[103] Zakaria A, Afifi SA, Elkhodairy K. Newly developed topical cefotaxime sodium hydrogels: antibacterial activity and in vivo evaluation. Biomed. Res. Int. 2016, 2016, 1–15.

[104] Vlaia L, Coneac G, Olariu I, Vlaia V, Lupuleasa D. Cellulose-derivatives-based Hydrogels as Vehicles for Dermal and Transdermal Drug Delivery, In: Majee SB, ed, Emerging concepts in analysis and applications of hydrogels, InTech, London, 2016, 159–200.

[105] Ciolacu D, Rudaz C, Vasilescu M, Budtova T. Physically and chemically cross-linked cellulose cryogels: Structure, properties and application for controlled release. Carbohydr. Polym. 2016, 151, 392–400.

[106] Chang C, Duan B, Cai J, Zhang L. Superabsorbent hydrogels based on cellulose for smart swelling and controllable delivery. Eur. Polym. J. 2010, 46, 92–100.

[107] El-Hag Ali A, Abd El-Rehim H, Kamal H, Hegazy D. Synthesis of carboxymethyl cellulose based drug carrier hydrogel using ionizing radiation for possible use as specific delivery system. J. Macromol. Sci. Pure Appl. Chem. 2008, 45, 628–634.

[108] Treesuppharata W, Rojanapanthua P, Siangsanohb C, Manuspiyac H, Ummartyotin S. Synthesis and characterization of bacterial cellulose and gelatin-based hydrogel composites for drug-delivery systems. J. Appl. Biotechnol. Rep. 2017, 15, 84–91.

[109] Pandey M, Mohamad N, Amin MC. Bacterial cellulose/acrylamide pH-sensitive smart hydrogel: development, characterization, and toxicity studies in ICR mice model. Mol. Pharm. 2014, 11, 3596–3608.

[110] Klemm D, Kramer F, Moritz S, Lindstroem T, Ankerfors M, Gray D, Dorris A. Nanocelluloses: A New Family of Nature-Based Materials. Angew. Chem. Int. Ed. Engl. 2011, 50, 5438–5466.

[111] Xu Q, Ji Y, Sun Q, Fu Y, Xu Y, Jin L. Fabrication of cellulose nanocrystal/chitosan hydrogel for controlled drug release. Nanomaterials. 2019, 9, 253.

[112] Guo X, Jia X, Du J, Xiao L, Li F, Liao L, et al. Host-guest chemistry of cyclodextrin carbamates and cellulose derivatives in aqueous solution. Carbohydr. Polym. 2013, 98, 982–987.

[113] Medronho B, Andrade R, Vivod V, Ostlund A, Miguel MG, Lindman B, et al. Cyclodextrin-grafted cellulose: physicochemical characterization. Carbohydr. Polym. 2013, 93, 324–330.

[114] Cova TFGG, Nunes SCC, Pais AACC. Free-energy patterns in inclusion complexes: the relevance of non-included moieties in the stability constants. Phys. Chem. Chem. Phys. 2017, 19, 5209–5221.

[115] Zhu C, Krumm C, Facas GG, Neurock M, Dauenhauer PJ. Energetics of cellulose and cyclodextrin glycosidic bond cleavage. React. Chem. 2017, 2, 201–214.

[116] Krukiewicz K, Zak JK. Biomaterial-based regional chemotherapy: local anticancer drug delivery to enhance chemotherapy and minimize its side-effects. Mater. Sci. Eng. Biol. Appl. 2016, 62, 927–942.

[117] Du J, Guo X, Tu J, Xiao L, Jia X, Liao L, Liu L. Biopolymer-based supramolecular micelles from beta-cyclodextrin and methylcellulose. Carbohydr. Polym. 2012, 90, 569–574.

[118] Wintgens V, Layre AM, Hourdet D, Amiel C. Cyclodextrin polymer nanoassembblies: strategies for stability improvement. Biomacromolecules. 2012, 13, 528–534.

[119] Iohara D, Okubo M, Anraku M, Uramatsu S, Shimamoto T, Uekama K, Hirayama F. Hydrophobically modified polymer/α-cyclodextrin thermoresponsive hydrogels for use in ocular drug delivery. Mol. Pharm. 2017, 14, 2740–2748.

[120] Blanchemain N, Karrout Y, Tabary N, Bria M, Neut C, Hildebrand HF, Siepmann J, Martel B. Comparative study of vascular prostheses coated with polycyclodextrins for controlled ciprofloxacin release. Carbohydr. Polym. 2012, 90, 1695–1703.

[121] Muankaew C, Jansook P, Loftsson T. Evaluation of γ-cyclodextrin effect on permeation of lipophilic drugs: application of cellophane/fused octanol membrane. Pharm. Dev. Technol. 2017, 22, 562–570.

[122] Cova TF, Murtinho D, Pais AACC, Valente AJM. Combining Cellulose and Cyclodextrins: Fascinating Designs for Materials and Pharmaceutics. Front Chem. 2018, 6, 271.

[123] Lin N, Dufresne A. Supramolecular Hydrogels from In Situ Host-Guest Inclusion between Chemically Modified Cellulose Nanocrystals and Cyclodextrin. Biomacromolecules. 2013, 14, 871–880.

[124] Malik N S, Ahmad M, Minhas M U. Cross-linked β-cyclodextrin and carboxymethylcellulose hydrogels for controlled drug delivery of acyclovir. PLoS ONE. 2017, 12, 1–17.

[125] Zhang L, Zhou J, Zhang L. Structure and properties of β-cyclodextrin/cellulose hydrogels prepared in NaOH/urea aqueous solution. Carbohydr. Polym. 2013, 94, 386–393.

[126] Dodov M G, Crcarevska M S, Goracinova K, Trajkovic-Jolevska S. Formulation and evaluation of diazepam hydrogel for rectal administration. Acta Pharm. 2005, 55, 251–261.

[127] Koffia A A, Agnely F, Ponchel G, Grossiordb J L. Modulation of the rheological and mucoadhesive properties of thermosensitive poloxamer-based hydrogels intended for the rectal administration of quinine. Eur. J. Pharm. Sci. 2006, 27, 328–335.

[128] El-Leithy ES, Shaker DS, Ghorab MK, Abdel-Rashid RS. Evaluation of mucoadhesive hydrogels loaded with diclofenac sodium–chitosan microspheres for rectal administration. AAPS PharmSciTech. 2010, 11, 1695–1702.

[129] Cappello B, Iervolino M, Miro A, Chetoni P, Burgalassi S, Saettone MF. Formulation and preliminary in vivo testing of rufloxacin-cyclodextrin ophthalmic solutions. J. Incl. Phenom. Macrocycl. Chem. 2002, 44, 173–176.

[130] Kirange RH, Chaudhari RB. Utilizing mucoadhesive polymers for nasal drug delivery system. Int. J. Pharm. Sci. Res. 2017, 8, 1012–1022.

[131] Zaki NM, Awad GA, Mortada ND, Abd ElHady SS. Enhanced bioavailability of metoclopramide HCl by intranasal administration of a mucoadhesive in situ gel with modulated rheological and mucociliary transport properties. Eur. J. Pharm. Sci. 2007, 32, 296–307.

[132] Vidgren P, Vidgren M, Arppe J, Hakuli T, Laine E, Paronen P. In vitro evaluation of spray-dried mucoadhesive microspheres for nasal administration. Drug Dev. Ind. Pharm. 1992, 18, 581–597.

[133] Ugwoke MI, Kaufmann G, Verbeke N, Kinget R. Intranasal bioavailability of apomorphine from carboxymethylcellulose-based drug delivery systems. Int. J. Pharm. 2000, 202, 125–131.

[134] Ikeda K, Murata K, Kobayashi M, Noda K. Enhancement of bioavailability of dopamine via nasal route in beagle dogs. Chem. Pharm. Bull. 1992, 40, 2155–2158.

[135] El-Gizawy SA, Osman MA, El-Hagaar SM, Hisham DM. Nasal drug delivery of a mucoadhesive oxybutynin chloride gel: in vitro evaluation and in vivo in situ study in experimental rats. J. Drug Deliv. Sci. Technol. 2013, 23, 569–575.

[136] Ozsoy Y, Akev N, Can A, Gerçeker AA. In vivo studies on nasal preparations of ciprofloxacin hydrochloride. Pharmazie. 2000, 55(8), 607–609.

[137] Kouchak M, Handali S. Effects of various penetration enhancers on penetration of aminophylline through shed snake skin. Jundishapur J. Nat. Pharm. Prod. 2014, 9, 24–29.

[138] Nandy BC, Gupta RN, Rai VK, Das S, Tyagi LK, Roy S, Meena KC. Transdermal iontophoretic delivery of atenolol in combination with penetration enhancers: optimization and evaluation on solution and gels. Int. J. Pharm. Sci. Drug Res. 2009, 1, 91–99.

[139] Tavakoli N, Minaiyan M, Heshmatipour M, Musavinasab R. Transdermal iontophoretic delivery of celecoxib from gel formulation. Res. Pharm. Sci. 2015, 10, 419–428.

[140] Bertsch P, Schneider L, Bovone G, Tibbitt MW, Fischer P, Gstöhl S. Injectable biocompatible hydrogels from cellulose nanocrystals for locally targeted sustained drug release. ACS Appl. Mater. Interfaces. 2019, 11, 38578–38585.

[141] Shahin M, Hady SA, Hammad M, Mortada N. Novel jojoba oil-based emulsion gel formulations for clotrimazole delivery. AAPS Pharm. Sci. Tech. 2011, 12, 239–247.

[142] Sabale V, Vora S. Formulation and evaluation of micro-emulsion-based hydrogel for topical delivery. Int. J. Pharm. Invest. 2012, 2, 140–149.

[143] Salerno C, Carlucci AM, Bregni C. Study of in vitro drug release and percutaneous absorption of fluconazole from topical dosage forms. AAPS Pharm. Sci. Tech. 2010, 11, 986–993.

[144] Marcos X, Pérez-Casas S, Llovo J, Concheiro A, Alvarez-Lorenzo C. Poloxamer-hydroxyethyl cellulose-α-cyclodextrin supramolecular gels for sustained release of griseofulvin. Int. J. Pharm. 2016, 500, 11–19.

Diana Ciolacu and Valentin I. Popa

Chapter 11
Nanocelluloses: preparations, properties, and applications in medicine

11.1 Introduction

In the drive toward sustainable development, the growing application of renewable, biodegradable, green, and environmental-friendly materials to substitute nonrenewable resources have roused substantial interest on a global scale [1]. The uses of materials from natural sources play an important role in obtaining sustainable materials due to their reinforcement capacity, biodegradability, and biocompatibility [2].

Cellulose, one of the most abundant natural polymers, has the potential to overcome challenges pertaining to material biodegradability, renewability, cost, and energy. Furthermore, cellulose obtained from wood and also from agricultural waste, crops, and forestry residues provides a renewable resource for the production of new materials due to its availability in large amounts and low cost [3–5]. The ability to control the material features at the nanoscale brings new and promising properties, such as high mechanical characteristics and low density, which provides the opportunity to develop new nanomaterials with various morphologies and various functions.

Nanocelluloses (NCs) have unique features, such as small dimensions, a variety of shapes and enhanced specific surface, high absorption capacity, high thermal stability and high chemical resistance to dilute solutions of acids and alkalis, organic solvents, proteolytic enzymes, antioxidants, and some other therapeutically active substances [6–8]. The diversity of properties together with a huge variety of cellulose sources used as a raw material lead to a vast array of industrial applications of NCs, like manufacturing of paper or packaging products, automobile or aerospace materials, textiles, paints and coatings, and in environmental treatments [9–13]. The use of nanoscale cellulose in composites will allow the production of much lighter weight materials, by replacing metals and plastics, making existing products much more effective and enabling the development of many new products with widespread application [14]. Cellulose nanocrystals (CNCs) have the ability to form a sol–gel and, thus, have been incorporated into polymers, resins, and biopolymers to improve the strength over other composite materials, while nanofibrillated cellulose (NFCs) are an ideal component in medical films and gels, largely because of its ability to absorb water and act as a prophylactic barrier [15]. Moreover, these specific features open

Diana Ciolacu, "Petru Poni" Institute of Macromolecular Chemistry, Iasi, Romania
Valentin I. Popa, "Gheorghe Asachi" Technical University, Iasi, Romania

https://doi.org/10.1515/9783110658842-011

new promising application areas, particularly in various branches of care and cure such as hygiene, cosmetics, pharmaceutics, medicine, and so on [16–19].

This chapter provides an overview of the recent advances in the preparation of nanoparticles obtained from cellulose and highlights the development of various NC-based materials. A detailed summary is given on various applications of NCs, including past achievements and the potential future directions, with emphasis on the use of NCs in medical applications.

11.2 Cellulose: structural aspects

Cellulose is a polydisperse linear homopolymer consisting of β-1,4-glycosidic linked D-glucopyranose units. The structural units are joined by single oxygen atoms between the C-1 of one pyranose ring and the C-4 of the next pyranose ring [20]. The glucose units in the cellulose polymer are referred to as *anhydroglucopyranose units* (AGU) and every second AGU ring is rotated with 180° in the plane. In this manner, two adjacent structural units are arranged to form a so-called cellobiose unit (Fig. 11.1a).

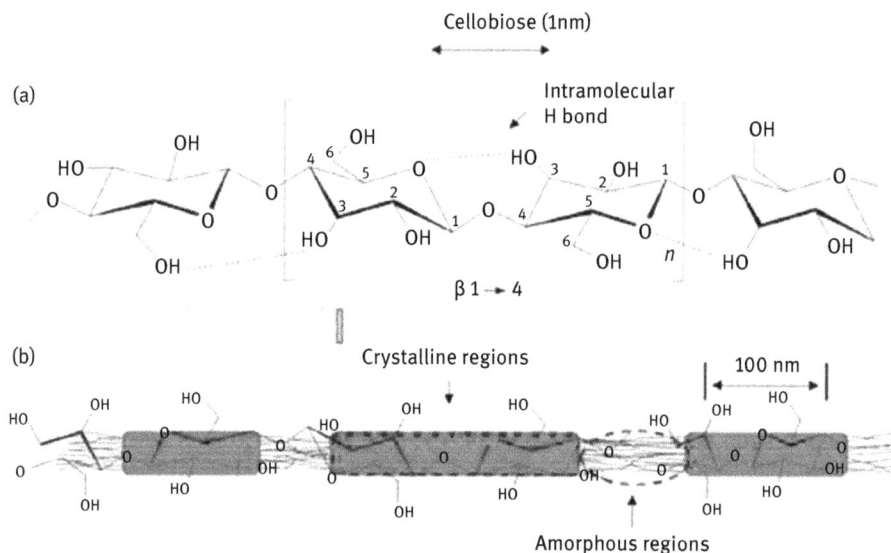

Fig. 11.1: (a) Molecular structure of cellulose; (b) configuration of crystalline and amorphous regions in cellulose. Reproduced with permission from Reference [20].

The cellulose chain consists at one end of a D-glucopyranose unit in which the anomeric carbon atom is involved in a glycosidic linkage (the nonreducing end), whereas the other end has a D-glucopyranose unit in which the anomeric carbon atom is free and is in equilibrium with an aldehyde structure (the reducing end). Thus, the cellulose chain has a chemical polarity. Determination of the relative orientation of cellulose chains in the three-dimensional structure has been and remains one of the major problems in the study of cellulose [21].

One of the most specific characteristics of cellulose is that each of its monomers bears three hydroxyl groups. These hydroxyl groups and their hydrogen bonding ability play a major role in directing crystalline packing and in governing important physical properties of these highly cohesive materials [22]. The intra- and intermolecular hydrogen bonds permit organization of the chains in highly ordered, crystalline structures. The chains are usually longer than crystalline regions, and pass through several different crystalline regions, with areas of disorder (amorphous structure) between them (Fig. 11.1b) [20].

The chain length of cellulose expressed in the number of constituents AGUs (degree of polymerization, DP) varies with the origin and treatment of the raw material. In the case of wood pulp, the values are typically 300 and 1,700. Cotton and other plant fibers have DP values in the 800–10,000 range depending on treatment; similar DP values are observed in bacterial cellulose. Regenerated fibers from cellulose contain 250–500 repeating units per chain, while by partial chain degradation yields powdery cellulose of microcrystalline cellulose (MCC) with a DP values between 150 and 300 [21].

Cellulose exists in several crystal modifications, differing in unit cell dimensions and chain polarity, such as cellulose I, II, III, and IV (Fig. 11.2).

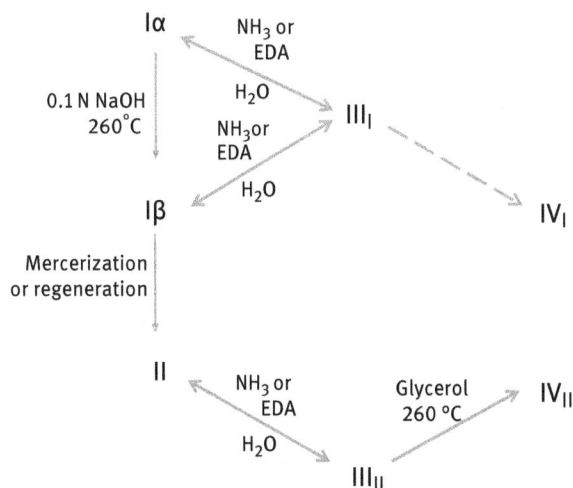

Fig. 11.2: The relationships between various cellulose allomorphs.

A major discovery due to cross-polarization magic angle spinning (CP-MAS) studies has shown that cellulose I (native cellulose, CI) is a mixture consisting of two distinct crystalline forms: cellulose Iα and Iβ, which can be found in different ratios, as a function of the cellulose origin [23].

CI can be converted to cellulose II (CII) by regeneration or mercerization processes. The transition from CI to CII is irreversible and this implies that CII is thermodynamically more stable in comparison with CI. By using X-ray and neutron fiber diffraction analysis, it was reported that CI and CII are characterized by a substantial difference in cellulose chain polarity in unit cells, parallel for CI (with their reducing ends always on the same side within the crystalline domains), and antiparallel for CII (with their reducing ends alternatively on either side within the crystalline domains), although arguments about this problem are still going on in the field of cellulose science [24, 25].

Cellulose III (CIII) allomorph is prepared by treating CI or CII with liquid ammonia or with certain organic amines, such as ethylenediamine (EDA), followed by washing with alcohol, when is obtained CIII$_I$ or CIII$_{II}$, respectively. These transformations are reversible, suggesting that the chain orientation is similar to the one of the starting materials. However, at the crystalline level, an extensive decrystallization and fragmentation of the cellulose crystals were observed during the conversion of CI to CIII$_I$ [22, 26, 27]. During the reverse transition to CI partial recrystallization takes place, but the distortion and fragmentation of the crystals are irreversible and restoration of the damage done to the morphological surface is incomplete.

Cellulose IV (CIV) is obtained by treating CIII at high temperature in glycerol. For this allomorph was also reported the obtaining of CIV$_I$ and CIV$_{II}$ starting from CI or CII via CIII$_I$ and CIII$_{II}$, respectively. An explanation of the necessity to convert CI to CIV$_I$ through CIII$_I$ is the fibrillation of cellulose microfibrils, which makes them more accessible for further transformation to CIV$_I$ [26, 28]. However, the fibrillation makes CIV$_I$ less suitable for crystallographic analysis and thus it makes it more difficult to interpret CIV$_I$ as a crystal. For these reasons, the issue about CIV$_I$, whether it is a crystal with an orthogonal unit cell or a less crystalline form of CI, is still debated. Based on X-ray diffraction method and solid-state ^{13}C NMR spectroscopy, it was suggested that CIV$_I$ is not a distinct polymorph, but is cellulose Iβ with various degrees of lateral disorder, which was obtained by fragmentation of crystallites during preparation of CIII and subsequent of thermal treatment [29].

The crystalline allomorphs of cellulose differ from each other, also by the shapes of the crystalline unit cells. The projection of the CI-unit cell has a parallelepiped shape, while CIV-unit cell has a square shape and unit cells of CII and CIII have a rhombic shape [16]. Regarding the cross-sectional shape of NC crystallites, these were depicted as a square or rectangle and in recent studies have shown that the most likely cross-sectional shape is a hexagon [30, 31].

11.3 Nanocelluloses: preparation and characterization

Recently, TAPPI (Technical Association of the Pulp and Paper Industry) has proposed to standard the terms for cellulose nanomaterials, due to the unclear nomenclature used to designate the crystalline cellulose nanoparticles (such as rod-like colloidal particles, nanocrystalline cellulose (NCC), cellulose whiskers, cellulose microcrystallites, microcrystals, etc.) [32–35].

Thus, cellulose nanomaterials (NCs) can be classified into two major subcategories based on their shape, dimension, function, and preparation method, which in turn primarily depend on the cellulosic source and processing conditions: (i) *cellulose nano-objects* and (ii) *cellulose nanostructured* materials (Fig. 11.3) [35, 36].

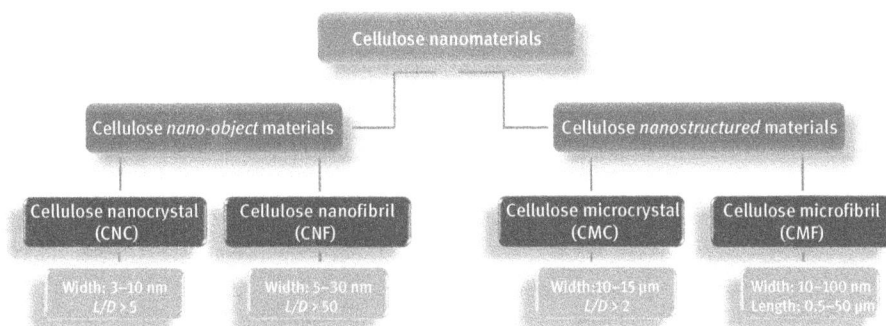

Fig. 11.3: Standard terms for cellulose nanomaterials.

The two subcategories can be divided into specific subgroups:
- cellulose nano-objects materials: (i) CNC and (ii) *cellulose nanofibril* (CNF),
- cellulose nanostructured materials: (i) *cellulose microcrystal* (CMC) and (ii) *cellulose microfibril* (CMF).

The main difference between cellulose nano-objects (10–100 μm) and cellulose nanostructured (1–50 nm) is the size of the cellulose nanomaterial, data summarized in Tab. 11.1 [37].

CMC is a commercially available cellulose material (named *microcrystalline cellulose* – MCC), obtained by hydrolysis of cellulose with mineral acid, a multisided aggregate of spherical or rodlike particles with dimensions of 10–200 μm, while CMF is produced by intensive mechanical refinement of cellulose pulp and consist of multiple aggregates of elementary nanofibrils with a width of 20–100 nm and a length of 500–2,000 nm.

Tab. 11.1: The dimensions of nanocelluloses obtained by different methods from various sources.

Cellulose nanofibers	Example	Dimensions		Reference
		Diameter, D	Length, L	
Cellulose nanocrystals	Microcrystalline cellulose	10 nm	200–400 nm	[38]
	Soft wood pulp	5–15 nm	100–250 nm	[39]
		4–5 nm	100–150 nm	[40, 41]
	Hardwood pulp	3–5 nm	100–300 nm	[30, 43, 44]
		4–5 nm	140–150 nm	[40, 41]
		19 ± 5 nm	151 ± 39 nm	[45]
	Bacterial cellulose	10–100 nm	0.1–1 μm	[46]
		5–50 nm	100 nm–several μm	[47]
	Cotton	7.3 nm,	135 nm	[48]
		10 nm	100–150 nm	[47]
	Tunicate	19 nm	201 nm	[49]
		15–30 nm	1–1.5 μm	[30, 43, 44]
		16 nm	1,160 nm	[41, 50]
	Ramie	6–8	150–250	[41, 51]
		5–10 nm	50–150 nm	[52]
	Sisal	3–5	100–500	[41]
	Valonia	10–20 nm	1,000 nm	[41]
	Kenaf bast	10–15 nm	hundreds nm	[53]
	Grass	15 ± 3 nm	120 ± 15 nm	[54]
Cellulose nanofibrils	Softwood pulp	3–5 nm	several μm	[38, 55]
	Hardwood pulp	20 ± 14 nm	1,030 ± 334 nm	[45]
		14 ± 0.7 nm	597 nm	[56]
	Wheat straw	30–40 nm	Several μm	[57]
	Flax bast	5–80 nm	Several μm	[58]
	Soybean stock	50–100 nm	Several μm	[59]
	Tunicate	54.24 nm	6.85 μm	[49]
	Alfa fibers	1.4–4.6 nm	Several μm	[42]
	Bacterial cellulose	70–140 nm	Several μm	[60]
		80–170 nm	Several μm	[61]

Cellulose nanofibers may be classified in two main classes, CNCs and CNFs, as has been recommended by TAPPI [1, 32, 62]. The terminology of CNCs goes by a few terms, such as NCC, nanowhisker (CNW), and nanorods, while for CNFs have been used different interchangeably terms, such as NFC, nanofibrillar cellulose, nanofibrous cellulose, and bacterial nanocellulose (BNC) [37, 63]. CNFs can be distinguished through their structure that is composed of stretched masses of elementary nanofibrils with alternating crystalline and amorphous domains. In contrast to CNF, CNC is less flexible due to its higher crystallinity and to an elongated rodlike shape made up of crystalline regions isolated from CNFs.

The variation of the lignocellulosic source, as well as the influence of the extraction process, bring to NC differences in the particle size and shape, crystal structure, morphology, crystallinity, and properties [63]. For example, CNCs extracted from tunicates and green algae have crystal lengths in the range of a few micrometers, while the crystallites from wood and cotton have lengths of the order of a few hundred nanometers [43].

The main extraction processes in the preparation of NCs are acid hydrolysis and mechanical treatments. Acid hydrolysis can be controlled by reaction temperature, agitation and time, as well as the nature of the acid, acid-to-pulp ratio, and cellulose source. Mechanical processes can be divided into high-pressure homogenization and refining, microfluidization, grinding, cryo-crushing, and high-intensity ultrasonication.

11.4 Cellulose nanocrystals (CNCs)

CNCs, also known as *cellulose whiskers*, are obtained by a well-known isolation process – acid hydrolysis. This procedure consists of an acidic attack through transverse hydrolysis along the amorphous regions of fibrils, followed by a longitudinal cutting of the microfibrils, thus leaving highly crystalline domains [64]. Compared to cellulose fibers, CNC possesses many advantages, such as nanoscale dimension, high specific strength and modulus, high surface area, and unique optical properties, and therefore has wide possibility of applications – a fact that has significantly attracted interest from both research scientists and industrialists [43].

The physical dimension of CNC, designated by length (L), diameter (D), and aspect ratio (L/D), is greatly dependent to the cellulosic source and preparation condition (Fig. 11.4). It has been reported that nanocrystalline particles extracted from tunicates and BC are usually larger when compared to CNCs obtained from wood or cotton. This is because tunicates and BC are highly crystalline and contain longer nanocrystallites [37]. Generally, CNCs obtained from cotton have a lower aspect ratio (L/D within 10–30) in comparison with an L/D around 70, corresponding to CNCs obtained from tunicates. The dimensions of CNCs obtained from wood, cotton, and tunicates are L/D = 100–200 nm/3–5 nm, L/D = 100–300 nm/5–10 nm,

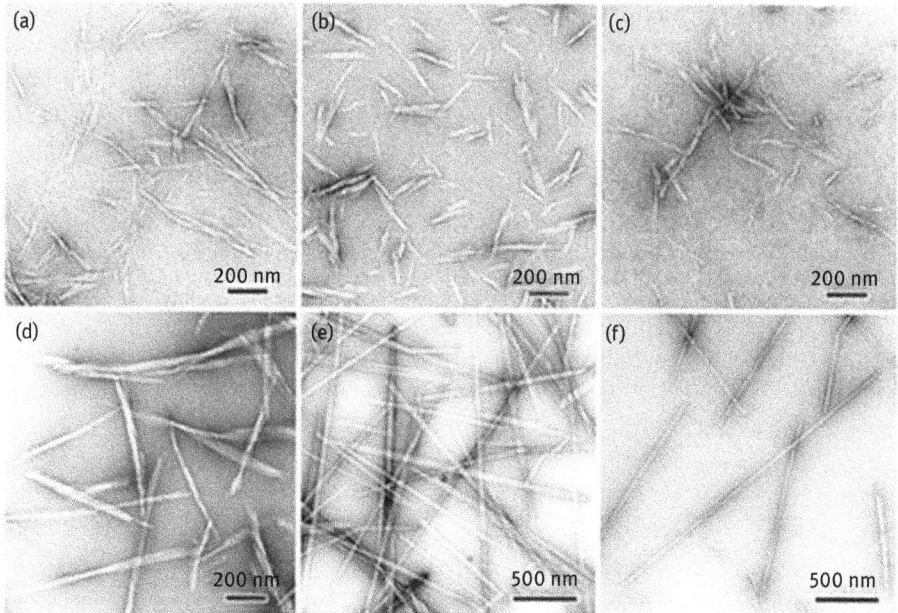

Fig. 11.4: TEM images of negatively stained preparations of CNCs of various origins: (a) wood; (b) cotton; (c) bamboo; (d) *Gluconacetobacter Xylinus*; (e) *Glaucocystis*; and (f) *Halocynthia papillosa*. Reproduced with permission from Reference [65].

and L/D = 500–2,000 nm/10–20 nm, respectively [64–66]. CNC crystals may also show different geometries, depending on their biological source; for example, algal cellulose membrane displays a rectangular structural arrangement, whereas both bacterial and tunicate cellulose chains have twisted-ribbon geometry [30].

The preparation of CNC depends on different factors, such as the concentration of cellulose, the type and concentration of the strong acid, the reaction time and temperature, and the time of ultrasonic treatment.

It was established that the reaction time is one of the most crucial factors of the acid hydrolysis process because a sufficient reaction time is required to break down the hierarchical structure of the fiber into a nanocrystalline structure [67]. In addition, if the hydrolysis is inadequate, incomplete removal of the amorphous regions can occur resulting in decreased crystallinity and a change in particle morphology. Likewise, by increasing the severity of the hydrolysis (increase reaction time), it is possible to depolymerize the crystalline cellulose and the length of the particles will decrease, which reduces the aspect ratio and can even result in spherical particles [68].

Different acids have been shown to successfully degrade cellulose fibers, such as phosphoric [69], hydrobromic [70], and nitric acids [71], but sulfuric [40] and hydrochloric acids [72] have been extensively used.

One of the main reasons for using sulfuric acid as hydrolyzing agent is its reaction with the hydroxyl groups from surface, via an esterification process, allowing the grafting of anionic sulfate ester groups. The presence of these negatively charged groups induces the formation of a negative electrostatic layer covering the nanocrystals and promotes their dispersion in water [73]. CNCs show some dispersibility in aqueous-based mixtures and in organic solvents with high dielectric constants, such as dimethyl sulfoxide (DMSO) and ethylene glycol, but tend to aggregate in highly hydrophobic solutions [74]. Several chemical modifications have been applied to CNCs surfaces in order to improve their stability in organic media or to make them compatible with hydrophobic matrices. These can be classified into three distinctive groups: (i) substitution of hydroxyl groups with small molecules, (ii) polymer grafting based on the "grafting onto" strategy with different coupling agents, and (iii) polymer grafting based on the "grafting from" approach with a radical polymerization involving ring opening polymerization, atom transfer radical polymerization, and single-electron transfer living radical polymerization [2, 73].

The main challenge with chemical modification is to choose a reagent and reaction medium that enable modification of the nanocrystal surface without the nanocrystal dissolving in the reaction medium and without undesired bulk changes. An alternative to chemical surface modification is the adsorption of surfactants at the colloid surface to improve nanoparticle stability in organic solvents [74].

11.5 Cellulose nanofibrils

CNFs, the so-called NFC, consists of a stretched bundle of cellulose nanofibers, with long, flexible, and entangled chains, which are formed as a result of cellulose chain-stacking, induced by hydrogen bonds [75].

The isolation process consists in fibrillation of woody or nonwoody plants raw materials through mechanical grinding. Since an energy-intensive mechanical grinding is required to separate the stacked fibrils, different types of pretreatments, such as enzymatic, chemical, alkaline, or radiation, are typically applied prior to the fibrillation process, in order to remarkably lower the cost and energy [20].

The mechanical treatments applied for the disintegration of cellulosic materials are high-pressure homogenizers, microfluidizers or grinders, cryo-crushing, and high-intensity ultrasonic treatments. High-pressure homogenization can be considered as an efficient method for refining of cellulosic fibers, because of its high efficiency, simplicity, and no need for organic solvents [76, 77]. Internal fibrillation happens as a result of breaking the hydrogen bonds due to mechanical actions, while external fibrillation happens on the surface of fiber by abrasive actions (Fig. 11.5).

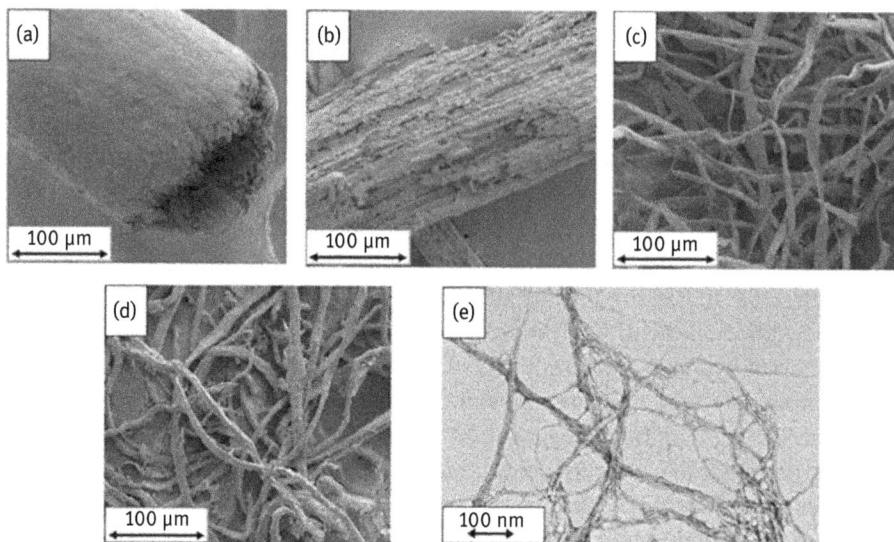

Fig. 11.5: FESEM micrograph of: (a) raw sugar palm fiber; (b) sugar palm bleached fiber (SPBF); (c) sugar palm cellulose, (d) PFI-refined fiber; and TEM nanograph of (e) sugar palm nanofibrillated cellulose (SPNFCs). Reproduced with permission from Reference [77].

All these mechanical methods involve high consumption of energy, which can cause a significant decrease in both the yield and fibril length. Thus, current research has been focused on finding environmental conservation, high efficiency, and low costs methods to isolate CNFs [75]. The combination of mechanical treatments with the pretreatments presented above, to either remove amorphous area or chemically functionalize the particle surface and to endow it with new properties, have brought positive results in this regard [38, 66].

Enzymatic hydrolysis, an environmental-friendly pretreatment, has the effect to reduce the number of passes through the homogenizer, by using commercial cellulase, which determines an enzymatic breakage of the cellulose networks, promoting the cell wall delamination and releasing the CNF [78]. A significantly lower environmental impact and low-energy consumption were recorded by using steam explosion process, which consist in a short time vapor phase cooking, at temperatures between 180 and 210 °C, followed by explosive decompression and sudden release of pressure, during which the flash evaporation of water exerts a thermomechanical force causing a substantial breakdown of lignocellulosic structure, hydrolysis of the hemicelluloses, depolymerization of the lignin, and defibrillation [79].

Surface carboxylation by 2,2,6,6-tetramethylpiperidine-1-oxyl radical (TEMPO)-mediated oxidation is a chemical pretreatment, which involves the addition of negatively charged entities at the microfibrils surface, facilitating the isolation of nanofibers and significantly decreasing the energy consumption [80, 81].

Alkaline-acid pretreatment is used before mechanical isolation of CNF in order to solubilize the lignin, hemicelluloses, and pectins [57, 75]. Other chemical pretreatment is the carboxymethylation process, which makes the surfaces negatively charged, promotes formation of stable suspension from carboxymethylated fibers, and increases the breakup of lignocellulosic fibers to nanosize [82].

NFCs display two main drawbacks, which are associated with its intrinsic physical properties. The first one is the high number of hydroxyl groups, which lead to strong hydrogen interactions between two nanofibrils and to the gellike structure once produced. The second drawback is the high hydrophilicity of this material, which limits its uses in several applications such as in paper coating (increase of dewatering effect) or composites (tendency to form agglomerates in petrochemical polymers). The most feasible solution to this is chemical surface modification to reduce the number of hydroxyl interactions and also to increase the compatibility with several matrices [2, 43, 83].

11.6 Applications of nanocelluloses

The current trend to develop new materials from renewable sources, as well as the versatility of NC-based materials, have opened up doors for a wide possibility of applications.

The use of NCs as reinforcing agent in nanocomposites presents great *advantages* such as renewable nature, biodegradability, high specific surface area, low density, wide availability of sources, low energy consumption, relatively reactive surface, and relatively easy processability due to their nonabrasive nature, which allows high filling levels; it also has some major *disadvantages*, as difficult compatibilization of highly hydrophilic NCs with most of polymeric matrices, high moisture absorption, and limitation of processing temperature [2, 84]. The strategies for improving the stability of cellulose nanoparticle dispersions in nonpolar or low polarity solvents are as follows: (i) physically coating of the surface with a surfactant or (ii) chemically grafting nonpolar moieties onto the surface. The chemical modification of NCs improves its dispersibility in organic solvents and thus greatly enlarges its potential applications in different areas [85].

In the recent years, cellulose nanocomposites have been extensively used in various applications, such paper and packaging products, light-responsive composites and other electronic devices, ultrafiltration membranes, pharmaceutical and medical products, as well as in construction (as structural composites), automotive (as parts based on micropatterns), and furniture (Fig. 11.6) [17, 86, 87].

CNF has been used in paper coatings for the formation of a continuous layer on the paper surfaces in order to induce suitable characteristics, such as improved mechanical strength and water/gas impermeability, or as a green binder not only

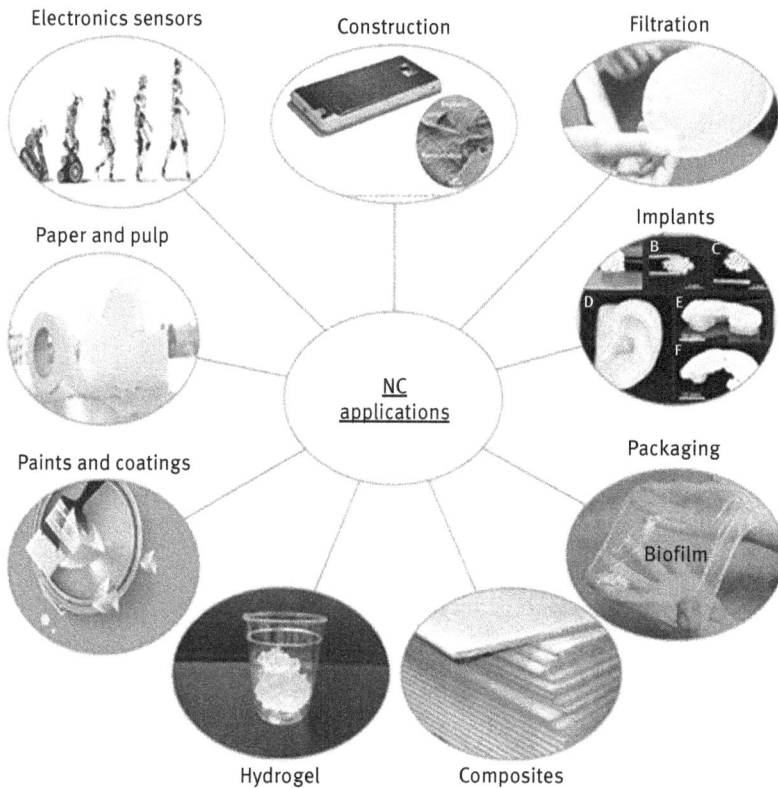

Fig. 11. 6: Potential applications of nanocellulose in various fields. Reproduced with permission from Reference [17].

to improve the properties of coating layers on paper/paperboard but also to reduce the level of latex [88–91]. Furthermore, CNF has a great potential for utilization as a binder in energy storage devices (e.g., electrodes in Li-ion batteries), alternative to petroleum-based polymers [92], but one of the most promising binder applications is in the production of particleboards and medium density fiberboards, as a replacement for urea-formaldehyde (UF) resin [20]. Cellulose nanomaterials can be a potential candidate in electronics industry, used to improve the conductivity and flexibility of materials, such as paper-based sensors, flexible electrode, electronic device, and conducting adhesives [93–95]. Moreover, cellulose nanofibers can be used for paint coatings and adhesives due to their three-dimensional network with high shear tolerance properties [96–98].

11.7 Nanocelluoses for biomedical applications

NCs are characterized by crucial properties for biomedical applications, such as biocompatibility, nontoxicity, tissue-friendliness, wound-healing properties, antimicrobial effects, as well as high binding potential through the available OHs and negative interfacial charges that facilitate the electrostatic adsorption on tissues. These properties make NCs adequate for different medical applications: wound dressing, cartilage/bone regeneration, dental repairs, and cancer curing drugs [20].

NCs have been used in *wound-dressing* applications due to its anti-infection features and its ability to increase tensile properties of the scaffolds. CNFs were tested for wound dressing in severely burned patients, where a successful grafting to the skin donor sites upon the cellulose dehydration was reported, and which could be easily attached to the wound bed and remain on the site until the donor site was renewed with no allergic reaction or inflammatory response [92, 99]. When an electrical charge is introduced into NC, it can be functionalized with various biomolecules, for example, cell adhesion peptides [100] and silk fibroin [101], which improve the capacity of NC for colonization with cells and for wound healing [102]. Coated CNF threads with stem cells created sutures with positive influence on postsurgery inflammation and wound-healing properties [55]. Positive effects regarding the adhesion and proliferation of human dermal fibroblasts were recorded when different NC solutions covered individual cellulose microfibers of a cellulose mesh and filled the wide spaces between them [103]. Chemical modifications of NC for further usage in wound-healing applications include converting it to cellulose acetate or to hydroxyethyl cellulose, fact which improve the electrospinnability of cellulose. The formulations of NCs and silver nanoparticles can be used as microbial medicaments, antibacterial agents in wound dressing, bandages, implants, skins replacements for burnings, face masks, artificial blood vessels, cuffs for nerve surgery, drug delivery, cell carriers and support matrices for enzyme immobilization, cosmetic tissues, and so on [2, 41].

NCs have been widely used in *tissue engineering* due to their capability to reinforce polymer matrices and promote cell proliferation, such as injectable tissue scaffolds, bone tissue regeneration, improving bone implant adhesion, artificial ligaments or tendon substitutes, blood vessels, neural tissue, vascular grafts, liver and adipose tissue, congenital heart defects, ophthalmologic applications, mainly construction of contact lenses, and so on [104, 105].

CNCs have low extension to break, high aspect ratios, high surface areas, high crystallinity, and apparent biocompatibility [106, 107]. Thus, CNCs can reinforce polymer matrices, such as alginate/gelatin, poly(lactic acid), silk fibroin, in conjunction with hydroxyapatite, or in electrospinning to produce strong fibers [20]. CNCs were used to reinforce collagen films, which supported the viability of mouse embryonic 3T3 fibroblasts and thereby had promising results for skin tissue engineering [106]. In another study, cotton-derived CNCs were electrospun together with poly(lactic-*co*-glycolic acid). The resulting scaffolds improved the adhesion, spreading, and proliferation of 3T3 fibroblasts in comparison with neat PLGA nanofiber membranes [108]. One of the new approaches for repair or regenerate the cartilage is three-dimensional bioprinting that needs a proper bioink and specific viscoelastic features. Recent studies on cartilage regeneration proposed a CNF-based bioink in which CNF and alginate were used to help bioprinting coupled with human bone marrow-derived stem cells and human nasal chondrocytes [109]. A new study used hydrogel scaffold based on gelatin-CNF and β-tricalcium phosphate displaying both osteoconductive and osteoinductive characteristics, in the search for the optimal bone substitute [110, 111]. The obtained data proved that the CNF reduced the degradation rate of the hydrogel matrix, which in turn favored the gradual and constant release of the simvastatin entrapped in the scaffold.

Even if materials based on NC are often considered without or with low cytotoxicity and immunogenicity, several studies demonstrated a considerable cytotoxicity and pro-inflammatory activity of NC in vitro and in vivo. It was established that the type of NCs has a great influence on them cytotoxicity. Thus, there are examples of nonimmunogenic NCs, like cotton-derived CNCs [48] or cellulose nanofibers isolated from Curauá leaf fibers (*Ananas erectifolius*) [112]. However, some types of wood-derived CNCs can induce a pulmonary inflammatory response in mice after aspiration, manifested by an increase in leukocytes and eosinophils in the lungs, and also induced oxidative stress and tissue damage [102, 113, 114]. The morphology of cellulose nanoparticles can also influence their cytotoxicity and immunogenicity. CNF showed more pronounced cytotoxicity and oxidative stress responses in human lung epithelial A549 cells than CNC. However, exposure to CNC caused an inflammatory response with significantly elevated proinflammatory cytokines and chemokines compared to CNF. Interestingly, cellulose staining indicated that CNC particles, but not CNF particles, were taken up by the cells [115]. In vivo experiments performed in mice also confirmed different immune responses to CNF and to CNC [102]. Another possibility to reduce cytotoxicity of NCs is their functionalization with specific chemical groups. Thus, the modification of CNF with carboxymethyl groups (anionic NC) or with hydroxypropyl trimethylammonium groups (cationic NC) leads to a lower proinflammatory effect in human dermal fibroblasts, in lung MRC-5 cells and in human macrophage-like THP-1 cells [116].

In *drug delivery*, the development of "smart" responsive NCs-based aerogels with favorable floatability and mucoadhesive properties, which can respond to changes in their surrounding environments, was provided for therapeutic applications [117]. Another hot topic is the application of CNFs for drug delivery in the form of tablet coating, membranes, bionanocomposite delivery systems [118], or films for sustained parenteral delivery of drugs, for example, analgesics, antiphlogistics, corticoids, and antihypertensives [119, 120]. In combination with silver nanoclusters, NFC was designed as a novel composite with fluorescence and antibacterial activity for potential wound dressings [121]. In combination with polypyrrole, NC was proposed for constructing hemodialysis membranes [122]. On the other hand, the fluorescent labeling of cellulosic nanofibers with various fluorophores is also of potential interest in bio-imaging, targeting and sensing uses [123, 124].

The applications of BNCs are strongly expanding fields of personal care and medicine due to the specific structure and properties, such as purity, high water retention capacity, shapeability, and biocompatibility. The development of implants ranges from the design of materials for bone and cartilage repair to the development of tubular prototypes as grafts for vascular surgery. For the repair of congenital heart defects, BNC was used as a new patch material for closing ventricular septal defects in a pig model. This material could serve as an alternative to materials currently used in clinical practice, namely, polyester, expanded polytetrafluoroethylene, and autologous or bovine pericardium, which are often associated with compliance mismatch and with a chronic inflammatory response [125]. Hydrogels based on BNC mimic basic living processes and are of growing importance as bioactive scaffolds. In all cases, BNC is active as a three-dimensional template for in vitro and in vivo tissue growth [2, 126].

11.8 Conclusions

This chapter describes some of the most recent advances regarding NCs, which represent a natural resource of green and sustainable materials with importance for the future of nanotechnologies. NCs present unique features, such as small dimensions, a variety of shapes and enhanced specific surface, high sorption ability, high thermal stability, and high resistance to dilute solutions of acids and good biocompatibility. These specific features open new and promising application areas, particularly in biomedical area, including tissue engineering and wound healing. Despite the great potential, the widespread use of various kinds of NCs is limited due to lack of information on their biological safety.

Even if NC materials are often considered without or with low cytotoxicity and immunogenicity, several studies demonstrated a considerable cytotoxicity and proinflammatory activity of NC in vitro and in vivo. It was demonstrated that the cytotoxicity and immunogenicity of NC can be modulated by the type of NCs (wood, cotton, hemp, flax, wheat straw, ramie, algae, tunicin, etc.), the morphology (NCF or CNC), or the physicochemical properties (by chemical modifications with specific chemical groups). The nondegradability of NCs in the human organism is another limitation for its direct use in tissue engineering. Degradability of cellulose can be induced by incorporating cellulase enzymes in conjunction with β-glucosidase, or partially, by making of oxidized NCs.

Certainly, NCs have a great potential for the development of a new generation of biomedical materials and the acceptance of NCs as commercially materials for biomedical application will be achieved by rigorously controlling them properties and by implementing of reliable and reproducible production techniques.

List of abbreviations

AGU	Anhydroglucopyranose unit
BC	Bacterial cellulose
BNC	Bacterial nanocellulose
CI	Cellulose I
CII	Cellulose II
CIII	Cellulose III
CIV	Cellulose IV
CMC	Cellulose microcrystal
CMF	Cellulose microfibril
CNC	Cellulose nanocrystal
CNF	Cellulose nanofibril
CNW	Cellulose nanowhisker
DMSO	Dimethyl sulfoxide
DP	Degree of polymerization
NC	Nanocelluloses
NCC	Nanocrystalline cellulose
NFC	Nanofibrillated cellulose
TEMPO	2,2,6,6-Tetramethylpiperidine-1-oxyl radical

Acknowledgments: This work was supported by a grant of the Romanian Ministry of Research and Innovation, CCCDI – UEFISCDI, project number PN-III-P1-1.2-PCCDI-2017-0697/13PCCDI/2018, within PNCDI III.

References

[1]   Shak KPY, Pang YL, Mah SK. Nanocellulose: Recent advances and its prospects in environmental remediation. Beilstein J. Nanotechnol. 2018, 9, 2479–2498.

[2]   Ciolacu D, Darie RN. Nanocomposites Based On Cellulose, Hemicelluloses, And Lignin. In: Visakh PM, Morlanes MJM., eds. Nanomaterials and Nanocomposites: Zero- to three-dimensional materials and their composites. Wiley-VCH Verlag GmbH, 2016, chap 11. 391–423.

[3]   Loow YL, New EK, Yang GH, Ang LY, Foo LYW, Wu TY. Potential use of deep eutectic solvents to facilitate lignocellulosic biomass utilization and conversion. Cellulose. 2017, 24, 3591–3618.

[4]   Gnanasekaran D. Green Biopolymers and its Nanocomposites in Various Applications: State of the Art. In: Gnanasekaran D., ed. Green biopolymers and their nanocomposites. chap 1. Springer, Singapore, 2019, 1–27.

[5]   Dharmalingam K, Anandalakshmi R. Polysaccharide-based Films for Food Packaging Applications. In: Katiyar V, Gupta R, Ghosh T., eds. Advances in sustainable polymers. chap 9. Springer, Singapore, 2019, 183–207.

[6]   Ioelovich M. Characterization of Various Kinds of Nanocellulose. In: Kargarzadeh H, Ahmad I, Thomas S, Dufresne A., eds.. Handbook of nanocellulose and cellulose nanocomposites, 2017, 1, chap 2. 51–100.

[7]   Wang J, Liu X, Jin T, He H, Liu L. Preparation of nanocellulose and its potential in reinforced composites: A review. J. Biomater. Sci. Polym. Ed. 2019, 30(11), 919–946.

[8]   Islam MT, Alam MM, Patrucco A, Montarsolo A, Zoccola M. Preparation of nanocellulose: A Review. AATCC J. Res. 2014, 1(15), 17–23.

[9]   Boufi S, González I, Delgado-Aguilar M, Tarrès Q, Pèlach MÀ, Mutje P. Nanofibrillated cellulose as an additive in papermaking process: A review. Carbohyd. Polym. 2016, 154, 151–166.

[10]   Li F, Mascheroni E, Piergiovanni L. The potential of nanocellulose in the packaging field: A review. Packag. Technol. Sci. 2015, 28, 475–508.

[11]   Mahfoudhi N, Boufi S. Nanocellulose as a novel nanostructured adsorbent for environmental remediation: A review. Cellulose. 2017, 24, 1171–1197.

[12]   Vilarinho F, Sanches Silva A, Vaz MF, Farinha JP. Nanocellulose in green food packaging. Crit. Rev. Food Sci. Nutr. 2018, 58, 1526–1537.

[13]   Balea A, Fuente E, Blanco A, Negro C. Nanocelluloses: Natural-based materials for fiber reinforced cement composites. A critical review. Polymers-Basel. 2019, 11(518), 1–32.

[14]   Shatkin JA, Wegner TH, Bilek EM, Cowie J. Market projections of cellulose nanomaterial-enabled products. Part 1. Appl. Tappi J. 2014, 13(5), 9–16.

[15]   Anderson SR, Esposito D, Gillette W, Zhu JY, Baxa U, Mcneil SE. Enzymatic preparation of nanocrystalline and microcrystalline cellulose. Tappi J. 2014, 13(5), 35–42.

[16]   Ioelovich M. Nanocellulose – Fabrication, Structure, Properties, and Application in the Area of Care and Cure. In: Grumezescu A.., ed. Fabrication and self-assembly of nanobiomaterials. Elsevier, 2016, chap 9. 243–288.

[17]   Sharma A, Thakur M, Bhattacharya M, Mandal T, Goswami S. Commercial application of cellulose nano-composites – A review. Biotechnol. Rep. (Amst). 2019, 21(e00316), 1–15.

[18]   Curvello R, Raghuwanshi VS, Garnier G. Engineering nanocellulose hydrogels for biomedical applications. Adv. Colloid Interface Sci. 2019, 267, 47–61.

[19]   Zhang Y, Chang PR, Ma X, Lin N, Huang J. Strategies to Explore Biomedical Application. In: Huang J, Dufresne A, Lin N., eds. Nanocellulose: From fundamentals to advanced materials, Wiley-VCH Verlag GmbH, 2019 chap 11, 349–395.

[20]   Tayeb AH, Amini E, Ghasemi S, Tajvidi M. Cellulose nanomaterials – binding properties and applications: A review. Molecules. 2684 2018, 23(10), 1–24.

[21]   Ciolacu D, Popa VI. Molecular structure of cellulose. In: Cellulose allomorphs: Structure, accessibility and reactivity. Nova Science Publisher, New York, 2011, chap1. 1–5.

[22]   Dufresne A. Cellulose and Potential Reinforcement. In: Nanocellulose: From nature to high performance tailored materials. Walter de Gruyter, Berlin, Germany, 2012, 1–46.

[23]   Atalla RH, VanderHart DL. Native cellulose: A composite of two distinct crystalline forms. Science. 1984, 223, 283–285.

[24]   Kim NH, Imai T, Wada M, Sugiyama J. Molecular directionality in cellulose polymorphs. Biomacromolecules. 2006, 7, 274–280.

[25]   Ciolacu D, Popa VI. Cellulose Allomorphs – Overview and Perspectives. In: Lejeune A, Deprez T., eds. Cellulose: Structure and property, derivatives and industrial uses. Nova Science Publishers, New York, 2010, chap 1. 1–38.

[26]   Roche E, Chanzy H. Electron microscopy study of the transformation of cellulose I into cellulose IIII in Valonia. Int. J. Biol. Macromol. 1981, 3, 201–206.

[27]   Reis D, Vian B, Chanzy H, Roland JC. Liquid crystal-type assembly of native cellulose-glucuronoxylans extracted from plant cell wall. Biol. Cell. 1991, 73, 173–178.

[28]   Chanzy H, Henrissat B, Vincendon M, Tanner SF, Belton PS. Solid-state 13C-NMR and electron microscopy study on the reversible cellulose I → cellulose IIII transformation in Valonia. Carbohydr. Res. 1987, 160, 1–11.

[29]   Wada M, Heux L, Sugiyama J. Polymorphism of cellulose I family: Reinvestigation of cellulose IVI. Biomacromol. 2004, 5, 1385–1391.

[30]   Elazzouzi-Hafraoui S, Nishiyama Y, Putaux JL, Heux L, Dubreuil F, Rochas C. The shape and size distribution of crystalline nanoparticles prepared by acid hydrolysis of native cellulose. Biomacromolecules. 2008, 9, 57–65.

[31]   Yang B, Dai Z, Din SY, Wyman CE. Enzymatic hydrolysis of cellulosic biomass. Biofuels. 2011, 2(4), 421–449.

[32]   TAPPI, ISO/TS 20477:2017: Standard terms and their definition for cellulose nanomaterial.

[33]   TAPPI, ISO/TR 19716:2016: Characterization of cellulose nanocrystals.

[34]   TAPPI, ISO/AWI TS 21346: Characterization of individualized cellulose nanofibrils.

[35]   TAPPI, WI 3021: Standard terms and their definition for cellulose nanomaterial.

[36]   Mariano M, El Kissi N, Dufresne A. Cellulose nanocrystals and related nanocomposites: Review of some properties and challenges. J. Polym. Sci. Pol. Phys. 2014, 52, 791–806.

[37]   Kargarzadeh H, Ioelovich M, Ahmad I, Thomas S, Dufresne A. In: Methods for Extraction of Nanocellulose from Various Sources. Kargarzedeh H, Ahmad I, Thomas S, Dufresne A., eds. Handbook of nanocellulose and cellulose nanocomposites. chap 1. Wiley-VCH, Weinheim, Germany, 2017, 1–50.

[38]   Lavoine N, Desloges I, Dufresne A, Bras J. Microfibrillated cellulose – its barrier properties and applications in cellulosic materials: A review. Carbohydr. Polym. 2012, 90, 735–764.

[39]   Pu Y, Zhang J, Elder T, Deng Y, Gatenholm P, Ragauskas AJ. Investigation into nanocellulosics versus acacia reinforced acrylic films. Compos. Part B Eng. 2007, 38, 360–366.

[40]   Beck-Candanedo S, Roman M, Gray DG. Effect of reaction conditions on the properties and behavior of wood cellulose nanocrystal suspensions. Biomacromolecules. 2005, 6, 1048–1054.

[41]   Rebouillat S, Pla F. State of the art manufacturing and engineering of nanocellulose: A review of available data and industrial applications. J. Biomater. Nanobiotechnol. 2013, 4, 165–188.

[42]   Kassab Z, Boujemaoui A, Youcef HB, Hajlane A, Hannache H, El Achaby M. Production of cellulose nanofibrils from alfa fibers and its nanoreinforcement potential in polymer nanocomposites. Cellulose. 2019, 26, 9567–9581.

[43]   Peng BL, Dhar N, Liu HL, Tam KC. Chemistry and applications of nanocrystalline cellulose and its derivatives: A nanotechnology perspective. Can. J. Chem. Eng. 2011, 89, 1191–1206.

[44]   Grishkewich N, Mohammed N, Tang J, Tam KC. Recent advances in the application of cellulose nanocrystals. Curr. Opin. Colloid In. 2017, 29, 32–45.

[45]   Xu X, Liu F, Jiang L, Zhu JY, Haagenson D, Wiesenborn DP. Cellulose nanocrystals vs. cellulose nanofibrils: A comparative study on their microstructures and effects as polymer reinforcing agents. ACS Appl. Mater. Interfaces. 2013, 5(8), 2999–3009.

[46]   Di Z, Shi Z, Ullah MW, Li S, Yang G. A transparent wound dressing based on bacterial cellulose whisker and poly (2-hydroxyethyl methacrylate). Int. J. Biol. Macromol. 2017, 105, 638–644.

[47]   Lin N, Huang J, Dufresne A. Preparation, properties and applications of polysaccharide nanocrystals in advanced functional nanomaterials: A review. Nanoscale. 2012, 4(11), 3274–3294.

[48]   Catalán J, Ilves M, Järventaus H, Hannukainen KS, Kontturi E, Vanhala E, Alenius H, Savolainen KM, Norppa H. Genotoxic and immunotoxic effects of cellulose nanocrystals in vitro. Environ. Mol. Mutagen. 2015, 56, 171–182.

[49]   Yu SI, Min SK, Shin HS. Nanocellulose size regulates microalgal flocculation and lipid metabolism. Sci. Rep. 2016, 6(35684), 1–9.

[50]   de Souza Lima MM, Wong JT, Paillet M, Borsali R, Pecora R. Translational and rotational dynamics of rodlike cellulose whiskers. Langmuir. 2003, 19(1), 24–29.

[51] Habibi Y, Goffin A, Schiltz N, Duquesne E, Dubois P, Dufresne A. Bionanocomposites based on poly(ε-caprolactone)-grafted cellulose nanocrystals by ring opening polymerization. J. Mater. Chem. 2008, 18(41), 5002–5010.
[52] de Menezes A, Siqueira G, Curvelo AAS. Dufresne A. Extrusion and characterisation of functionalised cellulose whiskers reinforced polyethylene nanocomposites. Polymer. 2009, 50, 4552–4563.
[53] Abdul Khalil HPS, Saurabh CK, Adnan AS, Nurul Fazita MR, Syakir MI, Davoudpour Y, Rafatullah M, Abdullah CK, Haafiz MKM, Dungani R. A review on chitosan-cellulose blends and nanocellulose reinforced chitosan biocomposites: Properties and their applications. Carbohydr. Polym. 2016, 150, 216–226.
[54] Zhao Y, Gao G, Liu D, Tian D, Zhu Y, Chang Y. Vapor sensing with color-tunable multilayered coatings of cellulose nanocrystals. Carbohydr. Polym. 2017, 174, 39–47.
[55] Basu A, Lindh J, Ålander E, Strømme M, Ferraz N. On the use of ion-crosslinked nanocellulose hydrogels for wound healing solutions: Physicochemical properties and application-oriented biocompatibility studies. Carbohydr. Polym. 2017, 174, 299–308.
[56] Gamelas JA, Pedrosa J, Lourenço AF, Mutjé P, González I, Chinga-Carrasco G, Singh G, Ferreira PJ. On the morphology of cellulose nanofibrils obtained by TEMPO-mediated oxidation and mechanical treatment. Micron. 2015, 72, 28–33.
[57] Alemdar A, Sain M. Isolation and characterization of nanofibers from agricultural residues-wheat straw and soy hulls. Bioresour. Technol. 2008, 99, 1664–1671.
[58] Bhatnagar A, Sain M. Processing of cellulose nanofiber-reinforced composites. J. Reinf. Plast. Comp. 2005, 24(12), 1259–1268.
[59] Wang B, Sain M. Isolation of nanofibers from soybean source and their reinforcing capability on synthetic polymers. Compos. Sci. Technol. 2007, 67(11–12), 2521–2527.
[60] Sano MB, Rojas AD, Gatenholm P, Davalos RV. Electromagnetically controlled biological assembly of aligned bacterial cellulose nanofibers. Ann. Biomed. Eng. 2010, 38, 2475–2484.
[61] Mahdavi M, Mahmoudi N, Rezaie Anaran F, Simchi A. Electrospinning of nanodiamond-modified polysaccharide nanofibers with physico-mechanical properties close to natural skins. Mar. Drugs. 128 2016, 14(7), 1–12.
[62] Carpenter AW, de Lannoy CF, Wiesner MR. Cellulose nanomaterials in water treatment technologies. Environ. Sci. Technol. 2015, 49, 5277–5287.
[63] Kaushik M, Moores A. Review: Nanocelluloses as versatile supports for metal nanoparticles and their applications in catalysis. Green Chem. 2016, 18, 622–637.
[64] Habibi Y, Lucia LA, Rojas OJ. Cellulose nanocrystals: Chemistry, self-assembly, and applications. Chem. Rev. 2010, 110, 3479–3500.
[65] Kaushik M, Fraschini C, Chauve G, Putaux JL, Moores A. In: Transmission Electron Microscopy for the Characterization of Cellulose Nanocrystals. Maaz K., ed. The transmission electron microscope. IntechOpen, 2015, chap 6. 129–163.
[66] Chakrabarty A, Teramoto Y. Recent advances in nanocellulose composites with polymers: A guide for choosing partners and how to incorporate them. Polymers. 2018, 10(517), 1–47.
[67] Razali N, Hossain M, Taiwo OA, Ibrahim M, Nadzri NWM, Razak N, Fazita N, Rawi M, Mahadar MM, Kassim MHM. Influence of acid hydrolysis reaction time on the isolation of cellulose nanowhiskers from oil palm empty fruit bunch microcrystalline cellulose. BioResources. 2017, 12(3), 6773–6788.
[68] Moon RJ, Martini A, Nairn J, Simonsen J, Youngblood J. Cellulose nanomaterials review: Structure, properties and nanocomposites. Chem. Soc. Rev. 2011, 40, 3941–3994.
[69] Espinosa SC, Kuhnt T, Foster EJ, Weder C. Isolation of thermally stable cellulose nanocrystals by phosphoric acid hydrolysis. Biomacromolecules. 2013, 14, 1223–1230.

[70]   Sadeghifar H, Filpponen I, Clarke SP, Brougham DF, Argyropoulos DS. Production of cellulose nanocrystals using hydrobromic acid and click reactions on their surface. J. Mater. Sci. 2011, 46, 7344–7355.

[71]   Habibi Y, Dufresne A. Highly filled bionanocomposites from functionalized polysaccharide nanocrystals. Biomacromolecules. 2008, 9, 1974–1980.

[72]   Yu H, Qin Z, Liang B, Liu N, Zhou Z, Chen L. Facile extraction of thermally stable cellulose nanocrystals with a high yield of 93% through hydrochloric acid hydrolysis under hydrothermal conditions. J. Mater. Chem. A. 2013, 1, 3938–3994.

[73]   Dufresne A. Nanocellulose: A new ageless bionanomaterial. Mater. Today. 2013, 16(6), 220–227.

[74]   Klemm D, Kramer F, Moritz S, Lindström T, Ankerfors M, Gray D, Dorris A. Nanocelluloses: A new family of nature-based materials. Angew. Chem. Int. Ed. 2011, 50, 5438–5466.

[75]   Khalil HPSA, Davoudpour Y, Islam MN, Mustapha A, Sudesh K, Dungani R, Jawaid M. Production and modification of nanofibrillated cellulose using various mechanical processes: A review. Carbohydr. Polym. 2014, 99, 649–665.

[76]   Abdul Khalil HP, Davoudpour Y, Islam MN, Mustapha A, Sudesh K, Dungani R, Jawaid M. Production and modification of nanofibrillated cellulose using various mechanical processes: A review. Carbohydr. Polym. 2014, 99, 649–665.

[77]   Ilyas RA, Sapuan SM, Ibrahim R, Abral H, Ishak MR, Zainudin ES, Atikah MSN, Nurazzi NM, Atiqah A, Ansari MNM, Syafri E, Asrofi M, Sari NH, Jumaidin R. Effect of sugar palm nanofibrillated cellulose concentrations on morphological, mechanical and physical properties of biodegradable films based on agro-waste sugar palm (Arenga pinnata (Wurmb.) Merr) starch. J. Mater. Res. Technol. 2019, 8(5), 4819–4830.

[78]   Cherian BM, Leao AL, de Souza SF, Thomas S, Pothan LA, Kottaisamy M. Isolation of nanocellulose from pineapple leaf fibres by steam explosion. Carbohydr. Polym. 2010, 81, 720–725.

[79]   Iwamoto S, Isogai A, Iwata T. Structure and mechanical properties of wet-spun fibers made from natural cellulose nanofibers. Biomacromolecules. 2011, 12, 831–836.

[80]   Isogai T, Saito T, Isogai A. Wood cellulose nanofibrils prepared by TEMPO electro-mediated oxidation. Cellulose. 2011, 18, 421–431.

[81]   Yuan Z, Wen Y. Enhancement of hydrophobicity of nanofibrillated cellulose through grafting of alkyl ketene dimer. Cellulose. 2018, 25, 6863–6871.

[82]   Plackett D, Iotti M. In: Preparation of Nanofibrillated Cellulose and Cellulose Whiskers. Dufresne A, Thomas S, Pothen L.A., eds. Biopolymer nanocomposites: Processing, properties, and applications. chap. 14. John Wiley & Sons, Inc., Hoboken, New Jersey, 2013, 309–338.

[83]   Pereda M, Dufresne A. In: Cellulose Nanocrystals and Related Polymer Nanocomposites. Thakur VK, Singha AS., eds. Biomass-based biocomposites. Smithers Rapra Technology, 2013, chap 15. 305–348.

[84]   Martínez-Sanz M, López-Rubio A, Lagarón JM. Cellulose Nanowhiskers: Properties and Applications as Nanofillers in Nanocomposites with Interest in Food Biopackaging Applications. In: Silvestre C, Cimmino S., eds. Ecosustainable polymer nanomaterials for food packaging. chap 8. CRC Press, New York, 2013, 199–230.

[85]   Dufresne A. Processing of polymer nanocomposites reinforced with polysaccharide nanocrystals. Molecules. 2010, 15(6), 4111–4128.

[86]   Siró I, Plackett D. Microfibrillated cellulose and new nanocomposite materials: A review. Cellulose. 2010, 17, 459–494.

[87]   Kalia S, Dufresne A, Cherian BM, Kaith BS, Avérous L, Njuguna J, Nassiopoulos E. Cellulose based bio- and nanocomposites: A review. Int. J. Polym. Sci. 2011, 2011, 1–36.

[88] Herrera MA, Mathew AP, Oksman K. Gas permeability and selectivity of cellulose
      nanocrystals films (layers) deposited by spin coating. Carbohydr. Polym. 2014, 112, 494–501.
[89] Gicquel E, Martin C, Garrido Yanez J, Bras J. Cellulose nanocrystals as new bio-based coating
      layer for improving fiber-based mechanical and barrier properties. J. Mater. Sci. 2017, 52,
      3048–3061.
[90] Vähä-Nissi M, Koivula HM, Räisänen HM, Vartiainen J, Ragni P, Kenttä E, Kaljunen T, Malm T,
      Minkkinen H, Harlin A. Cellulose nanofibrils in biobased multilayer films for food packaging.
      J. Appl. Polym. Sci. 2017, 134(44830), 1–8.
[91] Mousavi SMM, Afra E, Tajvidi M, Bousfield DW, Dehghani-Firouzabadi M. Application of
      cellulose nanofibril (CNF) as coating on paperboard at moderate solids content and high
      coating speed using blade coater. Prog. Org. Coat. 2018, 122, 207–218.
[92] Sabo R, Yermakov A, Law CT, Elhajjar R. Nanocellulose-enabled electronics, energy
      harvesting devices, smart materials and sensors: A review. J. Renew. Mater. 2016, 4,
      297–312.
[93] Müller D, Cercená R, Gutiérrez Aguayo AJ, Porto LM, Rambo CR, Barra GMO. Flexible PEDOT-
      nanocellulose composites produced by in situ oxidative polymerization for passive
      components in frequency filters. J. Mater. Sci. Mater. Electron. 2016, 27, 8062–8067.
[94] Guo X, Zhang Q, Li Q, Yu H, Liu Y. Composite aerogels of carbon nanocellulose fibers
      and mixed-valent manganese oxides as renewable supercapacitor electrodes. Polymers. 129,
      2019, 11(1), 1–13.
[95] Hsu HH, Zhong W. Nanocellulose-based conductive membranes for free-standing
      supercapacitors: A review. Membranes. 74, 2019, 9(6), 1–21.
[96] Ma IAW, Shafaamri A, Kasi R, Zaini FN, Balakrishnan V, Subramaniam R. Arof AK
      Anticorrosion properties of epoxy/nanocellulose nanocomposite coating. Bioresources.
      2017, 12, 2912–2929.
[97] Yabuki A, Fathona IW. In: Recent Trends in Nanofiber-Based Anticorrosion Coatings. Barhoum
      A, Bechelany M, Makhlouf A., eds. Handbook of nanofibers, 2018, 1–32.
[98] Bhat AH, Dasan YK, Khan I, Jawaid M. In: Cellulosic Biocomposites: Potential Materials For
      Future. Jawaid M, Salit MS, Alothman OY., eds. Green biocomposites, 2017, 69–100.
[99] Hakkarainen T, Koivuniemi R, Kosonen M, Escobedo-Lucea C, Sanz-Garcia A, Vuola J,
      Valtonen J, Tammela P, Mäkitie A, Luukko K, Yliperttula M, Kavola H. Nanofibrillar cellulose
      wound dressing in skin graft donor site treatment. J. Control Release. 2016, 244, 292–301.
[100] Trovatti E, Tang H, Hajian A, Meng Q, Gandini A, Berglund LA, Zhou Q. Enhancing strength
      and toughness of cellulose nanofibril network structures with an adhesive peptide.
      Carbohydr. Polym. 2018, 181, 256–263.
[101] Shefa AA, Amirian J, Kang HJ, Bae SH, Jung HI, Choi H, Lee SY, Lee BT. In vitro and in vivo
      evaluation of effectiveness of a novel TEMPO-oxidized cellulose nanofiber-silk fibroin
      scaffold in wound healing. Carbohydr. Polym. 2017, 177, 284–296.
[102] Bacakova L, Pajorova J, Bacakova M, Skogberg A, Kallio P, Kolarova K, Svorcik V. Versatile
      application of nanocellulose: From industry to skin tissue engineering and wound healing.
      Nanomaterials. 2019, 9(164), 1–39. 10.3390,
[103] Bacakova L, Pajorova J, Zikmundova M, Filova E, Mikes P, Jencova V, Kostakova EK, Sinica
      A. In: Nanofibrous Scaffolds for Skin Tissue Engineering and Wound Healing based on
      Nature-Derived Polymers. Khalil I., ed. Current and future aspects of nanomedicine.
      IntechOpen, 2019, 1–30.
[104] Domingues RMA, Gomes ME, Reis RL. The potential of cellulose nanocrystals in tissue
      engineering strategies. Biomacromolecules. 2014, 15, 2327–2346.

[105] Mathew AP, Oksman K, Pierron D, Harmand MF. Fibrous cellulose nanocomposite scaffolds prepared by partial dissolution for potential use as ligament or tendon substitutes. Carbohydr. Polym. 2012, 87, 2291–2298.

[106] Li W, Guo R, Lan Y, Zhang Y, Xue W, Zhang Y. Preparation and properties of cellulose nanocrystals reinforced collagen composite films. J. Biomed. Mater. Res. Part A. 2014, 102, 1131–1139.

[107] Supramaniam J, Adnan R, Mohd Kaus NH, Bushra R. Magnetic nanocellulose alginate hydrogel beads as potential drug delivery system. Int. J. Biol. Macromol. 2018, 118, 640–648.

[108] Mo Y, Guo R, Liu J, Lan Y, Zhang Y, Xue W, Zhang Y. Preparation and properties of PLGA nanofiber membranes reinforced with cellulose nanocrystals. Colloids Surf. B Biointerfaces. 2015, 132, 177–184.

[109] Nguyen D, Hägg DA, Forsman A, Ekholm J, Nimkingratana P, Brantsing C, Kalogeropoulos T, Zaunz S, Concaro S, Brittberg M, Lindahl A, Gatenholm P, Enejder A, Simonsson S. Cartilage tissue engineering by the 3D bioprinting of iPS cells in a nanocellulose/alginate bioink. Sci. Rep. 2017, 7(658), 1–10.

[110] Sukul M, Min YK, Lee SY, Lee BT. Osteogenic potential of simvastatin loaded gelatin-nanofibrillar cellulose-β tricalcium phosphate hydrogel scaffold in critical-sized rat calvarial defect. Eur. Polym. J. 2015, 73, 308–323.

[111] Rusu D, Ciolacu D, Simionescu BC. Cellulose Chem. Technol. 2019, 53(9–10), 907–923.

[112] Souza SF, Mariano M, Reis D, Lombello CB, Ferreira M, Sain M. Cell interactions and cytotoxic studies of cellulose nanofibers from Curauá natural fibers. Carbohydr. Polym. 2018, 201, 87–95.

[113] Yanamala N, Farcas MT, Hatfield MK, Kisin ER, Kagan VE, Geraci CL, Shvedova AA. In vivo evaluation of the pulmonary toxicity of cellulose nanocrystals: A renewable and sustainable nanomaterial of the future. ACS Sustain. Chem. Eng. 2014, 2, 1691–1698.

[114] Shvedova AA, Kisin ER, Yanamala N, Farcas MT, Menas AL, Williams A, Fournier PM, Reynolds JS, Gutkin DW, Star A, Reiner RS, Halappanavar S, Kagan VE. Gender differences in murine pulmonary responses elicited by cellulose nanocrystals. Part Fibre Toxicol. 2016, 13(28), 1–20.

[115] Menas AL, Yanamala N, Farcas MT, Russo M, Friend S, Fournier PM, Star A, Iavicoli I, Shurin GV, Vogel UB, Fadeel B, Beezhold D, Kisin ER, Shvedova AA. Fibrillar vs crystalline nanocellulose pulmonary epithelial cell responses: Cytotoxicity or inflammation? Chemosphere. 2017, 171, 671–680.

[116] Lopes VR, Sanchez-Martinez C, Strømme M, Ferraz N. In vitro biological responses to nanofibrillated cellulose by human dermal, lung and immune cells: Surface chemistry aspect. Part Fibre Toxicol. 2017, 14(1), 1–13.

[117] Bhandari J, Mishra H, Mishra PK, Wimmer R, Ahmad FJ, Talegaonkar S. Cellulose nanofiber aerogel as a promising biomaterial for customized oral drug delivery. Int. J. Nanomed. 2017, 12, 2021–2031.

[118] Trache D. Nanocellulose as a promising sustainable material for biomedical applications. AIMS Mat. Sci. 2018, 5(2), 201–205.

[119] Kolakovic R, Laaksonen T, Peltonen L, Laukkanen A, Hirvonen J. Spray-dried nanofibrillar cellulose microparticles for sustained drug release. Int. J. Pharm. 2012, 430, 47–55.

[120] Kolakovic R, Peltonen L, Laukkanen A, Hirvonen J, Laaksonen T. Nanofibrillar cellulose films for controlled drug delivery. Eur. J. Pharm. Biopharm. 2012, 82, 308–315.

[121] Díez I, Eronen P, Österberg M, Linder MB, Ikkala O, Ras RHA. Functionalization of nanofibrillated cellulose with silver nanoclusters: Fluorescence and antibacterial activity. Macromol. Biosci. 2011, 11, 1185–1191.

[122] Ferraz N, Carlsson DO, Hong J, Larsson R, Fellstrom B, Nyholm L, Stromme M, Mihranyan A. Haemocompatibility and ion exchange capability of nanocellulose polypyrrole membranes intended for blood purification. J. R. Soc. Interface. 2012, 9, 1943–1955.
[123] Navarro JRG, Conzatti G, Yu Y, Fall AB, Mathew R, Edén M, Bergström L. Multicolor fluorescent labeling of cellulose nanofibrils by click chemistry. Biomacromolecules. 2015, 16(4), 1293–1300.
[124] Li M, Li X, Xiao HN, James TD. Fluorescence sensing with cellulose-based materials. ChemistryOPEN. 2017, 6, 685–696.
[125] Lang N, Merkel E, Fuchs F, Schumann D, Klemm D, Kramer F, Mayer-Wagner S, Schroeder C, Freudenthal F, Netz H, Kozlik-Feldmann R, Sigler M. Bacterial nanocellulose as a new patch material for closure of ventricular septal defects in a pig model. Eur. J. Cardiothorac. Surg. 2015, 47, 1013–1021.
[126] Gatenholm P, Klemm D. Bacterial nanocellulose as a renewable material for biomedical applications. MRS Bull. 2010, 35, 208–213.

Dana Mihaela Suflet

Chapter 12
Ionic derivatives of cellulose: new approaches in synthesis, characterization, and their applications

12.1 Introduction

Cellulose is the most abundant polymer on Earth, being the most common and renewable organic compound. The annual cellulose synthesis of plants is close to 10^{12} tons. Most of the cellulose is utilized as a raw material in paper production, while only few tons are used to obtain new materials in special domains such as top technologies or the biomedical field.

The versatility and wide availability of cellulose could complement various challenges and requirements of modern life, placing this natural and biodegradable polymer in a special position compared to other polymers. As a consequence of its high capacity of chemical modification, cellulose can be transformed into a variety of useful products covering a wide range of applications, from traditional plastics and fibers to special devices and innovative products.

In this regard, the purpose of this chapter is to supplement the information presented in the first edition of this book [1], with new information on cellulose ionic derivatives, with emphasis on recent progress in the synthesis and characterization and new approaches in their applications.

12.2 Cellulose

Cellulose is the most widespread natural polymer, constituting the main component of the primary cell wall of green plants. It has a linear structure, consists from D-anhydroglucopyranose units (AGU), linked together by β-(1,4) glycosidic bonds (Fig. 12.1).

Each of the AGU units presents three hydroxyl (OH) groups at C-2, C-3, and C-6 positions, capable of initiating the typical reactions for primary and secondary alcohols. Terminal groups at the either end of the cellulose molecule are quite different in nature from each other. Thus, at the end of the polysaccharide chain at the C-1 atom is an aldehyde group with reducing properties, while at the opposite end of the C-4 atom there is an alcohol OH component, and the unit is called the "nonreducing end."

Dana Mihaela Suflet, "PetruPoni" Institute of Macromolecular Chemistry, Iasi, Romania

https://doi.org/10.1515/9783110658842-012

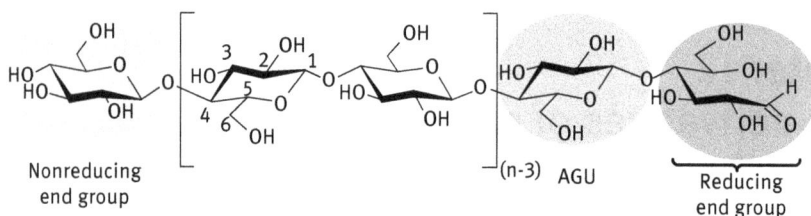

Fig. 12.1: Molecular structure of cellulose.

The chemical reactivity of cellulose is determined to a large extent by the supramolecular structure of its solid state. Cellulose contains 31.48% by weight of hydroxyl groups, which play the influential role in the preparation of cellulose derivatives. As shown in Fig. 12.1, cellulose possesses one primary and two secondary hydroxyl groups per AGU (at C-2, C-3, and C-6, respectively). Like any hydroxyl-containing compound, these hydroxyl groups can undergo addition, substitution, and oxidation reactions. As a result of the inductive effects of neighboring substituents, the acidity and the tendency for dissociation of hydroxyl groups increase in the order HO-6 < HO-3 < HO-2 [2, 3]. The reactivity of these hydroxyl groups depends on the acidity or alkalinity of the reaction medium. For example, for etherification of hydroxyl groups conducted in an alkaline condition, HO-2 is the most readily etherified of the three. Moreover, HO-2 groups often exhibit increased reactivity in the case of heterogeneous esterification or etherification reactions. On the contrary, for esterification, the primary hydroxyl group HO-6 is the most active. Although these hydroxyl groups are reactive, they cannot be completely accessible for the reaction, as hydroxyl groups frequently participate in intra- and intermolecular hydrogen bonds [3].

Once the hydroxyl groups AGU are accessible, they offer a variety of possibilities for making useful derivatives. The properties of the derivatives depend heavily on the type, distribution, and uniformity of the substitution groups. The average number of hydroxyl groups replaced by the substituents is the degree of substitution (DS); the maximum of the DS is 3.

12.3 Cellulose derivatives

The hydroxyl groups of AGU can enter into all classical reactions for alcoholic compounds, especially esterification, esterification, and oxidation reactions. Esterification is usually accomplished by reacting the cellulose with an acid, or with the acidic anhydride or the acid chloride of the acid, as an active agent with high reactivity. The most important cellulose esters are the cellulose sulfate, –phosphate, –nitrate, –acetate, –acetobutyrate, –acetophthalate, –xanthogenate, –carbamate, and cellulose formate.

The main routes to the synthesis the cellulose ethers consist in the reaction of cellulose: (i) with alkyl halides in the presence of strong base (the common Williamson ether reaction); (ii) with alkylene oxides in a weakly basic medium; and (iii) by Michael addition of acrylic or related unsaturated compounds like acrylonitrile. The most important cellulose ethers are methylcellulose, ethylcellulose, carboxymethyl cellulose, hydroxylethyl cellulose, hydroxylpropyl cellulose, trialkylsilylcellulose, mixed ethers of alkylhydroxyalkyl, benzyl cellulose, and triphenylmethyl cellulose. The chemical modification of cellulose leads to the widening of the cellulose applications as fibers, films, plastics, coatings, filtration media, as well as additives for food products, cosmetics, pharmaceuticals, extractive industry, and so on. According to the nature of substituent, the derivatives can be structured in nonionic and ionic derivatives.

12.3.1 Ionic derivatives of cellulose

Various ionic cellulose derivatives can be synthesized by the introduction of ionic groups into the cellulose backbone by esterification or etherification of OH groups, or by oxidation with or without cleavage of C–C bonds. The main classes of cellulose ionic derivatives are presented in Fig. 12.2.

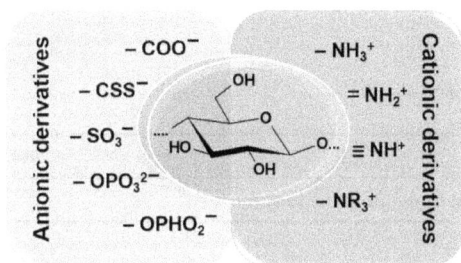

Fig. 12.2: Structure of ionic site of derivatives cellulose.

12.3.1.1 Anionic derivatives of cellulose

Sulfate derivatives: Cellulose sulfate is a semiester of cellulose, with much more beneficial properties than the parent cellulose, for example, cellulose sulfate can be obtained with a wide range of substitution degrees, the sulfate groups conferring water solubility, and unique biological properties (i.e., antivirus, anticoagulant, antibacterial).

The cellulose derivatives with sulfate groups (Fig. 12.3) can be prepared both in a heterogeneous and homogeneous system. Under heterogeneous conditions, cellulose

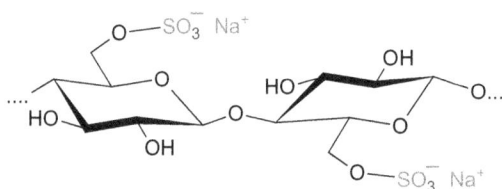

Fig. 12.3: Structure of cellulose sulfate.

in the presence of a mixture of sulfuric acid and isopropyl alcohol leads to a water-soluble derivate with a total DS of about 0.7, but with a severe chain degradation. Otherwise, the sulfation of cellulose under homogenous conditions resulted in water-soluble derivatives with only minor cellulose degradation. Furthermore, cellulose sulfate can also be synthesized quasi-homogeneously through direct sulfation or acetosulfation, which means that the suspension of cellulose turns into optically transparent solution during the reaction and the cellulose is dissolved in the reaction mixture [4–6].

Principle route of the synthesis of cellulose sulfates can be as follows:

a) *Sulfation of OH groups of unmodified cellulose in heterogeneous conditions*:
A large variety of sulfating agents for unmodified cellulose have been studied so far. The most used systems for synthesis of cellulose sulfate are listed in Tab. 12.1.

Tab. 12.1: The systems for the synthesis of cellulose sulfate.

System	DS	Ref.
Heterogeneous condition throughout the reaction		
H_2SO_4/DMF	–	
H_2SO_4/DMAC	0.45–2.21	[4]
H_2SO_4/DMSO	–	
H_2SO_4/NMP	–	
H_2SO_4/chlorinated hydrocarbons	0.3	
H_2SO_4/SO_2	~0.9	
H_2SO_4/diethyl ether	0.2–0.4	
H_2SO_4/low aliphatic alcohol	0.1–1.0	[7–9]
$ClSO_3H$/SO_2	~1.8	
$ClSO_3H$/pyridine/toluene	~2.8	

Tab. 12.1 (continued)

System	DS	Ref.
Heterogeneous condition throughout the reaction		
ClSO$_3$H/DMF	≤3.0	[10]
NH$_2$SO$_3$H/DMF	–	[10]
NH$_2$SO$_3$H/DMAC	–	[4]
NH$_2$SO$_3$H/NMP	0.31–1.04	[4]
SO$_3$/SO$_2$	~2.5	
SO$_3$/CS$_2$	≤3.0	
SO$_3$/diethyl ether	1.3–2.1	
SO$_2$Cl$_2$/DMF, formamide	0.2–0.5	[4, 11]
With transition from a heterogeneous to homogeneous system during the sulfation reaction		
H$_2$SO$_4$	1.0–2.0	[12]
ClSO$_3$H/pyridine	1.9–2.8	
ClSO$_3$H/formamide	≤3.0	
SO$_3$/DMSO	1.3–2.0	
SO$_3$/DMF	1.5–2.6	
SO$_3$/pyridine	≤2.5	[13, 14]
SO$_3$/triethyl phosphate	≤3.0	
Homogeneous condition throughout the reaction		
ClSO$_3$H/DMF into cellulose solution in DMAc/LiCl	–	[10]
N$_2$O$_4$/DMF/SO$_2$	–	[9, 15]

b) *Sulfation of free OH groups in partially functionalized cellulose esters or ethers with primary substituent acting as a protecting group*:
The free groups of various modified cellulose esters and ethers can be sulfated by conventional sulfating agents in a suitable dipolar aprotic medium, while the primary substituent acts as a protecting group.

The acetyl group of partially esterified cellulose proved to be a very suitable protecting group in a subsequent sulfation, as it is definitely stable in the anhydrous acid reaction medium, in contrast with the more mobile formyl group. Sulfation of cellulose acetates with DS between 0.35 and 1.5 is preferentially performed under homogeneous conditions using as solvent DMF and as sulfating agents: SO_3 or $ClSO_3H$/acetic anhydride [4, 10, 16], SO_2Cl_2 [4], NH_2SO_3H [9, 16]; H_2SO_3/pyridine/DMF [16], and $ClSO_3H$/pyridine [17]. The synthesis of cellulose sulfate through the acetosulfation is completed after the cleavage of the acetyl groups using NaOH.

c) *Sulfation by displacement of an ester or ether group present in the AGU*:
   Typical for this route is the formation of labile ester or ether groups that are easily displaced by sulfating agents from its position in the AGU. Such labile groups can be, for example, the nitrite group as an ester or the trialkysilyl group as an ether.

In heterogeneous conditions, the sulfation of cellulose leads to derivatives with DS of 0.2–2.8, with a severe depolymerization of the polysaccharide chain (e.g., in the presence of the mixture of sulfuric acid and isopropanol). In the case of the esterification reaction of hydroxyl groups in homogenous conditions, when obtained the water-soluble derivatives, generally in the acid half-ester form, the range of DS is also 0.2–2.59 and can be easily converted to a neutral sodium salt.

The number-average degrees of polymerization (DP_n) of these cellulose sulfates were determined to be in the range of 59 and 232 via size exclusion chromatography (SEC) [18]. FT Raman spectroscopy offers the information about the substitution of the hydroxyl groups and it is a promising method of characterizing cellulose derivatives. In this respect, the band between 825 and 847 cm^{-1} is attributed to stretching vibrations of C–O–S groups, while the bands at 1,068–1,075 cm^{-1} are attributed to stretching vibrations of O=S=O. Moreover, the change of the signals ascribed to stretching vibrations of CH and CH$_2$ groups in the range of 2,800–3,050 cm^{-1} is visible, after the sulfation of cellulose. The band at 2,896 cm^{-1} becomes weaker and a new band between 2,950 and 2,966 cm^{-1} is notable. With the increase of total DS, this new band shifts to higher wave numbers and becomes more intense. Within the spectra of CS with low total DS (e.g., below 0.47), this signal at 2,950–2,955 cm^{-1} appears only as a shoulder of the peak at 2,896 cm^{-1}. However, within the spectra of CS exhibiting high total DS, especially the total DS of more than 1, this signal has a strong intensity. Finally, within the spectrum of CS showing total DS of more than 1.5, this new band is the dominant one, while the signal at 2,896 cm^{-1} turns into a shoulder of this new band [10, 18].

^{13}C NMR (nuclear magnetic resonance) spectrometry can be used in order to confirm the synthesis of cellulose sulfate [10, 13, 14]. In Fig. 12.4, for example, the ^{13}C NMR spectra of CS samples recorded in D$_2$O are shown. A new signal at 66.3 ppm beside the signal at 60 ppm was visible. In this case, because both signals are derived from C-6 of AGU, the DS was calculated based on the integrals of both signals. Moreover, the signal of C-1 shifts from 102.5to 100.5 ppm as shown in Fig. 12.4; this is due to the influence of the sulfation at OH-2 position. Based on the integrals of the two peaks ascribed to C-1, the DS can be estimated. In addition, new peaks between 80 and 82 ppm are notable, which are ascribed to C-2 and C-3 with sulfate groups at

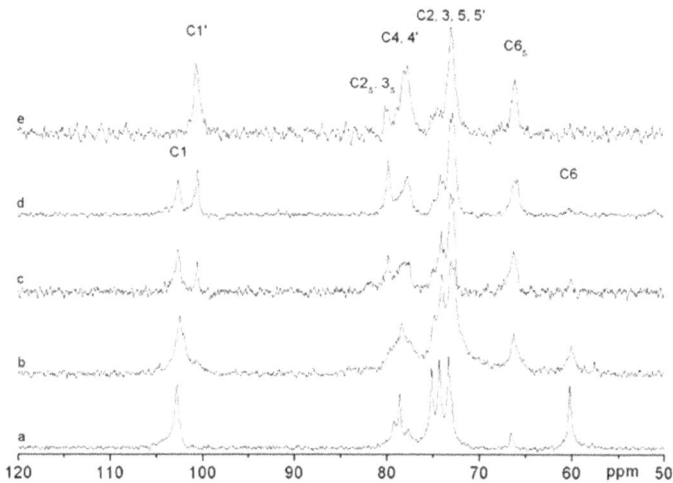

Fig. 12.4: ^{13}C NMR spectra (120–50 ppm) of (a) CS with DS = 0.2; (b) CS with DS = 0.66; (c) CS DS = 1.09; (d) CS with DS = 1.57; (e) CS with DS = 2.59, in D$_2$O at room temperature [10]. Reproduced by permission of Wiley, copyright 2019.

OH-2 and OH-3 positions. Other signals assigned C-2, 3, 4, and 5 are between 70 and 80 ppm in the CS spectra.

Wide-angle X-ray diffraction (WAXD) confirms the sulfatation reaction when the crystalline regions of cellulose are completely destroyed during the homogeneous sulfation reactions, and a part of them remains after the heterogeneous sulfation [6].

Phosphate derivatives: The phosphorylation of cellulose could be performed both in heterogeneous or homogeneous systems. The heterogeneous reactions, which will be discussed in this section, consist in the attachment of the phosphorous groups to the cellulose chain by reactions with the hydroxyl groups from the original polymer, when cellulose esters, as monobasic phosphate, dibasic phosphate, or phosphite, were obtained (Fig. 12.5). The most frequently used agents for the obtaining of the

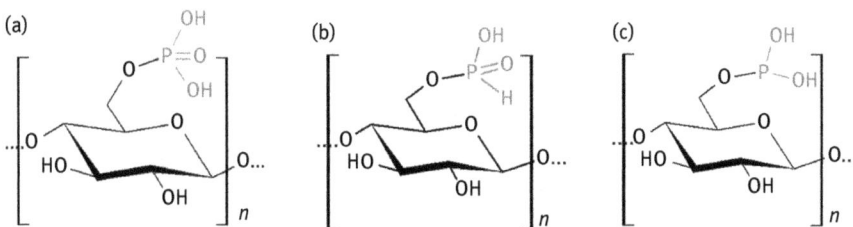

Fig. 12.5: Structure of phosphate cellulose: (a) monobasic; (b) dibasic; and (c) phosphite.

cellulose phosphates (CPs) are as follows: the phosphoric acid (H_3PO_4), the phosphorous acid (H_3PO_3), the phosphorous pentoxide (P_2O_5), the phosphorous oxychloride ($POCl_3$), and the phosphorous trichloride (PCl_3) [19]. These phosphorylated agents lead to inorganic cellulose derivatives, having a lower reactivity in esterification, but allow to obtain reaction products with much less degraded macromolecular chains.

Synthesis of the *dibasic* CPs using water-free orthophosphoric acid as a phosphating agent leads to obtain the soluble or insoluble derivatives. Water-soluble CP of high polymerization degree can be prepared with anhydrous phosphoric acid. The reactivity of the phosphorylation agent can be increased by using a mixture of H_3PO_4 with P_2O_5. In this case, the DS increases with the increase of molar ratio between the phosphorylation agent and AGU, but these hard conditions lead to the increase in the degradation of cellulosic chains. Water-soluble derivatives were also synthesized in ternary systems of $H_3PO_4/P_2O_5/DMSO$ or $H_3PO_4/P_2O_5/aliphatic$ alcohols, when the degree of phosphorylation was up to 0.2. The most used systems for synthesis of CP are listed in Tab. 12.2.

Tab. 12.2: The systems for synthesis of cellulose phosphate.

System	DS	Solubility	Ref.
H_3PO_4/urea	0.30–0.60	Water-insoluble	[20]
$H_3PO_4/P_2O_5/Et_3PO_4$/aliphatic alcohols	1.00–2.50	Gel	[21–24]
$(NH_4)_2HPO_4$/urea	0.30–0.60	Water-insoluble	[25]
$POCl_3/N_2O_4$/DMF system	up to 1.00 0.30–0.60	Soluble	[26]
PCl_3/DMF	–	–	[27]
Sodium trimetaphosphate (STMP)	–	Insoluble	
Anhydrides of hydrophosphorus and acetic acid	–	–	

Other cellulose derivatives with phosphorus groups were obtained by phosphorylation reactions of partially substituted cellulose derivatives such as carboxyalkyl cellulose, acetate, nitrate, and so on. Carboxymethylcellulose (CMC) with a DS of 0.8 was converted to a derivative with a DS in phosphorous of 0.3 using the H_3PO_4/urea system with the phosphate groups again preferentially located at C-6 [28]. CMC containing one phosphate group for each disaccharide unit (CMCP) was also prepared using a 1% solution of CMC in alkaline conditions (pH 12, 2M NaOH) and trisodium trimetaphosphate as phosphating agent, in sufficient amount to activate two hydroxyl groups per disaccharide unit, for example, to functionalize titanium oxide (TiO_2) surfaces in order to increase the osseointegration of the host bone tissue. The carboxyethylcellulose (CEC) was also phosphorylated when a considerable higher DS in phosphorous of 0.6 was obtained with the H_3PO_4/urea system [29].

The hydroxypropyl cellulose (HPC) was phosphorylated to introduce new functional groups that could promote and increase the adhesion on metallic substrates. The reaction was carried out using the H_3PO_4/DMF/ triethylamine system. The HPC was dissolved overnight in DMF. Triethylamine and a solution of poly(phosphoric acid) in DMF was added. The reaction mixture was heated slowly to 120 °C for 6 h.

The *monobasic cellulose phosphate* (PCell) (Fig. 12.5b), with DS up to 1, was obtained by reaction of cellulose with phosphorous acid in molten urea [30], through a hydrogen bond complex between H_3PO_3, urea, and cellulose. This method allows the synthesis of water-soluble ionic derivatives with a thermal stability up to 200 °C, which allows the use of monobasic CP in processes that reach higher temperatures. The chemical structure of PCell was investigated by FTIR and NMR spectroscopy and DS was calculated by electrochemical methods [30]. These PCell was also used as intermediary products in the synthesis of polyelectrolytes having two different ionization groups, P–OH and COOH. The reaction takes place in the presence of sodium ethoxide at room temperature by addition of acrylonitrile to P–H bonds from monobasic CP [31].

FTIR and NMR spectroscopy, X-ray diffraction (XRD), scanning electron microscopy (SEM), rheological measurements, viscosimetric behavior, and thermal analysis are the most widely used methods reported in the literature for the characterization of the CP. The main absorption bands shown in FTIR spectra of phosphorylated celluloses are listed in Tab. 12.3.

Tab. 12.3: The main absorption FTIR bands of phosphorylated cellulose [21, 30, 32].

wavelength (cm^{-1})	Characteristic link
3,400–3,500	O–H stretching
2,800–2,900	CH and CH$_2$ stretching
2,440–2,275	P–H
1,625	Absorbed H$_2$O
1,418–1,029	C–O–C stretching from the glucosidic units or from β-(1→4)-glucosidic bonds; C–H wagging; C–O stretching
1,260–855	P–O

Tab. 12.3 (continued)

wavelength (cm^{-1})	Characteristic link
1,200–1,383	P=O stretching
1,055–950	C–OP stretching
1,100–940	P–OH
803–754	O–P–O stretching
980–900	P–O–P
1,050–970; 820–740	P–OC stretching

XRD allowed the determination of the crystallinity index (CI). The results of the studies indicated a decrease in crystallinity with increasing DS, but at low DS the materials are already considerably amorphous when compared to the crystalline Avicel cellulose. The CI was found to be 14.5% for DS=1.4; 32.4% for DS = 0.8 compared with 82.9% for microcrystalline cellulose (Avicel) [21].

NMR spectroscopy provides clear evidence for derivative synthesis. The ^{31}P NMR/MAS spectrum exhibited a signal at 20.4 ppm. The regioselectivity of this reaction was elucidated by ^{13}C NMR/MAS studies, when the specific peak displacements were observed.

The *electrorheological studies* of the CP ester suspension showed a response and a behavior similar to a Bingham flow. The shear stress for the CP ester suspension showed a linear dependence on the volume fraction of particles and the dependence with a factor equal to 1.41 power on the electric field [33, 34].

The thermostability and thermal decomposition kinetics of phosphates cellulose were studied in order to obtain information on their thermal behavior for future applications. The thermal decomposition unfolds mainly in two steps: the first one occurs with a high rate of decomposition while the second phase unfolds more slowly. The temperature value corresponding to the maximum rate of weight loss (T_m) of the second step, which was attributed to the oxidative decomposition of the PCell, was found to be around 250–270 °C. It was observed that in the temperature range of 22–180 °C the weight loss of PCell is almost 9%, losses that were attributed to the moisture evaporation [30]. In the case of CP, the thermal decomposition in air showed the lowest activation energy (E = 60 kJ/mol) and frequency factor (Ln Z = 15 min^{-1}) values, suggesting that the CP exhibits the low thermostability compared with other cellulose derivatives. It is of interest that values of the order (n) remain almost constant in a narrow range of 0.8–1.2 and do not vary either with the chemical structure or with test atmosphere, except for that in the oxygen [35].

The presence of phosphate groups on the cellulose chain confers the polyelec-trolyte properties. From this point of view, the polyelectrolyte behavior of PCell was studied by electrochemical methods [36] using purified samples by diafiltration and transformation into acidic form by passing through a cationic exchange column filled with a sulfonic Dowex 50W × 8, 20–50 mesh resin. In Fig. 12.6, the potentio-metric titration curves of PCell with monovalent (KOH) and divalent bases (Ca(OH)$_2$), with and without added salt and their first derivate, are shown. All the titration curves with KOH (Fig. 12.6a) present only one inflexion point, correspond-ing to the dissociation of one kind of acidic groups –HP(O)(OH). The presence of salt in the system has a slight influence on the polyion dissociation in accordance to the case of strong polyelectrolytes [36, 37].

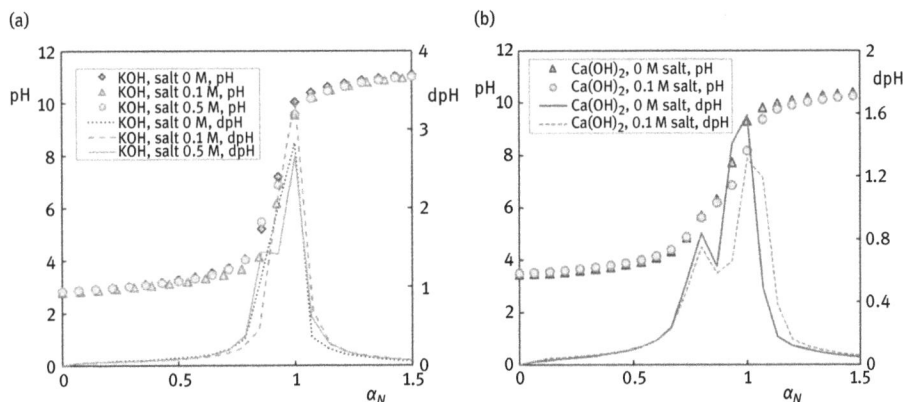

Fig. 12.6: Potentiometric titration of PCell. Influence of added salt on the potentiometric titration: (a) with monovalent bases (KOH) and (b) with divalent base (Ca(OH)$_2$).

In the case of titration with a bivalent base (Fig. 12.6b) the derived curve shows two very close peaks. The appearance of an additional peak on derivative curves corre-sponds to the presence of the more accessible acid groups to titration than the other acid groups. This could be due to the specific rigid conformation of cellulose chains or the different accessibility of the titrant ions related to their size and/or their hydration.

From the potentiometric data, the apparent dissociation constant, pK_a, have been calculated using the Henderson–Hasselbalch relation (12.1):

$$pK_a = pH + \log\left(\frac{1-\alpha_T}{\alpha_T}\right) \tag{12.1}$$

where pK_a is the apparent dissociation constant; $\alpha_T = \alpha_{H^+} + \alpha_N$ is the total fraction of ionized groups, α_{H^+} is the initial degree of dissociation, and α_N is the degree of neutralization. The intrinsic dissociation constant pK_0 was estimated from extrapolation of $pK_a = f(\alpha_T)$ curves. The $pK_0 \sim 3$ value obtained from the potentiometric titrations of PCell with KOH is in accordance with values given in literature for polyelectrolytes with moderately strong acid groups [36, 38].

The transport coefficient of monovalent counterions (f^{M^+}) and the equivalent conductance of polyanion (α_p) were calculated from conductometric data, at complete neutralization (Fig. 12.7).

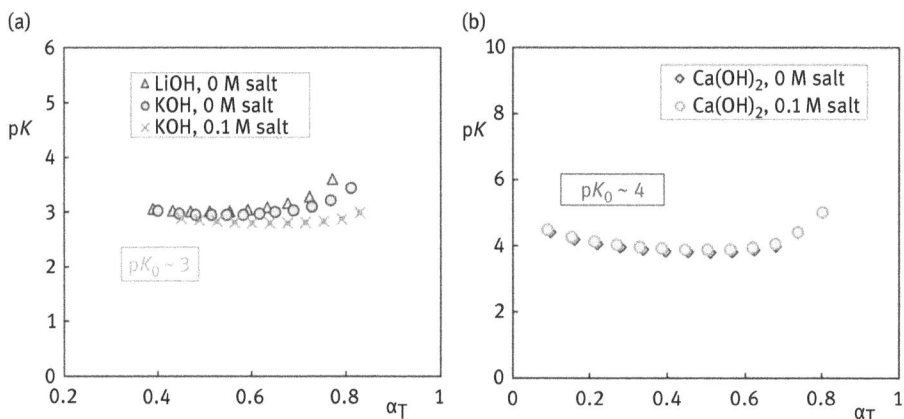

Fig. 12.7: Dependence of the apparent dissociation constant on the total ionization degree: (a) titration of PCell with monovalent bases and (b) titration of PCell with bivalent base.

The viscosimetric measurements of PCell carried out in pure water, with or without added salt (concentration from 5×10^{-4} to 1 M KCl), showed a typical polyelectrolyte behavior. The reduced viscosity increased sharply with the decrease of the polymer concentration (C_P) because in high diluted solutions the charges induce a fully stretched conformation of macroions.

Increasing the amount of added salt, the electrostatic repulsions are gradually screened, so that at high salt concentration the macroion behaves as neutral polymer (Fig. 12.8). The Fuoss relation (12.2) was used to evaluate the intrinsic viscosity of phosphorylated cellulose:

$$\frac{C_P}{\eta_{sp}} = \frac{1}{A} + \left(\frac{B}{A}\right) C_P^{1/2} \tag{12.2}$$

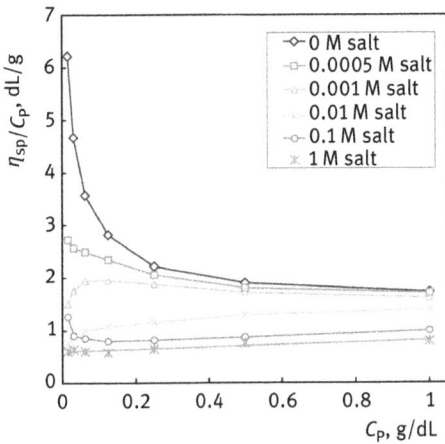

Fig. 12.8: η_{sp}/C_P dependence on polymer concentration of PCell in the absence or in the presence of KCl.

where A is the intrinsic viscosity and B is a constant. Figure 12.9 shows the Fuoss plot of PCell at various concentration of added salt. The value of $[\eta]_{water}$ = 4.59 dL/g was obtained from the extrapolation of Fuoss equation (12.2) at infinite dilution in aqueous solution without added salt (Fig. 12.9). This value had the same magnitudes with the value reported in literature [39], calculated by Wolf equation $[\eta]$ = 6.658 dL/g.

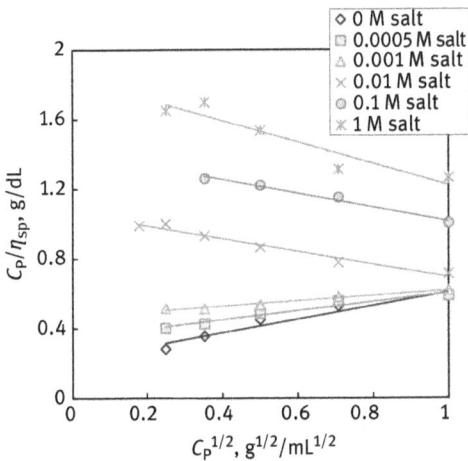

Fig. 12.9: C_P/η_{sp} dependence on the $C_P^{1/2}$ of PCell at various concentrations of added salt.

Carboxylic derivatives: *Carboxymethyl cellulose* (CMC) is the most important cellulose ionic derivative that finds applications in various fields. The CMC production is estimated to have a volume more than 300,000 t/year worldwide. CMC is a linear and water-soluble polymer and can be used as free acid or sodium salt, or mixtures (Fig. 12.10). CMC is usually produced in a heterogeneous process by Williamson etherification, when the cellulose suspended in a mixture of isopropanol/water (or ethanol/acetone) is treated with aqueous solution of NaOH (20–30%) and monochloroacetic acid or its sodium salt. Recently, it was demonstrated that the change of concentrations or the time of the reaction can only influence the DS, but not the substituents distribution within the AGU or along the cellulose chains [40]. Treatment of cellulose with aqueous NaOH solution has been shown to be a very effective activation method and, in addition, the high activation power before etherification leads to completely homogeneous carboxymethylation. Commercial CMCs are white, odorless, nontoxic powders, in the form of sodium salt, and usually have a DS between 0.5 and 1.0 with carboxylic groups fixated in all position of the AGU, but predominantly in positions C-2 and C-6. The sodium salt of CMC is a weak polyelectrolyte with a pK_a value of $3.0–4.0 \pm 0.2$. Homogeneous carboxymethylation results in a different substitution pattern, with a preference for the C-6 position and of course uniform distribution along the chain. The water solubility of CMC, the most relevant property, is mainly dependent on the DS, DP, and cellulose accessibility and etherification procedure.

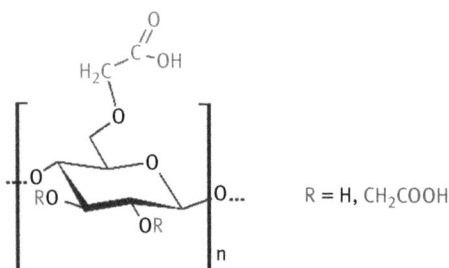

$R = H, CH_2COOH$

Fig. 12.10: Structure proposed for carboxymethyl cellulose.

The synthesis of CMC by homogeneous process was performed in various nonaqueous solvents as reaction media, for instance, *N,N*-dimethylacetamide/lithium chloride (DMAc/LiCl). The CMC prepared in the DMAc/LiCl system and solid NaOH showed a medium DS and exhibited a higher nonuniformity of substitution along the polymer chains.

CMC with DS values at higher than 2.2 is available in a one-step synthesis in *N*-methylmorpholine-*N*-oxide (NMMNO) and in dimethyl sulfoxide/tetrabutylammonium fluoride (DMSO/TBAF) [41]. A completely homogeneous cellulose carboxymethylation with sodium monochloroacetate is possible in the presence of an aqueous NaOH solution, for example, Ni(tren)(OH)$_2$ [tren = tris(2-aminoethyl)amine] system [42].

An efficient method for the synthesis of CMC is the conversion of cellulose into triacetate cellulose (CTA) intermediate, which then is converted to CMC by deprotection, followed by etherification using NaOH and ClCH$_2$COONa in the tetrahydrofuran-water (THF-H$_2$O) solvent system [43]. Using this method, CMC with a maximum DS of 1.02 can be obtained, at 50 °C in 1 h using 0.95 M and 1.1 M solutions of NaOH and ClCH$_2$COONa, respectively.

As with the other derivatives, various approaches have been proposed for the structural characterization of CMC. Thus, the structure–property relationship of the CMC is supported by determining DS and the distribution of the substituent in monomers. DS is easily determined by the direct titration method of the carboxylic groups [44] or by NMR spectroscopy. The properties of CMC, such as solubility and viscosity, are closely related to the individual DS at the C-2, C-3, and C-6 positions along the cellulose chains, as well as the total DS. Usually, a slight preference for the C-2 position of AGU unit was reported when compared with that of C-6 and a low substitution in the C-3 position, for samples with DS > 1, synthesized both homogeneous and heterogeneous conditions. CMC with a preferential C-6 substitution can be realized by a two-phase process, with benzene and ethanol as component of the system (in this case DS < 1).

Another approach for structural characterization of CMC polymeric chains is the quantitative NMR spectroscopic analysis of the derivatives of CMC. NMR is one of the most powerful and simple analytical methods for characterizing cellulose derivatives [45]. In Fig. 12.11 the ^{13}C NMR spectra of CMC with various DS are shown, and in Fig. 12.12 the HSQC and HSQC-TOCSY spectra of CMC with DS about 1 are shown.

In Tab. 12.4 the empirical equations used to calculate the molecular weight of CMC in different environments are listed.

The rheological properties of NaCMC solutions vary within wide limits with DP DS and the amount and the structure of supramolecular aggregates (gel particles) in systems. The thermal analysis of NaCMC shows a sharp peak at 297 °C, which is attributed to decarboxylation processes of carboxylated glucose units and/or carboxymethyl group decomposition. At 366 °C, another stage of maximum mass loss rate is observed. The solid residue at 700 °C was around 25% w/w [48].

Carboxyethyl cellulose (CEC) can be prepared in analogy to CMC by etherification with halogenpropionic acid, but also by homogeneous saponification of cyanoethylcellulose synthetized by Michael addition of acrylonitrile onto cellulose in weakly alkaline conditions. Due to the greater spacer length between the polymer backbone and carboxylic group, CEC synthesized by latter procedure becomes water soluble at DS between 0.15 and 0.20.

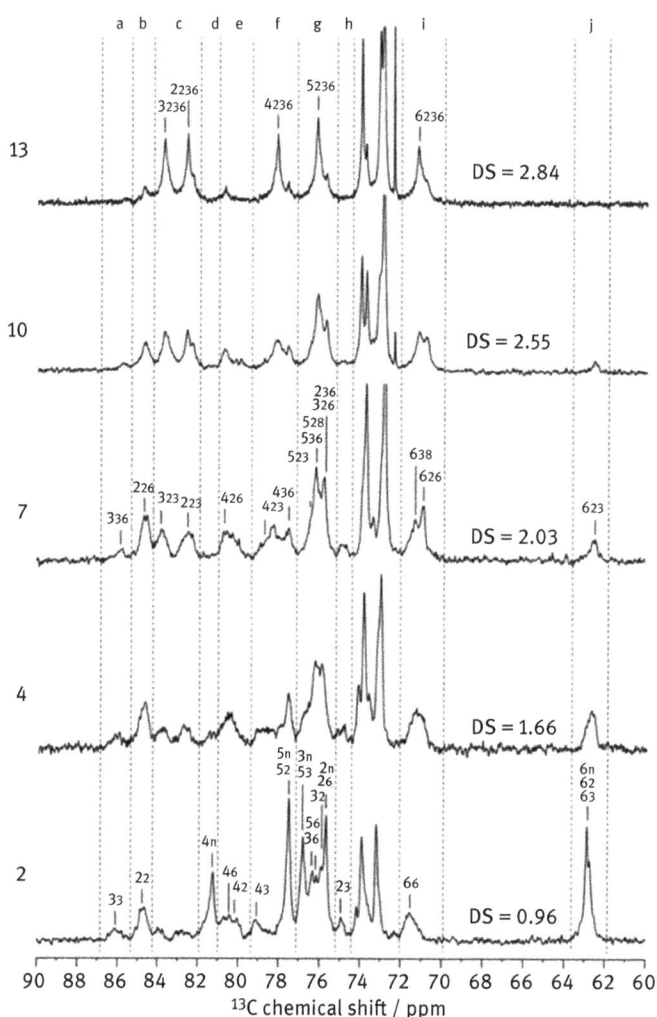

Fig. 12.11: Quantitative ^{13}C NMR spectra of CMC with various DS in D$_2$O at 363 K [45]. Reproduced by permission of Wiley, copyright 2019.

Dicarboxymethylcellulose (DCMC) carrying two carboxylic groups at same C atom was synthesized by etherification of cellulose dissolved in LiCl/DMAc system with bromomalonic acid. A sample with DS > 0.3 was completely soluble in water at neutral pH and had a pK_a = 2.1.

Cellulose xanthogenate: This derivative plays an important role in viscose fiber manufacturing processes, and is synthesized by reaction of alkalycellulose and

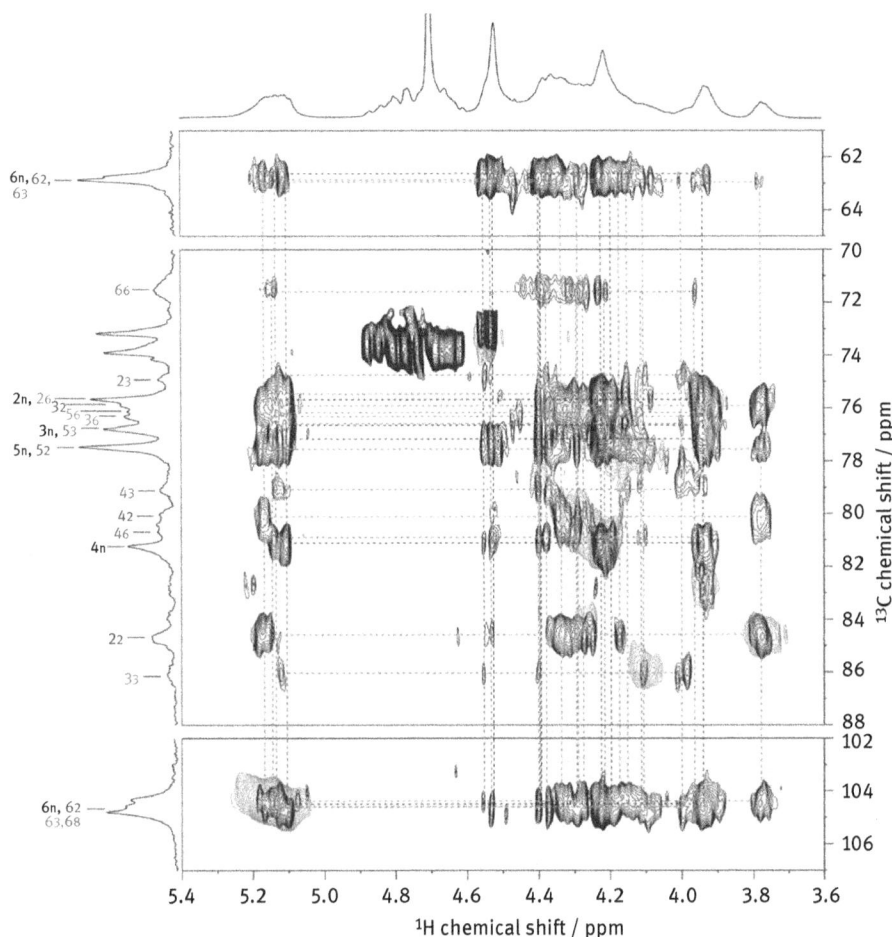

Fig. 12.12: 2D contour plots of the ^1H–^{13}C HSQC-TOCSY spectrum of CMC (DS = 0.96) recorded with a TOCSY mixing time of 120 ms [45]. Reproduced by permission of Wiley, copyright 2019.

carbon disulfide (CS$_2$) at 30 °C. The xanthogenate is unstable in aqueous medium over the whole range of pH and rapidly decomposes in an alkaline medium by losing slowly anionic groups and forming Na$_2$CS$_3$, Na$_2$S, and Na$_2$CO$_3$ [19].

12.3.1.2 Cationic derivatives of cellulose

Cationic cellulose derivatives are industrially very important compounds and have many applications in food, chemical, pharmaceutical, paper industry, cosmetics, and textiles, in flotation and flocculation, and in drilling fluids. The easiest way to

Tab. 12.4: Empirical equations used for calculate the molecular weight.

Solvent	Viscosity	Ref.
NaCl 0.2 M	$[\eta] = 4.3 \times 10^{-4} \times M_W^{0.74}$	[46]
NaCl 0.05 M	$[\eta] = 1.90 \times 10^{-4} \times M_W^{0.82}$	[46]
NaCl 0.01 M	$[\eta] = 8.10 \times 10^{-5} \times M_W^{0.92}$	[46]
NaCl 0.005 M	$[\eta] = 7.20 \times 10^{-5} \times M_W^{0.95}$	[46]
Infinite ionic strength	$[\eta] = 1.90 \times 10^{-3} \times M_W^{0.6}$	[46]
Water-cadoxene	$[\eta] = 1.93 \times 10^{-3} \times M_W^{1.0}$	[47]

Here, $[\eta]$ is the intrinsic viscosity, and M_w the molecular weight.

synthesize ethers with an amino group is the reaction of cellulose with ethylene imine in an aqueous alkaline medium.

Epoxidation of cellulose to an alkylcelluloses with quaternary ammonium end-groups were reported in the literature [49]. Cellulose was dissolved in the LiCl/DMAc system after an adequate preactivation and then etherified with glycidyl-trimethyl-ammonium chloride. The cationic ether precipitated from the reaction system and was purified by ultrafiltration. Water-soluble samples with DS values up to 1 were obtained. The *aminoalkylation* of cellulose also leads to also water-soluble cationic derivatives by reaction of chloroethyldiethyl when diethylaminoe-thylcellulose (DEAE-cellulose) was obtained. The DEAE-cellulose was synthetized in homogeneous conditions by dissolved cellulose at −10 °C in aqueous NaOH and then etherified at room temperature, when a sample with DS = 0.25–0.30 was obtained. The quaternized cellulose (QC) derivatives were synthesized in a homogeneous medium by dissolving the cellulose in the aqueous NaOH or NaOH/urea system and used various quaternization agents [50, 51]. These derivatives have been shown to be effective in pharmaceutical fields as carrier for anionic drugs, genes, protein, or can be used as flocculation agents in water purification process.

12.3.2 Nonionic derivatives of cellulose

The nonionic cellulose derivatives can be also synthesized by esterification and etherification reactions. The most known nonionic derivatives are cellulose acetate, methylcellulose, ethylcellulose, hydroxyethylcellulose, hydroxypropylmethylcellu-lose, silyl ethers of cellulose, and others.

12.4 Recent applications of ionic derivatives of cellulose

Cells encapsulation: The implantation of encapsulated cells overexpressing specific biomolecules offers a means to treat a variety of diseases, including cancer. Encapsulated cells have been evaluated in clinical trials for the treatment of diabetes as well as for the treatment of pancreatic and mammary cancer. A number of different materials can be used for encapsulation of cells, such as alginate, agarose, and poly-sulphone, but *cellulose sulfate* presents some advantages over other encapsulation materials. Cellulose sulfate was used for the encapsulation of mammalian cells and it has been shown to provide a means to protect cells from rejection by the immune system, to localize the cells at the site of implantation, to provide a long-term micro-environment for survival of the cells, and to allow release of biomolecules from the capsules, as well as entry of food and nutrients for the cells (Fig. 12.13) [52–55].

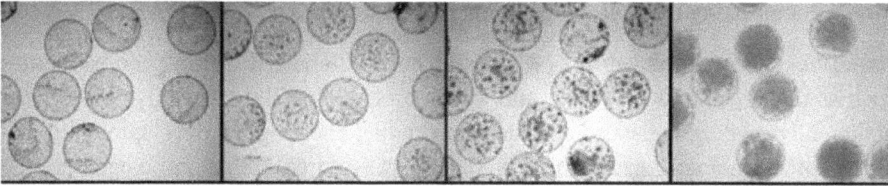

Fig. 12.13: Growth of Hut-78 cells within cellulose sulfate capsules over time. Photomicrographs taken on day 0, day 2, day 8, and day 17 (day of encapsulation) demonstrate continual growth of the cells within the capsules and good filling of the capsules by day 17. The capsules are in all cases regular with a measured diameter of 0.7 mm [54]. Reproduced by permission of Elsevier, copyright 2019.

Ortner et al. [56] succeeded encapsulating HEK293 cells with magnetic nanoparticles using a biologically inert polyanion (sodium cellulose sulfate) as a matrix. The capsule formation was enabled by a modified standard procedure based on gelating sodium cellulose sulfate in a solution of polycationic poly-diallyl dimethyl ammonium chloride. In addition, using the ability of cellulose ionic derivatives to form polyelectrolyte complexes with polymers of opposite charge, they can be successfully used in various biomedical applications such as cell immobilization and drug delivery [57]. A potential candidate for the encapsulation of cells, such as β-cells that form islets to activate the secretion of insulin, could be the alginate–cellulose composites, based on TEMPO-mediated oxidized bacterial cellulose (TOBC). Park et al. [58] confirmed that when TOBC are incorporated into alginate hydrogels, encapsulated cells are more viable and proliferate more easily due to the 3D fibrous contribution of TBCs, which is similar to the extracellular matrix.

3D printing: Three-dimensional (3D) printing is classified as a revolutionary, disruptive manufacturing technology, leading to major innovations in a broad range of areas, including energy, biotechnology, medical devices, and many more. Carbohydrate polymers from natural sources have been studied and used in the pharmaceutical and biotechnological industries for many years. Besides its desired gelling properties, these natural polysaccharides have been used as hydrogels for 3D bioprinting [59]. For example, in the presence of divalent cations, such as calcium, anionic polysaccharides are able to form gels because of associations between the calcium ion and the acid residues [60]. The CMC is used in many products such as stabilizer, emulsifier, tablet disintegrant, viscosity modifier, suspending aid, binder, film former, water absorbent, sizing aid, and metal ion adsorbent. In 3D printing the properties of inks are of great importance. Thus, CMC is often used as viscosity modifier in 3D printing inks, increasing the ink viscosity and slowing the flocculation kinetics, facilitating the fabrication of unsupported spanning structures.

CMC is a widely used matrix material due to its viscosity thickening capability and thixotropic rheology; thereby, Park et al. [61] prepared a silver nanowire-based 3D conductor using the CMC matrix (Fig. 12.14).

Nanofibers: Nanofibers are a new class of nanomaterials with inherited properties such as the large surface-to-area ratio, high porosity, flexibility, stability, and permeability. Electrospinning is also a modern technique, enabling the development of matrices with nanometer-sized fibers with similar features and morphologies to the extracellular matrix (ECM). The ECM is the noncellular component present within all tissues and organs, and provides not only essential physical scaffolding for the cellular constituents but also initiates crucial biochemical and biomechanical cues that are required for tissue morphogenesis, differentiation, and homeostasis. Materials with similar properties and structure are therefore believed to stimulate cell proliferation and encourage wound healing. Electrospun matrices have also been shown to provide a high-surface area and microporosity, making them ideal for loading of drugs or other biomolecules. Different fibers made from natural polymers are often used as drug carriers, absorbents, or moisturizers in wound dressings. The CMC as sodium salt (NaCMC) is one of the most common functional parts of different modern wound dressings, either alone or as part of multilayered wound dressings. CMC with different molecular weight and DS was electrospun in the presence of polyoxyethylene (PEO) as copolymer [62]. The nanofibers displayed homogeneous structures with mean diameters of 200–250 nm, regardless of the molecular weight and DS. The CMC/PEO/hydroxyapatite nanofibers were also obtained for use in regenerative medicine [63].

Gelatin/CMC mixture was used to produce nanofiber scaffold to reduce the cost of prosthetic skin. The scaffold with an 80/20 gelatin/CMC ratio presented the largest pores in the structure and also showed 44.67% swelling ratio [64]. Electrospinning process of nanofibers specially based on derivatives of cellulose presents high behest for developing several kinds of novel drug delivery systems (DDSs) due to their

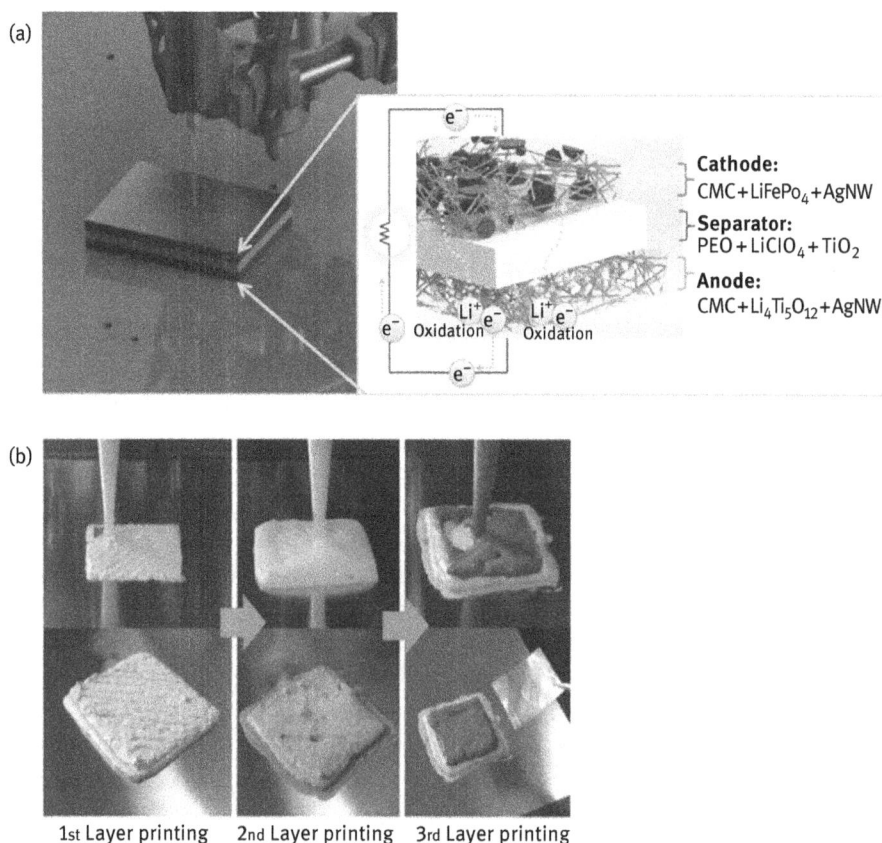

Fig. 12.14: Computer-aided design representation of three-layered 3D-printed battery and a detail schematic overview of layers (a); images of 3D printing process of the 3D printed battery (anode, electrolyte, and cathode are sequentially printed to form a three-layered battery structure). (b) Adapted with the permission from Ref. [61], Copyright© 2017.

specific characteristics, the simplicity, beneficial, and impressive fabricating procedure. Other applications of nanomaterials based on derivatives cellulose nanofibers are membrane for water treatment [65].

Drug delivery systems: Oral route is one of the most widely used noninvasive ways of administering medicines, avoiding the disturbance and pain of the patient. Oral controlled release systems were designed to deliver drugs at duration, sustained, and controlled frequency besides keeping plasma concentrations of drug at therapeutic levels. The efficiency of drug used by patients is often limited by delivery route, drug instability, immunogenicity, and physiological barriers. Consequently, continuous efforts are being made to develop DDSs. Various techniques such as emulsion, dry spraying, and extrusion are often used in the industry to produce drug particles encapsulated with various materials.

The ionic derivatives of cellulose, especially CMC, are frequently used in formulation of controlled drug release systems. CMC included in drug administration systems can prevent crystallization of the active principle and also protect it from degradation. Furthermore, CMC could improve the frequency of drug administration by increasing the drug diffusion rate from the polymer matrix, by the decrease erosion (degradation) rate of polymer, or by gels formed by CMC in contact with water. In this regard, the CMC was included in various oral controlled delivery as tablets, films, nanocomposite, or capsules as drug/cross-linked systems, thereby increasing the efficiency of administration compared with system without CMC [66]. Thus, CMC and Gum acacia were used as the adsorbent and solid carriers to prepare a dry emulsion based on indomethacin-5-fluorouracilmethyl ester. Studies have shown an improvement in oral drug administration and also a reduction in side effects of drugs [67]. Calcium alginate (CA)–CMC beads have been prepared as carriers by ionic gelation method for colon-specific oral drug delivery. The swelling, mucoadhesiveness, biodegradability, and pH-sensitive studies encourage of the use system in the administration of colon-specific drugs [68]. Another DDS based on CMC-grafted graphene oxide with pH-sensitive and controlled drug-release properties was developed for loaded of methotrexate [69]. This delivery system exhibited several advantages: high drug-loading efficiency (39.33%) and encapsulation efficiency (29.50%); good pH-dependent drug-release properties; reduced cytotoxicity against NIH-3T3 normal cells but high cytotoxicity against HT-29 cancer cells; higher plasma drug concentration and longer action time; and superior tumor inhibition activities and liver metastasis-inhibition activities. Recently, eco-friendly smart vehicles based on zinc oxide (ZnO) nanoparticles incorporated into CMC beads and then coated with a CS layer via a self-assembly technique to form core–shell polyelectrolyte complexes were also investigated for cancer colon-specific drug delivery [70].

The pH stimuli-responsive microspheres from monobasic CP (Fig. 12.15) were synthesized by chemical cross-linking with epichlorohydrin, using the water-in-oil inverse emulsion technique, in order to use as DDSs [71]. Nanoparticles based on organol-soluble ionic cellulose derivatives containing carboxyl or amino groups, with average diameters of 80–330 nm, were synthesized by Wang et al. [72] by nanoprecipitation. Based on the sustainable character of cellulose and reversible pH sensitivity, these nanoparticles could be used for nonadhesive coatings and targeted drug delivery.

The microparticles based on CS were prepared using micronization technique, so-called supercritical fluid-assisted atomization introduced by hydrodynamic cavitation mixer. By using this procedure, the uniform CS microspheres with narrow (0.3–2.0 μm) and perfect spherical morphology were obtained. In addition, the pharmaceutical composites based on protein-Na-CS with controlled size distributions for sustained release in various drug delivery strategies can be performed

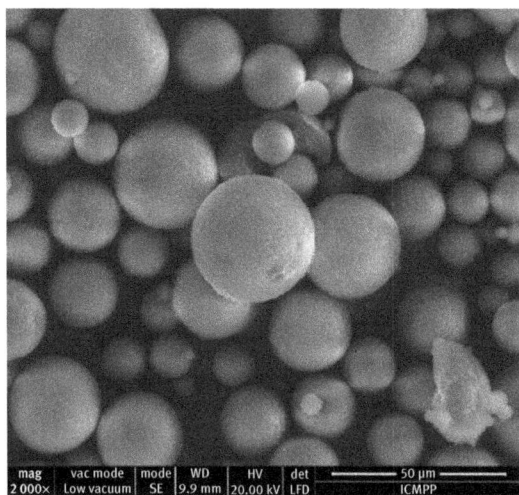

Fig. 12.15: SEM photograph of microspheres from monobasic cellulose phosphate.

using this supercritical fluid process without any residual organic solvent [73]. The cationic cellulose derivatives were also used to prepare nanoparticles. The quaternized cellulose with 3-chloro-2-hydroxypropyltrimethylammonium chloride (DS = 0.46 or 0.63) was used to obtain nanoparticles by ionic crosslinking with sodium tripolyphosphate (TPP). The obtained nanoparticles showed high loading efficiency of the bovine serum albumin (BSA) [74].

Anticoagulant product: Polysaccharide sulfates in organisms exhibit biological activities, such as antiviral coagulation and blood coagulation. Thus, the anticoagulant efficacy was investigated by Wang et al. [75] using in vitro and in vivo coagulation assays and amidolytic tests in comparison with heparin. The results indicated that Na-MCS exhibited higher anticoagulation activity based on activated partial thromboplastin time assay and prolonged the thrombin time to a lesser extent than heparin. Subcutaneous administration of Na-MCS to mice increased the clotting time in a moderate dose-dependent manner with a longer duration. Na-MCS exhibited anticoagulation activity mainly by accelerating the inhibition of antithrombin III on coagulation factors FIIa and FXa in plasma. The effects of molecular weight of CS and sulfate group distribution on anticoagulation were also investigated and reported in literature [76].

Contraceptives: In the medical field, Na-CS has been used as a contraceptive and vaginal microbicide. In vitro antifertility and antimicrobial effects studies have been performed for rabbits. In addition, in clinical trials, the CS with a lower molecular weight than the current clinical formulation showed a high degree of contraceptive efficacy [77].

Antiviral HIV treatments: CS with high DS values showed a high antihuman-immunodeficiency virus (anti-HIV) activity. The anti-HIV test was expressed as the number of viable cells on the third and sixth days of incubation of partially infected MT-4 cells in the presence of CS. It is suggested that the anti-HIV activity of these sulfated compounds is due to the inhibition of viral binding with MT-4 cells [17]. Research has shown that CS is one of the most widely used cellulose derivatives for antiviral treatment, being marketed through the Ushercell™ drug. This drug based on CS with high molecular weight, developed by Polydex (Canada), was found active against HIV, herpes simplex virus, *Neisseria gonorrhoeae*, *Chlamydia trachomatis*, papillomaviruses, and *Gardnerella vaginalis* [78, 79].

Orthopedic biomaterials: CP can promote the growth of calcium phosphate crystals, being used as an implantable orthopedic biomaterial [21, 80]. Once implanted, PC could promote the formation of calcium phosphates, having therefore closer resemblance to bone functionality and assuring a satisfactory bonding at the interface between hard tissue and biomaterial [81]. The PCell can influence the crystallization/separation of slightly soluble salts such as calcium carbonate and calcium oxalate with an increase of their stability in aqueous dispersions [30]. These properties, in connection with the biocompatibility and lack of toxicity of the parent polymer, are promising for the use of monobasic CP in such medical applications as additive by the preparation of hydroxyapatite or hydroxyapatite-based organoinorganic composites by hydrothermal method, and in renal excretion therapy of oxalate by crystallization of calcium oxalate in urine in patients with absorptive hypercalciuria [82].

Membrane/film: Incorporation of CS (with high anticoagulant activity) in membranes or biomaterials from cellulose increases the hemocompatibility of dialysis membranes or other biomaterials [83]. In addition, CP incorporated into hemodialysis membranes prohibits the activation of detrimental blood proteins in hemodialysis. The CMC films were prepared with the aim to obtain a human skin equivalent for testing the adhesive properties of medical patches. The films exhibited hydrophilic character and acceptable mechanical properties. Moreover, their surface roughness and friction coefficient values were in reasonably close range to human skin parameters [84]. The CMC films are sensitive to moisture and exhibit both tribological and morphological properties similar to human skin. The biodegradable films prepared by cross-linking CMC and polyvinyl alcohol (PVA) showed that the increase of the CMC content induces the increase of water absorption and permeability. The CMC/PVA films has a smooth surface and the good biodegradability that recommends the CMC/PVA blend films to be used as a coating material in agriculture [85].

The CP is used as an efficient and nonconventional polyelectrolyte in preparation of films by layer-by-layer (LbL) procedure. Ullah et al. [86] demonstrated the ability of CP to be used as a binding and dispersing agent for mounting photoactive TiO_2 on glass, quartz, silicon, and bacterial cellulose by LbL technique. Moreover, they have introduced polyoxometalates such as phosphotungstic acid (HPW) for expanding the application of films as photochromic and photocatalytic materials. Due to the high affinity of CP for TiO_2 via Ti–O–P bonds, the CP/TiO_2 films showed good photoactivity, while $CP/HPW/TiO_2$ films showed much improved photochromic behavior. The proposed method for the obtaining of these types of films is simple, economical, and environmentally safe and offers more practicability and versatility for the use of TiO_2 and HPW in various applications such as photocatalysis, photochromism, as well as self-cleaning devices.

HPCs with phosphate groups are used for the deposition of ultra-thin layers on metal surfaces like aluminum, titanium, or steel for adhesion promotion and corrosion inhibition [87] with application in medical field.

Separation process: Increasing environmental degradation has motivated many researchers to develop new low-cost adsorbents derived from renewable resources. From this point of view, the cellulose derivatives find applications as adsorption materials for removal of heavy metal ions from waste streams and industrial effluents. The ion-exchange capacity of the CP increases with a decrease in hydrate ionic radii, and with an increase in quantity and concentration of metal ions in the following order, for divalent ions: $Ba^{2+} > Sr^{2+} > Ca^{2+} > Mg^{2+}$. In addition, these derivatives have shown a high affinity for certain cations, principally Th^{4+}, Ti^{4+}, U^{4+}, Ce $^{4+}$, Fe^{3+}, ZrO^{2+}, and UO_2^{2+} [1]. CP and CMC have also been used as a *cation-exchange material* in the treatment of calcium-related diseases or complexation of heavy metals ions such as Cd(II), Ni(II), Cu(II), Mn(II), and Pb(II) [88, 89].

Nadagouda et al. reported a simple and convenient approach to obtain nanocomposites based on CMC, at room temperature in aqueous medium, for the retention of various metal salts such as Cu, In, Fe, and Ag. These CMC nanocomposites could be used in biolabeling for ultrasensitive detection of biological species such as antibodies, DNA, and cells, thus replacing other toxic semiconducting nanoparticles-based biolabeling [90].

Ecological problems recommend the use of natural polyelectrolytes in the treatment of domestic and industrial wastewater. Since most of the pollutants are negatively charged (clays, metal oxides, dyes, emulsions, etc.) in the separation/purification processes, cationic polyelectrolytes are frequently used. Thus, the use of PCell as a flocculant agent for retention of zinc and ferric oxide particles is presented in the literature [39].

The anionic microspheres based on monobasic CP were obtained by chemical cross-linking with epichlorohydrin and could be used as *dyes absorbent materials*. The efficiency of PCell microspheres as adsorbent for the retention of cationic dyes (methylene blue and rhodamine 6G) increases with an increase in the pH of the medium, reaching high values at pH 6, when all phosphorous groups of the sorbent are dissociated [71]. Carboxylate-functionalized adsorbent based on cellulose nanocrystals was prepared and used for the adsorption and removal of multiple cationic dyes (crystal violet, methylene blue, malachite green, and basic fuchsin). The maximum cationic dyes uptake range was found to be from 30.0 to 348.9 mg/g. Adsorption kinetic studies revealed a second-order pseudo-model and the thermodynamic analysis revealed that the adsorption process was spontaneous and exothermic [91]. Cationic cellulose derivatives are also used to remove acid dyes from the aqueous effluent produced by the textile industry. The maximum color removal efficiency measured was found 99.2% at pH 3 [92].

Flame retardant additive: CP was first formulated as a flame retardant additive for textiles due to its inherent flame resistance [1]. Since then, the number of applications for this derivative has increased. Due to its excellent ion-exchange properties, CP has been extensively used for the separation of trace metals [89]. Its biocompatibility and bioactivity have also generated substantial attention for the potential applications of CP in biomedical research. The ion-exchange ability of phosphorylated cellulose and its strong affinity for divalent cations means that it can inhibit the absorption of dietary calcium, which is beneficial for kidney stone treatment.

Applications in foods: Cellulose derivatives have been used in the manufacture of processed foods since a long time. Cellulose derivatives (in particular CMC) have five important roles in the food industry, such as the organization of flow properties, emulsification and foam stability, alteration of the formation of ice crystals, growth, and the ability to bind water. The ability of the derivatives to form thin layers has led to their use in coating foodstuffs to improve the fresh quality of frozen and processed meat, poultry, and seafood, reducing fat oxidation and discoloration, thus improving the appearance of products in retail packages by eliminating the drop and sealing the volatile aromas. Moreover, it functions as a support for food additives, such as antimicrobial agents and antioxidants [93, 94].

12.5 Conclusions

Generally, the properties of cellulose derivatives are mainly determined by the type of the functional group, but they can be significantly modified by changing the DS and DP, as well as by different patterns of substitution both within anhydroglucose unit and along the macromolecular chain.

The water insolubility of cellulose, generally attributed to the existence of extended/intermolecular hydrogen bonds, determines that most chemical modification reactions occur through heterogeneous processes.

The introduction of ionic groups on the cellulosic chain leads to obtain soluble derivatives, thus expanding the cellulose application fields. The production of water-soluble cellulose derivatives generally proceeds without residues of hazardous monomers and has excellent biocompatibility, which can be considered as a decisive advantage in biomedical or nutritional use.

Based on numerous research efforts emphasized by recent literature, it may be assumed that important future trends in cellulose esters chemistry will be closely related to the developments of modern top nanotechnologies.

List of abbreviations

AGU	D-Anhydrogluco pyranose unit
$Ca(OH)_2$	Calcium hydroxide
CEC	Carboxyethylcellulose
$ClSO_3H$	Chlorosulfonic acid
CMC	carboxymethyl cellulose
CMCP	CMC containing phosphate group
CP	Cellulose phosphate
CS	Cellulose sulfate
CS_2	Carbon disulfide
CTA	Cellulose triacetate
D_2O	Deuterium oxide
DCP	Dibasic cellulose phosphates
DDSs	Drug delivery systems
DMAc	*N,N*-dimethylacetamide
DMF	*N,N*-dimethylformamide
DMSO	Dimethyl sulfoxide
DP	Degree of polymerization
DS	Degree of substitution
DS_{Ac}	Degree of substitution of cellulose acetate
ECM	Extracellular matrix
Et_3PO_4	Triethylphosphate
FTIR	Fourier-transform infrared spectroscopy
H_2SO_4	Sulfuric acid
H_3PO_3	Phosphorous acid
H_3PO_4	Phosphoric acid
HPC	Hydroxypropyl cellulose
HSQC	Heteronuclear single quantum correlation
KCl	Potassium chloride
KOH	Potassium hydroxide
LbL	Layer-by-layer
LiCl	Lithium chloride
N_2O_4	Dinitrogen tetroxide
Na-CMC	Carboxymethyl cellulose in sodium salt
Na-CS	Sodium cellulose sulfate
NaOH	Sodium hydroxide
$(NH_4)_2HPO_4$	Ammonium phosphate dibasic
NH_2SO_3H	Sulfamidic acid/amidosulfonic acid

NMP	n-Methyl-2-pyrrolidone
NMR	Nuclear magnetic resonance
P_2O_5	Phosphorus pentaoxide
PCell	Monobasic cellulose phosphate
PEO	Polyoxyethylene
$POCl_3$	Phosphorus (v) oxychloride
PVA	Polyvinyl alcohol
QC	Quaternized cellulose
SEM	Scanning electron microscopy
SO_2	Sulfur dioxide
SO_2Cl_2	Sulfuryl chloride
SO_3	Sulfur trioxide
STMP	Sodium trimetaphosphate
TOBC	Tempo-mediated oxidized bacterial cellulose
WAXD	Wide-angle X-ray diffraction
XRD	X-ray diffraction

References

[1] Ciolacu D, Olaru L, Suflet D, Olaru N. Cellulose ester – from Traditional Chemistry to Modern Approaches and Applications. In: Popa VI., ed. Pulp production and processing: From papermaking to high-tech products, iSmithers Rapra, 2013, 253–283.

[2] Hon DNS. Cellulose and its Derivatives: Structures, Reactions, and Medical Uses. In: Dumitriu S., ed. Polysaccharides in medicinal applications. New York, Marcel Dekker, Inc., 1996, 87–105.

[3] Klemm D, Philipp B, Heinze T, Heinze U, Wagenknecht W. General considerations on structure and reactivity of cellulose. Comprehensive cellulose chemistry: Fundamentals and analytical methods, 1, Wiley-VCH Verlag GmbH, Weinheim, Germany, 1998, 9–165.

[4] Hettrich K, Wagenknecht W, Volkert B, Fischer S. New possibilities of the acetosulfation of cellulose. Macromol. Symp. 2008, 262, 162–169.

[5] Zhang K, Peschel D, Brendler E, Groth T, Fischer S. Synthesis and bioactivity of cellulose derivatives. Macromol. Symp. 2009, 280, 28–35.

[6] Zhang K, Peschel D, Baucker E, Groth T, Fischer S. Synthesis and characterisation of cellulose sulfates regarding the degrees of substitution, degrees of polymerisation and morphology. Charbohydr. Polym. 2011, 83, 1659–1664.

[7] Yao S. An improved process for the preparation of sodium cellulose sulphate. Chem. Eng. J. 2000, 78(2–3), 199–204.

[8] Chen G, Zhang B, Zhao J, Chen H. Improved process for the production of cellulose sulfate using sulfuric acid/ethanol solution. Carbohydr. Polym. 2013, 95(1), 332–337.

[9] Gohdes M, Mischnick P. Determination of the substitution pattern in the polymer chain of cellulose sulfates. Carbohydr. Res. 1998, 309, 109–115.

[10] Zhang K, Brendler E, Geissler A, Fischer S. Synthesis and spectroscopic analysis of cellulose sulfates with regulable total degrees of substitution and sulfation patterns via ^{13}C NMR and FT Raman spectroscopy. Polymer. 2011, 52, 26–32.

[11] Hweige RG. Polysaccharide sulfates. I. Cellulose sulfate with a high degree of substitution. Carbohydr. Res. 1972, 21, 219–228.

[12]   Thomas M, Chauvelon G, Lahaye M, Saulnier L. Location of sulfate groups on sulfoacetate derivatives of cellulose. Carbohydr. Res. 2003, 338, 761–770.

[13]   Muhitdinov B, Heinze T, Turaev A, Koschella A, Normakhamatov N. Homogenous synthesis of sodium cellulose sulfates with regulable low and high degree of substitutions with SO₃/Py in N,N-dimethylacetamide/LiCl. Eur. Polym. J. 2019, 119, 181–188.

[14]   Muhitdinov B, Heinze T, Normakhamatov N, Turaev A. Preparation of sodium cellulose sulfate oligomers by free-radical depolymerization. Carbohydr. Polym. 2017, 173, 631–637.

[15]   Schweiger RG. New cellulose sulfate derivatives and applications. Carbohydr. Res. 1979, 70, 185–198.

[16]   Groth T, Wagenknecht W. Anticoagulant potential of regioselective derivatized cellulose. Biomaterials. 2001, 22, 2719–2729.

[17]   Yamamoto I, Takayama K, Honma K, et al. Synthesis, structure and antiviral activity of sulfates of cellulose and its branched derivatives. Carbohydr. Polym. 1991, 14, 53–63.

[18]   Zhang K, Brendler E, Fischer S. FT Raman investigation of sodium cellulose sulphate. Cellulose. 2010, 17, 427–435.

[19]   Klemm D, Philipp B, Heinze T, Heinze U, Wagenknecht W. General considerations on structure and reactivity of cellulose. Comprehensive cellulose chemistry: Fundamentals and analytical methods. Wiley-VCH Verlag GmbH, Weinheim. Germany, 1998, 2, 133–160.

[20]   Reid JD, Mazzeno LW. Preparation and properties of cellulose phosphates. Ind. Eng. Chem. 1949, 41(12), 2828–2831.

[21]   Granja PL, Pouysegu L, Petraud M, De Jeso B, Baquey C, Barbosa MA. Cellulose phosphates as biomaterials. I. Synthesis and characterization of highly phosphorylated cellulose gels. J. Appl. Polym. Sci. 2001, 82, 3341–3353.

[22]   Granja PL, Pouysegu L, Deffieux D, et al. Cellulose phosphates as biomaterials. II. Surface chemical modification of regenerated cellulose hydrogels. J. Appl. Polym. Sci. 2001, 82, 3354–3365.

[23]   Wanrosli WD, Rohaizu R, Ghazali A. Synthesis and characterization of cellulose phosphate from oil palm empty fruit bunches microcrystalline cellulose. Carbohydr. Polym. 2011, 84, 262–267.

[24]   Wanrosli WD, Zainuddin Z, Ong P, Rohaizu R. Optimization of cellulose phosphate synthesis from oil palm lignocellulosics using wavelet neural networks. Ind. Crop. Prod. 2013, 50, 611–617.

[25]   Nuessle AC, Ford FM, Hall WP, Lippert AL. Some aspects of the cellulose-phosphate-urea reaction. Text. Res. J. 1956, 26(1), 32–39.

[26]   Wagenknecht W, Nehls I, Philipp B, Schnabelrauch M, Klemm D, Hartmann M. Untersuchungen zur bildung wasserloslicher cellulose phosphate durch homogene acylierung im system cellulose/N₂O₄/DMF. Acta. Polym. 1991, 42(11), 554–560.

[27]   Wagenknecht W, Philipp B, Schleicher H. Zur veresterung und auflosung der cellulose mit saureanhydriden und saurechloriden des schwefels und phosphors. Acta. Polym. 1979, 30, 108–112.

[28]   Nehls I, Loth F. ¹³C-NMR-spektroskopische untersuchungen zur phosphatierung von cellulose produkten im system H₃PO₄/Harnstoff. Acta. Polym. 1991, 42(5), 233–235.

[29]   Pasqui D, Rossi A, Di Cintio F, Barbucci R. Functionalized titanium oxide surfaces with phosphated carboxymethyl cellulose: Characterization and bonelike cell behavior. Biomacromolecules. 2007, 8(12), 3965–3972.

[30]   Suflet DM, Chitanu GC, Popa VI. Phosphorylation of polysaccharides: New results on synthesis and characterisation of phosphorylated cellulose. React. Funct. Polym. 2006, 66 (11), 1240–1249.

[31] Inagaki N, Nakamura S, Asai H, Katsuura K. Phosphorylation of cellulose with phosphorous acid and thermal degradation of the product. J. Appl. Polym. Sci. 1976, 20, 2829–2836.

[32] Nam S, Condon BD, White RH, Zhao Q, Yao F, Cintron MS. Effect of urea additive on the thermal decomposition kinetics of flame retardant greige cotton nonwoven fabric. Polym. Degrad. Stabil. 2012, 97, 738–746.

[33] Choi U-S, Ahn B-G. Electrorheology of cellulose phosphate ester suspension as a new anhydrous ER fluid. Colloids Surf. A. 2000, 168, 71–76.

[34] Park DP, Hwang JY, Choi HJ, Kim CA, Jhon MS. Synthesis and characterization of polysaccharide phosphates based electrorheological fluids. Mat. Res. Innovat. 2003, 7, 161–166.

[35] Huang MR, Li XG. Thermal degradation of cellulose and cellulose esters. J. Appl. Polym. Sci. 1998, 68, 293–304.

[36] Suflet DM, Nicolescu A, Popescu I, Chitanu GC. Phosphorylated polysaccharides. 3. Synthesis of phosphorylated curdlan and its polyelectrolyte behaviour compared with other phosphorylated polysaccharides. Carbohyd. Polym. 2011, 84(3), 1176–1181.

[37] Nagaya J, Minakata A, Tanioka A. Effects of the charge density and counterion species on the conductance of ionene solutions. Colloids Surf. A. 1999, 148, 163–169.

[38] Minakata A, Takayama K, Yano S, Tanaka Y, Ariki T, Shimizu T. Polyelectrolytic behavior of a novel fluorine-containing ionomer, PPFA. J. Phys. Chem. B 2003, 107, 8146–8151.

[39] Ghimici L, Suflet DM. Phosphorylated polysaccharide derivatives as efficient separation agents for zinc and ferric oxides particles from water. Sep. Purif. Technol. 2015, 144, 31–36.

[40] Heinze T, Pfeiffer K. Studies on synthesis and characterization of carboxymethylcellulose. Angew. Makromol. Chem. 1999, 266, 37–45.

[41] Ramos LA, Frollini E, Heinze T. Carboxymethylation of cellulose in the new solvent dimethylsulfoxide/tetrabutylammonium fluoride. Carbohydr. Polym. 2005, 60, 259–267.

[42] Heinze T, Liebert T, Klufers P, Meister F. Carboxymethylation of cellulose in unconventional media. Cellulose. 1999, 6(2), 153–165.

[43] Bisht SS, Pandey KK, Joshi G, Naithani S. New route for carboxymethylation of cellulose: Synthesis, structural analysis and properties. Cell. Chem. Technol. 2017, 51, 609–619.

[44] Eyler RW, Klug ED, Diephuis F. Determination of degree of substitution of sodium carboxymethylcellulose. Anal. Chem. 1947, 19, 24–27.

[45] Kono H, Oshima K, Hashimoto H, Shimizu Y, Tajima K. NMR characterization of sodium carboxymethylcellulose: Substituent distribution and mole fraction of monomers in the polymer chains. Carbhydr. Polym. 2016, 146, 1–9.

[46] Brown W, Henley D. Studies on cellulose derivatives. Part IV. The configuration of the polyelectrolyte sodium carboxymethyl cellulose in aqueous sodium chloride solutions. Die Makromolekulare Chemie. 1964, 79(1), 68–88.

[47] Okatova OV, Lavrenko PN, Dautzenberg H, Filipp BN, Tsvetkov VN. Polyelectrolyte effects in diffusion and viscosity phenomena in water-cadoxene solutions of carboxymethylcellulose. Polym. Sci. U.S.S.R. 1990, 32(3), 533–539.

[48] Yanez-S M, Matsuhiro B, Maldonado S, et al. Carboxymethylcellulose from bleached organosolv fibers of Eucalyptus nitens: Synthesis and physicochemical characterization. Cellulose. 2018, 25, 2901–2914.

[49] Ott G, Schempp W, Krause T. Kationisierung von zellstoff unter homogenen reaktionsbedingungen. Die Angew. Makromol. Chem. 1989, 113, 213–218.

[50] Zamana M, Xiao H, Chibante F, Ni Y. Synthesis and characterization of cationically modified nanocrystalline cellulose. Carbohydr. Polym. 2012, 89, 163–170.

[51] Sirvio J, Honka A, Liimatainen H, Niinimaki J, Hormi O. Synthesis of highly cationic water-soluble cellulose derivative and its potential as novel biopolymeric flocculation agent. Carbohydr. Polym. 2011, 86, 266–270.

[52] Dangerfield JA, Salmons B, Corteling R, Abastado JP, Sinden J, Gunzburg WH, Brandtner EM. The diversity of uses for cellulose sulphate encapsulation. In: Brandtner EM, Dangerfield JA., eds. Bioencapsulation living cells diverse medical application, Bentham Science Publishers, 2013, 70–92.

[53] Lohr JM, Haas SL, Kroger JC, et al. Encapsulated cells expressing a chemotherapeutic activating enzyme allow the targeting of subtoxic chemotherapy and are safe and efficacious: Data from two clinical trials in pancreatic cancer. Pharmaceutics. 2014, 6, 447–466.

[54] Salmons B, Gunzburg WH. Release characteristics of cellulose sulphate capsules and production of cytokines from encapsulated cells. Int. J. Pharm. 2018, 548, 15–22.

[55] Weber W, Rinderknecht M, Daoud-El Baba M, de Glutz FN, Aubel D, Fussenegger M. CellMAC: A novel technology for encapsulation of mammalian cells in cellulose sulfate/pDADMAC capsules assembled on a transient alginate/Ca^{2+} scaffold. J. Biotechnol. 2004, 114, 315–326.

[56] Ortner V, Kaspar C, Halter C, et al. Magnetic field-controlled gene expression in encapsulated cells. J. Control. Release 2012, 158, 424–432.

[57] Zhang Q, Lin D, Yao S. Review on biomedical and bioengineering applications of cellulose sulfate. Carbohydr. Polym. 2015, 132, 311–322.

[58] Park M, Lee D, Hyun J. Nanocellulose-alginate hydrogel for cell encapsulation. Carbohydr. Polym. 2015, 116, 223–228.

[59] Tai C, Bouissil S, Gantumur E, et al. Use of anionic polysaccharides in the development of 3D bioprinting technology. Appl. Sci. 2019, 9, 2596–2609.

[60] Dai L, Cheng T, Duan C, et al. 3D printing using plant-derived cellulose and its derivatives: A review. Carbohydr. Polym. 2019, 203, 71–86.

[61] Park JS, Kim T, Kim WS. Conductive cellulose composites with low percolation threshold for 3D printed electronics. Sci. Rep. 2017, 7, 3246–3256.

[62] Maver T, Kurecic M, Smrke DM, Kleinschek KS, Maver U. Electrospun nanofibrous CMC/PEO as a part of an effective pain relieving wound dressing. J. Sol-Gel Sci. Technol. 2016, 79, 475–486.

[63] Gasparic P, Kurecic M, Kargl R, et al. Nanofibrous polysaccharide hydroxyapatite composites with biocompatibility against human osteoblasts. Carbohydr. Polym. 2017, 177, 388–396.

[64] Jongwuttanaruk K, Surin P, Wiwatwongwana F. Characterization of Gelatin/CMC scaffolds by electrospinning and comparison with freeze dry techniques. Int. J. Mater. Mech. Manuf. 2019, 7(2), 91–94.

[65] Mokhena TC, Jacobs V, Luyt AS. A review on electrospun bio-based polymers for water treatment. eXPRESS Polym. Lett. 2015, 9(10), 839–880.

[66] Javanbakht S, Shaabani A. Carboxymethyl cellulose-based oral delivery systems. Int. J. Biol. Macromol. 2019, 133, 21–29.

[67] Wang J, Hu Y, Li L, Jiang T, Wang S, Mo F. Indomethacin-5-fluorouracil-methyl ester dry emulsion: A potential oral delivery system for 5-fluorouracil. Drug Dev. Ind. Pharm. 2010, 36 (6), 647–656.

[68] Agarwal T, Narayana SGH, Pal K, Pramanik K, Giri S, Banerjee I. Calcium alginate-carboxymethyl cellulose beads for colon-targeted drug delivery. Int. J. Biol. Macromol. 2015, 75, 409–417.

[69] Jiao Z, Zhang B, Li C, et al. Carboxymethyl cellulose-grafted graphene oxide for efficient antitumor drug delivery. Nanotechnol. Rev. 2018, 7, 291–301.

[70] Sun X, Liu C, Omer A, et al. pH-sensitive ZnO/carboxymethyl cellulose/chitosan bio-nanocomposite beads for colon-specific release of 5-fluorouracil. Int. J. Biol. Macromol. 2019, 128, 468–479.

[71] Suflet DM, Popescu I, Pelin IM. Preparation and adsorption studies of phosphorylated cellulose microspheres. Cell. Chem. Technol. 2017, 51(1–2), 23–34.

[72] Wang Y, Heinze T, Zhang K. Stimuli-responsive nanoparticles from ionic cellulose derivatives. Nanoscale. 2016, 8(1), 648–657.

[73] Wang Q, Guan YX, Yao SJ, Zhu ZQ. Microparticle formation of sodium cellulose sulfate using supercritical fluid assisted atomization introduced by hydrodynamic cavitation mixer. Chem. Eng. J. 2010, 159, 220–229.

[74] Song Y, Zhou J, Li Q, Guo Y, Zhang L. Preparation and characterization of novel quaternized cellulose nanoparticles as protein carriers. Macromol. Biosci. 2009, 9, 857–863.

[75] Wang ZM, Li L, Zheng BS, Normakhamatov N, Guo SY. Preparation and anticoagulation activity of sodium cellulose sulfate. Int. J. Biol. Macromol. 2007, 41, 376–382.

[76] Wang ZM, Xiao KJ, Li L, Wu JY. Molecular weight-dependent anticoagulation activity of sulfated cellulose derivatives. Cellulose. 2010, 17(5), 953–961.

[77] Mauck C, Weiner DH, Ballagh S, et al. Single and multiple exposure tolerance study of cellulose sulfate gel: A Phase I safety and colposcopy study. Contraception. 2001, 64, 383–391.

[78] Rohan LC, Sassi AB. Vaginal drug delivery systems for HIV prevention. AAPS J. 2009, 11(1), 78–87.

[79] Anderson RA, Feathergill K, Diao XH, et al. Contraception by Ushercell™ (cellulose sulfate) in formulation: Duration of effect and dose effectiveness. Contraception. 2004, 70, 415–422.

[80] Fricain JC, Granja PL, Barbosa MA, de Jeso B, Barthe N, Baquey C. Cellulose phosphates as biomaterials. In vivo biocompatibility studies. Biomater. 2002, 23, 971–980.

[81] Pelin IM, Suflet DM. Guided bone repair using synthetic apatites-biopolymer composites, In: Aflori M., ed., Intelligent polymers for nanomedicine and biotechnologies, USA, CRC press Taylor & Francis Group, 2018, 111–143.

[82] Hayashi Y, Kaplan RA, Pak CYC. Effect of sodium cellulose phosphate therapy on crystallization of calcium oxalate in urine. Metabolism. 1975, 24(11), 1273–1278.

[83] Lukas J, Richau K, Schwarz HH, Paul D. Surface characterization of polyelectrolyte complex membranes based on sodium cellulose sulfate and poly (dimethyldiallylammonium chloride). J. Membrane Sci. 1995, 106, 281–288.

[84] Antosika AK, Piatek A, Wilpiszewska K. Carboxymethylated starch and cellulose derivatives-based film as human skin equivalent for adhesive properties testing. Carbohydr. Polym. 2019, 222, UNSP 115014. Doi: org/10.1016/j.carbpol.2019.115014.

[85] Zhang L, Zhang G, Lu J, Liang H. Preparation and characterization of carboxymethyl cellulose/polyvinyl alcohol blend film as a potential coating material. Polym. Plast. Technol. Eng. 2013, 52(2), 163–167.

[86] Ullah S, Acuna JJS, Pasa AA, et al. Photoactive layer-by-layer films of cellulose phosphate and titanium dioxide containing phosphotungstic acid. Appl. Surf. Sci. 2013, 277, 111–120.

[87] Kowalik T, Adler HJ, Plagge A, Stratmann M. Ultrathin layers of phosphorylated cellulose derivatives on aluminium surfaces. Macromol. Chem. Phys. 2000, 201, 2064–2069.

[88] Chauhan S. Use of cellulose and its derivatives for metal ion sorption. J. Chem. Pharm. Res. 2016, 8(4), 416–420.

[89] Rocha JC, Toscano IAS, Burba P. Lability of heavy metal species in aquatic humic substances characterized by ion exchange with cellulose phosphate. Talanta 1997, 44(1), 69–74.

[90] Nadagouda MN, Varma RS. Synthesis of thermally stable carboxymethyl cellulose/metal biodegradable nanocomposites for potential biological applications. Biomacromolecules. 2007, 8, 2762–2767.

[91] Qiao H, Zhou Y, Yu F, et al. Effective removal of cationic dyes using carboxylate-functionalized cellulose nanocrystals. Chemosphere. 2015, 141, 297–303.

[92] Kono H, Kusumoto R. Removal of anionic dyes in aqueous solution by flocculation with cellulose ampholytes. J. Water Process Eng. 2015, 7, 83–93.

[93] Khan MI, Adrees MN, Tariq MR, Sohaib M. Application of edible coating for improving meat quality: A review. Pak. J. Food Sci. 2013, 23(2), 71–79.

[94] Hamad AMA, Ates S, Durmaz E. Evaluation of the possibilities for cellulose derivatives in food products. Kastamonu University Journal of Forestry Faculty 2016, 16(2), 383–400.

Masayuki Yamaguchi, Shogo Nobukawa, Mohd Edeerozey
Abd Manaf, Kultida Songsurang, and Hikaru Shimada

Chapter 13
Novel methods to control the optical
anisotropy of cellulose esters

13.1 Introduction

One of the most important applications for cellulose esters must be optical films
used for liquid crystal display (LCD), in which anisotropy in refractive index, that
is, birefringence, must be precisely controlled. The consumption of such applications had a tremendous increase in accordance with the rapid growth of LCD in the
last two decades [1–3]. Among optical functional films, polarizer protective films
and optical retardation films are the main targets for cellulose esters.

In the case of a polarizer protective film, the birefringence has to be free not to
change the polarizing state of the light. For this purpose, a solution-cast method is
usually employed for cellulose triacetate (CTA). Because the solution-cast film has
no molecular orientation in the film plane, it does not provide the orientation birefringence in the film plane. However, the refractive index in the thickness direction
is usually lower than those in the film plane. Therefore, birefringence occurs to
some degree when seeing from a tilted angle, that is, not from the normal direction,
as explained later. Furthermore, a recent trend of a protective film is to reduce the
film thickness to decrease the thickness of a display [4]. Therefore, the moisture
transmittance has to be controlled [5]. Substitution by a longer ester group such as
butyryl, instead of acetyl, is one way to reduce the moisture transmittance [6].

For a retardation film, a precise control of optical retardation, that is, product of
birefringence and film thickness is required. Although the main target was to compensate the retardation that occurred in the liquid crystal layer [7], various properties are
required for the retardation films recently due to the advanced technology in LCD,
such as retardation film showing extraordinary wavelength dispersion (the absolute
value of birefringence increases with increase in the wavelength), three-dimensional
refractive index control, low photoelastic birefringence in the glassy state, and

Masayuki Yamaguchi, Kultida Songsurang, Hikaru Shimada, School of Materials Science, Japan
Advanced Institute of Science and Technology
Shogo Nobukawa, School of Materials Science, Japan Advanced Institute of Science and
Technology; Department of Life Science and Applied Chemistry, Nagoya Institute of Technology
Mohd Edeerozey Abd Manaf, School of Materials Science, Japan Advanced Institute of Science
and Technology; Faculty of Manufacturing Engineering, Universiti Teknikal Malaysia Melaka

https://doi.org/10.1515/9783110658842-013

high heat resistance with reduced thermal expansion. Among them, the addition of an antiplasticizer, that increases the modulus in the glassy state, was found to be an effective method to reduce the photoelastic birefringence under stress in the glassy state [8, 9] and the thermal expansion [10]. The reduced thermal expansion with a small level of photoelastic birefringence under stress widens the service temperature without generating excess birefringence. Furthermore, a multiband quarter-wave plate is strongly required for organic electroluminescence display to reduce the reflection and also for optical pickup lenses. As explained later, cellulose esters are appropriate materials to produce the extraordinary wavelength dispersion, although conventional polymeric materials exhibit ordinary wavelength dispersion.

Because cellulose esters have unique characteristics besides high transparency with high heat resistance, they are good candidates to be employed for advanced displays. In this section, notable characteristics of cellulose esters for optical films are introduced, such as (1) material design by chemical modification, (2) utilization of crystalline parts for good processability, and (3) capability to accept various organic compounds with low molecular weight and its applications. Prior to explanation about them, the basics of optical properties, especially birefringence, are mentioned briefly.

13.2 Basics of birefringence

The characteristics of a polarized light are usually modified after passing through an optical anisotropic material, which is determined by the optical retardation Γ given by the product of birefringence Δn and film thickness d as follows [11–14];

$$\Gamma = d\,\Delta n \tag{13.1}$$

The origins of birefringence are classified into orientation birefringence, photoelastic birefringence in the glassy state, and form birefringence. Among them, the orientation birefringence, Δn_O, is usually employed to provide birefringence for a retardation film, which is determined by the intrinsic birefringence, Δn^0, and the Hermans orientation function F [15] as follows;

$$\Delta n_O = F\,\Delta n^0 \tag{13.2}$$

$$F = \frac{3\langle \cos^2 \theta \rangle - 1}{2} \tag{13.3}$$

Therefore, the intrinsic birefringence can be considered as the birefringence at $F = 1$, that is, perfect orientation. The value is independent of the molecular weight and decided by the chemical structure of a repeat unit, that is, monomer. It can be expressed by the following equation.

$$\Delta n^0 = \frac{2\pi}{9} \frac{\left(\bar{n}^2 + 2\right)^2}{\bar{n}} N \Delta\alpha \tag{13.4}$$

where $\bar{n}$ is the average refractive index, N is the number of chains per unit volume, and $\Delta\alpha$ is the polarizability anisotropy.

Figure 13.1 illustrates the contribution of intrinsic birefringence and the degree of orientation. Each ellipsoid represents the refractive index of a repeating unit, denoting its intrinsic birefringence. Therefore, a slender ellipsoid with a large aspect ratio has a large intrinsic birefringence. As seen in the figure, a high level of orientation birefringence is obtained when long ellipsoids exhibit marked orientation (F is closer to unity).

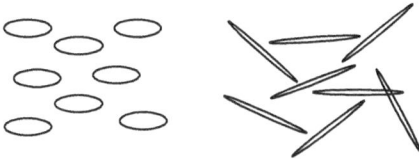

Fig. 13.1: Schematic illustration of orientation birefringence using refractive index ellipsoids. (Left) Low intrinsic birefringence with high orientation function and (right) high intrinsic birefringence with low orientation birefringence.

Another important expression of the orientation birefringence is the stress-optical rule. Since the stress in the rubbery state, σ, is provided by the decrease in the entropy owing to the molecular orientation, it can be expressed using a draw ratio λ_D as follows;

$$\sigma = N k_B T \left(\lambda_D{}^2 - \frac{1}{\lambda_D}\right) \tag{13.5}$$

where k_B is the Boltzmann constant.

The orientation birefringence is also a function of the draw ratio because it is determined by the orientation function of polymer chains.

$$\Delta n_O = \frac{2\pi}{45} \frac{\left(\bar{n}^2 + 2\right)^2}{\bar{n}} N \Delta\alpha \left(\lambda_D{}^2 - \frac{1}{\lambda_D}\right) \tag{13.6}$$

As a result, the orientation birefringence is proportional to the applied stress, which is known as the stress-optical rule.

$$\Delta n_O = C_R \, \sigma \tag{13.7}$$

In the equation, C_R is called the stress-optical coefficient, given by the following equation.

$$C_R = \frac{2\pi}{45 k_B T} \frac{\left(\bar{n}^2 + 2\right)^2}{\bar{n}} \Delta\alpha \tag{13.8}$$

It should be noted that the birefringence generated in the glassy state, Δn_G, is also proportional to the stress, which was firstly reported by Brewster.

$$\Delta n_G = C_G\, \sigma \tag{13.9}$$

where C_G is also called as the stress-optical coefficient. In some literatures, "Brewster" was used as a unit of C_G, which is $1 \times 10^{-12}\ \mathrm{Pa}^{-1}$.

Although the formula is the same with that in the rubbery state, the proportional constant, that is, stress-optical coefficient, C_G, is, of course, different from that in the rubbery state C_R. This is reasonable because the origin of the birefringence is different. The photoelastic birefringence in the glassy state is originated from the polarizability anisotropy generated by the localized deformation such as distortion/twisting of chemical bonds in the glassy state. In general, it is decided by the chemical structure of a repeat unit. The birefringence in the glassy state is an unfavorite property that should be reduced to keep the polarized state of a light [14, 16]. In some cases, thermal expansion results in the stress generation, leading to excess photoelastic birefringence. Therefore, the reduction of the thermal expansion is also required for films used in LCD [7].

The third one is the birefringence originated from the ordered arrangement of refractive index fluctuation, known as form birefringence [17]. The size of the structure must be smaller than the wavelength of light, which is, however, larger than the origin of the other birefringences. It can be detected in nanocomposites, block-copolymer having phase separation, foam, and rough surface with periodical groove, because these materials and structures give periodical change in refractive index in the submicron scale. Assuming that a system is composed of a polymeric material and voids, the refractive index of the void, that is, air, is much lower than that of a polymer. In this case, the form birefringence, Δn_F, ascribed to the void structure is given by the following equation [18, 19],

$$\Delta n_F = n_{2,//} - n_{2,\perp} \tag{13.10}$$

$$n_{2,//} = \left[n_{0,//}^2 + \frac{1}{3} \left(\frac{D}{\lambda}\right)^2 \pi^2 f^2 (1-f^2)(n^2 - n_0^2)^2 \right]^{1/2} \tag{13.11}$$

$$n_{2,\perp} = \left[n_{0,\perp}^2 + \frac{1}{3} \left(\frac{D}{\lambda}\right)^2 \pi^2 f^2 (1-f^2) \left(\frac{1}{n^2} - \frac{1}{n_0^2}\right)^2 n_{0,//}^6 n_{0,\perp}^2 \right]^{1/2} \tag{13.12}$$

$$n_{0,//} = \left[f n^2 + (1-f) n_0^2 \right]^{1/2} \tag{13.13}$$

$$n_{0,\perp} = \left[\frac{n^2 n_0^2}{fn^2 + (1-f)n_0^2}\right]^{1/2}$$ (13.14)

where f is the filling factor shown in Fig. 13.2.

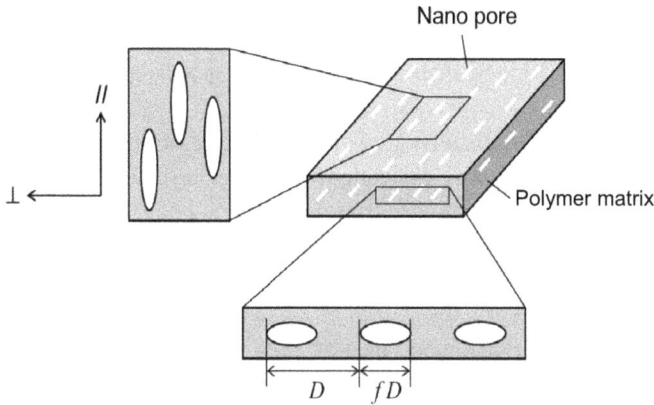

Fig. 13.2: Schematic illustration of nano-pore structure showing form birefringence. Reproduced with permission from S. Nobukawa, H. Shimada, Y. Aoki, A. Miyagawa, V. A. Doan, H. Yoshimura, Y. Tachikawa, and M. Yamaguchi, *Polymer*, **55**, 3247 (2014), Elsevier [18].

Birefringence of polymeric materials is originated from these three mechanisms, that is, orientation birefringence, Δn_O; glass birefringence, Δn_G; and form birefringence, Δn_F. The actual birefringence observed can be written as the sum of them.

$$\Delta n = \Delta n_F + \Delta n_O + \Delta n_G$$ (13.15)

The size of each origin is different, which can be summarized in Fig. 13.3 [20].

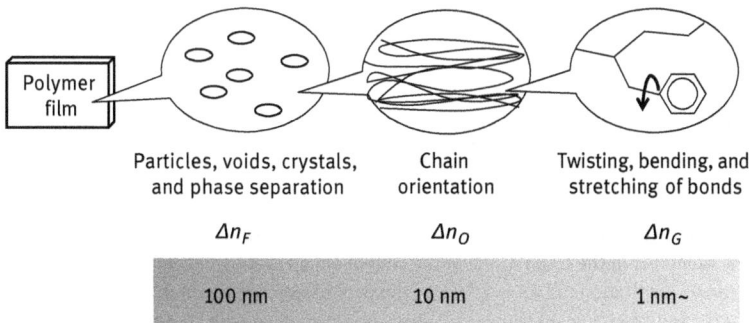

Fig. 13.3: Origin of birefringences [20].

13.3 Effect of chemical structure in cellulose esters

As explained in the previous section, the orientation birefringence is provided by the flow/deformation in the rubbery state, at which the degree of molecular orientation as well as the polarizability anisotropy decides the birefringence. The latter one is dependent upon the chemical structure. In the case of cellulose esters, it has been known that the pyranose-ring, that is, main chain of cellulose, has a low level of intrinsic birefringence [21]. Therefore, the species and their contents of ester groups greatly affect the orientation birefringence.

Figure 13.4 shows the wavelength dispersion of orientation birefringence for various cellulose esters [21]. CDA is the cellulose diacetate with the degree of acetyl substitution (S_{Ac}) of 2.41, whereas CTA is the cellulose triacetate with S_{Ac} of 2.96. CAB and CAP are the cellulose acetate butyrate and cellulose acetate propionate, respectively. The numerals in the sample code in CAB represent the butyryl content in the weight fraction. The samples were stretched at a draw ratio of 2.0, which was carried out at the temperature where the tensile storage modulus is 10 MPa at 10 Hz, that is, in the rubbery state. It should be noted that not only the magnitude of birefringence but the wavelength dispersion is dependent upon the species and their contents of ester groups (including hydroxyl group). Except for CTA, all CAB, CAP, and CDA samples employed in this experiment exhibited positive birefringences with extraordinary wavelength dispersion; that is, the absolute value of the orientation birefringence increases with increasing the wavelength, which is an inevitable property to prepare a

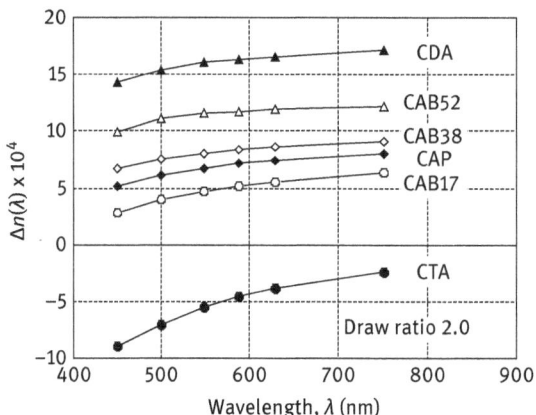

Fig. 13.4: Wavelength dependence of orientation birefringence $\Delta n(\lambda)$ for stretched films of various cellulose esters. The numerals in the CAB represent the weight concentration of the butyryl group. The degree of propionyl substitution in CAP is 2.58 and those of butyryl group are 0.73, 1.74, and 2.64 for CAB17, CAB38, and CAB52, respectively. The degrees of acetyl substitution are 2.41 for CDA and 2.96 for CTA. Reproduced with permission from M. Yamaguchi, K. Okada, M. E. A. Manaf, Y. Shiroyama, T. Iwasaki, and K. Okamoto, *Macromolecules*, **42**, 9034 (2009), ACS Publications [21].

multiband wave plate [22]. Considering that the refractive index usually decreases with increasing the wavelength dispersion, which is expressed by the Sellmeier equation (eq. (13.16)), it is not usual for conventional plastic films to show extraordinary wavelength dispersion.

$$n(\lambda)^2 = 1 + \sum_i \frac{A_{S-i}}{\lambda^2 - B_{S-i}} \tag{13.16}$$

where λ is the wavelength and A_{S-i} and B_{S-i} are the Sellmeier's coefficients.

The mechanism to show the extraordinary wavelength dispersion was explained by the contributions of at least two origins [22]. In the case of cellulose esters employed in Fig. 13.4, one is the orientation birefringence from the acetyl group, which is negative with strong dependence on the wavelength. Another one is that from the hydroxyl, propionyl, and butyryl group, which is positive with weak wavelength dispersion. Since the total birefringence is given by the sum of them, the orientation birefringence shows extraordinary wavelength dispersion as exemplified in Fig. 13.5. Among CDA, CAP, and CAB samples in Fig. 13.4, CDA shows the largest positive value of orientation birefringence, indicating the significant contribution of the positive orientation birefringence for the hydroxyl group. Since the demand for retardation films showing an extraordinary wavelength dispersion increasing rapidly these days for advanced systems, such as optical pick-up lens and anti-reflection films of electroluminescent display, the phenomenon is focused in industry.

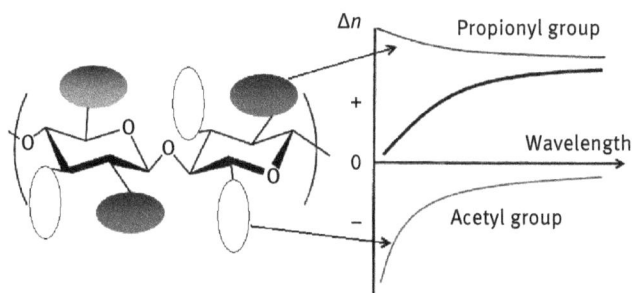

Fig. 13.5: Illustration of CAP molecules and the contribution of each ester group to the orientation birefringence. The bold line in the right figure represents the birefringence for the oriented film, which shows extraordinary wavelength dispersion. Reproduced with permission from M. Yamaguchi, K. Okada, M. E. A. Manaf, Y. Shiroyama, T. Iwasaki, and K. Okamoto, *Macromolecules*, **42**, 9034 (2009), ACS Publications [21].

Furthermore, the magnitude and wavelength dispersion are dependent upon the stretching condition, because the orientation of acetyl group shows the different tendency from that of propionyl group [23]. In the case of CAP (the degree of propionyl substitution, $S_{Pr} = 2.58$ and $S_{Ac} = 0.19$), a high level of stretching provides the strong wavelength dispersion. Low temperature stretching also gives the strong wavelength dispersion. This is an anomalous behavior because the stretching condition cannot alter the wavelength dispersion for most polymers [22]. Moreover, the substitution site of ester groups must affect the birefringence, because the C-6 site is in the different situation from the C-2 and C-3 sites. Although the regioselective cellulose acetate can be prepared using ionic liquids [24] or the specific preparation method [25] these days, Nobukawa et al. clarified this effect by a simple method [26]. They added xylan acetate having two substitution sites, that is, C-2 and C-3, in CTA, and evaluated the orientation birefringence as compared with that of pure CTA. It was revealed that acetyl groups in the C-2 and C-3 sites provide the negative orientation birefringence, while that in the C-6 site gives the positive one. Later, Hayakawa et al. revealed that the intrinsic birefringence and its wavelength dispersion are dependent upon the conformations of the acetyl group at the C-6 site using the density fluctuation theory [27]. These results demonstrate that the precise control of substitution sites can enhance the optical performance.

The orientation birefringence is detected even in a solution-cast film. Although the polymer chains show random orientation in the film plane (in-plane birefringence is zero), they tend to exist in the film plane due to the compressive stress applied at the evaporation process of solvent. Therefore, the refractive index in the thickness direction is different from those in the film plane (see Fig. 13.6), which is expressed by out-of-plane birefringence Δn_{th} defined by the following equation.

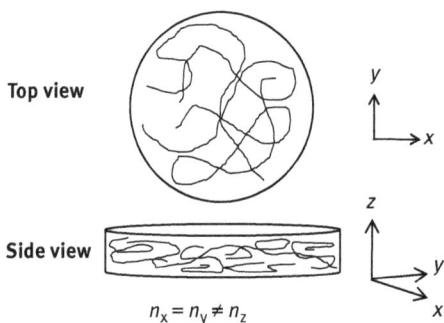

Fig. 13.6: Schematic illustration of molecular orientation in a solution-cast film.

$$\Delta n_{th} = \frac{n_x + n_y}{2} - n_z \qquad (13.17)$$

Since in-plane orientation of polymer chains is pronounced at a high evaporation rate, the absolute value of the out-of-plane birefringence increases with the evaporation rate. In fact, Songsurang et al. found that the carbonyl orientation is enhanced in the film plane [2, 28], which was also predicted by another research group using computer simulation [27]. Figure 13.7 shows the wavelength dispersion of the out-of-plane birefringence for the solution-cast CTA (S_{Ac} = 2.96) films obtained by various evaporation rates shown in the left figure. Even though the out-of-plane birefringence of CTA is not so large, further efforts will be required to improve the quality of display [29]. Later, we will show one of the material designs to reduce the out-of-plane birefringence.

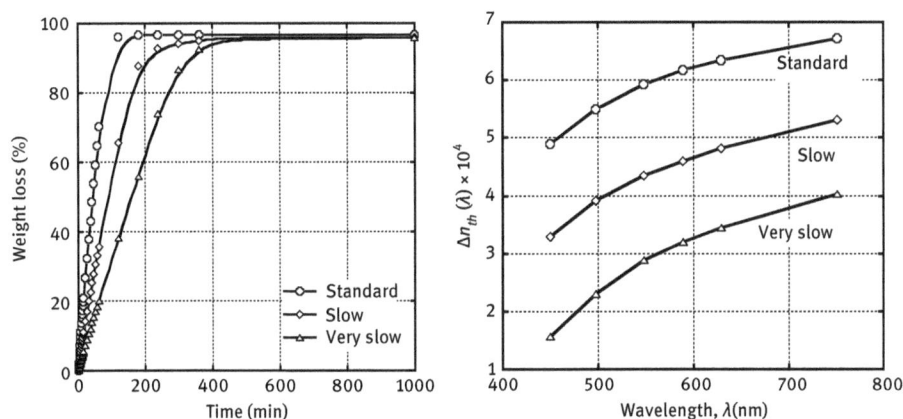

Fig. 13.7: (Left) Time variation of weight loss at the solution-cast processes for CTA (S_{Ac} = 2.96) at room temperature. The solvent employed was dichloromethane and the initial concentration of CTA in the solution was 4 wt.%. The evaporation rate was controlled by covering an aluminum foil having holes. (Right) Wavelength dispersion of the out-of-plane birefringence $\Delta n_{th}(\lambda)$ for the solution-cast films. Reproduced with permission from K. Songsurang, A. Miyagawa, M. E. A. Manaf, P. Phulkerd, S. Nobukawa, and M. Yamaguchi, *Cellulose*, **20**, 83 (2013), Springer [28].

For the photoelastic birefringence in the glassy state, CTA (S_{Ac} = 2.96) shows a relatively low value. The stress-optical coefficient C_G was reported to be ca. 7.9×10^{-12} Pa^{-1} at 633 nm [30]. The value is much lower than bisphenol-A polycarbonate (80×10^{-12} Pa^{-1}) [9] and polystyrene (32×10^{-12} Pa^{-1}) [31], and higher than poly(methyl methacrylate) (-4.1×10^{-12} Pa^{-1}) [32]. The C_G value increases with increasing the hydroxyl group in cellulose acetate [30]. Therefore, C_G of CDA (8.6×10^{-12} Pa^{-1}, S_{Ac} = 2.44) is higher than that of CTA. Moreover, cellulose esters having longer ester groups, such as propionyl and butyryl, show high C_G values [33].

13.4 Role of crystals in cellulose esters

A retardation film is usually produced by hot-stretching process. During the stretching, polymer chains orient to the stretching direction, which provides the orientation birefringence. Then, the molecular orientation has to be "frozen" by cooling below the glass transition temperature. Therefore, the solidification time has to be much shorter than the characteristic time for orientation relaxation [34]. Because the orientation relaxation usually occurs in a short period for a conventional thermoplastic material, it is not so easy to control the birefringence and thus the retardation precisely at an actual processing operation. In the case of cellulose esters, however, the situation is much different from conventional thermoplastic materials.

Figure 13.8 shows the growth curves of the stress and birefringence at hot-stretching process, which were evaluated by the simultaneous measurements of stress and birefringence for CTA (S_{Ac} = 2.96) and CDA (S_{Ac} = 2.41) beyond their glass transition temperatures [30]. It was found that both stress and birefringence grow monotonically, and the birefringence is almost proportional to the stress for CTA. This is a typical behavior of a rubbery polymer as shown in eq. (13.7), that is, stress-optical rule. In the figure, furthermore, the time variation of the stress and birefringence after the cessation of the stretching, that is, stress relaxation process, is shown. As seen in the figure, the stress rapidly decreases in a short period, as similar to conventional thermoplastics. However, surprisingly, the orientation birefringence of CTA shows a slight decrease and a large value even after 1 min. Moreover, the birefringence of CDA keeps an almost constant value after the cessation of the stretching, although the stress decays rapidly. Even at the hot-stretching process, the birefringence is not proportional to the stress for CDA. These results demonstrate that the stress-optical rule is not applicable to the cellulose acetates. A similar behavior was also detected in CAP [18]. From the viewpoint of processing conditions at hot-stretching process, it is a great benefit for cellulose esters because the cooling condition does not play an important role on the retardation control.

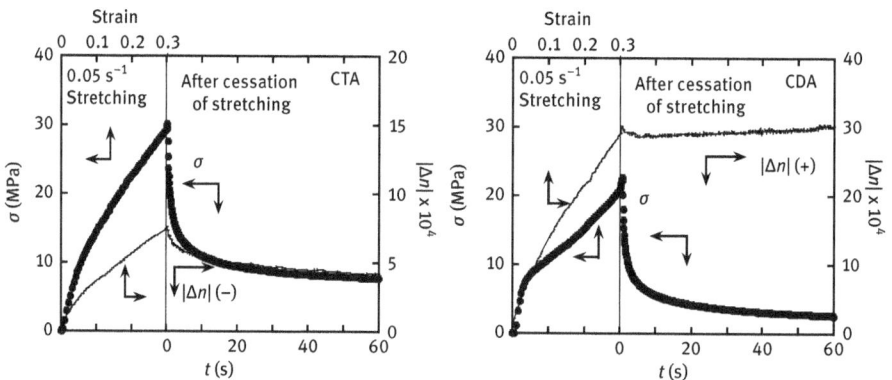

Fig. 13.8: Growth curves of stress σ and the absolute values of birefringence |Δn| at hot-stretching process and their relaxation curves after the cessation of the stretching for (left) CTA (S_{Ac} = 2.96) and (right) CDA (S_{Ac} = 2.44). The sign of the birefringence is denoted in the parenthesis of Δn. Reproduced with permission from K. Hatamoto, H. Shimada, M. Kondo, S. Nobukawa, and M. Yamaguchi, *Cellulose*, **25**, 4453 (2018), Springer [30].

This anomalous behavior was explained by a strong contribution of the crystalline part to decide the birefringence. Although cellulose esters show a high level of transparency, they are known to have a small content of crystallites [35, 36]. Their melting point is higher than the glass transition temperature and closer to (or even higher than) the degradation temperature. Therefore, the hot-stretching was performed for the sample film having crystallites. When using a solution-cast film, its crystallinity is enhanced by preheating and hot-stretching processes as shown in Fig. 13.9 [25]. The crystallites have high intrinsic birefringence, which was revealed by the simulation [27]. Moreover, the relaxation time for orientation must be much longer than that of amorphous chains. As a result, the orientation of crystallites is clearly detected even after stress relaxation [23, 30, 37]. Figure. 13.10 shows the two-dimensional wide-angle X-ray diffraction pattern for the CTA (S_{Ac} = 2.96) film at various stages of hot-stretching, that is, (a) as solution-cast, (b) after heating for hot-stretching, (c) quenched film immediately after hot-stretching, and (d) quenched film after stress relaxation for 1 min. The strong diffraction peak of the (110) plane [38], detected at 2 theta = 7.8°, appears on equator of the sample after the hot-stretching, suggesting that chain axis of CTA crystals orients to the stretching direction. The diffraction peaks on equator are also clearly detected after the stress relaxation, demonstrating that the crystalline parts do not show orientation relaxation. In other words, the crystalline parts mainly decide the orientation birefringence. Since the stress is generated by the orientation of amorphous chains, the birefringence has a different tendency from the stress at hot-stretching and relaxation processes.

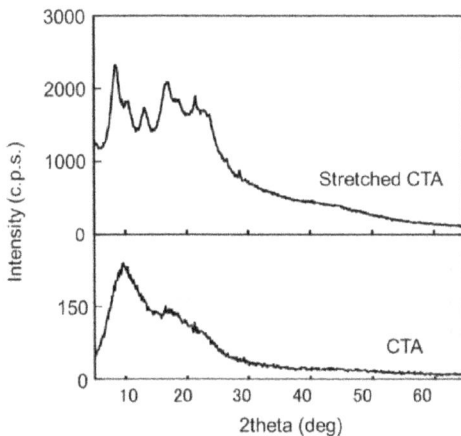

Fig. 13.9: WAXD profiles of a solution-cast CTA (S_{Ac} = 2.96) films (bottom) prior to heating and (top) after hot-stretching. Reproduced with permission from K. Songsurang, A. Miyagawa, M. E. A. Manaf, P. Phulkerd, S. Nobukawa, and M. Yamaguchi, *Cellulose*, **20**, 83 (2013), Springer [28].

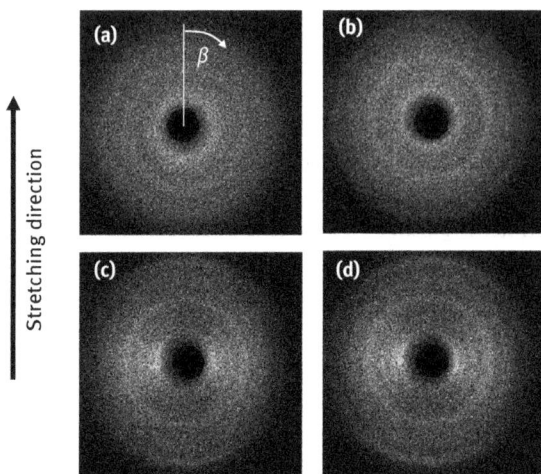

Fig. 13.10: 2D-WAXD profiles for CTA (S_{Ac} = 2.96): (a) solution-cast film, (b) after heating, (c) after hot-stretching, and (d) after stress relaxation. Reproduced with permission from K. Hatamoto, H. Shimada, M. Kondo, S. Nobukawa, and M. Yamaguchi, *Cellulose*, **25**, 4453 (2018), Springer [30].

The crystallites act as branch/crosslink points in the rubbery state. Therefore, the sample shows marked melt elasticity including the strain-hardening in the transient elongational viscosity. In general, the strain-hardening in elongational viscosity is detected in a polymer melt with long-chain branches, and provides good processability [39–43]. Figure 13.11 shows the growth curves of uniaxial elongational viscosity for cellulose acetate (S_{Ac} = 2.18) plasticized by 20 wt.% of dimethyl phthalate (DMP). The measurement was performed at 190°C, that is, much higher temperature than the glass transition temperature. In the figure, the solid line $3\eta^+$ represents three times of transient shear viscosity calculated from the linear viscoelastic properties [44] and the numerals denote the elongational strain rates. The upward deviation of the viscosity curve from the solid line is called strain-hardening, which is responsible for a small level of neck-in and reduced draw resonance at hot-stretching [45, 46], and uniform film thickness after hot-stretching [47]. Although various techniques have been proposed to provide the strain-hardening for conventional thermoplastics to improve the processability [43, 47–50], cellulose esters do not need such a technique due to the existence of the crystalline parts. For CTA, conventional melt processing operations such as T-die extrusion are not applicable because of a large content of crystallites, while commercialized CAP and CAB are used as thermoplastics. Considering that the solution-cast method needs a huge investment for a production plant, it is desired for CTA to be available for melt-processing operations. The authors expect that this problem will be solved in near future using some specific additives. Recently, the crystallinity of polyamide and polyester was found to be greatly reduced by the salt addition technique [51, 52]. In fact, the addition of lithium bromide was found to reduce the

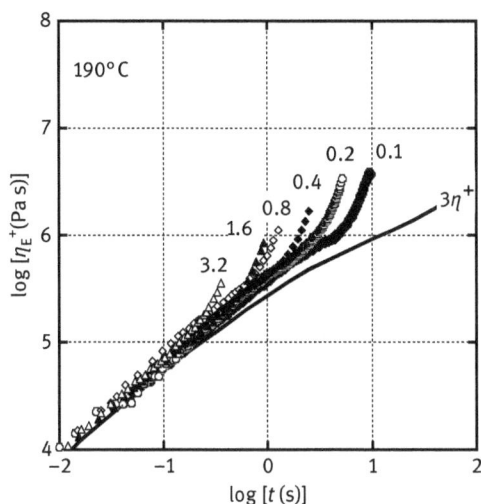

Fig. 13.11: Growth curves of uniaxial elongational viscosity $\eta_E{}^+$ at various strain rates denoted as numerals for plasticized cellulose acetate (S_{Ac} = 2.18) at 190 °C. The amount of dimethyl phthalate (DMP), as a plasticizer, was 20 wt.%.

crystallinity of a solution-cast CTA (S_{Ac} = 2.85) film [53]. Furthermore, it was reported that the addition of triacetin reduces the crystallinity of CDA (S_{Ac} = 2.4) [54].

13.5 Modification by low-molecular-weight compound

Miscibility of cellulose esters with another polymer has been studied for a long time. It has been known that poly(epichlorohydrin) [55], poly(vinyl acetate) [56], poly(butylene succinate) [57], and poly(3-hydroxyalkanoate) [58, 59] are miscible with some cellulose esters. A small amount of poly(lactic acid) is also miscible with CAP [60]. Among them, the blends of CAP and poly(vinyl acetate) show a low level of orientation birefringence even after stretching. This is attributed to the opposite sign of intrinsic birefringence for CAP and poly(vinyl acetate). Moreover, the addition of a small amount of poly(lactic acid) can enhance the orientation birefringence of CAP greatly, even though the characteristic time of orientation relaxation for poly(lactic acid) is significantly shorter than that of CAP. These modification techniques of orientation birefringence are originated from the orientation correlation; that is, one polymer species with a short relaxation time is oriented by the topological interaction with another polymer species having a long relaxation time. This is called nematic interaction [61–64]. It should be noted that the nematic interaction is expected for not only polymer species but also low-molecular-weight compounds (LMCs) including plasticizers [65–71]. Because cellulose esters are miscible with various conventional plasticizers, this technique provides the capability of the precise control of orientation birefringence.

The nematic interaction with LMCs is illustrated in Fig. 13.12, in which small molecules, for example, plasticizer, orient to the stretching direction along with surrounding polymer chains. When LMCs have a high level of intrinsic birefringence, the orientation birefringence is greatly enhanced. In contrast, once the sign of the orientation birefringence is opposite to that of the polymer chains, the birefringence can be reduced. Figure 13.13 shows the wavelength dispersion of orientation birefringence for CAP (S_{Pr} = 2.58 and S_{Ac} = 0.19) and CAP containing 10 wt.% of bisphenoxyethanolfluorene (BPEF) [67]. Both films were prepared by hot-stretching process at the same stress level, that is, the same degree of molecular orientation of

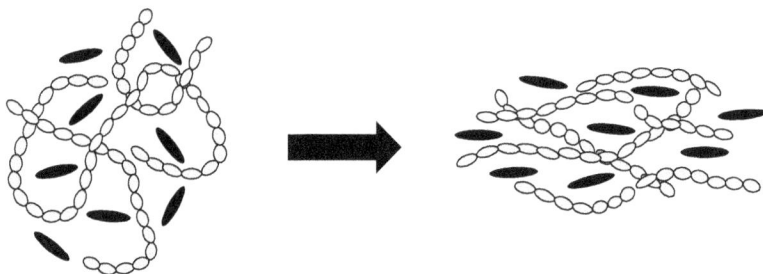

Fig. 13.12: Schematic illustration of orientation correlation between polymer chains and LMCs.

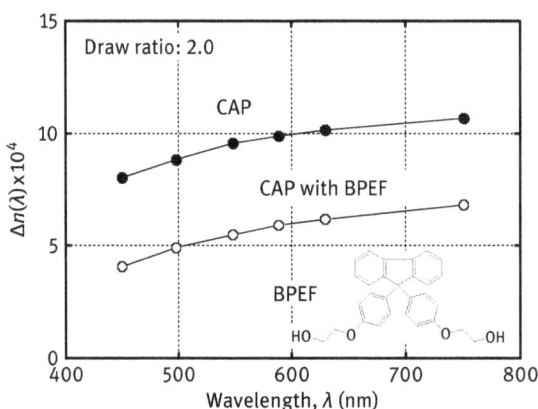

Fig. 13.13: Wavelength dependence of orientation birefringence $\Delta n(\lambda)$ of hot-stretched films of CAP (S_{Pr} = 2.58 and S_{Ac} = 0.19) and CAP/BPEF (90/10). Reproduced with permission from M. E. A. Manaf, A. Miyagawa, S. Nobukawa, Y. Aoki, and M. Yamaguchi, *Opt. Mater.*, **35**, 1443 (2013), Elsevier [67].

polymer chains. Because BPEF has negative intrinsic birefringence, the blend shows lower orientation birefringence than the pure CAP.

Furthermore, extraordinary wavelength dispersion can be provided by using this technique [22, 66]. Figure 13.14 shows the orientation birefringence for CTA (S_{Ac} = 2.96) and CTA containing tricresyl phosphate (TCP). Although CTA exhibits the negative orientation birefringence with ordinary wavelength dispersion, the blends show the positive orientation birefringence with extraordinary wavelength dispersion. This is attributed to the great contribution of the TCP molecules having positive orientation birefringence with weak wavelength dependence. In the figure, the dotted line represents the ideal birefringence for a multiband quarter-wave plate. The experimental values of the sample containing 10% of TCP almost correspond with the ideal ones.

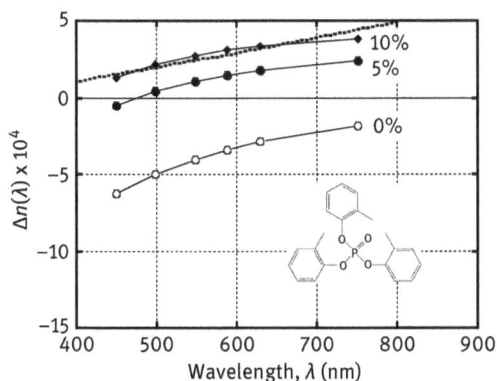

Fig. 13.14: Wavelength dependence of orientation birefringence $\Delta n(\lambda)$ for (open circles) CTA (S_{Ac} = 2.96), (closed circles) CTA/TCP (95/5), and (closed diamonds) CTA/TCP (90/10), stretched at a draw ratio of 1.5. In the figure, the straight dotted line represents the ideal wavelength dispersion for a multiband quarter-wave plate with a thickness of 400 μm. Reproduced with permission from M. E. A. Manaf, M. Tsuji, Y. Shiroyama, and M. Yamaguchi, *Macromolecules*, **44**, 3942 (2011), ACS Publications [66].

The orientation birefringence is also detected in a solution-cast film [28], because the polymer chains in a film show in-plane orientation. Owing to the nematic interaction, a disk-shaped compound tends to stay in the film plane, which usually enhances the refractive index in the in-plane direction. As a result, the out-of-plane birefringence is enhanced in general (Fig. 13.15). However, when the direction in the polarizability anisotropy is perpendicular to the long axis of the LMC, the refractive index in the out-of-plane direction (normal to film plane) is enhanced, leading to a small level of the out-of-plane birefringence, which has a strong demand in the

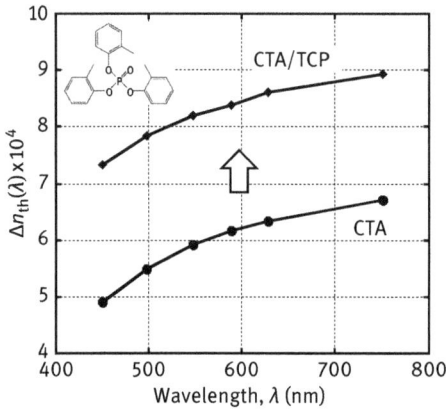

Fig. 13.15: Wavelength dependence of out-of-plane birefringence $\Delta n_{th}(\lambda)$ for solution-cast films of CTA (S_{Ac} = 2.96) and CTA/TCP (95/5). Reproduced with permission from K. Songsurang, A. Miyagawa, M. E. A. Manaf, P. Phulkerd, S. Nobukawa, and M. Yamaguchi, *Cellulose*, **20**, 83 (2013), Springer [28].

field of optical films. Recently, Kiyama et al. reported that the addition of ferrocene molecules can reduce the out-of-plane birefringence because the direction of the polarizing anisotropy is perpendicular to the long axis of ferrocene molecules as illustrated in Fig. 13.16 [68]. Furthermore, they clarified that the ferrocene addition reduces the stress-optical coefficient C_G in the glassy state [68].

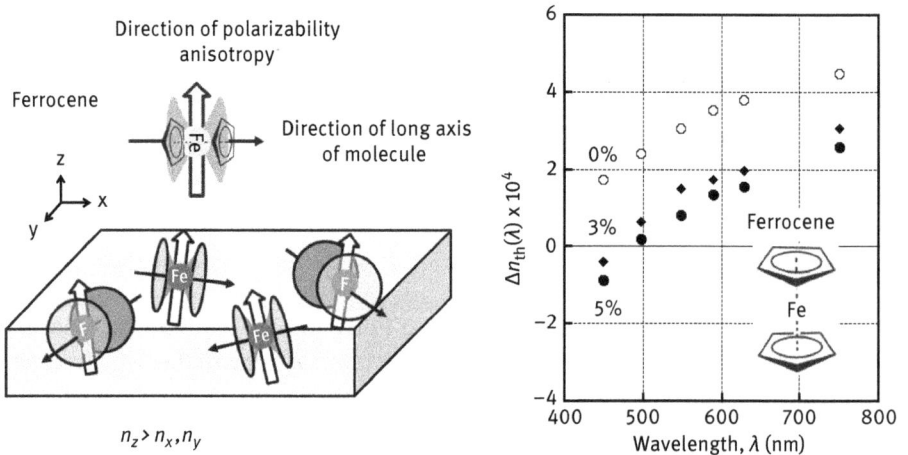

Fig. 13.16: Schematic illustration and the experimental results of out-of-plane birefringence $\Delta n_{th}(\lambda)$ for solution-cast films of CTA (S_{Ac} = 2.96) and CTA/Ferrocene. Reproduced with permission from A. Kiyama, S. Nobukawa, and M. Yamaguchi, *Opt. Mater.*, **72**, 491 (2017), Elsevier [71].

As explained previously, most cellulose esters show strain-hardening behavior in the elongational viscosity, which results in a small level of neck-in at hot-stretching. The reduced neck-in leads to the plane stretching mode, that is, one directional stretching with a constant width. Therefore, the stretching in the transversal direction is slightly applied even at uniaxial stretching, which results in the in-plane orientation of disk-shaped molecules [70]. In the case of rod-shaped molecules, however, the situation is different as shown in Fig. 13.17; that is, three dimensional control of refractive index can be available by using an appropriate LMC. Figure 13.18 shows the example of the wavelength dispersion of refractive indices in three directions, i.e., n_x, n_y, and n_z, for films of pure CTA and CTA containing 5wt% of disk-shaped molecule (TCP) and rod-shaped molecules (5CB) [70]. The order in the refractive indices is greatly affected by the LMC addition, which is predictable from the shape of LMC as shown in Fig. 13.17.

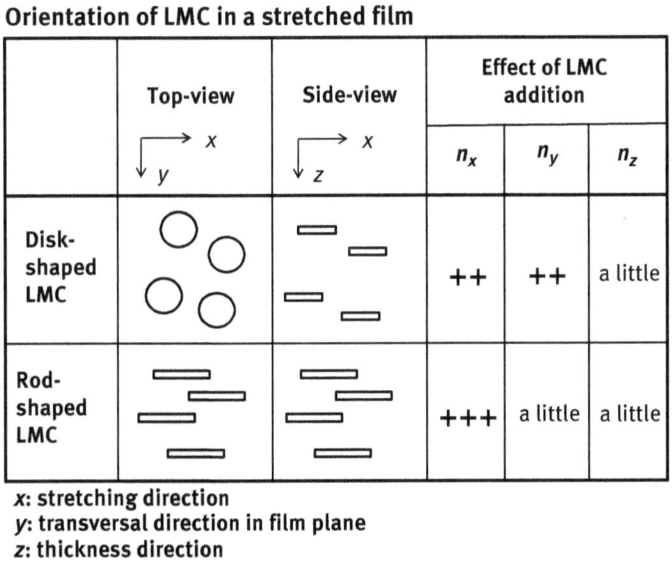

Orientation of LMC in a stretched film

	Top-view	Side-view	Effect of LMC addition		
	$\rightarrow x$ $\downarrow y$	$\rightarrow x$ $\downarrow z$	n_x	n_y	n_z
Disk-shaped LMC			++	++	a little
Rod-shaped LMC			+++	a little	a little

x: stretching direction
y: transversal direction in film plane
z: thickness direction

Fig. 13.17: Orientation of LMCs in a stretched film with planar elongation mode and the effect on the refractive index in each direction.

Some plasticizers are, of course, immiscible with cellulose esters. The miscibility can be roughly predicted by the difference in the solubility parameter [72]. Nobukawa et al. [18] and Shimada et al. [19, 20] proposed a unique material design to have extraordinary wavelength dispersion using in an immiscible blend with a plasticizer. Figure 13.19 shows the wavelength dependence of the orientation birefringence for CTA (S_{Ac} = 2.96)/DIDP (di-isodecyl phthalate) (90/10). Since the blend has phase-separated structure, the orientation of DIDP molecules relaxes immediately. This is reasonable because no nematic interaction is expected in a

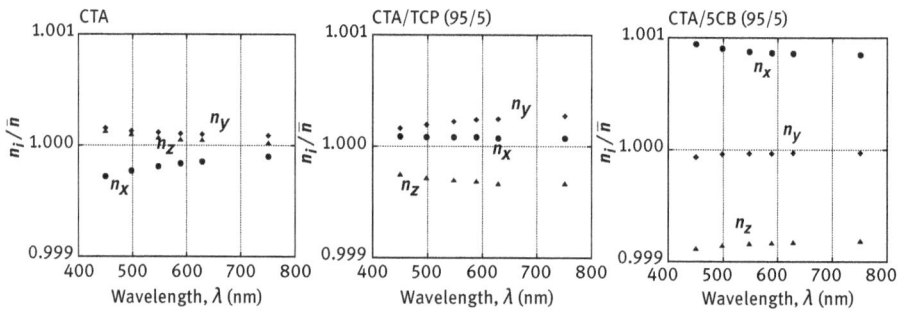

Fig. 13.18: Refractive index at three principal directions for the hot-stretched films. (left) CTA (S_{Ac} = 2.96), (center) CTA/TCP (tricresyl phosphate as a disk-shaped molecule (95/5), and (right) CTA/5CB (4-cyano-4′-pentylbiphenyl as a rod-shaped molecule) (95/5). × is the stretching direction and y is perpendicular to the stretching direction in the film, and z is the thickness direction. Reproduced with permission from K. Songsurang, H. Shimada, S. Nobukawa, and M. Yamaguchi, *Eur. Polym. J.*, **59**, 105 (2014), Elsevier [70].

phase-separated blend. As a result, the orientation birefringence for CTA/DIDP is the same as that of pure CTA. Then the film was immersed into methanol to remove DIDP from the film. After dying, the film shows extraordinary wavelength dispersion as seen in Fig. 13.19 (open circle) without losing the light transmittance. This is originated from the form birefringence. The prolonged voids appear after removal of

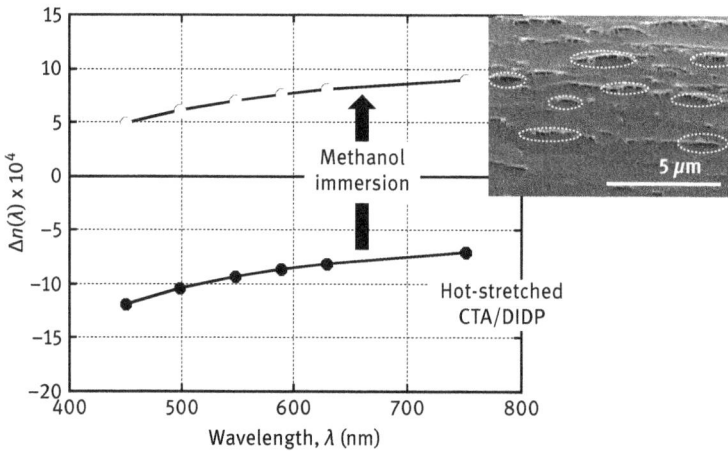

Fig. 13.19: Wavelength dispersion of birefringence for hot-stretched CTA (S_{Ac} = 2.96)/DIDP (90/10) films: (closed symbols) prior to methanol immersion and (open symbols) after methanol immersion. The SEM image in the right picture is the cut surface of the film after methanol immersion to remove DIDP. Reproduced with permission from H. Shimada, S. Nobukawa, and M. Yamaguchi, *Carbohydr. Polym.*, **120**, 22 (2015), Elsevier [19].

the DIDP phase, as denoted in the SEM image in the figure, and provides the form birefringence because of the great difference in the refractive index between CTA and voids (air). Since the film is thin with a small diameter of pores, light scattering is minimized. The schematic illustration of the structure development and the components of birefringence are shown in Fig. 13.20.

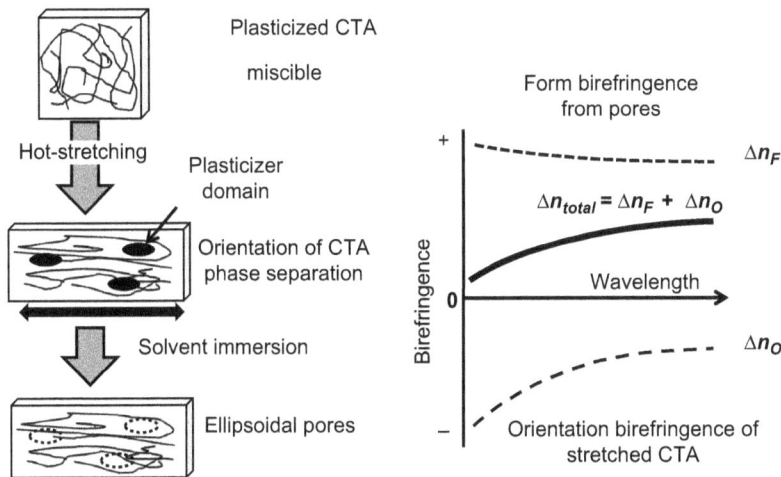

Fig. 13.20: Mechanism to show the extraordinary wavelength dispersion. (Left) Structure development and (right) the contribution of the form and orientation birefringences [20].

13.6 Conclusion

Cellulose esters have a great potential to be employed in advanced display systems as an optical functional film because of its good transparency and heat resistance. Furthermore, it should be notified that various methods can be applicable to modify the optical anisotropy. For example, it is not difficult to modify the ester group to adjust the optical properties such as stress-optical coefficients in both rubbery and glassy regions. Furthermore, the chemical modification is a good method to control the amount of crystallinity, which plays an important role on film processability and hot-stretching process. The capability to accept various types of LMCs should be also notified. The orientation birefringence and its wavelength dispersion can be controlled greatly by the addition of an appropriate compound with manipulated processing condition. Considering they are biomass-derived materials, further effort should be performed to be widely employed as optical functional films.

Acknowledgement: The authors would like to express their gratitude to Ms. Ayumi Kiyama, Mr. Kenji Masuzawa, Mr. Takuya Iwasaki, Ms. Kyoko Yokokura, Mr. Yasuhiko Shiroyama, Ms. Manami Mieda, Mr. Hiroki Hayashi, Mr. Akichika Nakao, Mr. Yoshihiko Aoki, Mr. Kazuya Hatamoto, Mr. Takeyoshi Kimura, and Ms. Maho Miyashita for their great efforts to accomplish this work.

References

[1]   Edgar KJ, Buchanan CM, Debenham JS, Rundquist PA, Seiler BD, Shelton MC, Tindall
       D. Advances in cellulose ester performance and application. Prog. Polym. Sci. 2001, 26,
       1605–1688.
[2]   Songsurang K, Shimada H, Nobukawa S, Yamaguchi M. Optical anisotropy of cellulose esters
       and its application to optical functional films. In: Thakur VK, Thakur MK., eds. Handbook of
       sustainable polymers: Processing and applications. Chap.9, Pan Stanford Publishing,
       Singapore, 2015, 341–384
[3]   Parker PM. The 2019–2024 world outlook for cellulose acetate. ICON Group International, Las
       Vegas, 2018.
[4]   Suzuki R, Nagura M, Sasada Y, Fukagawa N, Kawao N, Ito Y. Evolution of cellulose triacetate
       (TAC) films for LCDs: Novel technologies for high hardness, durability, and dimensional
       stability. SID Symp. Dig. Tech. Paper. 2015, 46, 446.
[5]   Manaf MEA, Tsuji M, Nobukawa S, Yamaguchi M. Effect of moisture on the orientation
       birefringence of cellulose esters. Polymers. 2011, 3, 955–966.
[6]   MaKeen LW. Film properties of plastics and elastomers. 3rd. ed. Chap. 14, Elsevier,
       Amsterdam, 2012.
[7]   Scharf T. Polarized light in liquid crystals and polymers. Wiley, Hoboken, 2007.
[8]   Miyagawa A, Korkiatithaweechai S, Nobukawa S, Yamaguchi M. Mechanical and optical
       properties of polycarbonate containing p-terphenyl. Ind. Eng. Chem. Res. 2013, 52,
       5048–5053.
[9]   Nobukawa S, Hasunuma S, Sako T, Miyagawa A, Yamaguchi M. Reduced stress-optical
       coefficient of polycarbonate by antiplasticization. J. Polym. Sci. B Polym. Phys. Ed. 2017, 55,
       1837–1842.
[10]   Miyagawa A, Nobukawa S, Yamaguchi M. Thermal expansion behavior of antiplasticized
       polycarbonate. Nihon Reoroji Gakkaishi. 2014, 42, 255–259.
[11]   Kuhn W, Grün F. Beziehungen zwischen elastischen konstanten und
       dehnungsdoppelbrechung hochelastischer stoffe. Kolloid-Z. 1942, 101, 248–271.
[12]   Treloar LRG. The physics of rubber elasticity. Clarendon Press, Oxford, UK, 1958.
[13]   Harding GF. Optical properties of polymers, Applied and Science, London, UK, 1986.
[14]   Born M, Wolf E. Principles of optics. 4th ed., Cambridge Univ. Press, Cambridge, UK, 2006.
[15]   Hermans PH, Platzek P. Beiträge zur Kenntnis des Deformationsmechanismus und der
       Feinstruktur der Hydratzellulose. Kolloid-Z. 1939, 88, 68–72.
[16]   Inoue T, Osaki K. On the strain-induced birefringence of glassy polymers. Polym. J. 1996, 28,
       76–79.
[17]   Richter I, Sun PC, Xu F, Fainman Y. Design considerations of form birefringent
       microstructures. Appl. Opt. 1995, 34, 2421–2429.

[18] Nobukawa S, Shimada H, Aoki Y, Miyagawa A, Doan VA, Yoshimura H, Tachikawa Y, Yamaguchi M. Extraordinary wavelength dispersion of birefringence in cellulose triacetate film with anisotropic nanopores. Polymer. 2014, 55, 3247–3253.

[19] Shimada H, Nobukawa S, Yamaguchi M. Development of microporous structure and its application to optical film for cellulose triacetate containing diisodecyl adipate. Carbohyd. Polym. 2015, 120, 22–28.

[20] Shimada H. PhD thesis, Japan Advanced Institute of Science and Technology, 2017.

[21] Yamaguchi M, Okada K, Manaf MEA, Shiroyama Y, Iwasaki T, Okamoto K. Extraordinary wavelength dispersion of orientation birefringence for cellulose esters. Macromolecules. 2009, 42, 9034–9040.

[22] Yamaguchi M, Manaf MEA, Songsurang K, Nobukawa S. Material design of retardation films with extraordinary wavelength dispersion of orientation birefringence – A review. Cellulose. 2012, 19, 601–613.

[23] Nobukawa S, Nakao A, Songsurang K, Phulkerd P, Shimada H, Kondo M, Yamaguchi M. Birefringence and strain-induced crystallization of stretched cellulose acetate propionate films. Polymer. 2017, 111, 53–60.

[24] Thomas R, Buchanan NL, Buchanan CM, Lambert JL, Donelson ME, Gorbunova MG, Kuo T, Wang B. WO2010019244A1, Regioselectively substituted cellulose esters produced in a carboxylated ionic liquid process and products produced therefrom, 2010.

[25] Kono H, Oka C, Kishimoto R, Fujita S. NMR characterization of cellulose acetate: Mole fraction of monomers in cellulose acetate determined from carbonyl carbon resonances. Carbohydr. Polym. 2017, 170, 23–32.

[26] Nobukawa S, Rogers YE, Shimada H, Iwata T, Yamaguchi M. Effect of acetylation site on orientation birefringence of cellulose triacetate. Cellulose. 2015, 22, 3003–3012.

[27] Hayakawa D, Gouda H, Hirono S, Ueda K. DFT study of the influence of acetyl groups of cellulose acetate on its intrinsic birefringence and wavelength dependence. Carbohydr. Polym. 2019, 207, 122–130.

[28] Songsurang K, Miyagawa A, Manaf MEA, Phulkerd P, Nobukawa S, Yamaguchi M. Optical anisotropy in solution-cast film of cellulose triacetate. Cellulose. 2013, 20, 83–96.

[29] Soeta H, Fujisawa S, Saito T, Berglund L, Isogai A. Low-birefringent and highly tough nanocellulose-reinforced cellulose triacetate. ACS Appl. Mater. Interfaces. 2015, 7, 11041–11046.

[30] Hatamoto K, Shimada H, Kondo M, Nobukawa S, Yamaguchi M. Effect of acetyl substitution on the optical anisotropy of cellulose acetate films. Cellulose. 2018, 25, 4453–4462.

[31] Inoue T, Onogi T, Tao ML, Osaki K. Viscoelasticity of low molecular weight polystyrene. Separation of rubbery and glassy components. J. Polym. Sci. B Polym. Phys. Ed. 1999, 37, 389–397.

[32] Tsugawa N, Ito A, Yamaguchi M. Effect of lithium salt addition on the structure and optical properties of PMMA/PVB blends. Polymer. 2018, 146, 242–248.

[33] Yamaguchi M. Optical properties of cellulose esters and their blends. In: Lejeune A, Deprez T eds., Cellulose, structure and properties, derivatives and industrial uses. 11, pp.325–340, New York, Nova, 2010.

[34] Meijer HEH, Janssen JMH, Anderson PD. Mixing of immiscible liquids. In: Manas-Zloczower I., ed. Mixing and compounding of polymers. 2nd ed., Chap. 3, pp.43–182, Hanser, Munich, Germany, 2009.

[35] Sata H, Murayama M, Shimamoto S. Properties and applications of cellulose triacetate film. Macromol Symp. 2004, 208, 323–333.

[36] Roche E, Chanzy H, Boudeulle M, Marchessault RH, Sundarajanid E. Three-dimensional crystalline structure of cellulose triacetate II. Macomolecules. 1978, 11, 86–94.

[37] Shimada H, Kiyama A, Phulkerd P, Yamaguchi M. Anomalous optical anisotropy of oriented cellulose triacetate film. Nihon Reoroji Gakkaishi. 2017, 45, 19–24.

[38] Sikorski P, Wada M, Heux L, Shintani H, Stokke BT. Crystal structure of cellulose triacetate I. Macromolecules. 2004, 37, 4547–4553.

[39] Yamaguchi M, Takahashi M. Rheological properties of low density polyethylenes produced by tubular and vessel processes. Polymer. 2001, 42, 8663–8670.

[40] Wagner MH, Yamaguchi M, Takahashi M. Quantitative assessment of strain hardening of low-density polyethylene melts by the molecular stress function model. J. Rheol. 2003, 47, 779–793.

[41] Yamaguchi M, Wakabayashi T. Rheological properties and processability of chemically modified poly(ethylene terephthalate-co-ethylene isophthalate). Adv. Polym. Technol. 2006, 25, 236–241.

[42] Yamaguchi M, Wagner MH. Impact of processing history on rheological properties for branched polypropylene. Polymer. 2006, 47, 3629–3635.

[43] Mieda N, Yamaguchi M. Flow instability for binary blends of linear polyethylene and long-chain branched polyethylene. J. Non-Newtonian Fluid Mech. 2011, 166, 231–240.

[44] Osaki K. Linear viscoelastic relation concerning shear stresses at the strat and cessation of shear flow. Nihon Reoroji Gakkaishi. 1976, 4, 166–168.

[45] Satoh N, Tomiyama H, Kajiwara T. Viscoelastic simulation of film casting process for a polymer melt. Polym. Eng. Sci. 2001, 41, 1564–1579.

[46] Kouda S. Prediction of processability at extrusion coating for low-density polyethylene. Polym. Eng. Sci. 2008, 48, 1094–1102.

[47] Yamaguchi M, Suzuki K. Enhanced strain-hardening in elongational viscosity for HDPE/crosslinked HDPE blend: 2. Processability of thermoforming. J. Appl. Polym. Sci. 2002, 86, 79–83.

[48] Yamaguchi M, Miyata H. Strain hardening behavior in elongational viscosity for binary blends of linear polymer and crosslinked polymer. Polym. J. 2000, 32, 164–170.

[49] Yokohara T, Nobukawa S, Yamaguchi M. Rheological properties of polymer composites with flexible fine fiber. J. Rheol. 2011, 55, 1205–1218.

[50] Fujii Y, Nishikawa R, Phulkerd P, Yamaguchi M. Modifying the rheological properties of polypropylene under elongational flow by adding polyethylene. J. Rheol. 2019, 63, 11–18.

[51] Tomie S, Tsugawa N, Yamaguchi M. Modifying the thermal and mechanical properties of poly(lactic acid) by adding lithium trifluoromethanesulfonate. J. Polym. Res. 2018, 25, 206.

[52] Sato Y, Ito A, Maeda S, Yamaguchi M. Structure and optical properties of transparent polyamide 6 containing lithium bromide. J. Polym. Sci. B Polym. Phys. Ed. 2018, 56, 1513–1520.

[53] Hatamoto H, Yamaguchi M. Improvement of processability for cellulose acetate by the salt addition. Proc. Autumnal Meeting of the Japan Soc. Plast. Eng. 2018, 1, 68.

[54] Phuong VT, Verstichel S, Cinelli P, Anguillesi I, Coltelli MB, Lazerri A. Cellulose acetate bends – Effect of plasticizers on properties and biodegradability. J. Renew. Mater. 2014, 2, 35–41.

[55] Yamaguchi M, Masuzawa K. Transparent polymer blends composed of cellulose acetate propionate and poly(epichlorohydrin). Cellulose. 2008, 15, 17–22.

[56] Tatsushima T, Ogata N, Nakane K, Ogihara T. Structure and physical properties of cellulose acetate butyrate/poly(butylene succinate) blend. J. Appl. Polym. Sci. 2005, 96, 400–406.

[57] Yamaguchi M, Masuzawa K. Birefringence control for binary blends of cellulose acetate propionate and poly(vinyl acetate). Eur. Polym. J. 2007, 43, 3277–3282.

[58] El-Shafee E, Saad GR, Fahmy SM. Miscibility, crystallization and phase structure of poly(3-hydroxybutyrate)/cellulose acetate butyrate blends. Eur. Polym. J. 2001, 37, 2091–2104.

[59] Yamaguchi M, Arakawa K. Enhancement of melt elasticity for poly(3-hydroxybutyrate-co-3-hydroxyvalerate) by addition of weak gel. J. Appl. Polym. Sci. 2007, 103, 3447–3452.

[60] Yamaguchi M, Lee S, Manaf MEA, Tsuji M, Yokohara T. Modification of orientation birefringence of cellulose ester by addition of poly(lactic scid). Eur. Polym. J. 2010, 46, 2269–2274.

[61] Merrill WW, Tirrell M, Tassin JF, Monnerie L. Diffusion and relaxation in oriented polymer media. Macromolecules. 1989, 22, 896–908.

[62] Doi M, Watanabe H. Effect of nematic interaction on the Rouse dynamics. Macromolecules. 1991, 24, 740–744.

[63] Urakawa O, Ohta E, Hori H, Adachi K. Effect of molecular size on cooperative dynamics of low mass compounds in polystyrene. J. Polym. Sci. B Polym. Phys. Ed. 2006, 44, 967–974.

[64] Nobukawa S, Urakawa O, Shikata T, Inoue T. Evaluation of nematic interaction pParameter between polymer segments and low-mass molecules in mixtures. Macromolecules. 2010, 43, 6099–6105.

[65] Yamaguchi M, Iwasaki T, Okada K, Okamoto K. Control of optical anisotropy of cellulose esters and their blends with plasticizer. Acta. Mater. 2009, 57, 823–829.

[66] Manaf MEA, Tsuji M, Shiroyama Y, Yamaguchi M. Wavelength dispersion of orientation birefringence for cellulose esters containing tricresyl phosphate. Macromolecules. 2011, 44, 3942–3949.

[67] Manaf MEA, Miyagawa A, Nobukawa S, Aoki Y, Yamaguchi M. Incorporation of low-mass compound to alter the orientation birefringence in cellulose acetate propionate. Opt. Mater. 2013, 35, 1443–1448.

[68] Nobukawa S, Aoki Y, Yoshimura H, Tachikawa Y, Yamaguchi M. Effect of aromatic additives with various alkyl groups on orientation birefringence of cellulose acetate propionate. J. Appl. Polym. Sci. 2013, 130, 3463–3472.

[69] Nobukawa S, Hayashi H, Shimada H, Kiyama A, Yoshimura H, Tachikawa Y, Yamaguchi M. Strong orientation correlation and optical anisotropy in blend of cellulose ester and poly(ethylene 2,6-naphthalate) oligomer. J. Appl. Polym. Sci. 2014, 131, 40570.

[70] Songsurang K, Shimada H, Nobukawa S, Yamaguhic M. Control of three-dimensional refractive indices of uniaxially-stretched cellulose triacetate with low-molecular-weight compounds. Eur. Polym. J. 2014, 59, 105–112.

[71] Kiyama A, Nobukawa S, Yamaguchi M. Birefringence control of solution-cast film of cellulose triacetate. Opt. Mater. 2017, 72, 491–495.

[72] Kugimoto D, Kouda S, Yamaguchi M. Improvement of mechanical toughness of poly(lactic acid) by addition of ethylene-vinyl acetate copolymer. Polym. Test. 2019, 80, 106021.

Index

https://doi.org/10.1515/9783110658842-014